The Neuroimmunological Basis of Behavior and Mental Disorders

Allan Siegel • Steven S. Zalcman
Editors

The Neuroimmunological Basis of Behavior and Mental Disorders

Springer

Editors
Allan Siegel
Departments of Neurology and Neuroscience
and Psychiatry
University of Medicine and Dentistry of New
Jersey — NJ Medical School,
Newark, NJ 07103, USA
siegel@umdnj.edu

Steven S. Zalcman
University of Medicine and Dentistry
of New Jersey — NJ Medical School,
Newark, NJ 07103, USA
zalcmass@umdnj.edu

ISBN 978-0-387-84850-1 ISBN 978-0-387-84851-8 (eBook)
DOI 10.1007/978-0-387-84851-8

Library of Congress Control Number: 2008937579

Printed on acid-free paper

springer.com

For Carla Siegel, and Anna and Mier Zalcman

"Who is wise? One who learns from every person."

Contents

Contributors

William A. Banks Departments of Internal Medicine, Geriatric Division and Pharmacological and Physiological Science, Saint Louis University School of Medicine, 1402 South Grand Blvd, St. Louis, MO 63104, bankswa@slu.edu

Francesca Baracchi Department of Anesthesiology, 7422 Medical Sciences Building I, 1150 West Medical enter Drive, University of Michigan, Ann Arbor, MI 48109–5615, USA

Arlene Bardeguez Department of Obstretrics, UMDNJ – NJ Medical School 185 South Orange Avenue, Newark, NJ 07103, USA

Rekha Bhatt Departments of Neurology & Neuroscience and Psychiatry, UMDNJ – New Jersey Medical School, MSB Room H-512, 185 South Orange Avenue, Newark, NJ 07103, USA

Suresh Bhatt Departments of Neurology & Neuroscience and Psychiatry, UMDNJ – New Jersey Medical School, MSB Room H-512, 185 South Orange Avenue, Newark, NJ 07103, USA

Q. Chang Integrative Immunology and Behavior Program, Department of Animal Sciences, College of Medicine, University of Illinois at Urbana-Champaign, 212 Edward R. Madigan Laboratory, 1201 West Gregory Drive, Urbana, IL 61801, USA

Robert Dantzer Integrative Immunology and Behavior Program, Department of Animal Sciences, College of Medicine, University of Illinois at Urbana-Champaign, 212 Edward R. Madigan Laboratory, 1201 West Gregory Drive, Urbana, IL 61801, USA

Daniel Dauer McKnight Brain Institute, Departments of Psychiatry, Neuroscience, and Pharmacology & Therapeutics, College of Medicine, University of Florida, Box 100256, Gainesville, FL 32610–0256, USA

Heidi E.W. Day Department of Integrative Physiology, Center for Neuroscience, University of Colorado-Boulder, Clare Small Building, Boulder, CO 80309–0354, USA

Adrian J. Dunn Department of Psychology, Pacific Biosciences Research Center, University of Hawaii, 1993 East-West Road, Honolulu, HI 96822–2321, USA, ajdunn@hawaii.edu

Monika Fleshner Department of Integrative Physiology, Center for Neuroscience, University of Colorado-Boulder, Clare Small Building, Boulder, CO 80309–0354, USA, fleshner@colorado.edu

Ron P.A. Gaykema Laboratory of Neuroimmunology & Behavior Program in Sensory and Systems Neuroscience, Department of Psychology & Neuroscience Graduate Program, University of Virginia, 102 Gilmer Hall, P.O. Box 400400, Charlottesville, VA 22904–4400, USA

Lisa E. Goehler Laboratory of Neuroimmunology and Behavior, Center for the Study of Complementary and Alternative Therapies, School of Nursing, University of Viginia, Charlottesville, VA 22908, USA, goehler@virginia.edu

Benjamin N. Greenwood Department of Integrative Physiology, Center for Neuroscience, University of Colorado-Boulder, Clare Small Building, Boulder, CO 80309–0354, USA

Grace K. Ha McKnight Brain Institute, Departments of Psychiatry, Neuroscience, and Pharmacology & Therapeutics, College of Medicine, University of Florida, Box 100256, Gainesville, FL 32610–0256, USA

Joanne M. Hash-Converse Behavioral Neuroscience Program, Department of Psychology, Rutgers University, Piscataway, NJ 08855, USA

Steven A. Hoffman. Neuroimmunology Labs, School of Life Sciences, College of Liberal Arts and Sciences, Arizona State University, P.O. Box 874501, Tempe, AZ 85287–4501, USA, SteveHoffman@asu.edu

Zhi Huang McKnight Brain Institute, Departments of Psychiatry, Neuroscience, and Pharmacology & Therapeutics, College of Medicine, University of Florida, Box 100256, Gainesville, FL 32610–0256, USA

John D. Johnson Department of Integrative Physiology, Center for Neuroscience, University of Colorado-Boulder, Clare Small Building, Boulder, CO 80309–0354, USA

Rodney W. Johnson Integrative Immunology and Behavior Program, Department of Animal Sciences, University of Illinois, 1207 West Gregory Drive, Urbana, IL 61801, USA, rwjohn@uiuc.edu

Keith W. Kelley Integrative Immunology and Behavior Program, Department of Animal Sciences, College of ACES, Department of Pathology, College of Medicine, University of Illinois at Urbana-Champaign, Urbana, IL 61801, USA, kwkelley@illinois.edu

Sarah L. Kennedy Department of Integrative Physiology, Center for Neuroscience, University of Colorado-Boulder, Clare Small Building, Boulder, CO 80309–0354, USA

Martin Klein Department of Medical Psychology – D349, VU University Medical Center, Van der Boechorststraat 7, 1081 BT Amsterdam, The Netherlands, M.Klein@vumc.nl

Rachel A. Kohman Joint Graduate Training Program in Toxicology, Department of Psychology, Rutgers University and University of Medicine and Dentistry of New Jersey, 152 Frelinghuysen Road, Piscataway, NJ 08855, USA

Alexander W. Kusnecov Joint Graduate Training Program in Toxicology, Rutgers University and University of Medicine and Dentistry of New Jersey, 152 Frelinghuysen Road, Piscataway, NJ 08855, USA; Behavioral Neuroscience Program, Department of Psychology, Rutgers University, Piscataway, NJ 08855, USA, kusnecov@rci.rutgers.edu

Jessica L. Lynch Departments of Internal Medicine, Geriatric Division and Pharmacological and Physiological Science, Saint Louis University School of Medicine, 1402 South Grand Blvd, St. Louis, MO 63104

Ruchika Mohla Department of Psychiatry, UMDNJ – New Jersey Medical School, Behavioral Health Science Building, Room F-1559, 183 South Orange Avenue, Newark, NJ 07103, USA

Norbert Müller Department of Psychiatry and Psychotherapy, Ludwig-Maximilians-University Munich, Nussbaumstrasse 7, D-80336 Munich-Germany, norbert.mueller@med.uni-muenchen.de

J.C. O'Conner Integrative Immunology and Behavior Program, Department of Animal Sciences, College of Medicine, University of Illinois at Urbana-Champaign, 212 Edward R. Madigan Laboratory, 1201 West Gregory Drive, Urbana, IL 61801, USA

Mark R. Opp Department of Anesthesiology, University of Michigan, 7422 Medical Sciences Building I, 1150 West Medical enter Drive, Ann Arbor, MI 48109–5615, USA; Department of Molecular & Integrative Physiology, University of Michigan, 7422 Medical Sciences Building I, 1150 West Medical enter Drive, Ann Arbor, MI 48109–5615, USA; Neuroscience Graduate Program, University of Michigan, 7422 Medical Sciences Building I, 1150 West Medical enter Drive, Ann Arbor, MI 48109–5615, USA, mopp@med.umich.edu

Paul H. Patterson Biology Division, California Institute of Technology, 216–76, Caltech Pasadena, CA 91125, USA, php@its.caltech.edu

Bradley D. Pearce Department of Psychology, Emory University, 532 North Kilgo Circle, Atlanta, GA 30322, USA, bpearce@emory.edu

John M. Petitto McKnight Brain Institute, Departments of Psychiatry, Neuroscience, and Pharmacology & Therapeutics, College of Medicine, University of Florida, Box 100256, Gainesville, FL 32610–0256, USA, jpetitto@UFL.EDU

Tulin O. Price Departments of Internal Medicine, Geriatric Division and Pharmacological and Physiological Science, Saint Louis University School of Medicine, 1402 South Grand Blvd, St. Louis, MO 63104

Pranela Rameshwar Department of Medicine, UMDNJ – New Jersey Medical School, 185 South Orange Avenue, Newark, NJ 07103, USA, rameshwa@umdnj.edu

Amy F. Richwine Integrative Immunology and Behavior Program, Department of Animal Sciences, University of Illinois, 1207 West Gregory Drive, Urbana, IL 61801, USA

Boris Sakic Department of Psychiatry and Behavioral Neurosciences, McMaster University, Hamilton, Ont., Canada L8N 3Z5.

Steven J. Schleifer Department of Psychiatry, UMDNJ – NJ Medical School, Behavioral Health Science Building, Room F-1430, 183 South Orange Avenue, Newark, NJ 07103, USA, schleife@umdnj.edu

Markus J. Schwarz Department of Psychiatry and Psychotherapy, Ludwig-Maximilians-University Munich, Nussbaumstrasse 7, D-80336 Munich-Germany

Allan Siegel Departments of Neurology & Neuroscience and Psychiatry, UMDNJ – New Jersey Medical School, MSB Room H-592, 185 South Orange Avenue, Newark, NJ 07103, USA, Siegel@UMDNJ.edu

Stephen Smith Biology Division, California Institute of Technology, Pasadena, CA 91125, USA

S.S. Szegedi Integrative Immunology and Behavior Program, Department of Animal Sciences, College of Medicine, University of Illinois at Urbana-Champaign, 212 Edward R. Madigan Laboratory, 1201 West Gregory Drive, Urbana, IL 61801, USA

Jianping Wang Division of Pharmacology and Toxicology, School of Pharmacy, University of Missouri, Kansas City, 2464 Charlotte Street, Kansas City, MO 64108, USA, wangjp@umkc.edu

Randall T. Woodruff Department of Psychiatry, UMDNJ – New Jersey Medical School, Behavioral Health Science Building, Room F-1559, 183 South Orange Avenue, Newark, NJ 07103, USA

Steven S. Zalcman Department of Psychiatry, UMDNJ – New Jersey Medical School, Behavioral Health Sciences Building, Room F-1559, 183 South Orange Avenue, Newark, NJ 07103, USA, zalcmass@umdnj.edu

Introduction

For many years, the immune and central nervous systems were thought to function independently with little or no interaction between the two. This view has undergone dramatic changes over the past three decades. Indeed, we now know that there exists various feedback loops between the brain and immune systems that impact significantly upon different behavioral processes, including normal behavior and mental disorders. Pioneering efforts in generating this change were initiated by a number of early investigators. Included were those whose efforts were directed at establishing neuroimmune connections as well as others whose research focused upon the relationship between immunity, cytokines, and behavior.

This book brings together outstanding scientists and clinicians who have made major contributions to the rapidly developing field investigating the relationship between immunity and behavior. The book is divided into three parts. The first part describes pathways by which the brain and immune systems communicate and interact with each other. In the chapter "Cytokines and the Blood–Brain Barrier" provides insight into interactions between the blood–brain barrier and cytokines. Such interactions underlie basic communication between the immune system and brain that are present in normal as well as in disease conditions. In the chapter "Neurochemical and Endocrine Responses to Immune Activation: The Role of Cytokines," the neurochemical and endocrine consequences of immune challenge and cytokine administration on central neurotransmitter activity are discussed. In the chapter "Neural Pathways Mediating Behavioral Changes Associated with Immunological Challenge," the authors identify mechanisms by which pathogens or mediators derived from the immune system interface with peripheral neural pathways to influence brain function. Two chapters describe cytokine-molecular signaling pathways: a general overview of the molecular basis of cytokine function is presented in the chapter "Molecular Basis of Cytokine Function," while an analysis of molecular signaling pathways underlying the behavioral effects of interferon-α is the focus of the chapter "Interferon-α, Molecular Signaling Pathways and Behavior." The brain-to-immune component of the neuroimmune axis is discussed in the chapter "Exercise and Stress Resistance: Neural-Immune Mechanisms" in the context of stress and exercise.

The second part of the book focuses upon the neuroimmunological basis of behavior. Topics include the neural and behavioral consequences of different types of immunological challenges as well as the role of cytokine mediators. The chapters "Alteration of Neurodevelopment and Behavior by Maternal Immune Activation" and "Interleukin-2 and Septohippocampal Neurons: Neurodevelopment and Autoimmunity" highlight how development in brain and behavior are regulated by cytokines. In the chapter "Cytokine-Induced Sickness Behavior and Depression," the mechanisms governing cytokine-induced sickness behavior and depression are analyzed. The effects of bacterial toxins upon behaviors relevant to mood, anxiety, and cognition are discussed in the chapter "Effect of Systemic Challenge with Bacterial Toxins on Behaviors Relevant to Mood, Anxiety and Cognition." Physiological processes with respect to cytokine activity with respect to sleep are reviewed in the chapter "Cytokines, Immunity, and Sleep". The roles of peripheral and central cytokines in the regulation of aggressive behavior are discussed in the chapter "Cytokines and Aggressive Behavior." The final chapter in this part, "Neurochemical and Behavioral Changes Induced by Interleukin-2 and Soluble Interleukin-2 Receptors," discusses neurochemical and behavioral changes induced by interleukin-2. In several chapters in this part, the effects of cytokines upon feeding, anorexia, and anxiety are also described.

Part 3 of the book focuses upon the neuroimmunological basis of mental disorders. The chapter "Immunity and Depression: A Clinical Perspective" discusses the linkage between immunity and depression. The role of cytokines and immunity in schizophrenia are considered in the chapter "Cytokines, Immunity and Schizophrenia with Emphasis on Underlying Neurochemical Mechanisms." The evidence implicating systemic autoimmunity in the etiology of selected forms of mental illness is reviewed in the chapter "Immunobiological and Neural Substrates of Cancer-Related Neurocognitive Deficits." Insights into the relationship between viral infections and forms of psychiatric illness are provided in the chapter "Autoimmunity and Brain Dysfunction." The immunobiological and neural substrates of cancer-related cognitive deficits are discussed in the chapter "Viruses and Psychiatric Disorders." The final chapter, "Microglial Cells and Inflammatory Cytokines in the Aged Brain," provides an understanding of the role of microglial cells and neuroinflammation in behavioral pathology of the aged.

As indicated above, this book describes landmark studies produced over a relatively short period of time. In summarizing the contributions of the research contained in this book, we can summarize the achievements resulting from this research. These include in part: (1) Characterization of neuroimmune feedback loops, demonstrating that the brain is not completely immune privileged; (2) development of the model of sickness behavior showing that adaptive behavioral changes, and in some cases abnormal responses, are induced by a variety of immunological challenges and by peripheral and central cytokine administration; (3) that cytokines are potent neuromodulators that may play important roles in underlying a variety of behaviors independent of an ongoing immune response; and (4) that molecular substrates governing the neuroimmunological basis of behavior and mental disorders have begun to be explored and systematically examined. From in utero development to neuroinflammation in the aged, there is still much to be discovered about the impact of immunological challenge and cytokines on brain and behavior. The future appears bright for investigators in this field.

The editors also wish to thank the editorial staff of Springer-Verlag directed by Ann Avouris and her crew in providing excellent support, direction, and guidance in the preparation of this book.

Newark, NJ Allan Siegel
Newark, NJ Steven S. Zalcman

Part I
Neuroimmune Interactions

Cytokines and the Blood–Brain Barrier

William A. Banks, Jessica L. Lynch, and Tulin O. Price

Abstract The blood-brain barrier (BBB) mediates interactions between the immune and central nervous systems in several ways and is central to many mechanisms that form communication pathways within the neuroimmune axis. Here, we briefly review the chief types of interactions. Cytokines and immune cells cross the BBB by regulated mechanisms. Cytokines alter BBB characteristics, including the integrity of the BBB, its transport systems, and its ability to control immune cell trafficking. The cells that comprise the BBB secrete cytokines, prostaglandins, nitric oxide, and other immune-active factors. Such secretion is both constituitive and inducible, depending on the substance secreted. Secretion is also polarized; that is, secretion can be from either the luminal or abluminal membrane. This raises the possibility that the BBB may recieve signal at one membrane and secrete cytokine from the other as a mechanism of communication within the neuroimmune axis. In brief, the BBB is a central player in a number of mechanisms and pathways that comprise the neuroimmune axis.

Keywords Cytokine · Blood-brain barrier · Brain · Immune cell · Interleukin · Tumor necrosis factor

1 Introduction

The cells which comprise the blood–brain barrier (BBB) are simultaneously both inside the central nervous system (CNS) and outside it, with a luminal surface facing into the blood stream and an abluminal surface facing into the brain interstitial fluid. Circulating immune cells that can cross the BBB are capable of being either inside the CNS or outside of it. Nearly all other cell types are permanently fixed in locations either inside or outside the CNS. Therefore, it is perhaps not so surprising

W.A. Banks (✉)
Division of Geriatrics, Department of Internal Medicine, GRECC, Veterans Affairs Medical Center-St. Louis and Saint Louis University School of Medicine, WAB, 915 N. Grand Blvd, St. Louis, MO 63106, USA
e-mail: bankswa@slu.edu

A. Siegel, S.S. Zalcman (eds.), *The Neuroimmunological Basis of Behavior and Mental Disorders*, DOI 10.1007/978-0-387-84851-8_1,

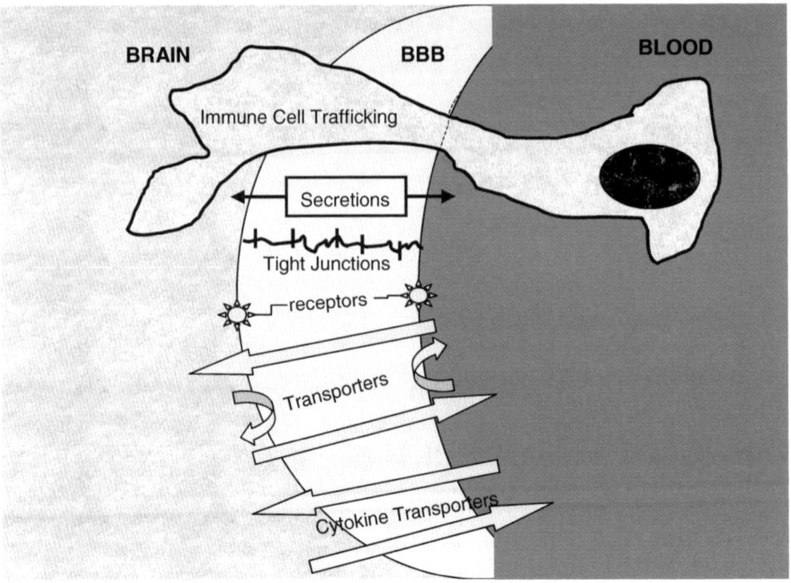

Fig. 1.1 Interactions between cytokines and the blood–brain barrier (BBB): pleuripotent proteins meet regulatory interface. Starting from bottom: cytokines can cross the BBB in the brain-to-blood or blood-to-brain directions; cytokines can affect influx and efflux transporters; cytokines can act at their own receptors on BBB cells or influence the action of other receptors located at the BBB; cytokines can open the tight junctions of the BBB; cytokines can be secreted by the cells of the BBB, including brain endothelial cells and the epithelium of the choroids plexus; cytokines influence immune cell trafficking

that the BBB and the neuroimmune system, both intimately involved in mediating interactions between the CNS and peripheral tissues, interact with each other.

The BBB is involved in a variety of ways with the neuroimmune system and the cytokines which mediate much of neuroimmune function (Fig. 1). The first interaction discovered was that the BBB can be disrupted by cytokines. However, more subtle aspects of BBB function are even more readily altered by cytokines, probably mediated through the cytokine receptors located on the cells which comprise both the vascular BBB and the blood–cerebrospinal fluid (BCSF) barrier. For example, some transport systems of the BBB are altered by neuroimmune stimulation, whereas others are not. Immune cell trafficking across the BBB, which requires orchestrated interactions between the cells which comprise the BBB and the immune cells, is likely mediated and certainly modulated by cytokines and cytokine receptors. Many cytokines cross the BBB in either the brain-to-blood or blood-to-brain direction by way of specific saturable transporters. This provides a direct way in which blood-borne cytokines can access tissue deep within the CNS. Additionally, cytokines can be secreted by the cells which comprise the BBB. Because the BBB is polarized with its luminal (blood side) membrane having different lipids, receptors, and transporters from that of its abluminal (brain side) membrane, the BBB can receive signals from one compartment (e.g., the blood stream) and secrete substances into the other (e.g., the CNS). Indeed, the vascular BBB is known to be able to secrete cytokines from either its luminal or abluminal surface and to respond to immune stimulation in a polarized fashion. Together, these interactions between the BBB and cytokines provide multilayered mechanisms which mediate interactions between the CNS and immune system. Here, we will briefly explore some of these interactions.

2 Barriers of the Brain

The brain uses barriers to separate itself from the bloodstream and so achieve the rigorous control of the brain microenvironment that is needed for complex neural signaling. There are two main physiological brain barriers and they differ in location, size, morphology, and function. The vascular BBB comprises endothelial cells and constitutes the interface between the blood and the interstitial fluid of the CNS tissue (Rapoport, 1976). The other blood barrier is the BCSF, which comprises a single layer of epithelial cells at the choroid plexus. Epithelial cells separate the plexus blood from the cerebrospinal fluid (CSF). The BCSF governs much of the exchange of water, ions, and other substances that occurs between CSF and blood (Spector and Johanson, 1989). There are a few, small localized brain regions (less than 1% of total brain weight), such as the area postrema and pineal, called circumventricular organs (CVOs), that lack the vascular BBB, but have a barrier of ependymal cells between the CVO tissue and CSF and of tanycytes between the CVO and adjacent brain tissue. Thus, a regulatory interface comprising a monolayer of cells separates the blood from the fluids of the CNS.

3 Morphologic Aspects of the Blood–Brain Barrier

Historically, the concept of the BBB developed from observations that were initiated in the 1880s. The German bacteriologist Paul Ehrlich and others demonstrated the existence of the BBB through the intravenous (i.v.) injections of vascular dyes (Bradbury, 1979). However, research into the BBB took over 80 years before electron microscopic studies convinced many that the BBB exists at the level of brain capillary endothelial cells (BCECs) and epithelial cells (Reese and Karnovsky, 1967). The vascular BBB is formed by a monolayer of endothelial cells that is different from other capillary beds in three fundamental ways: (i) high electrical resistance tight junctions cement together adjoining brain endothelial cells (BEC) to eliminate intercellular gaps; (ii) the BECs have very few pinocytotic vesicles; and (iii) BECs have few fenestrae. Taken together, these modifications eliminate the production of an ultrafiltrate from the plasma (Abbott, 2005).

The morphological properties of the BBB are present in the early embryonic mouse cerebral cortex. The first known marker of brain endothelial cells appears in the mice embryo, at day 10.5, before astrocytes are present (Qin and Sato, 1995).

Microvascular endothelium of the BBB has a close cellular connection with both astrocytic end-feet and pericytes. They both are closely applied to a continuous basement membrane surrounding the abluminal (brain) surface of the cerebral capillaries. All these elements with neurons are part of a functional neurovascular unit. Astrocyte–endothelial cell interactions are important in the maintenance of BBB tightness and function and are necessary for a functional neurovascular unit (Abbott et al., 2006; Walz, 2000). Pericytes are contractile connective tissue cells on the abluminal capillary walls and are pleuripotent (Dore-Duffy et al., 2000). The neurovascular unit (NVU) plays an important role in maintaining the structural integrity and the vasodynamic capacity of the BBB (Lai and Kuo, 2005). Impairments of the NVU and the BBB are present in the pathogenesis of many neurodegenerative CNS diseases (Alzheimer's disease, Parkinson's disease) and inflammation-related diseases in the brain (infections, stroke, vascular dementia, and multiple sclerosis; Avison et al., 2004; Persidsky et al., 2006; Zlokovic, 2005).

4 Properties of the Blood–Brain Barrier

The BBB functions as a dynamical physical barrier (tight junctions) and also as a metabolic barrier (enzymes, diverse transport systems) to the brain. The complex cellular system of the BBB both restricts and regulates exchange of substances between the blood and the brain. The tight junctions between endothelial cells form a diffusion barrier. This prevents the leakage of many substances from circulating blood into the brain via the paracellular route. Under physiological conditions, the BBB eliminates (toxic) substances from the endothelial compartment and supplies the brain's nutritive needs, and plays a role in communication between the brain and peripheral tissues (Banks, 1999).

In addition to the function of structural elements at the cerebral endothelia, drug-metabolizing enzymes and transport systems provide an enzymatic barrier (Lee et al., 2001; Pardridge, 2005). Transmembrane diffusion not only depends on a substance's lipid solubility but is also influenced by its molecular weight, charge, and other physicochemical properties. There are several types of selective, saturable mechanisms that transport substances (e.g., drugs) into and out of the brain at the BBB. These include carrier-mediated transport, receptor-mediated transcytosis, and efflux transporters. Carrier-mediated transport relies on molecular carriers present at both the luminal and the abluminal membranes of the BBB and may or may not require energy. They transport small molecules such as ions, energy sources, amino acids, and peptides.

Unidirectional systems are energy-dependent (active transport), whereas nonenergy-dependent transport (facilitated diffusion) is bidirectional. Receptor-mediated transcytosis involves the vesicular trafficking system of the brain endothelium and so requires energy in the transport process. Extremely large molecules and viruses use vesicular-dependent, or transcytotic, mechanisms to cross the BBB, but some small molecules are also vesicular-dependent. Several classes of vesicular transport likely occur at the BBB: podocytosis, clathrin-dependent, caveolar, and adsorptive endocytosis. Adsorptive endocytosis relies on nonspesific charge-based interactions. It is a mechanism by which glycoproteins can cross the BBB. Viruses use this mechanism to enter cells, including brain endothelial cells. Diapedesis, the process by which immune cells cross the BBB, has similarities to adsorptive endocytosis. Glycoproteins on the immune cell bind to glycoproteins on the brain endothelial cell. This initiates a series of events involving cytokines and other messengers to communicate between the immune and endothelial cells.

The BBB is also a polarized barrier; that is, a polarity exists between luminal and abluminal membrane surfaces of the endothelial cells contributing to the barrier function. Compositions of lipids, receptors, enzymes, ion channels, and transporters differ between the luminal and abluminal sides. Some substances are transported only in one direction or the other. For example, the P-glycoprotein (P-gp) system plays a unique role among efflux systems (brain-to-blood) in the BBB. One of the most important transporters, P-gp is expressed in the luminal membrane of the endothelial cells and actively excludes toxins and many xenobiotics from the brain (Begley, 2004).

The BBB has a secretor capacity besides transport function. The choroid plexus and the BECs are able to produce and secrete neuroactive and immunoactive substances, such as prostaglandins, nitric oxide, and cytokines.

5 Cytokine Release from BBB Endothelial Cells: Major Properties

Cytokines at the BBB can act in autocrine-, paracrine-, and hormone-like fashions. Cytokines are pleiotropic, redundant, and multifunctional. It is known that cytokines have effects on cells outside the immune system, and that non-immune cells can synthesize and secrete cytokines to regulate the immune response to injury and infection. Cytokines generally have short half-lives in the circulation, usually measured in minutes when they are injected i.v. (Vilcek, 2003). They can also act antagonistically and synergistically; that is, a cytokine may increase or decrease the production of another cytokine. A variety of immune cells secrete cytokines, such as monocytes, macrophages, activated T cells, B cells, natural killer (NK) cells, and fibroblasts in the periphery (Abbas et al., 2000). Also, cytokine production has been described in many other cell types such as smooth cells, muscle cells, endothelial cells, fibroblasts, keratinocytes, cardiac myocytes, and eccrine sweat glands.

In the CNS, cytokines are produced in a variety of cells including microglia, astrocytes, fibroblasts, and vascular endothelial cells (Vilcek, 2003). BECs, the major component of the NVU of the BBB, are themselves capable of secreting several cytokines either spontaneously or with stimulation that can act at both peripheral tissues and within the CNS (Fabry et al., 1993; Frigerio et al., 1998; Hofman et al., 1999; Quan and Banks, 2007; Reyes et al., 1999; Vadeboncoeur et al., 2003). Cytokine secretion is variable, inducible, and polarized in both the sense of receiving immune signals and of cytokine secretions.

6 Polarization of the Cytokine Secretion from BECs

Cytokine secretion from BECs is polarized; that is, endothelial cells can secrete substances into the blood or the CNS. This is because of the unique feature of the BBB having both a blood and a CNS side, facing simultaneously into the blood and into the CNS. The luminal (blood-side) and abluminal (brain-side) membranes of the BECs have unevenly distributed receptor and transporter proteins and lipid composition, which helps the polarization of the BBB properties (Banks and Broadwell, 1994; Betz and Goldstein, 1978; Davson and Segal, 1996; Taylor, 2002). Thus, BECs can receive stimulation from one side (e.g., luminal cell membrane surface), but released a substance into its opposite side (e.g., abluminal cell membrane surface) in response. As an example, this function of the endothelial cells of the BBB has been shown after application of lipopolysaccharide (LPS) to the abluminal membrane of the endothelial cells. IL-6 secretion then increased from its luminal membrane by about 10-fold. IL-6 was preferentially secreted from the luminal surface and this secretion was found to be more robust than from that of the abluminal surface (Verma et al., 2006). Also, when brain endothelial cells are exposed to luminal gp120 (HIV-1 viral coat protein), they secrete the cytokine endothelin-1 (ET-1) into the abluminal compartment (Didier et al., 2002). As yet another example, a feeding-related peptide, adiponectin, can interact with the luminal surface of the BBB to modify the secretions of cytokines into the CNS (abluminal surface) and can reduce the release of IL-6 by an immortalized cell line of rat brain endothelial cells (Qi et al., 2004; Spranger et al., 2006).

7 Modulators of the Cytokine Secretion from BBB Endothelial Cells

Cytokines are important mediators in physiologic and pathophysiologic processes affecting the CNS. Various and multiple stimuli regulate the production of cytokines. Invasive pathogens or their products, such as LPS (derived from the cell walls of gram-negative bacteria), cause the secretion of most proinflammatory cytokines. Also, other cytokines (such as TNF-α or IL-1), cocaine, HIV-1 related proteins, and other immune-active substances are inducers of proinflammatory cytokine production (Hofman et al., 1999; Lee et al., 2001; Reyes et al., 1999). BECs express proinflammatory cytokines during bacterial meningitis, HIV-associated dementia, and after traumatic or ischemic brain injury. For example, the meningeal pathogen, *Streptococcus suis* serotype 2, induces release of the proinflammatory cytokines tumor necrosis factor-alpha (TNF-α), interleukin-1 (IL-1), IL-6, and the chemokines IL-8 and monocyte chemotactic protein-1 (MCP-1) by human BECs (Vadeboncoeur et al., 2003). Hypoxia/ischemia also can be a stimulus for cytokine release by BECs (Reyes et al., 1999). In hypoxia, upregulation of ET-1 secretion induces MCP-1 release from human brain-derived endothelial cells to mediate the damage to hypoxic brain tissue (Chen et al., 2001).

BECs are known to promote inflammatory cascades in Alzheimer's disease (AD). It has been shown that brain microvessels obtained from AD patients release high levels of inflammatory proteins such as IL-1β, IL-6, TNF-α, and cerebrovascular transforming growth factor-β, after exposure of the cerebral vasculature to beta-amyloid peptides. These secretory products may together support inflammatory cascades, killing neurons and damaging the integrity of the NVU (Basu et al., 2004; Christov et al., 2004; Grammas and Ovase, 2001; Koistinaho and Koistinahi, 2005; Liao et al., 2004; Zhao et al., 2006). It has also been demonstrated that human BECs have the potential to contribute to coordinated dysfunction of the NVU. For example, hypoxia-stressed human BECs secrete IL-1β at physiological concentrations that are known to induce significant release of Aβ42 peptides from human neural cells in primary culture (Liao et al., 2004; Zhao et al., 2006).

Substances that effect feeding (such as LPS and adiponectin) can be modulators of BEC secretion. LPS applied to monolayer cultures of BECs enhanced the release of IL-6, IL-10, granulocyte-macrophage colony stimulating factor (GM-CSF), and TNF-α from these cells (Verma et al., 2006). Adiponectin can modulate the release of cytokines and reduce the secretion of IL-6 by BECs (Spranger et al., 2006; Verma et al., 2006). BECs are known to release IL-6, and its release is influenced by proinflammatory events (Reyes et al., 1999). IL-6 is important in lipid and carbohydrate metabolism (Di Gregorio et al., 2004; Wallenius et al., 2002) and can exert effects on appetite (Larson and Dunn, 2001). Cerebral IL-6 levels are inversely correlated with body fat mass, and IL-6 knock out mice have an obesity that is reversed by central administration of IL-6 (Stenlof et al., 2003; Wallenius et al., 2002). Therefore, IL-6 can mediate potential effects of adiponectin on energy expenditure. Adiponectin could modify the release of cytokines from BBB cells. Although BECs do not secrete adiponectin, they do express adiponectin receptors, and adiponectin inhibits the secretory profile of these cells. Adiponectin-induced cytokine secretion by BECs represents a potential mechanism linking circulating adiponectin and CNS pathways involved in energy homeostasis. This also suggests that blood-borne substances can interact with the luminal surface of the BBB to modify secretions into the CNS (Qi et al., 2004; Spranger et al., 2006).

One of the other activators of cytokine release is viral infection. Synthesis and release of the cytokine ET-1 by human BECs can be stimulated by HIV-1 and its viral coat protein (gp120) in HIV infection. Human BECs had increased ET-1 mRNA expression and secretion of the ET-1 peptide when infected with HIV-1, and gp120 caused a dose-dependent increase in ET-1 mRNA synthesis. Therefore, it has been suggested that ET-1 produced by BECs under HIV/gp120 stimulation may be a cause of the brain injury seen in AIDS dementia complex (Didier et al., 2002). ET-1 is the most potent vasoconstrictor peptide in humans (Yanagisawa et al., 1988) and it has neuroregulatory and physiologic functions. ET-1 has been implicated as a mediator of the cerebrovascular responses seen with ischemic strokes and subarachnoid hemorrhages (Lampl et al., 1997; Suzuki et al., 1990). Also, HIV-infected individuals show elevated ET-1 levels in the CSF which is correlated with the degree of encephalopathy (Rolinski et al., 1999). An interaction between gp120 and the endothelial cells may contribute to HIV-1 encephalopathy (Banks et al., 1997; Stins et al., 2001).

Microvessel endothelial cells, when treated exogenously with the HIV-derived Tat protein, can increase their mRNA expression and protein secretion of IL-8. Furthermore, Tat regulation of IL-8 production is modulated by cytokines; TNF-α upregulates, whereas TGF-β downregulates, Tat-induced IL-8 synthesis (Hofman et al., 1999).

Examples of cytokines acting on BEC to release other cytokines exist. IL-1β stimulates the release of IL-6 from brain BEC and smooth muscle pericytes (Fabry et al., 1993). Stimulation of isolated monkey BECs with IL-1β caused a significant release of IL-6 (Reyes et al., 1999). Taken together, cytokine release from the cells which form the BBB can provide a mechanism for communication between CNS and peripheral tissues.

8 Transport of Cytokines Across the BBB

Originally, it was assumed that peptides, including cytokines, did not cross the BBB. In analogy to albumin, cytokines were thought to be too large and too hydrophobic to cross the BBB by membrane diffusion. However, this early assumption did not take into account the possibility that saturable transport systems for cytokines could exist.

Several cytokines have been studied for their ability to cross the BBB. Typically in these studies, a radiolabeled cytokine is injected into the jugular vein. High-performance liquid chromatography (HPLC) is used to measure the cytokine stability in the blood and the brain. The influx rate and initial volume of distribution are calculated by the multiple-time regression analysis method developed by Blasberg et al. (1983), Kastin et al. (2001), and Patlak et al. (1983). Few studies have used species specific immunoassays to determine whether a cytokine can cross the BBB. For example, we have injected human IL-1α i.v. into mice to demonstrate that the blood-borne cytokine can cross the BBB (Banks and Kastin, 1997) in amounts sufficient to affect cognition (Banks et al., 2001).

9 Cytokines that Cross the Blood–Brain Barrier

9.1 Interleukins

The first cytokines to be studied were the IL-1s. Human IL-1α, murine IL-1α, and murine IL-1β are transported by a saturable mechanism in the blood-to-brain direction (Banks et al., 1991). We reported an influx transfer constant of 0.25–0.43 $\mu l\ g^{-1}\ min^{-1}$ for IL-1α and 0.47 $\mu l\ g^{-1}\ min^{-1}$ for IL-1β. The initial volume of distribution was 20.1 $\mu l\ g^{-1}$ and 16.5 $\mu l\ g^{-1}$, respectively. These levels are similar to or exceed the level of uptake of many other blood-borne substances that affect brain function, consistent with IL-1 being an important mediator between the CNS and periphery.

There are differences in the rate of transport among brain regions that are likely physiologically relevant to CNS function. IL-1α is transported into the hypothalamus more rapidly than most other parts of the CNS (Moinuddin et al., 2000), although there is an especially high rate of its uptake into the posterior division of the septum (Maness et al., 1995). Uptake at the posterior division of the septum is important in the effects of blood-borne IL-1α on memory (Banks et al., 2001). We have shown that the impairment of memory in mice induced by i.v. injected human IL-1α can be prevented by injecting a blocking antibody specific for human IL-1α into the posterior division of the septum. These results could only be achieved if the cytokine mediating cognitive impairment was acting in the posterior division of the septum and was derived from blood-borne IL-1α.

Similar to IL-1, IL-6 also has a saturable blood-to-brain transport system (Banks et al., 1994). Murine and human IL-6 have a unidirectional influx constant of 3.05–4.54 (10^{-4}) $\mu l\ g^{-1}\ min^{-1}$ respectively. After i.v. injection, intact IL-6 can be recovered from the CSF after 10 and 30 min.

Unlike the IL-1s and IL-6, IL-2 is not transported by a saturable mechanism in the blood-to-brain direction of normal mice (Banks et al., 2004). In fact, there are several mechanisms at work to prevent IL-2 from crossing the BBB in the blood-to-brain direction. IL-2 is rapidly degraded in the brain or at the BBB. There is also a circulating substance, possibly a soluble receptor for IL-2, which further retards IL-2 blood-to-brain transport. Most significantly, IL-2 is transported by a saturable system in the brain-to-blood direction and is the only cytokine to date known to have a saturable efflux system. However, even in the absence of an efflux system, significant amounts of cytokine can enter the blood from the brain with the reabsorption of CSF into the blood (Chen et al., 1997; Chen and Reichlin, 1998).

9.2 Tumor Necrosis Factor α

Peripherally administered TNF-α exhibits an entry rate through the BBB 10–100 times the rate of the vascular marker albumin (Gutierrez et al., 1993). This indicates that although large doses of TNF-α can disrupt the BBB, it is able to cross without BBB disruption. Self-inhibition with mTNF-α showed that this transport system was saturable. Similar to IL-1α, there is regional differences in TNF-α transport, such that the spinal cord has higher permeability than the brain. In the spinal cord, there is a greater volume of distribution and a faster influx rate in the cervical and lumbar segments. In brain, regional uptake differs by about 10-fold in young, healthy mice.

9.3 Interferons

Similar to TNF-α, there is saturable blood-to-brain transport of interferon γ(IFN-γ). Pan et al. reported increased permeability of IFN-γ in the cervical and lumbosacral spinal cord when compared to the brain and thoracic spinal cord (Pan et al., 1997). Interestingly, there is little to no transport of IFN-α, a glycoprotein of similar weight with IFN-γ.

Table 1 Permeability of the BBB to Representative Cytokines

Cytokine or related substances	Permeability	Reference
Adiponectin	NI	Spranger et al., (2006)
Brain-derived Neurotrophic Factor	SI	(120)
Ciliary Neurotrophic Factor	SI	(121;122)
Cytokine-induced Neutrophil Chemoattractant-1	NI	(10)
Epidermal Growth Factor	SI	(123)
Epogen	NI	(124;125)
Fibroblast Growth Factor	SI	(126)
Glial Cell Line-derived Neurotrophic Factor	NT	(127)
Interferons	SI	(94;128)
Interleukin-1alpha	SI, NE	(1;2;129)
Interleukin-1beta	SI	(2)
Interleukin-1 receptor anatagonist	SI	(130)
Interleukin-2	NT, SE	(11;12)
Interleukin-6	SI, NE	(4;131) (22)
Interleukin-10	NT	(132)
Leptin	SI	(72)
Leukemia Inhibitory Factor	SI	(133)
MIP's	NT	(134)
Nerve Growth Factor	SI	(95;135)
Neurotrophin 3	SI	(95;121)
Soluble Receptors	ST	(136)
Transforming Growth Factor alpha	SI	(137)
Transforming Growth Factor beta	NT	(138)
Tumor Necrosis Factor alpha	SI, NE	(5;84) (5;21;86)

SI = Saturable blood-to-brain transport (Influx); SE = Saturable brain-to-blood transport (Efflux); NI = Nonsaturable blood-to-brain transport; NE = Nonsaturable brain-to-blood transport; NT = No blood-to-brain transport.

9.4 Other Cytokines

Since the discovery that cytokines cross the BBB, over a dozen cytokines have been assessed for their blood-to-brain transport (Table 1). Several of these cytokines are transported across the BBB, including leukemia inhibitory factor, ciliary neurotrophic factor, and epidermal growth factor. Other cytokines, such as IL-2 and IFN-α, have been shown to not cross well in the blood-to-brain direction.

10 Transport Systems Are Specific

The cytokine transport systems are specific for closely related cytokines (Banks, 2005; Banks et al., 1995; Xiang et al., 2005) For example, whereas there is self-inhibition of TNF-α transport, there is a lack of inhibition by IL-1α, IL-1β, IL-6, or MIP-1α; thus the transporter is highly specific for TNF. The IL-1 transport system does not transport IL-6 or TNF. Each of these cytokines has its own distinct system. Furthermore, there is species specificity. The IL-1 transporter in the mouse will transport both human and murine IL-1α and murine IL-1β, but not human IL-1β. Additionally, the murine IL-1 transporter favors murine IL-1α over human IL-1α and murine IL-1α over murine IL-1β. Similarly, human TNF is not transported in all mouse strains or in the rat.

11 Modulation of Cytokine Transport

11.1 Morphine

It has long been appreciated that morphine is a potent neuroimmune modulator. Peripherally, chronic morphine treatment promotes a Th2 environment suppressing Th1 cytokine production (Roy et al., 2001, 1995). Morphine treatment has also been shown to alter cytokine transport across the BBB. Studies by Lynch and Banks (2008) established that unlike peripheral interactions, acute and chronic morphine treatment and withdrawal from morphine alters IL-1α, IL-2, and TNF-α transport across the BBB differentially. Acute morphine treatment increases blood-to-brain transport of IL-1α, whereas there is no change in blood-to-brain transport of IL-2 and TNF-α. Chronic morphine treatment and withdrawal from morphine did not alter blood-to-brain transport of IL-1α and TNF-α, but did increase blood-to-brain transport of IL-2. As previously mentioned, the permeability of the BBB to IL-2 is dominated by brain-to-blood transport (Banks et al., 2004), with only limited blood-to-brain transport (Waguespack et al., 1994). In this study, chronic morphine and withdrawal from morphine did not alter brain-to-blood efflux, but induced a novel saturable blood-to-brain transport system for IL-2.

11.2 Spinal Cord Injury

Leukemia inhibitory factor (LIF) has an important role in spinal cord regeneration. For example, LIF-secreting fibroblasts significantly increase axonal sprouting of the corticospinal tract in rats after spinal cord injury (SCI; Blesch et al., 1999). In studies by Pan et al., LIF transport was increased in the spinal cord of SCI mice when compared with controls 1 week after injury (Pan et al., 2006). Furthermore, enhanced LIF transport can be suppressed by both unlabeled LIF and a blocking antibody against its specific receptor (Pan et al., 2000). This indicates that enhanced LIF transport is not caused by barrier disruption but involves receptor-mediated transport across the BBB. Spinal cord and brain injury has also been shown to alter TNF transport (Pan et al., 1997). The upregulation of TNF transport differs both temporally and regionally from the BBB disruption that can also occur with SCI (Pan and Kastin, 2001).

11.3 Experimental Autoimmune Encephalomyelitis

Transport of TNF is enhanced in experimental autoimmune encephalomyelitis (EAE), an animal model of multiple sclerosis. This enhancement is to the saturable component of TNF transport and is not dependent on BBB disruption

(Pan et al., 1996). The rate of TNF transport returns to normal with the resolution of clinical signs of EAE.

In summary, many cytokines have been shown to be transported across the BBB by saturable transport systems, the only mechanism by which exogenous, peripherally administered cytokines directly interact with the brain. Furthermore, these cytokines have been shown to exert their effects on the brain. The transport of cytokines across the BBB is highly selective for their ligands and not all cytokines are transported similarly for a given brain region. Transporters are also affected by numerous physiological and pathological conditions. These complexities of BBB cytokine transport need to be further studied in both health and disease to fully understand how they can affect many neuropathological events.

12 Conclusions

This review has highlighted some of the ways in which cytokines interact with the BBB. These mechanisms are important to BBB functioning and disease. They also form the basis by which the immune system and CNS can interact and affect one another. These interactions are important in the pathological aspects of neuroimmune diseases, but are also likely to be highly relevant in normal physiological functions.

References

Abbas, KA, Lichtman, A, Prober, JS. Effector mechanisms of immune responses. 2000; In: Cellular and Molecular Immunology 4th Edition (Editors: KA Abbas, JS Prober) Saunders, Philadelphia.

Abbott, NJ. Dynamics of CNS barriers: evolution, differentiation, and modulation. Cell Mol Neurobiol 2005; 25:5–23.

Abbott, NJ, Ronnback, L, Hansson, E. Astrocyte-endothelial interactions at the blood-brain barrier. Nat Rev 2006; 7:41–53.

Avison, MJ, Nath, A, Greene-Avison, R, Schmitt, FA, eenberg, RN, rgar, JR. Neuroimaging correlates of HIV-associated BBB compromise. J Neuroimmunol 2004; 157:140–146.

Banks, WA. Physiology and pathophysiology of the blood-brain barrier: Implications for microbial pathogenesis, drug delivery and neurodegenerative disorders. J Neurovirology 1999; 5:538–555.

Banks, WA. Blood-brain barrier transport of cytokines: A mechanism for neuropathology. Curr Pharm Design 2005; 11:973–984.

Banks, WA and Broadwell, RD. Blood to brain and brain to blood passage of native horseradish peroxidase, wheat germ agglutinin and albumin: pharmacokinetic and morphological assessments. J Neurochem 1994; 62:2404–2419.

Banks, WA, Farr, SA, La Scola, ME, Morley, JE. Intravenous human interleukin-1α impairs memory processing in mice: Dependence on blood-brain barrier transport into posterior division of the septum. J Pharmacol Exp Ther 2001; 299:536–541.

Banks, WA and Kastin, AJ. Relative contributions of peripheral and central sources to levels of IL-1α in the cerebral cortex of mice: assessment with species-specific enzyme immunoassays. J Neuroimmunol 1997; 79:22–28.

Banks, WA, Kastin, AJ, Akerstrom, V. HIV-1 protein gp120 crosses the blood-brain barrier: role of adsorptive endocytosis. Life Sci 1997; 61:L119–L125.
Banks, WA, Kastin, AJ, Broadwell, RD. Passage of cytokines across the blood-brain barrier. Neuroimmunomodulation 1995; 2:241–248.
Banks, WA, Kastin, AJ, Gutierrez, EG. Penetration of interleukin-6 across the blood-brain barrier. Neurosci Lett 1994; 179:53–56.
Banks, WA, Niehoff, ML, Zalcman, S. Permeability of the mouse blood-brain barrier to murine interleukin-2: Predominance of a saturable efflux system. Brain Behav Immun 2004; 18:434–442.
Banks, WA, Ortiz, L, Plotkin, SR, Kastin, AJ. Human interleukin (IL) 1α, murine IL-1α and murine IL-1β are transported from blood to brain in the mouse by a shared saturable mechanism. J Pharmacol Exp Ther 1991; 259:988–996.
Basu, A, Krady, JK, Levison, SW. Interleukin-1: a master regulator of neuroinflammation. J Neurosci Res 2004; 78:151–156.
Begley, DJ. ABC transporters and the blood-brain barrier. Curr Pharm Design 2004; 10:1295–1312.
Betz, AL and Goldstein, GW. Polarity of the blood-brain barrier: neutral amino acid transport into isolated brain capillaries. Science 1978; 202:225–227.
Blasberg, RG, Fenstermacher, JD, Patlak, CS. Transport of α-aminoisobutyric acid across brain capillary and cellular membranes. J Cereb Blood Flow Metab 1983; 3:8–32.
Blesch, A, Uy, HS, Grill, RJ, Cheng, JG, Patterson, PH, Tuszynski, MH. Leukemia inhibitory factor augments neurotrophin expression and corticospinal axon growth after adult CNS injury. J Neurosci 1999; 19:3356–3366.
Bradbury, M. The Concept of a Blood-Brain Barrier. 1979; John Wiley and Sons LTD, New York.
Chen, G, Castro, WL, Chow, HH, Reichlin, S. Clearance of ^{125}I-labelled interleukin-6 from brain into blood following intracerebroventricular injection in rats. Endocrinology 1997; 138:4830–4836.
Chen, G and Reichlin, S. Clearance of [^{125}I]-tumor necrosis factor-α from the brain into the blood after intracerebroventricular injection into rats. Neuroimmunomodulation 1998; 5:261–269.
Chen, P, Shibata, M, Zidovetzki, R, Fisher, M, Zlokovic, BV, Hofman, FM. Endothelin-1 and monocyte chemoattractant protein-1 modulation in ischemia and human brain-derived endothelial cell cultures. J Neuroimmunol 2001; 116:62–73.
Christov, A, Ottman, JT, Grammas, P. Vascular inflammatory, oxidative and protease-based processes: implications for neuronal cell death in Alzheimer's disease. Neurol Res 2004; 26:540–546.
Davson, H and Segal, MB. Special aspects of the blood-brain barrier. 1996; 303–485.
Di Gregorio, GB, Hensley, L, Lu, T, Ranganathan, G, Kern, PA. Lipid and carbohydrate metabolism in mice with a targeted mutation in the IL-6 gene: absence of development of age-related obesity. Am J Physiol-Endoc Metab 2004; 287:E182–E187.
Didier, N, Banks, WA, Creminon, C, Dereuddre-Bosquet, N, Mabondzo, A. HIV-1-induced production of endothelin-1 in an in vitro model of the human blood-brain barrier. Neuroreport 2002; 13:1179–1183.
Dore-Duffy, P, Owen, C, Balabanov, R, Murphy, S, Beaumont, T, Rafols, JA. Pericyte migration from the vascular wall in response to traumatic brain injury. Microvasc Res 2000; 60:55–69.
Fabry, Z, Fitzsimmons, KM, Herlein, JA, Moninger, TO, Dobbs, MB, Hart, MN. Production of the cytokines interleukin 1 and 6 by murine brain microvessel endothelium and smooth muscle pericytes. J Neuroimmunol 1993; 47:23–34.
Frigerio, S, Gelati, M, Ciusani, E, Corsini, E, Dufour, A, Massa, G, Salmaggi, A. Immunocompetence of human microvascular brain endothelial cells: cytokine regulation of IL-1beta, MCP-1, IL-10, sICAM-1 and sVCAM-1. J Neurol 1998; 245:727–730.
Grammas, P and Ovase, R. Inflammatory factors are elevated in brain microvessels in Alzheimer's disease. Neurobiol Aging 2001; 22:837–842.
Gutierrez, EG, Banks, WA, Kastin, AJ. Murine tumor necrosis factor alpha is transported from blood to brain in the mouse. J Neuroimmunol 1993; 47:169–176.
Hofman, F, Chen, P, Incardona, F, Zidovetzki, R, Hinton, DR. HIV-tat protein induces the production of interleukin-8 by human brain-derived endothelial cells. J Neuroimmunol 1999; 94:28–39.

Kastin, AJ, Akerstrom, V, Pan, W. Validity of multiple-time regression analysis in measurement of tritiated and iodinated leptin crossing the blood-brain barrier: meaningful controls. Peptides 2001; 22:2127–2136.

Koistinaho, M and Koistinahi, J. Interactions between Alzheimer's disease and cerebral ischemia--focus on inflammation. Brain Res Brain Res Rev 2005; 48:240–250.

Lai, CH and Kuo, KH. The critical component to establish in vitro BBB model: pericyte. Brain Res Rev 2005; 50:258–265.

Lampl, Y, Fleminger, G, Gilad, R, Galron, R, Sarova-Pinhas, I, Sokolovsky, M. Endothelin in cerebrospinal fluid and plasma of patients in the early stages of ischemic stroke. Stroke 1997; 28:1951–1955.

Larson, SJ and Dunn, AJ. Behavioral effects of cytokines. Brain Behav Immun 2001; 15:371–387.

Lee, G, Dallas, S, Hong, M, Bendayan, R. Drug transporters in the central nervous system: brain barriers and brain paranchyma considerations. Pharmacol Rev 2001; 53:569–596.

Lee, YW, Hennig, B, Fiala, M, Kim, KS, Toborek, M. Cocaine activates redox-regulated transcription factors and induces TNF-alpha expression in human brain endothelial cells. Brain Res 2001; 920:125–133.

Liao, YF, Wang, BJ, Cheng, HT, Kuo, LH, Wolfe, MS. Tumor necrosis factor-alpha, interleukin-1-beta, and interferon-gamma stimulate gamma-secretase-mediated cleavage of amyloid precursor protein through a JNK-dependent MAPK pathway. J Biol Chem 2004; 279:49523–49532.

Lynch, JL and Banks, WA. Opiate modulation of IL-1alpha, IL-2, and TNF-alpha transport across the blood-brain barrier. Brain Behav Immun 2008 Oct; 22(7):1096–102.

Maness, LM, Banks, WA, Zadina, JE, Kastin, AJ. Selective transport of blood-borne interleukin-1α into the posterior division of the septum of the mouse brain. Brain Res 1995; 700:83–88.

Moinuddin, A, Morley, JE, Banks, WA. Regional variations in the transport of interleukin-1α across the blood-brain barrier in ICR and aging SAMP8 mice. Neuroimmunomodulation 2000; 8:165–170.

Pan, W, Banks, WA, Kastin, AJ. BBB permeability to ebiratide and TNF in acute spinal cord injury. Exp Neurol 1997; 146:367–373.

Pan, W, Banks, WA, Kastin, AJ. Permeability of the blood-brain barrier and blood-spinal cord barriers to interferons. J Neuroimmunol 1997; 76:105–111.

Pan, W, Banks, WA, Kennedy, MK, Gutierrez, EG, Kastin, AJ. Differential permeability of the BBB in acute EAE: enhanced transport of TNF-α. Am J Physiol 1996; 271:E636–E642.

Pan, W, Cain, C, Yu, Y, Kastin, AJ. Receptor-mediated transport of LIF across the blood-spinal barrier is upregulated after spinal cord injury. J Neuroimmunol 2006; 174:119–125.

Pan, W and Kastin, AJ. Increase in TNF alpha transport after SCI is specific for time, region, and type of lesion. Exp Neurol 2001; 170:357–363.

Pan, W, Kastin, AJ, Brennan, JM. Saturable entry of leukemia inhibitory factor from blood to the central nervous system. J Neuroimmunol 2000; 106:172–180.

Pardridge, WM. Molecular biology of the blood-brain barrier. Mol Biotechnol 2005; 30:57–70.

Patlak, CS, Blasberg, RG, Fenstermacher, JD. Graphical evaluation of blood-to-brain transfer constants from multiple-time uptake data. J Cereb Blood Flow Metab 1983; 3:1–7.

Persidsky, Y, Ramirez, SH, Haorah, J, Kanmogne, GD. Blood-brain barrier: Structural components and function under physiologic and pathologic conditions. Neuroimmune Pharmacol 2006; 1:223–236.

Qi, Y, Takahashi, N, Hileman, SM, Patel, HR, Berg, AH, Pajvani, UB, Scherer, PE, Ahima, S. Adiponectin acts in the brain to decrease body weight. Nature Medicine 2004; 10:524–529.

Qin, Y and Sato, TN. Mouse multidrug resistance 1a/3 gene is the earliest known endothelial cell differentiation marker during blood-brain barrier development. Dev Dynamics 1995; 202: 172–180.

Quan, N and Banks, WA. Brain-immune communication pathways. Brain Behav Immun 2007; 21:727–735.

Rapoport, SI. Blood Brain Barrier in Physiology and Medicine. 1976; Raven Press, New York.

Reese, TS and Karnovsky, MJ. Fine structural localization of a blood-brain barrier to exogenous peroxidase. J Cell Biol 1967; 34:207–217.

Reyes, TM, Fabry, Z, Coe, CL. Brain endothelial cell production of a neuroprotective cytokine, interleukin-6, in response to noxious stimuli. Brain Res 1999; 851:215–220.
Rolinski, B, Heigermoser, A, Lederer, E, Bogner, JR, Loch, O, Goebel, FD. Endothelin-1 elevated in the cerebrospinal fluid of HIV-infected patients with encephalopathy. Infection 1999; 27:244–247.
Roy, S, Balasubrmanian, S, Sumandeep, S, Charboneau, R, Wang, J, Melnyk, D, Beilman, GJ, Vatassery, R, Barke, RA. Morphine directs T cells towards T(H2) differentiation. Surgery 2001; 130:304–309.
Roy, S, Loh, HH, Barke, RA. Morphine-induced suppression of thymocyte proliferation is mediated by inhibition of IL-2 synthesis. Adv Exp Med Biol 1995; 373:41–48.
Spector, R and Johanson, CE. The mammalian choroid plexus. Sci Am 1989; 261:68–74.
Spranger, J, Verma, S, Gohring, I, Bobbert, T, Seifert, J, Sindler, AL, Pfeiffer, A, Hileman, SM, Tschop, M, Banks, WA. Adiponectin does not cross the blood-brain barrier, but modifies cytokine expression of brain endothelial cells. Diabetes 2006; 55:141–147.
Stenlof, K, Wernstedt, I, Fjallman, T, Wallenius, V, Wallenius, K, Jansson, JO. Internleukin-6 levels in the central nervous system are negatively correlated with fat mass in overwieght/obese subjects. J Clin Endocrinol Metab 2003; 88:4379–4383.
Stins, MF, Shen, Y, Huang, SH, Gilles, F, Kalra, VK, Kim, KS. Gp120 activates children's brain endothelial cells via CD4. J Neurovirol 2001; 7:125–134.
Suzuki, H, Sato, S, Suzuki, Y, Takekoshi, K, Ishihara, N, Shimoda, S. Increased endothelin concentration in CSF from patients with subarachnoid hemorrhage. Acta Neurol Scand 1990; 81:553–554.
Taylor, EM. The impact of efflux transporters in the brain on the development of drugs for CNS disorders. Clin Pharmacokinet 2002; 41:81–92.
Vadeboncoeur, N, Segura, M, Al-Numani, D, Vanier, G, Gottschalk, M. Proinflammatory cytokine and chemokine release by human brain microvascular endothelial cells stimulated by Streptococcus suis serotype 2. FEMS Immunol Med Mic 2003; 35:49–58.
Verma, S, Nakaoke, R, Dohgu, S, Banks, WA. Release of cytokines by brain endothelial cells: a polarized response to lipopolysaccharide. Brain Behav Immun 2006; 20:449–455.
Vilcek, J. The cytokines: an overview. 2003; The Cytokine Handbook, 4th Edition Editor MT Thompson, Elsevier, Amsterdam.
Waguespack, PJ, Banks, WA, Kastin, AJ. Interleukin-2 does not cross the blood-brain barrier by a saturable transport system. Brain Res Bull 1994; 34:103–109.
Wallenius, V, Wallenius, K, Ahren, B, Rudling, M, Carlsten, H, Dickson, SL, Ohlsson, C, Jansson, JO. Interleukin-6-deficient mice develop mature-onset obesity. Nat Med 2002; 8:75–79.
Walz, W. Role of astrocytes in the clearance of excess extracellular potassium. Neurochem Int 2000; 36:291–300.
Xiang, S, Pan, W, Kastin, AJ. Strategies to create a regenerating environment for the injured spinal cord. Curr Pharm Design 2005; 11:1267–1277.
Yanagisawa, M, Kurihara, H, Kimura, S, Tomobe, Y, Kobayashi, M, Mitsui, Y, Yazaki, Y, Goto, K, Masaki, T. A novel potent vasoconstrictor peptide produced by vascular endothelial cells. Nature 1988; 332:411–415.
Zhao, Y, Cui, JG, Lukiw, WJ. Natural secretory products of human neural and microvessel endothelial cells: implications in pathogenic "spreading" and Alzheimer's disease. Mol Neurobiol 2006; 34:181–192.
Zlokovic, BV. Neurovascular mechanisms of Alzheimer's neurodegeneration. Trends Neurosci 2005; 28:202–208.

Neurochemical and Endocrine Responses to Immune Activation: the Role of Cytokines

Adrian J. Dunn

Abstract This chapter reviews the experimental evidence that activation of the immune system, e.g., following infection or challenge (for example, with viruses), alters the metabolism of certain neurotransmitters in the brain, most notably serotonin, and the catecholamine, norepinephrine, and the amino acid tryptophan, as well as activating the hypothalamo-pituitary-adrenal (HPA) axis. There may be causal relationships between the noradrenergic activation and the HPA axis, and the neurochemical changes are implicated in the behavioral responses, in particular the sickness behaviors associated with injuries and infections. Nevertheless, there are many gaps in our knowledge, and we do not yet have a detailed understanding of the relationships between the immune activation and the brain responses.

Keywords Cytokine · Interleukin · Interferon · Dopamine · Norepinephrine · Serotonin · Acetylcholine · Tryptophan · Fos · Neurochemistry · Behavior · HPA axis · Cyclooxygenase

1 Introduction

The immune system is able to detect environmental threats to the organism that may not be recognized by the classic six senses. For optimal survival, animals need to detect such threats, and to mount appropriate responses. There is thus a need for communication between the nervous system and the immune system, two rather different bodily systems. This chapter reviews our present understanding of the mechanisms involved in immune system signaling to the brain, indicating which brain systems are known to respond to immune system signals and how. At our present level of understanding, this is heavily focused on cytokines, the hormones of the immune system, that have the ability to signal the brain. We will also discuss the ways in which the

A.J. Dunn (✉)
Department of Psychology and Pacific Biosciences Research Center, University of Hawaii, Honolulu, HI, USA
e-mail: ajdunn@hawaii.edu

A. Siegel, S.S. Zalcman (eds.), *The Neuroimmunological Basis of Behavior and Mental Disorders*, DOI 10.1007/978-0-387-84851-8_2,

brain responds to those threats, by altering behavior, and the other bodily systems necessary to maintain homeostasis, and thus support the survival of the organism.

2 Brain Responses to Immune System Activation

It has been known for some considerable time that stressful situations in animals and man cause a co-activation of the sympatho-adrenal system (the sympathetic nervous system plus the adrenal medulla), and of the hypothalamo-pituitary-adrenocortical (HPA) axis. This is the classical physiological stress response. Activation of the sympatho-adrenal system elevates circulating concentrations of the catecholamine, norepinephrine (NE) from terminals of sympathetic nerves, and of NE and epinephrine (Epi) from the adrenal medulla. The HPA axis activation is initiated by the secretion of corticotropin-releasing factor (CRF) from cells in the paraventricular nucleus (PVN) of the hypothalamus that project to the median eminence (Fig. 1). The CRF is carried in the portal blood to the anterior pituitary gland where it stimulates the secretion of adrenocorticotrophic hormone (ACTH) and β-endorphin. ACTH enters the general circulation reaching the adrenal cortex where it stimulates the secretion of glucocorticoid hormones (cortisol in most animals; corticosterone in rats and mice; Fig. 1). The catecholamines (NE and Epi) circulating in the blood increase heart rate and blood pressure enabling the blood to supply more nutrients to tissues such as muscles which are likely to be needed for the "fighting or fleeing" associated with stress. The glucocorticoids, as their name implies, shift metabolism to mobilize glucose, elevating plasma glucose concentrations. The latter is complemented by a catecholamine-enhanced degradation of glycogen to glucose.

Many laboratories have studied the effect of various stressors, such as electric footshock and short-term restraint on various chemical constituents of the brain. The results using a variety of different techniques have shown clearly that footshock and restraint both activate catecholamine-containing neurons in the brain, primarily NE, but probably also dopamine (3,4 dihydroxyphenylethylamine, DA) and Epi. Specifically, the release and metabolism of NE is enhanced throughout the brain, induced by activation of several brain stem nuclei, such as the locus coeruleus (LC) which innervates much of the cerebral cortex, parts of the diencephalon, and the cerebellum, and the nucleus of the solitary tract (NTS; see Chapter 3 by Goehler). Epi-containing neurons are believed to be activated also, although there have been relatively few studies of the epi-containing systems. The secretion of DA is also activated to differing extents in various regions of the brain. The metabolism of the indoleamine, serotonin (5-hydroxytryptamine, 5-HT) is also increased throughout the brain, as are the concentrations of tryptophan (Trp), an essential amino acid for protein synthesis that is also an essential precursor for the synthesis of serotonin. Subsequent experiments using more sophisticated techniques, such as in vivo microdialysis and in vivo voltammetry have indicated that the increased metabolism of NE, DA, and 5-HT reflects increased secretion of these neurotransmitters in the brain. The release of NE is ubiquitous in the brain, reflecting the widespread distribution of axons and terminals of NE from brain stem nuclei, such as the LC and the

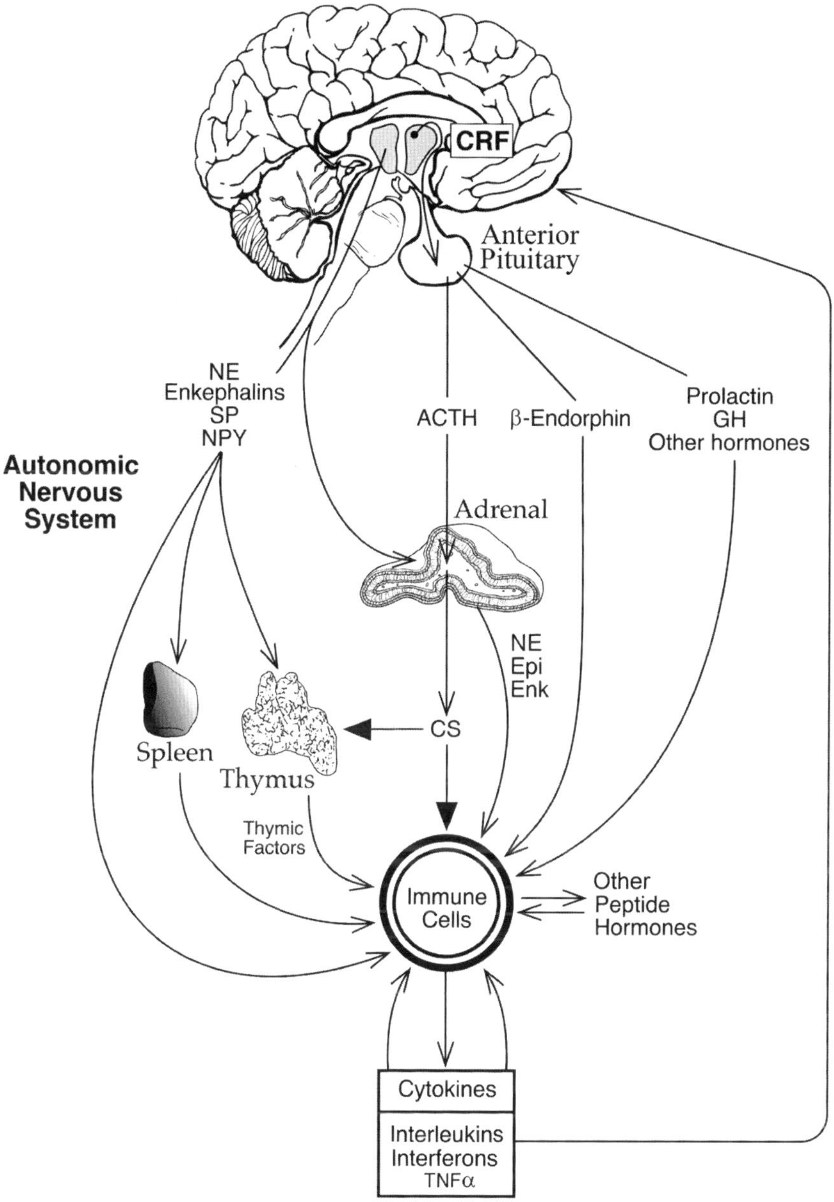

Fig. 1 Schematic of the interactions between the brain and components of the endocrine and immune systems. The ability of the brain to alter immune system function via a variety of endocrine pathways and the autonomic nervous system, and conversely the routes by which peptides and cytokines produced by cells of the immune system act on the brain are indicated. Abbreviations: E, epinephrine; ACTH, adrenocorticotropic hormone; CRF, corticotropin-releasing factor; CS, corticosteroids; Enk, enkephalins; GH, growth hormone; NE, norepinephrine; NPY, neuropeptide Y; SP, substance P; TNF, tumor necrosis factor (Modified from Dunn and Wang 1999)

NTS (Chapter 3). Similarly, the effects of stress on 5-HT are widespread reflecting the global distribution of terminals of projections from the raphe nuclei in the brain stem. However, dopaminergic systems are selectively activated in the mesolimbic and mesocortical projection systems (to the prefrontal cortex), and little, if at all, in the nigrostriatal system.

The response of the immune system during stress has long been considered anomalous, because immune system functions appeared to be inhibited during stress. This is primarily because glucocorticoids have long been known to have potent anti-inflammatory effects suppressing immune function (Munck and Guyre, 1986), even though it would be expected that the immune system would be important during stress, for example to coagulate blood, and to expel or destroy potential pathogens. A major factor contributing to this was a misinterpretation of the observation that during stress the thymus and the spleen were depleted of immune cells, and there were fewer immune cells in the circulation. These responses are now considered to reflect the mobilization (and hence apparent depletion) of immune cells to attack invading pathogens and/or repair wounds. The analogy is sending the soldiers from the barracks to the battle front (Dhabhar, 2002). The prevailing dogma that the immune system is inhibited when the organism is under stress has also been challenged, because many of the anti-inflammatory effects reflect the use of high doses of exogenous steroids, and/or potent synthetic glucocorticoids that far exceed concentrations achieved physiologically. Moreover, recently it has been shown that glucocorticoids are not exclusively anti-inflammatory, and can be immunoenhancing in some circumstances in the brain (e.g., Sorrells and Sapolsky, 2007).

There were a few reports in the literature that sickness might be associated with an activation of the HPA axis as indicated by increased concentration of glucocorticoids (Yelvington et al. 1987). We ourselves had noted increased plasma concentrations of corticosterone in mice that appeared to be sick or were wounded. Thus we decided to study the effect of infection of mice with influenza virus which was being studied in a nearby laboratory. The mice were infused intranasally with the virus, which caused an infection in the lungs, the normal site of infection for influenza. The dose chosen was such that the mice would become sick after about 2 days, and would normally die starting around 7 days. Mice were sacrificed at various times following infection with the virus, and HPA axis function was assessed by measuring plasma concentrations of ACTH and corticosterone. It was clear that as the mice became sick, plasma concentrations of ACTH and corticosterone increased progressively (Fig. 2; Dunn et al., 1989). Because there was no acute stimulus, the HPA axis was apparently chronically activated in contrast with the HPA responses to footshock or brief restraint after which plasma concentrations of ACTH and corticosterone normally return to baseline within an hour. This HPA axis activation extended the validity of its use to define stress, because influenza virus infection would clearly be regarded as stressful in man.

We also examined the neurochemical responses of the catecholamines and serotonin in the brains of the influenza virus-infected mice. Most interestingly, the influenza virus infection activated the brain noradrenergic and serotonergic systems as determined by measurement of their catabolites in a pattern resembling that

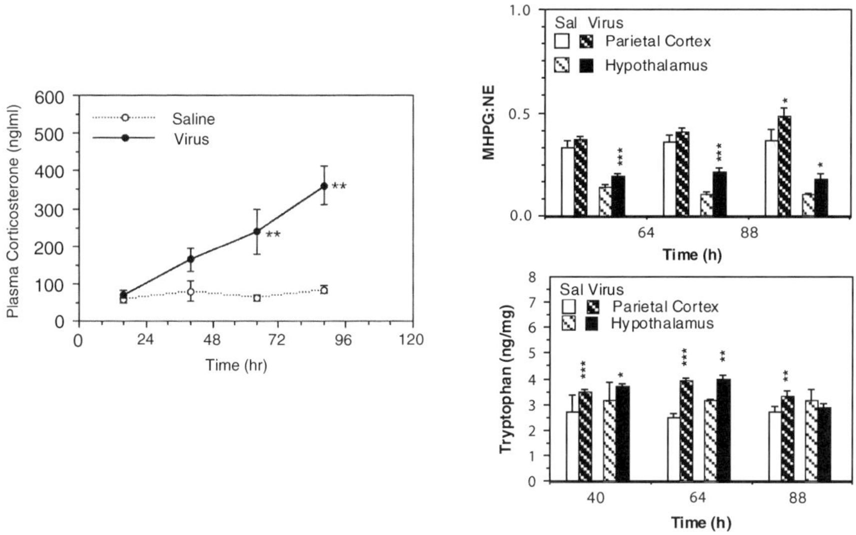

Fig. 2 The hypothalamo-pituitary–adrenocortical (HPA) and neurochemical responses at various times following influenza virus infection in mice. (Modified from Dunn et al., 1989)

observed following footshock and restraint, although there was little or no response in dopamine (Fig. 2; Dunn et al., 1989). Moreover, Trp concentrations were also elevated in a regionally nonselective manner. We had previously shown that treatment of mice with Newcastle disease virus (NDV) activated the HPA axis and brain noradrenergic and indoleaminergic systems (Dunn et al., 1987), so that it is likely that the responses we observed were a rather general response to viral infections (see review by Silverman et al., 2005). Similar findings have subsequently been made for a variety of different infectious agents (viruses, bacteria, protozoa, etc.) confirming the generality of these responses (see review by Besedovsky and del Rey, 1996). Thus two very different stressful treatments induced very similar physiological and neurochemical responses. These results support Selye's much maligned "nonspecificity" of the stress response. There was indeed similarity in the patterns of the physiological responses to different stressors.

The significance of the various physiological responses is partially understood. As mentioned above, the peripheral noradrenergic response serves to increase bloodflow to muscles and other organs that need more energy. The role of the glucocorticoid response is still controversial, but clearly it complements the sympathetic activation in generating more glucose as fuel for fighting and/or fleeing. The significance of the central noradrenergic response is thought to be to alert the brain and focus attention on novel stimuli in the environment that are likely to be the source of the stress, and hence target the response appropriately (Mason, 1980). The role of Trp and the indoleamines is unclear. Increasing brain concentrations of Trp may well be a precautionary measure, to prevent the brain running short of Trp, essential for the synthesis of proteins and serotonin.

3 The Mechanism(s) of Infection-Related Activation of the Stress Axis – The Involvement of Interleukin-1 (IL-1)

So what is the mechanism by which the body detects infections and initiates the stress responses? It seemed likely, a priori, that it would involve the immune system, because the immune system is the one responsible for surveillance of foreign antigens. We already had a clue, because in earlier work with NDV, we sought immune factors that might be responsible for its HPA and neurochemical effects. An important key was the seminal finding of Besedovsky et al. (1986) that peripheral administration of the cytokine, interleukin-1 (IL-1) potently stimulated the HPA axis in rats. Thus we injected mouse IL-1 which we prepared ourselves from stimulated mouse spleen cells, and recombinant human IL-1α intraperitoneally (ip) into mice, and observed not only a substantial HPA axis activation (increases in plasma ACTH and corticosterone), but also increases in brain 3-methoxy,4-hydroxyphenylethylene glycol (MHPG, the major brain catabolite of NE) and 5-hydroxyindoleacetic acid (5-HIAA, the major catabolite of 5-HT; Dunn 1988, see Fig. 3). Thus the mechanism appeared to be that immune cells recognized the administered pathogens, and initiated the synthesis and secretion of IL-1. The IL-1 then somehow signaled the brain to initiate the secretion of CRF necessary to initiate the activation of the HPA axis (Fig. 1). IL-1 also activated brain noradrenergic systems, and increased brain Trp and serotonin secretion.

IL-1 also elevates body temperature and is believed to be the major mediator of the fever associated with bacterial and viral infections (see review by Dinarello 1992). This fever is believed to be another aspect of the defensive mechanisms associated with sickness behavior, because viral replication is typically reduced at higher body temperatures (Hart, 1988; Dinarello, 1992).

4 Brain Responses to Other Cytokines

There is now a substantial literature on the neurochemical responses to immune activation in general, and in response to administration of cytokines and other immune factors. This literature has been reviewed relatively recently (Dunn, 2006), and the interested reader is referred to that source which will not be repeated here in detail.

Various results have been reported with IL-2, which is normally considered a growth-promoting cytokine for the immune system. There are reports of increases in NE and DA metabolism in mice (Zalcman et al., 1994), but decreased DA metabolism has also been reported (Dunn, 2006).

IL-6 is a cytokine that responds rapidly to almost any kind of infection or tissue damage. Like IL-1, it can activate the HPA axis, although it is far less potent than IL-1 (Wang and Dunn 1998). Peripherally administered mouse IL-6 is not pyrogenic (Wang et al., 1997), but IL-6 within the brain can induce fever. The

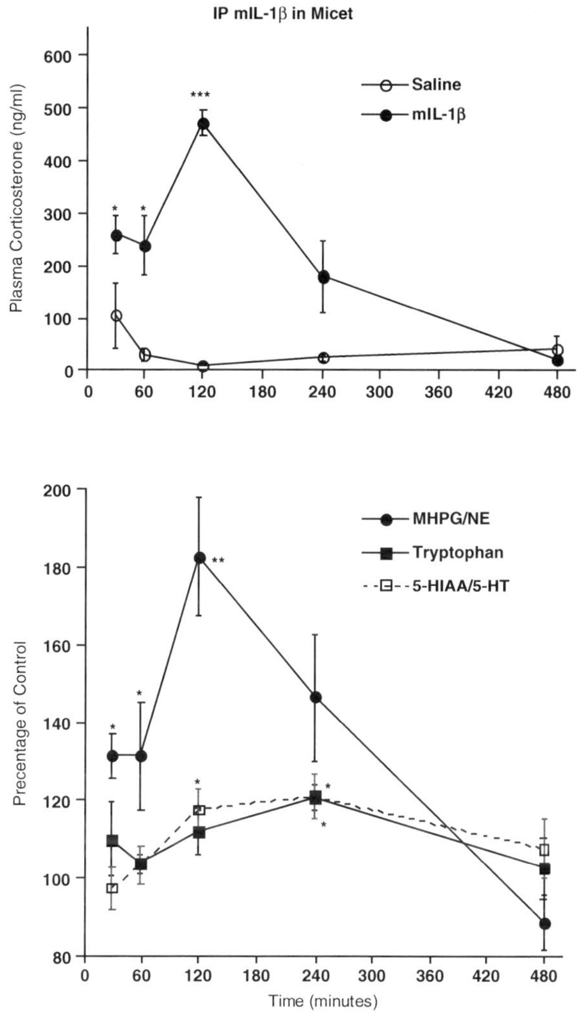

Fig. 3 The responses in plasma corticosterone (i.e., the HPA response) and certain neurochemical measures at various times following intraperitoneal (ip) administration of saline or 100 ng mouse interleukin-1β (IL-1β) to mice. Upper figure: Open circles – plasma concentrations of corticosterone (ng/ml) after injection of saline; filled circles – plasma corticosterone after ip injection of 100 ng IL-1β. Lower figure: Neurochemical measures made in the hypothalamus of the same animals as the upper figure: Filled circles: (3-methoxy,4-hydroxyphenylethylene glycol : norepinephrine) MHPG:NE ratios (an index of NE release); Open squares: (5-hydroxyindoleacetic acid : 5-hydroxytryptamine) 5-HIAA:5-HT ratios (an index of 5-HT release); Filled squares: Tryptophan concentrations, all expressed as the percentage of control (saline-injected) mice. Note the parallelism between the plasma corticosterone concentrations and the noradrenergic responses, while the tryptophan and 5-HIAA:5-HT responses parallel each other, but with a very different time course (Modified from Dunn 1988)

neurochemical responses to IL-6 reported have been somewhat variable. Zalcman et al. (1994) reported increases in prefrontal cortex DA and 5-HT metabolism, but Wang and Dunn (1998) observed effects on only Trp and 5-HT. Microdialysis and amperometric studies have also revealed activation by IL-6 of serotonergic systems (Barkhudaryan and Dunn, 1999; Zhang et al., 2001).

Tumor necrosis factor-α (TNF-α) has also been studied, but the interpretation of the literature is complicated by the failure of some studies to use homologous TNF, important because not all forms of TNF bind to receptors across species. At high doses, mouse TNF-α increased MHPG and Trp in mice (Ando and Dunn, 1999), along with a modest HPA activation which has been observed in several other reports (Dunn, 2006). TNF-α has little effect on body temperature in mice, although there was a small transient decrease at high doses (Wang et al., 1997).

The literature on the responses to the administration of interferons (IFNs) is markedly contradictory. IFN-α is used clinically to treat various forms of cancer, and it is well known to induce fever and HPA axis activation in man. However, studies in rodents have generated diverse results (Dunn, 2006). We have failed to observe any changes in body temperature in rats and mice using central or peripheral administration of homologous IFN-α or IFN-β. Likewise, we have not observed HPA activation in rats or mice, nor consistent effects on catecholamines or serotonin. We have, however, observed behavioral responses in tests for depression using recombinant rat IFN-α in rats, and natural mouse IFN-α in mice.

IFN-γ can profoundly affect indoleamine metabolism, although largely in the periphery. Its administration induces indoleamine-2,3-dioxygenase which converts Trp to kynurenine, and quinolinic acid. This can deplete circulating concentrations of Trp thus limiting the availability of this amino acid for protein and serotonin synthesis.

5 The Role of Catecholamines in the HPA Activation

The time courses of the brain noradrenergic response and the HPA response to IL-1 were very similar, suggesting that they might be related. This was also true for lipopolysaccharide (LPS) and viral infection. Because it was already known that noradrenergic neurons could activate the CRF-containing neurons in the PVN to initiate the HPA axis cascade (Saphier and Feldman, 1989; Al-Damluji, 1993), it seemed very likely that IL-1 stimulation of noradrenergic neurons was responsible for the CRF secretion. This was tested using the catecholamine-selective neurotoxin, 6-hydroxydopamine, by injecting it into the ascending noradrenergic bundle of rats to lesion the noradrenergic projection from the brain stem to the hypothalamic PVN, the nucleus in which the CRF-containing neurons considered critical for activating the secretion of ACTH are located (Saphier and Feldman, 1989; Al-Damluji, 1993). The results showed that such lesions in rats substantially reduced the IL-1-induced increases in plasma corticosterone, although there was a small residual response (Chuluyan et al., 1992). However in mice, whole brain depletion of NE resulted in only a very modest reduction in IL-1-induced plasma corticosterone, not evident in all experiments (Swiergiel et al., 1996). This suggested that NE may be involved in the IL-1-induced HPA axis activation, but that it was not the only mechanism.

A study by Nance's group (Wan et al., 1993), followed by studies from Bluthé et al. (1994) and Watkins et al. (1994), indicated that the vagus nerve was involved in the brain's response to LPS and suggested a potential route by which peripheral agents could signal the brain. Subsequent experiments verified that the HPA axis activation induced by ip IL-1 was indeed mediated at least in part by the vagus nerve (Fleshner et al., 1995). Our experiments in mice showed that subdiaphragmatic vagotomy attenuated, but did not prevent the neurochemical changes (NE and 5-HT) nor the HPA activation induced by ip administration of IL-1 or LPS (Wieczorek et al., 2005).

More recent experiments in rats have indicated that subdiaphragmatic vagotomy (Wieczorek and Dunn 2006a) and indomethacin pretreatment (Wieczorek and Dunn 2006b) largely prevented the increases in NE secretion in the hypothalamus induced by ip administered IL-1, but failed to block the increases in plasma ACTH and corticosterone. However, the combination of subdiaphragmatic vagotomy and indomethacin completely blocked both the noradrenergic activation, and the HPA axis activation to ip IL-1 in both rats and mice (Wieczorek and Dunn unpublished data). These results are consistent with the previously observed failures to block completely the neurochemical and endocrine responses with either treatment alone, and suggest strongly that there are indeed redundant pathways for the IL-1- and LPS-induced activation of the central noradrenergic system and the HPA axis in both species. Redundancy in what appears to be a critical defensive response is to be expected in biological systems.

6 The Significance of the Indoleamine Responses

The time courses of the brain indoleamine responses to IL-1 and LPS are distinct from those of the noradrenergic responses. As indicated above, the noradrenergic response in mice and rats peaks around 2 h, and has dissipated by 4 h, whereas the increases in Trp and 5-HIAA concentrations do not peak until 4–8 h, and dissipate slowly after that. This suggests that the two neurochemical responses involve different mechanisms and probably serve distinct functions. This conclusion is reinforced by the fact that whereas the noradrenergic responses are sensitive to cyclooxygenase (COX) inhibitors, the indoleamine responses are not (Dunn and Chuluyan, 1992). Moreover, the mechanism does not appear to involve the vagus, because the increases in Trp and 5-HIAA were not prevented in vagotomized mice (Wieczorek et al., 2005).

The brain content of Trp increases in response to a large variety of stimuli, including several psychotropic drugs, increases in body temperature and several different stressors (Dunn, 2006). The increases in Trp and 5-HIAA in response to footshock, restraint, IL-1, and LPS appear to depend upon peripheral sympathetic activity, because they can be blocked by pretreatment with the autonomic ganglionic blocker, chlorisondamine, and largely prevented by the β-adrenergic receptor antagonist, propranolol, but not by the α-adrenergic receptor antagonist, phentolamine, or the muscarinic receptor antagonist, scopolamine (Dunn and Welch, 1991). So the increases in brain Trp appear to reflect sympathetic activation. This is consistent with the ability of β_2-adrenergic agonists, such as clenbuterol, to increase brain concentrations of Trp (Edwards et al., 1989). However, β_2-adrenergic antagonists do not prevent the IL-1-induced increases in brain Trp, although some attenuations have been observed (unpublished observations). Recent studies in our laboratory have shown that both β_2- and β_3-adrenergic agonists increase net concentrations of brain Trp, and that β_3-adrenergic agonist administration can double the brain concentrations of Trp in mice (Lenard et al., 2003). Moreover, a β_2-adrenergic agonist increases brain Trp in β_3-knockout mice.

Nitric oxide synthase (NOS) also appears to be involved in the immune-induced activation of serotonergic neurons. Nonselective inhibitors of NOS attenuate or prevent the responses to IL-1 and LPS (Dunn, 1993), as well as to footshock (Dunn, 1998). Studies with selective NOS inhibitors indicate that iNOS is the principal form of NOS involved in the responses to IL-1 and LPS. However, the indoleamine responses in knockout mice lacking each of the various forms of NOS were not impaired, suggesting that there may be redundancy among the various forms of NOS (Dunn unpublished observations). The precise mechanism and significance of the NOS involvement is unclear. In particular the location of the NOS involved has not been identified.

7 The Relationship of the Neurochemical Responses to the Behavioral Responses

Infections and immune activation can profoundly influence behavior. It is a commonplace that this is also the role of brain neurotransmitters. The focus of behavioral studies has long been on sickness behavior. Sickness behavior was coined by Hart as a loose collection of behavioral responses that benefit a sick animal allowing it to avoid threatening situations (e.g., predators), while it is impaired by the illness. Hart (1988) wrote "The behavior of a sick individual is not a maladaptive and undesirable effect of illness but rather a highly organized strategy that is at times critical to the survival of the individual if it were living in the wild state." The concept of sickness behavior truly derived from Selye who called it "the syndrome of just being sick." ("Sick people are all indisposed, they look tired, have no appetite, gradually lose weight, do not feel like going to work, lie down rather than stand up. They all present a syndrome simply indicative of being ill" (Selye, 1979). Sickness behaviors include hypomotility (lethargy), hyperthermia, hypophagia (anorexia), decreased interest in exploring the environment, decreased libido, and increased sleep time.

To a first approximation, there are parallels between sickness behavior induced by immune activation and those induced by IL-1. One would anticipate that there would be clear relationships between the neurochemical changes associated with immune activation, but there are very few data to support this, even though the temporal aspects of the noradrenergic response and sickness behavior are very similar. Destruction of noradrenergic neurons in the brain, or the use of adrenergic receptor antagonists do little to alter behavioral responses to IL-1 administered either peripherally or centrally (Swiergiel et al., 1999). The only treatments identified that do impair IL-1-induced behavioral responses are COX inhibitors, such as indomethacin (Dunn and Swiergiel, 2000). When we studied the consumption of sweetened milk by mice (it may be thought of as the murine equivalent of ice cream!), we found that both COX1 and COX2 appear to be involved in the anorexic effects of IL-1, but at different times following its administration. In the first 30 min to 1 h, COX1 is involved, and the anhedonic effects of IL-1 can be inhibited by nonselective COX inhibitors, such as indomethacin, or SC-560, a COX1-selective inhibitor (Swiergiel and Dunn, 2002). In the second hour, indomethacin still inhibits the response, but SC-560 is not effective, whereas the COX2-selective inhibitors celecoxib and piroxicam are. Consistent with this, COX1-knockout mice do not show anorexia in the first hour, whereas COX2-knockout mice are not affected in the second (Swiergiel and Dunn 2002). Thus we conclude that the anorexic effects are mediated via prostaglandins or other eicosanoids.

Surprisingly, a host of other selective inhibitors for potential mediators had no significant amelioration of the depression of milk drinking by IL-1: including antagonists of dopaminergic receptors (haloperidol), α-adrenergic receptors (prazosin and phentolamine), β-adrenergic receptors (propranolol), 5-HT_1-, 5-HT_2- 5-HT_3-receptors; muscarinic cholinergic receptors (scopolamine), H_1-, H_2- and H_3-receptors, the opiate-receptor antagonist (naloxone), neurokinin-receptors (L659,877 and L703,606), CRF-receptors (alpha-helical CRF_{9-41}), melanocortin-4-receptors (SHU9119), NPY1 (BIBP3226), Substance P receptor antagonist (L703,606) 5-HT_{1A}- and 5-HT_{1B}-receptor agonists, selective neurotoxins for depleting DA and NE (6-hydroxydopamine) and NE (DSP-4), 5-HT (5,7-dihydroxytryptamine), the histamine synthesis inhibitor (α-fluoromethylhistidine), and NOS inhibitors (L-NAME, L-NMMA).

8 Does IL-1 Mediate the Neurochemical, Endocrine and Behavioral Responses to Immune Activation?

The foregoing has indicated that many of the responses to viral stimulation and immune activation are shared by IL-1, specifically the activation of the HPA axis, the activation of the central noradrenergic and serotonergic systems, the increase in brain Trp, and the behavioral effects. Because the synthesis of IL-1 is a ubiquitous response to immune stimulation, whether associated with pathogens or stimulants, such as LPS, is it the mediator of these responses? The answer is not simple. We have conducted a number of experiments to determine the involvement of IL-1 and certain other cytokines, such as IL-6, TNF-α, and the interferons. We have also studied responses to LPS, which is a relatively straightforward activator of the immune system, as well as number of challenges with pathogens.

8.1 Endotoxin (Lipopolysaccharide, LPS)

Peripheral administration of low doses of LPS is a very useful model for immune activation. It acts by binding to TLR-4 receptors in various organs, including endothelia. This ultimately results in the synthesis and secretion of IL-1. Not surprisingly, LPS elicits very similar neurochemical, endocrine, and behavioral responses to those of influenza virus and IL-1. It elevates plasma concentrations of ACTH and corticosterone, indicating HPA axis activation. It also increases brain MHPG, indicating activation of the brain noradrenergic systems, preferentially in the ventral (diencephalic) system, and increases brain 5-HIAA, indicating activation of brain serotonergic systems, and increasing brain concentrations of Trp (Dunn, 1992a, b). Each of these responses resembles those to IL-1. The differences are that the increases in plasma ACTH and corticosterone and the neurochemical responses were significantly slower, perhaps reflecting the delay involved in the induction of IL-1 by LPS. The distinction between the MHPG responses in the dorsal and ventral projection systems is less marked with LPS than with IL-1, and LPS also exhibits a significant activation of dopaminergic systems (i.e., increases in 3,4-dihydroxyphenylacetic acid (DOPAC) and homovanillic acid (HVA) not normally observed with IL-1. Whereas, the responses to icv IL-1 were quite similar to those of ip IL-1, icv LPS was less effective than ip. Thus it appears that induction of IL-1 cannot explain all the neurochemical effects of LPS.

LPS not only induces IL-1, but also IL-6 and TNF-α (IL-1 itself also induces IL-6; Zuckerman et al., 1989). Ip administration of IL-6 causes HPA axis activation in mice, but the effect is short lived, and IL-6 is far less potent than IL-1 (Wang and Dunn, 1998). IL-6 administration does not increase brain MHPG, but it does elevate brain concentrations of 5-HIAA and Trp, although it is significantly less potent in this respect than IL-1. Pretreatment with an antibody to mouse IL-6 did not alter the responses to IL-1, indicating that IL-6 does not the mediate the HPA or indoleaminergic effects of IL-1 (Wang and Dunn, 1999). However, the anti-IL-6 treatment did diminish the HPA response to LPS at later times, and attenuated the serotonin/Trp responses, suggesting that IL-6 contributes to the later responses to LPS. Administration of mouse TNF-α to mice also elevates MHPG and 5-HIAA, but only at relatively high doses (Ando and Dunn, 1999).

We have not observed significant anorexic effects of mouse IL-6 in mice, and mouse TNF-α induced anorexia only at very high doses (Swiergiel et al., 1997). Administration of the natural IL-1-receptor antagonist (IL-1ra) to mice at doses that substantially attenuated the anorexic response to mouse IL-1β significantly attenuated, but did not block, the anorexic response to low doses of LPS (Swiergiel et al., 1997; Swiergiel and Dunn 1999). Whereas the combination of IL-1ra with TNF-binding protein (TNFbp) significantly attenuated LPS-induced anorexia, the combination of IL-1ra, with the TNFbp, and a monoclonal antibody to IL-6 did prevent the anorexic response to low doses of LPS (Swiergiel and Dunn, 1999).

IL-1ra failed to alter the anorexic response to influenza virus infection when it was injected every 4 h for 5 days, however, when introduced continuously into mice using osmotic minipumps primed to deliver 50 μg/h, a statistically significant amelioration of the anorexia was observed, indicating that IL-1 participated in this response, although the anorexic effects were not prevented, perhaps because of the inability to block completely the actions of IL-1 (Swiergiel et al., 1997). In a subsequent series of experiments we tested the effects of continuous infusion of IL-1ra, repeated injection of TNFbp, and an IL-6 antibody. Statistically significant effects were observed on food pellet and sweetened milk intake, the anorexic effects, but the progression of the disease was not prevented (Swiergiel and Dunn, 1999).

9 Conclusions

Immune activation in the periphery causes profound changes in the brain. There is a sustained activation of noradrenergic and serotonergic neurons throughout the brain, resulting in an outpouring of NE and serotonin, however, the serotonergic response occurs much later than the noradrenergic response. There are also increases in brain concentrations of the essential amino acid, Trp. CRF-containing neurons are also activated resulting in an activation of the HPA axis, increasing the secretion of ACTH from the anterior pituitary gland, and consequent elevation of glucocorticoids from the adrenal cortex. This activation is mediated by the activation of brain stem noradrenergic neurons, and also by eicosanoids synthesized by COX in the hypothalamus. IL-1 secreted by macrophages and other immune cells is a major factor in the activation of the brain noradrenergic and serotonergic systems, but other immune factors (e.g., other cytokines) may also be involved. The glucocorticoid hormones alter metabolism and mobilize glucose to fuel the functions of the immune system. They also provide negative feedback limiting the activation of the immune system. IL-1 also initiates mechanisms, most notably behavioral ones, to limit activity of the animal (e.g., sickness behavior), and conserve energy for defensive fighting and/or escape. IL-1 (via eicosanoids) also increases body temperature inducing fever that limits viral replication, and enhances immune activity to neutralize pathogens.

References

Al-Damluji S (1993) Adrenergic control of the secretion of anterior pituitary hormones. Bailliere's Clin Endocrinol Metab 7:355–392.

Ando T, Dunn AJ (1999) Mouse tumor necrosis factor α increases brain tryptophan concentrations and norepinephrine metabolism while activating the HPA axis in mice. Neuroimmunomodulation 6:319–329.

Barkhudaryan N, Dunn AJ (1999) Molecular mechanisms of actions of interleukin-6 on the brain, with special reference to serotonin and the hypothalamo-pituitary-adrenocortical axis. Neurochem Res 24:1169–1180.

Besedovsky HO, A del Rey (1996) Immune-neuro-endocrine interactions: facts and hypotheses. Endocrine Rev 17:64–102.

Besedovsky HO, del Rey A, Sorkin E, Dinarello CA (1986) Immunoregulatory feedback between interleukin-1 and glucocorticoid hormones. Science 233:652–654.
Bluthé R-M, Walter V, Parnet P et al. (1994) Lipopolysaccharide induces sickness behaviour in rats by a vagal mediated mechanism. C.R. Acad. Sci. Paris 317:499–503.
Chuluyan H, Saphier D, Rohn, WM and Dunn, AJ (1992) Noradrenergic innervation of the hypothalamus participates in the adrenocortical responses to interleukin-1. Neuroendocrinol 56:106–111.
Dhabhar F (2002) Stress-induced augmentation of immune function -The role of stress hormones, leukocyte trafficking and cytokines. Brain Behav Immun 16:785–798.
Dinarello, C. (1992) Role of interleukin-1 in infectious diseases. Immunol Rev 127:119–146.
Dunn AJ, Powell, ML, Moreshead WV et al. (1987) Effects of Newcastle disease virus administration to mice on the metabolism of cerebral biogenic amines, plasma corticosterone, and lymphocyte proliferation. Brain Behav Immun 1:216–230.
Dunn AJ (1988) Systemic interleukin-1 administration stimulates hypothalamic norepinephrine metabolism parallelling the increased plasma corticosterone. Life Sci 43:429–435.
Dunn AJ, Powell ML, Meitin C et al. (1989) Virus infection as a stressor: influenza virus elevates plasma concentrations of corticosterone, and brain concentrations of MHPG and tryptophan. Physiol Behav 45:591–594.
Dunn AJ, Welch J (1991) Stress- and endotoxin-induced increases in brain tryptophan and serotonin metabolism depend on sympathetic nervous system activation. J Neurochem 57: 1615–1622.
Dunn AJ (1992a) Endotoxin-induced activation of cerebral catecholamine and serotonin metabolism: comparison with interleukin-1. J Pharmacol Exptl Therap 261:964–969.
Dunn AJ (1992b) The role of interleukin-1 and tumor necrosis factor α in the neurochemical and neuroendocrine responses to endotoxin. Brain Res Bull 29:807–812.
Dunn AJ Chuluyan H (1992) The role of cyclo-oxygenase and lipoxygenase in the interleukin-1-induced activation of the HPA axis: dependence on the route of injection. Life Sci 51:219–225.
Dunn AJ (1993) Nitric oxide synthase inhibitors prevent the cerebral tryptophan and serotonergic responses to endotoxin and interleukin-1. Neurosci Res Commun 13:149–156.
Dunn AJ (1998) Brain catecholaminergic and tryptophan responses to restraint are attenuated by nitric oxide synthase inhibition. Neurochem Intl 33:551–557.
Dunn AJ Wang JP (1999) Cytokines and the Brain. The Encyclopedia of Neuroscience (Ed: Adelman G, Smith B) Elsevier, 2nd edition, 506–509.
Dunn AJ, Swiergiel AH (2000) The role of cyclooxygenases in endotoxin- and interleukin-1-induced hypophagia. Brain Behav Immun 14:141–152.
Dunn AJ (2006) Effects of cytokines and infections on brain neurochemistry. Clin Neurosci Res 6:52–68.
Edwards DJ, Sorisio DA, Knopf S (1989) Effects of the β2-adrenoceptor agonist clenbuterol on tyrosine and tryptophan in plasma and brain of the rat. Biochem Pharmacol 38:2957–2965.
Fleshner M, Goehler LE, Hermann, J, et al. (1995) Interleukin-1β induced corticosterone elevation and hypothalamic NE depletion is vagally mediated. Brain Res Bull 37:605–610.
Hart BL (1988) Biological basis of the behavior of sick animals. Neurosci Biobehav Rev 12:23–137.
Lenard NR, Gettys TW, Dunn AJ (2003) Activation of β_2- and β_3-adrenergic receptors increases brain tryptophan. J Pharmacol Exptl Therap 305:653–659.
Mason ST (1980) Noradrenaline and selective attention: a review of the model and the evidence. Life Sci 27:617–631.
Munck A, Guyre PM (1986) Glucocorticoid physiology, pharmacology and stress. Adv Expt/Med Biol 196:81–96.
Saphier D, Feldman S (1989) Adrenoreceptor specificity in the central regulation of adrenocortical secretion. Neuropharmacol 28:1231–1237.
Silverman MN, Pearce BD, Biron CA Miller A.H (2005) Immune modulation of the hypothalamic-pituitary-adrenal (HPA) axis during viral infection. Viral Immunol 18:41–78.

Sorrells SF, Sapolsky RM. (2007) An inflammatory review of glucocorticoid action in the CNS. Brain Behav Immun 21:259–272.
Selye H (1979) The Stress of My Life. Van Nostrand Reinhold, New York.
Swiergiel AH, Dunn AJ, Stone EA (1996) The role of cerebral noradrenergic systems in the Fos response to interleukin-1. Brain Res Bull 41:61–64.
Swiergiel AH, Smagin GN Dunn AJ (1997) Influenza virus infection of mice induces anorexia: comparison with endotoxin and interleukin-1 and the effects of indomethacin. Pharmacol Biochem Behav 57:389–396.
Swiergiel AH, Dunn AJ (1999) The roles of IL-1, IL-6 and TNFα in the feeding responses to endotoxin and influenza virus infection in mice. Brain Behav Immun 13:252–265.
Swiergiel AH, Burunda T, Patterson B. et al. (1999) Endotoxin- and interleukin-1-induced hypophagia are not affected by noradrenergic, dopaminergic, histaminergic and muscarinic antagonists. Pharmacol Biochem Behav 63:629–637.
Swiergiel AH, Dunn AJ (2002) Distinct roles for cyclooxygenases 1 and 2 in interleukin-1-induced behavioral changes. J Pharmacol Exptl Therap 302:1031–1036.
Wan WH, Vriend CY, Wetmore L et al. (1993) The effects of stress on splenic immune function are mediated by the splenic nerve. Brain Res Bull 30:101–105.
Wang J-P, Ando T, Dunn AJ (1997) The effect of homologous interleukin-1, interleukin-6 and tumor necrosis factor-α on the core body temperature of mice Neuroimmunomodulation 4: 230–236.
Wang JP Dunn AJ (1998) Mouse interleukin-6 stimulates the HPA axis and increases brain tryptophan and serotonin metabolism. Neurochem Intl 33:143–154.
Wang J-P Dunn AJ (1999) The role of interleukin-6 in the activation of the hypothalamo-pituitary-adrenocortical axis and brain indoleamines induced by endotoxin and interleukin-1β. Brain Res 815: 337–348.
Watkins LR, Wiertelak EP, Goehler LE et al. (1994) Characterization of cytokine-induced hyperalgesia. Brain Res 654:15–26.
Wieczorek M, Swiergiel AH, Pournajafi Nazarloo et al. (2005) Physiological and behavioral responses to interleukin-1β and LPS in vagotomized mice. Physiol Behav 85:500–511.
Wieczorek M, Dunn AJ (2006a) Effect of subdiaphragmatic vagotomy on the noradrenergic and HPA axis activation induced by intraperitoneal interleukin-1 administration in rats. Brain Res 1101:73–84.
Wieczorek M, Dunn AJ (2006b) Relationships among the behavioral, noradrenergic and pituitary-adrenal responses to interleukin-1 and the effects of indomethacin. Brain Behav Immun 20:477–487.
Yelvington DB, Rosenthal MJ, Ratner A (1987) Effect of illness on hormonal response to footshock stress. Proc Soc Exptl Biol Med 184:239–242.
Zalcman S, Green-Johnson JM, Murray L, et al. 1994. Cytokine-specific central monoamine alterations induced by interleukin-1, -2 and -6. Brain Res 643:40–49.
Zhang J-J, Terreni L, De Simoni M-G, Dunn AJ (2001) Peripheral interleukin-6 administration increases extracellular concentrations of serotonin and the evoked release of serotonin in the rat striatum. Neurochem Intl 38:303–308.
Zuckerman SH, Shellhaas J, Butler LD (1989) Differential regulation of lipopolysaccharide-induced interleukin 1 and tumor necrosis factor synthesis: effects of endogenous and exogenous glucocorticoids and the role of the pituitary-adrenal axis. Europ J Immunol 19:301–305.

Neural Pathways Mediating Behavioral Changes Associated with Immunological Challenge

Lisa E. Goehler and Ron P.A. Gaykema

Abstract Peripheral infection or inflammation influences behavior by interacting with neural systems in the periphery and central nervous system that mediate pain, arousal and behavior. Multiple parallel pathways convey information relevant to sickness behavior, local inflammation and pain. For instance, vagal sensory neurons seem to contribute to systemic, brain-mediated responses, whereas spinal viscerosensory nerves modulate local inflammation and pain. Neural pathways of immune-to-brain communication drive projection neurons in the brainstem that potently influence forebrain regions, including the paraventricular thalamus, much of the hypothalamus, the basal forebrain, amygdala and bed nucleus of the stria terminalis. Cortical components of the immune-responsive network include the anterior cingulate, medial prefrontal and insular cortices. This immune "sensory system" provides the means by which the brain can monitor peripheral immune challenges and carry out relevant behavioral responses.

Keywords Viscerosensory · Inflammation · Infection · Primary afferents · Arousal · Brainstem · C-fibers · Vagus

1 Introduction

Peripheral infection and inflammation leads to the production of pro-inflammatory mediators locally in the infected or inflamed tissue. These pathogens, and/or immune-derived mediators (such as cytokines), influence mood, cognition, and behavior, and to do so they must interface with brain regions involved in these functions. The major pathways for access of pathogens to internal tissues are via inhalation or ingestion. Consequently, lungs and gastrointestinal tract are most often the initial sites of colonization for bacteria and parasites. Although inflammation in other places such as skin occurs, immune activation in internal tissues more reliably activates "systemic" as

L.E. Goehler (✉)
Laboratory of Neuroimmunology and Behavior, Center for the Study of Complementary and Alternative Therapies, School of Nursing, University of Virginia, Charlottesville, VA, 22908, USA
e-mail: goehler@virginia.edu

A. Siegel, S.S. Zalcman (eds.), *The Neuroimmunological Basis of Behavior and Mental Disorders*, DOI 10.1007/978-0-387-84851-8_3,

opposed to "local" physiological responses. In particular, activation of vagal afferents is more associated with "systemic" (Holzer 2007) responses to inflammation than activation of other neural pathways.

In this chapter, we will focus on mechanisms by which pathogens or mediators derived from the immune system interface with peripheral neural pathways influence brain functions. We note functional differences between peripheral neural pathways (vagal vs. spinal; visceral vs. somatic). We suggest that the basis for these differences follows from the fact that vagal input transmits directly to brain stem viscerosensory regions that serve as interfaces between signals arising from the body, and the neurocircuitry that modulates fundamental and widespread features of brain functioning, including arousal, affect and cognition.

2 Types of Immune-Sensitive Nerves

Peripheral nerves provide information to the central nervous system (CNS) pertinent to conditions in bodily tissues. Peripheral sensory nerves that respond to infection or inflammation belong to either general visceral or somatic sensory categories and are either lightly myelinated (Aδ) or unmyelinated (C) fibers. General visceral sensory fibers innervate blood vessels, and thoracic, abdominal, and pelvic viscera as they run with the vagus (parasympathetic) and sympathetic nerves, whereas somatic sensory neurons target the skin via the spinal segmental nerves, and mucosal tissues of the head (mouth, nose) via the trigeminal cranial nerve. In this way, peripheral neurons monitor both internal and external sites of body–pathogen interfaces (Fig. 1).

One of the important features of C-fibers relevant to inflammation and immune function is that they are bidirectional i.e., they perform both afferent and efferent functions. When stimulated they release neuropeptides, most notably substance

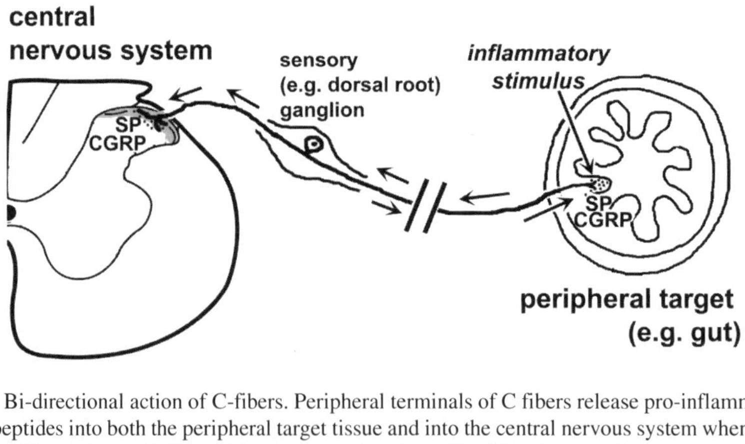

Fig. 1 Bi-directional action of C-fibers. Peripheral terminals of C fibers release pro-inflammatory neuropeptides into both the peripheral target tissue and into the central nervous system when stimulated. Abbreviations: CGRP, calcitonin gene-related peptide; SP, substance P

P and/or calcitonin gene-related peptide (CGRP), from their peripheral terminals in tissue. These neuropeptides, among other things, induce redness and swelling (plasma extravasation). In this way C-fibers are able to modulate conditions within peripheral tissues as well as signal activation to the CNS.

3 Cranial Nerve Viscerosensory Pathways

Among the twelve cranial nerves, two have been implicated in immune-to-brain signaling: the glossopharyngeal and the vagus. Together these two general viscerosensory nerves innervate most of the alimentary canal, as well as many other important visceral tissues including lung, liver, and lymph nodes that are notable as major points of immune pathogen interface. The cell bodies of vagal and glossopharyngeal sensory neurons comprise the vagal–glossopharyngeal complex, which contains three fused ganglia: the petrosal, which contains glossopharyngeal neurons, and the nodose (or inferior vagal) and jugular (or superior vagal) ganglia. This ganglion complex lies just outside the caudal cranium. The central projections of these neurons terminate in the dorsal vagal complex of the caudal brain stem (Fig. 2A; see below).

3.1 *The Glossopharyngeal Nerve*

The glossopharyngeal nerve (the ninth cranial nerve, which also carries special visceral gustatory signals) innervates the posterior two-thirds of the tongue and other posterior oral structures. This region constitutes the initial entry point for ingested pathogens, and contains specialized immune structures including the tonsils. Evidence for a functional role in immune surveillance by the glossopharyngeal nerve was reported by Romeo et al., showing that application of either lipopolysaccharide (LPS) or interleukin (IL)-1 into the soft palate (receptive field of the glossopharyngeal nerve) induces a fever that can be blocked by prior section of the glossopharyngeal nerve (Romeo et al., 2001). Sectioning the glossopharyngeal nerve did not block fever induced by systemic (intraperitoneal) injections of LPS or IL-1, supporting the idea that this nerve signals immune activation locally within the oral cavity.

Besides innervating the oral cavity, sensory fibers of the glossopharyngeal nerve innervate the carotid bodies. The carotid bodies are located at the carotid bifurcation and consist of a collection of chemosensory glomus cells, which are sensitive to blood gasses and likely other chemical stimuli in the general circulation (Matsuura 1973). The carotid bodies express IL-1 receptor type 1 immunoreactivity (Wang et al., 2002a), indicating that in addition to monitoring stimuli relevant to respiratory reflexes, these structures may also transduce systemic immune-related signals.

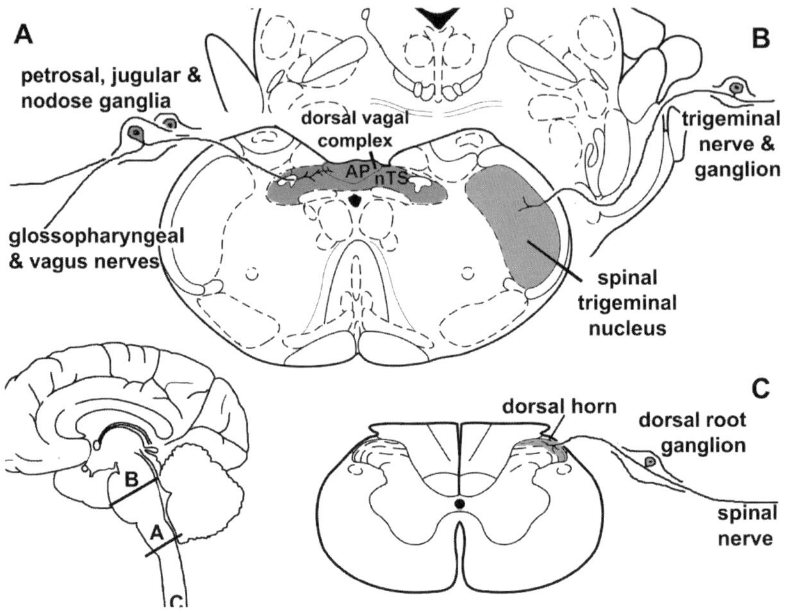

Fig. 2 Schematic view of relationships of the peripheral nerves and sensory ganglia to the primary sensory regions of the central nervous system where immune-related information is processed. The cell bodies of immune-responsive vagal and glossopharyngeal sensory neurons reside in the jugular, petrosal, and nodose ganglia of the vagal ganglion complex (depicted on the left, A). The central projections of these neurons enter the caudal medulla dorsally and run in the solitary tract to their targets, primarily in the medial and commissural subnuclei of the nucleus of the solitary tract (nTS) and area postrema (AP). The cell bodies of immune-responsive neurons of the trigeminal system are located in the trigeminal ganglia (depicted on the right side, B), enter with other trigeminal system nerve fibers but turn caudally to run in the spinal trigeminal tract to terminate in the spinal trigeminal nucleus of the caudal medulla. Spinal viscerosensory and somatic sensory neurons reside in the dorsal root ganglia and send their central processes to the superficial layers of the dorsal horn of the spinal cord. The transverse sections through the pons, medulla (A,B), and spinal cord (C) were modified after G. Paxinos and C. Watson, The Rat Brain Atlas, Fourth Edition, 1998. Academic Press, San Diego). The lower left image shows the lateral view of the human brain to indicate the relative levels of entry by the above-mentioned cranial and spinal nerves

3.2 The Vagus Nerve

The vagus nerve (the tenth cranial nerve), is nicknamed "the wanderer," because it innervates most internal structures of the neck, and thoracic and abdominal cavities, with the exception of the spleen (Cano et al., 2001, Nance and Burns 1989). Thus, the vagus nerve is particularly well-situated to serve as a conduit for immunosensory information arising from internal bodily structures. Among the neural pathways that communicate immunosensory information to the CNS, the vagus has received the bulk of experimental attention.

3.2.1 Targets of Vagal Sensory Nerves

In order to interface with peripheral nerves, pathogens must cross the epithelial barrier in the lung and gut, and sensory nerves must be present in the epithelial or subepithelial tissues. Berthoud and Neuhuber (2000) reported that anterograde tracing of vagal sensory neurons revealed vagal nerves that innervate the submucosal and epithelial regions of the intestine that are thus situated to respond promptly to pathogens in the gastrointestinal system. Abundant immune-type tissue and cells are found throughout the gastrointestinal tract, which is not surprising as this is a barrier site for infectious agents. Specialized immune tissue, including lymphoid nodules (which are organized somewhat like lymph nodes) and Peyer's patches of the small intestine, reside directly beneath the epithelium, and macrophages and dendritic cells line the epithelium, and Peyer's patches (Nagura et al., 1991). Berthoud et al. reported that vagal sensory fibers in the submucosa are closely associated with a cell type described as possessing several long processes, and were found in close association with mast cells (Berthoud and Neuhuber 2000, Goehler et al., 1999). These findings indicate that vagal sensory neurons occupy a position in which they might be sensitive to mediators produced by immune cells responding to local infection. In addition, peripheral nerves including the vagus are enriched with several types of immune cells (Goehler et al., 1999). Most of these cells are myeloid cells of the monocyte, macrophage, and dendritic cell family, based on morphological features and markers such as constitutive expression of major histocompatibility complex (MHC-II), a protein that enables antigen presentation to T cells, a critical step in the induction of systemic immune responses. Mast cells also occupy the vagus nerve (Goehler et al., 1999), and are potential sources of cytokines and other pro-inflammatory mediators including histamine and substance P. Numerous dendritic-like cells are interspersed among vagal nerve fibers, and within the paraganglia their processes encircle adjacent chemosensory (glomus) cells. These dendritic-like cells are also found among the cell bodies within the vagal sensory ganglia. In addition, the connective tissue surrounding the nerve contains myeloid cells, mostly ED-1 (CD68) and complement receptor-3 positive macrophages, as well as mast cells and possible lymphoid cells. When treated with i.p. LPS, dendritic-like cells, as well as some macrophages, express IL-1 immunoreactivity (Bronzino et al., 1976). The sensitivity of these cells to LPS suggests that they may serve a sentinel function, alerting the brain, via cytokine expression in the vagus nerve, regarding the presence of infectious microorganisms.

The potential contribution of vagal afferents innervating the lung to behavioral responses to immune challenge should not be underappreciated. The lung is richly innervated by both myelinated and unmyelinated vagal afferent fibers from both vagal sensory ganglia that innervate the airway epithelium and the neuroepithelial bodies (Adriaensen et al., 2006). These fibers express EP2 and EP4 prostaglandin and the P2X3 adenosine receptors, Thus vagal sensory fibers are adventitiously located to detect inhaled pathogens and respond to local immune activation in the lung. And given the access of the lung to circulating blood and lymph (pumped in by the right ventricle of the heart), vagal afferents could well be exposed to cytokines or pathogen products from "non-visceral" regions of the body.

In contrast to the mucosal surfaces of the lung and alimentary canal, as well as the other visceral including liver and pancreas, the spleen does not receive viscerosensory innervation from either the vagus or the spinal viscerosensory nerves (Cano et al., 2001, Opp 2005). The reason for this is unknown but may be related to the fact that the spleen is not quite on the first line of defense. Rather, pathogens typically arrive at lymph nodes first before the general circulation. Lymph nodes thus provide a site of early immune activation, as these are the major locations in which antigen presenting cells interface with the T cells that serve to co-ordinate immune responses. The lymphatic system comprises an interconnecting network of conducting vessels that carry immune cells and antigens, including microorganisms, from lymph node to node, progressively to the heart. Lymph nodes are innervated by both sympathetic and sensory neuropeptide-containing nerve fibers (Felton et al., 1984, Fink and Weihe 1988, Popper et al., 1988). Much of the lymphatic system, notably the pelvic, mesenteric, deep cervical, and mediastinal ducts and nodes, lies within the range of vagal afferent peripheral terminal fields as well. Vagal sensory neurons likely innervate these lymph nodes, based on findings that injections of the retrograde tracer Fluorogold into cervical and pelvic lymph nodes labeled neurons in the nodose and jugular ganglia (Goehler et al., 2000). These observations are consistent with a role for vagal afferents in monitoring early stage activation in immune-related tissues.

3.2.2 Experimental Evidence for the Vagus in Immune–Brain Signaling

Initial experimental evidence supporting a functional role for the vagus in immunosensory signaling was provided by studies that involved cutting the vagus nerve in the abdomen, below the diaphragm (subdiaphragmatic vagotomy). This is a partial lesion, leaving thoracic structures, notably lung and lymph nodes with intact vagal innervation. Unfortunately, sectioning the vagus above these structures is not compatible with life. Thus, negative findings from vagotomy studies can be interpreted as negating a role for subdiaphragmatic vagal afferents only. Findings from these studies have generally shown that vagotomy can block or attenuate a wide range of illness responses, including hyperalgesia, fever, hypersomnelence, hypothalamo-pituitary–adrenal activation, conditioned taste aversion, and social withdrawal (Bret-Dibat et al., 1995, Bluthe et al., 1994, Fleshner et al., 1995, 1998, Gaykema et al., 1995, 2000, Goehler et al., 1995, Hansen and Krueger 1997, Kapcala et al., 1996, Opp 2005, Opp and Toth 1998, Romanovsky et al., 1997, Sehic and Blatteis 1996, Wan et al., 1994, Watkins et al., 1994a, b, 1995a, b). In general, the effects of vagotomy are most pronounced when the immune stimulus is presented to peritoneal cavity, and when the dose of stimulant is low (Bluthe et al., 1996, Gaykema et al., 2000, Hansen et al., 2001). These latter findings suggest that the vagus nerve may contribute to the signaling of immune activity locally in visceral tissues and that higher doses of immune stimulants such as cytokines recruit additional immunosensory pathways associated with the brain, e.g., brain barrier tissues. Additionally, vagal sensory nerves left intact (innervating thoracic structures such as lung and lymph nodes) may contribute to immune signaling following subdiaphragmatic vagotomy.

Although the results from vagotomy studies support a role for the vagus in immune sensory signaling, the conclusions from these studies have been complicated by the fact that the vagus carries both sensory and motor nerve fibers. Thus, cutting the vagus may inhibit sickness responses not because it interrupts sensory signaling, but because it produces side effects or impairs immune functioning as a result of interrupting parasympathetic outflow, thus impairing vagal cholinergic anti-inflammatory mechanisms (Pavlov and Tracey 2005). However, vagotomized animals develop fevers identical to controls when the thermogenic stimulus is not associated with immune challenge (Milligan et al., 1997, Sugimoto et al., 1999), and vagotomy does not impair cytokine expression following LPS treatment, or the entry of cytokines or LPS into the systemic circulation (Gaykema et al., 2000, Hansen et al., 2000), two factors that might otherwise lead to impaired illness responses. In fact, vagotomy prevents fever responses to low doses of i.p. injected IL-1, even when the injected IL-1 reaches the systemic circulation (Gaykema et al., 2000). Taken together, findings from vagotomy studies support the idea that vagotomy effects follow primarily from interruption of vagal sensory transmission of cytokine or pathogen-derived signals.

Activation of vagal sensory neurons following systemic treatment with cytokines or pathogen products (such as LPS) provides evidence that these neurons indeed respond to immune challenge. For instance, IL-1 induces c-fos mRNA (Ek et al., 1998) and c-Fos protein (Goehler et al., 1997) in vagal sensory neurons, and increases electrically recorded neural firing in hepatic and gastric vagal sensory fibers (Ek et al., 1998, Niijima 1996). Similarly, peripheral injections of LPS induce c-Fos immunoreactivity in vagal sensory neurons, as does staphylococcal enterotoxin B (SEB), a product of gram-positive bacteria (Gaykema et al., 1998, 1999). Mesenteric vagal afferents respond rapidly to local (ex vivo) or systemic LPS administration (Liu et al., 2006, Wang et al., 2005). Evidence that this activation indeed results in signaling to the brain was provided by findings that systemic injections of IL-1 or LPS provoke glutamate release (the primary vagal sensory neurotransmitter) by central terminals of vagal neurons (Mascarucci et al., 1998). Consistent with these findings, subclinical gastrointestinal infection with live, gram negative bacteria (*Campylobacter jejuni* or *Citrobacter rodentium*) induces anxiety-like behavior in mice, concomitant with c-Fos protein induction in vagal sensory neurons, in the absence of circulating cytokines (Goehler et al., 2005, Lyte et al., 1998, 2006). Overall, these findings provide evidence for the idea that vagal sensory nerves innervating the gastrointestinal tract serves as a conduit by which such infections influence behavior.

3.2.3 Vagal Transduction of Pathogen Signals

The expression of c-Fos protein in vagal sensory neurons, and the prompt response of vagal nerve fibers following peripheral immune challenge implies that these neurons express receptors for pathogens and inflammatory mediators. For instance the rapid increase in mesenteric vagal afferent firing following LPS treatment (Liu et al., 2006, Wang et al., 2005) implies that these nerves express receptors, such as TLR 4/CD14 that transduce LPS signals. Receptors for both IL-1 and tumor necrosis factor (TNF) have been demonstrated on neurons in the vagal ganglia, as well as in other cells of these structures. Besides neurons, sensory ganglia contain satellite cells, which may be analogous to CNS glia, as well as macrophage/dendritic-like immune cells and endothelial cells associated with the vasculature that are potentially sources of either cytokine or cytokine receptors. Ek et al. (1998) demonstrated mRNA expression for type 1 IL-1 receptors in the primary sensory neurons and satellite cells in the vagal ganglia and Emch et al. (2001) reported TNF receptor 1 (TNFR1) immunoreactivity on the central projection of vagal sensory nerves within the brain stem, as well as on the cell bodies within the vagal sensory ganglia. Interestingly, TNFR1 was absent on the peripherally projecting fibers. Activation of TNFR1 receptors enhances glutamate release from vagal terminals (Herman et al., 2005), suggesting a role for TNF primarily as a neuromodulator, enhancing immune related signaling to the brain. Levels of TNF rapidly rise in the general circulation following peripheral administration of LPS (Hansen et al., 2000), providing a potential source of circulating cytokine possibly relevant to receptors expressed in the ganglia and brain stem (Hermann et al., 2005). The vagal response to LPS may have followed from circulating IL-l generated macrophages in response to the LPS injection, or by cells in the vagal ganglia, which express the LPS receptor, Toll like receptor (TLR)-4 (Hosoi et al., 2005) as well as TLR-9 (Sako et al., 2005), which responds to bacterial DNA. Vagal sensory neurons also express receptors for prostaglandins and nucleotides (Goehler et al., 2006), important intermediary molecules for a wide variety of inflammatory stimuli that constitute parts of the inflammatory cascade leading to brain responses such as fever. Thus, vagal sensory nerves may signal the presence of local tissue cytokines in the lung and gut epithelium, as well as cytokines and other mediators in the blood perfusing the nerve or ganglia. Taken together, findings from receptor localization and expression studies support a role of vagal sensory neurons in signaling peripheral immune activation.

Besides activating vagal afferent neurons directly, pathogens or cytokines may activate vagal afferents pathways via the chemoreceptive cells located in paraganglia or neuroepithelial bodies, as mentioned above. The vagal paraganglia are collections of glomus cells interspersed throughout the vagus nerve. These glomus cells cluster around blood and lymph vessels, suggesting that these cells are likely monitoring substances circulating in body fluids. Glomus cells of vagal paraganglia, like those of the carotid bodies, express IL-1 receptors (Goehler et al., 1998, Wang et al., 2000). Immune cells co-distribute with these glomus cells (Goehler et al., 1999). These cells express MHC-II constitutively, and LPS treatment induces IL-1-like immunoreactivity in them, suggesting that they may serve sentinel or surveillance functions. The paraganglia are innervated by vagal afferents (Berthoud and Neuhuber 2000, Berthoud et al., 1995), thus this arrangement may allow the vagus to monitor immune-related stimuli circulating in either blood or lymph. The neuroepithelial bodies are clusters of neuroendocrine cells that occupy the epithelium in lung airways (Ek et al., 1998). These cells are innervated by vagal afferent Aδ and spinal viscerosensory C fibers that express adenosine and transient receptor potential vanilloid (TRPV; capsaicin-sensitive) receptors. Thus, paraganglia and neuroepithelial bodies could conceivably expand the range of immune related signals to vagal afferents.

The intrinsic (enteric) neurons that reside throughout the extent of the gastrointestinal tract and control secretion and motility are sensitive to immune activation within the gut (Sharkey and Kroese 2001, Sharkey and Mawe 2002). Indeed, enteric responses directed toward expelling pathogens provide a critical initial host defense mechanism. Enteric neurons express receptors for IL-1 and TNF, and the activation of these receptors facilitates enteric neuronal excitability (Sharkey and Mawe 2002). Vagal sensory neurons monitor enteric ganglia (Berthoud and Neuhuber 2000) and may provide a pathway by which mediators induced by infection or inflammation, such as cytokines, may indirectly signal the brain.

4 Spinal and Trigeminal Nerves, Inflammation, and Pain Modulation

4.1 Spinal Viscerosensory Nerves

Spinal viscerosensory neurons co-habit dorsal root ganglia with somatic sensory neurons (Fig. 2C). Their peripheral projections run out with the spinal nerves to the sympathetic chain ganglia. Some travel with the sympathetic nerves to innervate blood or with the splanchnic and mesenteric nerves to innervate internal visceral tissues, including heart, lung, lymph nodes, gastrointestinal tract, and abdominal and pelvic viscera. These fibers respond to (among many other stimuli) inflammatory conditions within the gut, and seem to play a role in visceral pain and hypersensitivity (Holzer 2006, 2007, Kirkup et al., 2001).

Although spinal viscerosensory nerves are acutely sensitive to inflammation, they do not seem to play a role directly in the induction of sickness responses (Dogan et al., 2003). Rather, as Holzer (2007) has described, the spinal viscerosensory fibers in the gut, at least, serve as a "local emergency system." When sensory nerves are stimulated by disturbance of the mucosal barrier, the neuropeptide CGRP is released from their peripheral terminals. CGRP acts to induce hyperemia, increase prostacyclin (PGI_2), and decrease tissue TNF-α release. In this way, spinal visceral sensory fibers act rapidly to dilute gastric acid and inhibit pro-inflammatory cytokine release. On the other hand, sensory neurons can contribute to inflammation under certain circumstances, which seems to be mediated by neurons expressing vanilloid-1 (capsaicin receptors), and may involve endocannabinoids as well. Thus, whereas the precise role of spinal visceral afferents in local inflammation of the gut is not completely elucidated, they clearly interact with inflammatory and immune-related mechanisms in response to pathogenic conditions.

4.2 Spinal and Trigeminal Somatic Nerves

Activation of spinal and trigeminal nociceptive nerves follow release of chemicals from damaged tissues or breach of skin. The cell bodies of these "somatic" c-fiber sensory neurons that respond to inflammation or immune activation reside in the dorsal root ganglia, associated with the spinal nerves that exit each spinal segment. The trigeminal sensory neurons occupy the trigeminal ganglia (Fig. 2B), which lie beneath base of the brain, and send their peripheral terminals to tooth pulp, mucosal surfaces of mouth (including the tongue) and nose, mouth and skin of the head.

In contrast to cytokine or pathogen signaling to cranial viscerosensory nerves, which contribute to the familiar constellation of brain mediated illness responses (as noted above), an emerging theme is that cytokines produced in damaged or inflamed trigeminal or spinal nerves contribute to mechanisms of pain transmission. From the perspective of influences on behavior, it is important to note that modulation of pain transmission likely influences stress responses and affective states (Craig 2002), and thereby influence ongoing sickness responses.

Like the vagus, the spinal nerves contain immune cells, and MHC-II positive dendritic-like cells are interspersed among the nerve fibers (Goehler et al., 1999). These cells express TNF and IL-1 during inflammatory neuritis (Gazda et al., 2001). The release of TNF, in particular has been shown to dramatically facilitate pain transmission in the spinal cord (Scafers et al., 2003, Sorkin and Doom 2000), possibly via enhancement of glutamate release from primary afferent terminals (58). Epineurial treatment with TNF produces behavioral allodynia (Sorkin and Doom 2000), an enhanced pain state whereby normally non-painful stimuli become painful. Blocking the actions of TNF in models of neuritis prevent enhanced pain sensitivity (hyperalgesia; Sorkin and Doom 2000). Similar findings were obtained using IL-1, where exogenous IL-1 enhances pain states (Clark et al., 2006) that can be blocked by treatment with the IL-1 receptor antagonist. Indeed, responses to neuropathic pain were dramatically reduced in animals lacking IL-1 R1 (Wolf et al., 2006), providing strong evidence for a role of IL-1 in inflammatory pain. Taken together, current findings indicate that immune cells in peripheral nerves can potently modulate signals relevant to pain and inflammation.

4.2.1 Receptors in Spinal and Trigeminal Ganglia

Cells located within the dorsal root ganglia, which could be immune cells or satellite cells, express IL-1, TNF, and IL-6 in models of inflammatory or neuropathic pain (Cunha et al., 2005, Hou et al., 2003, Hwang et al., 2003, Lee et al., 2004, Li et al., 2004, 2005, Liu et al., 2006, Ozaktay et al., 2006, Zhang et al., 2002), as well as their receptors. Sensory neurons in the dorsal root ganglia express TNFR1 (Hermann et al., 2005, Li et al., 2004) and TNF activates second messenger systems (p38 mitogen-activated kinase (Scafers et al., 2003), protein kinase A (Zhang et al., 2002)) in these sensory neurons. Blocking the actions of TNF prevented the activation of p38 (Scafers et al., 2003), and blocking protein kinase A blocks the effects of TNF on sensory neuron excitability (Zhang et al., 2002). Endogenous TNF is produced by immune cells in the ganglia (Li et al., 2004), and taken together, these findings support a role for locally (within the ganglion) generated TNF (and possibly IL-1 and IL-6), in the pathogenesis of neuropathic pain. In addition, sensory ganglia express receptors for purines, tachykinins, serotonin, prostaglandins, and TRPV receptors (capsaicin etc.), which have all been implicated in signaling inflammation and/or immune activation.

4.2.2 Peripheral Actions of Trigeminal and DRG C-Fibers

The interaction of cytokines and spinal nerve sensory neurons is, unsurprisingly, bi-directional. Acute intraplantar administration of capsaicin, to *activate* TRPV receptor expressing fibers innervating the hindpaw, induced hyperalgesia and the expression of cytokines (IL-1, IL-6, TNF) in the hindpaw skin (Saade et al., 2002). However, the expression of cytokines was absent in animals that had been previously treated with capsaicin (to *lesion* TRPV receptor expressing fibers in the sciatic nerve; Hou et al., 2003). This finding indicates that peripheral nerves, besides responding to cytokine signals, can influence the expression of these same cytokines, as well as the release of pro-inflammatory peptides, such as substance P (SP; Hou et al., 2003). This indicates that a potentially important efferent function of C-fibers is to modulate peripheral cytokine expression, and raise the possibility that pathological pain states may be complicated by a positive feedback loop in which cytokines can activate pain-transmitting neurons, which can in turn upregulate the expression of the cytokines. Interestingly, SP and CGRP may have contrasting effects on immune cell function. Whereas SP is generally pro-inflammatory, CGRP seems to exert suppressive effects on pro-inflammatory cytokine expression.

5 Central Projections of Immune-Responsive Nerves: Interface with Brain Regions Involved in Behavior

As noted previously, immune mediators produce physiological and behavioral illness responses by activating brain neurocircuitry that mediate these responses. The identity of this neurocircuitry has been probed using expression of the activation marker c-Fos following treatment with cytokines or LPS. These studies have shown that immune-responsive brain nuclei are associated with autonomic functions, including the hypothalamus, amygdala, visceral thalamus, periaquiductal gray, and cingulate and infralimbic cortex (Elmquist et al., 1997, Elmquist and Saper 1996, Gaykema et al., 2007a, Goehler et al., 2000, Konsman et al., 1999, Wan et al., 1993; Fig. 3). Whereas the neurocircuitry mediating some autonomic functions, including fever (Elmquist et al., 1997, Zhang et al., 2000), and activation

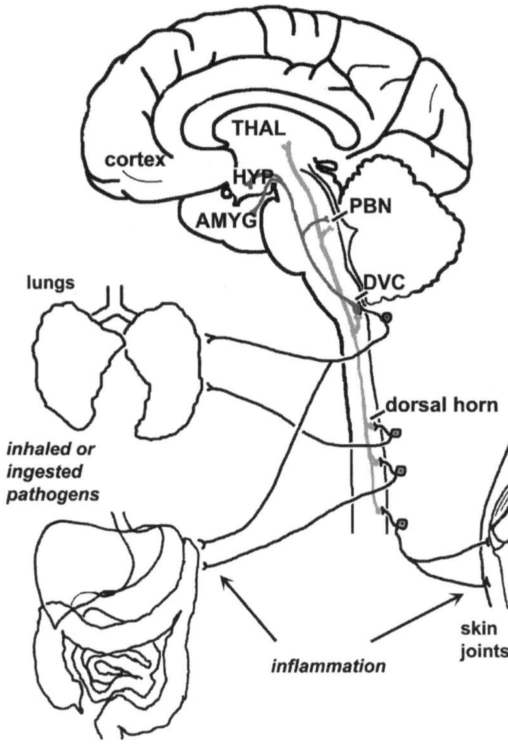

Fig. 3 Schematic and simplified depiction of the neural pathways by which peripheral immune activation is signaled to the brain regions that influence behavior. Spinal pathways (*green*) signal pain and inflammation, and vagal pathways (*red*) signal presence of pathogens to brain stem and forebrain regions that initiate integration of homeostatic information with ongoing behavior (glossopharyngeal and trigeminal contributions were omitted to preserve clarity of the figure). Abbreviations: AMYG, amygdala, DVC, dorsal vagal complex; PBN, parabrachial nucleus; THAL thalamus

of the hypothalamo-pituitary–adrenal axis (Ericsson et al., 1994, 1997) have been described in at least rough detail, neurocircuitry driving other responses, notably behavioral ones, are less clear.

5.1 Immune-Neural Interface in the Brain Stem Ventrolateral Medulla and Dorsal Vagal Complex: Cranial Nerve Viscerosensory Projections

Sensory fibers associated with the vagal and glossopharyngeal ganglia collect signals from the tissues that they innervate, and convey this information to brain stem dorsal vagal complex : the nucleus of the solitary tract (nTS) and the area postrema (a circumventricular organ). These nuclei coordinate local, protective reflexes, such as emesis and gastric retention. In addition they relay a wide variety of viscerosensory signals to forebrain regions concerned with integration of visceral information with ongoing behavior and other sensory inputs. Notably, brain regions driven by ascending pathways emanating from the dorsal vagal complex (Gaykema et al., 1998, Ricardo and Koh 1978) overlaps significantly with those shown to respond to peripheral immune stimulation (Brady et al., 1994, Elmquist et al., 1997). This arrangement is consistent with the idea that one pathway by which cytokines signal the brain to activate illness responses is via peripheral nerves that in turn drive the dorsal vagal complex and its projections to higher brain regions.

5.1.1 The Nucleus of the Solitary Tract (nTS)

The nTS is well known for its role as the primary sensory relay nucleus for the vagus, glossopharyngeal, and facial cranial nerves, which carry taste and general visceral sensory information. With the area postrema (below), the nTS forms the sensory dorsal vagal complex (DVC). Also, the nTS receives ascending input from spinal visceral and somatic sources ("bottom-up") and descending input from multiple forebrain regions, including the paraventricular hypothalamus and central extended amygdala ("top-down"). Information available to the nTS is not limited to neural input, as the nTS contains receptors for glucocorticoids (Roozendaal et al., 1999) and, via the area postrema and the vagus nerve (see below), senses circulating catecholamines and other hormones that signal psychological stress (Miyashita and Williams 2004) or visceral challenge. By its ascending projections, directly or via interface with the ventrolateral medulla (VLM), the nTS contributes to the central autonomic network of brain nuclei that are integrated to provide autonomic control and adjustments in concert with behavioral demands. As the primary viscerosensory relay nucleus, the nTS also contributes to the mediation of sickness symptoms (Marvel et al., 2004), and plays a critical role in the pathway by which peripheral arousal facilitates memory (Miyashita and Williams 2004, Williams and McGaugh 1993). Taken together, these features place the nTS in a fortuitous position to monitor information about bodily conditions, and interface with the forebrain regions charged with the integration of cognitive, behavioral, and neuroendocrine responses to challenges.

5.1.2 The Area Postrema

The area postrema is a circumventricular organ, in which, like the other circumventricular organs, the blood–brain barrier is weak. This allows resident cells (including neurons) within the area postrema access to substances in the blood that are excluded from the rest of the brain. Such substances include cytokines, which are large, lipophobic molecules that do not readily pass the blood–brain barrier, as well as pathogens or pathogen products such as LPS. Like other circumventricular organs, the area postrema, harbors immune cells that produce cytokines likely important for sickness behavior and cohabit with typical brain neurons that innervate brain regions "behind" the blood–brain barrier (Goehler et al., 2006). Unlike the other circumventricular organs, however, the area postrema receives direct viscerosensory input via the vagus nerve, which terminates extensively throughout the area postrema (Shapiro and Miselis 1985). This arrangement allows the area postrema access to a uniquely wide variety of peripheral signals: those present in the general circulation, in the cerebrospinal fluid, and, carried by vagal sensory nerves, those arising from distant viscera, e.g., related to local inflammation. These signals can then be propagated to neurons in the nTS, and in the lateral parabrachial nucleus in the pons, the principal targets of area postrema projection neurons (Shapiro and Miselis 1985). Whereas the area postrema is most famous as an "emetic" center, it apparently also contributes to EEG synchronization and slow wave sleep (Bronzino et al., 1976a, b), which are consistent features of the sickness syndrome.

5.1.3 The Ventrolateral Medulla

The VLM is part of the reticular formation that harbors a mixture of large neurons that express catecholamines or other substances including glutamate and neuropeptides. Via input from the dorsal vagal complex, the VLM modulates, as part of the bodily homeostatic control, pulmonary and cardiovascular function. Also, neurons in the VLM projecting to more rostral brain regions play an important role in responses to immune and other visceral challenges (Elmquist and Saper 1996, Ericsson et al., 1994, Bluthe et al., 1994, Gaykema et al., 2007b). In particular, the catecholaminergic neurons that target the hypothalamus respond to circulating cytokines, and provide the primary drive on hypothalamic neuroendocrine systems under conditions of sickness (Ericsson et al., 1994), and other viscerosensory challenges (Rinaman 1999, 2004).

5.1.4 Functional Evidence of a Role of the DVC in Behavior Modulation

Whereas the DVC/VLM seem to contribute to a range of brain-mediated sickness responses (albeit variably), a particularly salient role seems to be related to behavioral responses, notably those associated with behavioral depression such as social withdrawal and psychomotor retardation (Marvel et al., 2004) as well as anxiety-like behavior (Lyte et al., 2006). Temporary inactivation (using microinjection of local anesthetic to produce a "reversible lesion") of the DVC dramatically prevents social withdrawal and psychomotor retardation, as well as the characteristic pattern of induction of a neuronal activation marker protein, c-Fos, normally seen following peripheral administration of LPS (Marvel et al., 2004). These findings provide functional evidence for the DVC as a crossroads of signals relevant to sickness induced behavioral depression.

5.1.5 Projections from the DVC and VLM Link Brain Stem Immunosensory Regions with Those Mediating Behavior

As mentioned above, the brain regions activated by immune-derived stimuli are tied into the constellation of interconnected nuclei known to play key roles in the coordination of autonomic and neuroendocrine functions (output) in support of behavioral responses to external ("psychological") and internal ("visceral") demands (Capuron et al., 2005, Elmquist et al., 1997, Goehler et al., 2005, Herman et al., 2005, Park et al., 2006, Sawchenko et al., 2000). These include components of the ascending viscerosensory pathways, the central autonomic network, the nucleus accumbens as well as cortical areas including the insular, anterior cingulate, and medial prefrontal cortex.

Ascending projections from the VLM/DVC derive from at least two neurochemically distinct groups of neurons. The largest group comprises the noradrenergic and adrenergic neurons that innervate structures distributed more rostrally along the neuraxis, in particular the parabrachial, periaqueductal and dorsal raphe nuclei, hypothalamus, basal forebrain, including the amygdala and bed nucleus of the stria terminalis (Buller et al., 2001, Cunningham and Sawchenko 1988, Espana and Berridge 2006, Ericson et al., 1989, Gaykema et al., 2007b, Hajszan and Zaborszky 2002, Herbert and Saper 1992, Peyron et al., 1996). The VLM, along with the DVC, seems to provide most of the noradrenergic, and all of the adrenergic innervation of the hypothalamus, including the paraventricular nucleus (PVN; contains corticotropin releasing hormone, CRH, neurons driving corticosteroid responses) and tuberomammillary neurons (histaminergic neurons), as well as most of the innervation to the basal forebrain (including cholinergic neurons; Hajszan and Zaborszky 2002). Thus VLM and DVC neurons make compelling candidates as links between visceral challenges, arousal, and thus potentially, affective states. Many of these adrenergic and noradrenergic neurons located in the VLM and DVC become strongly activated by systemic challenge with immune stimulants (Elmquist and Saper 1996, Gaykema et al., 2007b, Goehler et al., 2005). The other group of ascending projection neurons reside in the DVC and express a variety of peptides (e.g., glucagon-like peptide-1, cholecystokinin, galanin; Herbert and Saper 1990, Rinaman, 1999, 2004). These nerve fibers project to the parabrachial nucleus and hypothalamic structures in particular (see below), and, like the catecholaminergic neurons, many respond to immune activation.

Besides direct input from the DVC and VLM, several forebrain regions involved in arousal and responses to stress (including the amygdala, BST and hypothalamus) receive additional drive via the lateral parabrachial nucleus (PBN), often referred to as the second-order viscerosensory relay nucleus because it receives direct input from the DVC (Krout and Loewy 2000). Many of these regions receive input from both the PBN and the DVC/VLM. In this way, the DVC/VLM is in a prime position to influence in neurochemically distinct and complex ways, the targets implicated in behavioral responses to pathogens.

Clues to the neural substrates that represent targets through which DVC and VLM likely influence behavioral changes associated with infection derive from studies investigating brain activation patterns when sick animals are challenged with behavioral tasks, such as exploring a novel environment, social interaction or exposure to a palatable sweetened milk solution (Krout and Loewy 2000). LPS-treated rats show reduced locomotor speed and occasional bouts of immobility, indicative of psychomotor retardation and/ or fatigue, on the elevated plus maze, and show reduced social interaction (Gaykema et al., 2007a, Marvel et al., 2004) and consumption of the sweetened milk (Gaykema et al., 2007a, Park et al., 2008). This reduction in behavior is associated with a reduced expression of the activation marker protein c-Fos in brain regions involved in arousal and behavior, including the histaminergic tuberomamillary nucleus, dorsal hippocampus, the striatum, nucleus accumbens, cingulate cortex, lateral and medial septum, and the diagonal band. These findings suggest that interference with these systems may contribute to neurovegetative syndrome-like symptoms seen in sickness (see below). This idea is supported by the further finding that DVC inactivation abolished the suppressive effects of LPS on behavior and c-Fos induction (Gaykema et al., 2007a, Marvel et al., 2004), indicating that the DVC contributes to the pathway by which immune activation inhibits behavior. Taken together, these findings illuminate an interaction between the DVC and brain arousal systems in aspects of sickness behavior.

Besides input to hypothalamic brain regions that control or influence sleep, wakefulness, and behavioral arousal, the DVC has been reported to synchronize EEG (Golanov and Reis 2001), which would support to typically increase non-REM sleep in acute illness (Opp 2005). In case of the nTS, this effect was mediated via projections to the rostral portion of the VLM, although further neural pathway(s) subserving this effect were not delineated (Golanov and Reis 2001). Since EEG is determined by the discharge of thalamo-cortical projectons, these findings suggest that a viscerosensory drive, perhaps via the parabrachial nucleus (Krout and Loewy 2000), on the thalamus may contribute to somnolence and the increase in SWS observed during sickness (Opp 2005). As yet however, such a sickness-responsive pathway has not been identified. Nonetheless, these effects are consonant with the demonstration that sectioning of the vagus nerves influence sleep and sleep/waking behavior following immune challenge (Opp and Toth 1998).

5.2 *Inflammation and Potentiation of Pain States: Trigeminal and Spinal Pathways*

Neurons of the trigeminal ganglia send their terminals into the pons, and those fibers carrying information related to inflammation continue caudally to the spinal trigeminal nucleus of the caudal medulla and cervical spinal cord (Fig. 2B). The central projections of spinal C-fibers terminate primarily in the superficial laminae of the spinal dorsal horn (lamina 1; Fig. 2C).

Besides enhancing pain states by modulating peripheral nerve function, cytokines including IL-1 are produced by glia within the spinal cord, and play a critical role in modulation of pain states (Samad et al., 2001, Wolf et al., 2006). The induction of spinal cytokines may follow the release from primary sensory neurons of the mediators, such as the chemokine fractalkine, which seems to serve as a pain-related signal to spinal cord glia critical for the induction of enhanced pain states, such as allodynia (Milligan et al., 2005). Also, circulating or induced TNF potently enhances pain signal transmission via enhancement of sensory neurotransmitter (glutamate) release into the CNS (Hermann et al., 2005). Prostaglandins likely play a role in pain transmission as well (Samad et al., 2001).

Pain-related and viscerosensory signals from the spinal dorsal horn and trigeminal nuclei of the brain stem are propagated via ascending neural projections in the spinothalamic and trigeminothalamic tracts to brain stem regions, including the nTS (Gamboa-Esteves et al., 2001, Menetry and Basbaum 1987) and lateral parabrachial nucleus (Hwang et al., 2003), as well as the thalamus. These brain regions begin the process by which information associated with neuroendocrine, physiological, and emotional responses to challenges, including host defense (as above) is integrated. In this way, pain-related signals can potently modulate behavior, by increasing drive on stress-related brain functions and sickness behaviors.

6 Summary and Perspectives

Based on its role in detecting pathogens and alerting the brain to their presence, the immune system has been described as "diffuse sensory system" (Besedovsky and del Rey 1992). This sensory system comprises multiple (humoral, neural) pathways that each may be relatively important for specific aspects of nervous system responses to infection. A hallmark of sensory systems in general and physiological regulatory (homeostatic) sensory systems in particular involves a redundancy in the pathways that carry relevant signals to the brain (i.e., parallel pathways). For instance, vagal and spinal (sympathetic) visceral afferents respond to different stimuli and contribute to distinct aspect of host defense. Vagal sensory nerve fibers respond rapidly to bacterial LPS, and this pathway contributes to the induction of acute phase responses such as fever and neuroendocrine activation. Spinal visceral fibers respond to inflammation, and like vagal fibers (e.g., see inflammatory reflex Pavlov and Tracey 2005) act to locally protect internal tissues, but their contribution to CNS responses appears to be more directed toward modulation and enhancement of pain states (visceral hypersensitivity). This arrangement helps to ensure that signals critical to appropriate responses to dangerous physiological conditions will be conveyed to the CNS regions involved in coordinating such responses. In the context of infection, peripheral nerves, by virtue of their extensive innervation of bodily locations serve as early warning systems (or "smoke detectors"), by alerting the CNS to the presence of possible danger while the infection is still localized, and thus easier to combat than when it has become systemic.

When compared with the other neural pathways that respond to immune activation, the vagus contributes primarily (though not exclusively) to "systemic" (i.e., non-local brain mediated) responses to infection or inflammation. In this way, the vagus serves as a mind–body link. For example, animal and human studies implicate the vagus in the modulation of affective states (Zagon 2006), and the vagus also carries feedback signals regarding peripheral responses (such as circulating epinephrine) to behavioral arousal (Miyashita and Williams 2004). These feedback signals have been shown to facilitate memory (Clark et al., 1998, Miyashita and Williams 2004, Williams and McGaugh 1993). Thus, several lines of enquiry implicate the vagus, especially the sensory component, in the regulation or modulation of cognitive and affective functions (Craig 2002). However, although the major sensory modality in the spinal dorsal root ganglia influenced by immune activation seems to be pain, it is important to note that pain (like sickness) has profound influences on behavior, cognition, and affect. As an example, inflammation or infection in the gastrointestinal tract (associated with inflammatory bowel disorders or irritable bowel syndrome) is associated with increased expression of pro-inflammatory cytokines and other mediators, as well as mood symptoms, notably anxiety and depression (Addolorato et al., 1997, Simrin et al., 2002). These affective symptoms likely follow from a confluence of signals including the cognitive response to a painful and stressful disorder, as well as from mediators induced during inflammation carried by vagal and spinal viscerosensory fibers that ultimately induce components of sickness behavior. Thus recognition of the multiple and largely parallel neural pathways that signal inflammation and infection to the brain will be critical to the development of comprehensive therapeutic interventions for concomitant behavioral sequelae.

Acknowledgments This work was supported by NIH grants MH 55283, MH 64648, MH 50431, and MH 68834.

References

Addolorato G, Capristo E, Stafanini GF, Gasbarrini G (1997) Inflammatory bowel disease: a study of the association between anxiety and depression, physical morbidity, and nutritional status. Scand J Gastroenterol 32: 1013–1021

Adriaensen D, Brouns I, Pintalon I, De Proost I, Timmermans J-P (2006) Evidence for a role of neuroepithelial bodies as complex airway sensors: comparison with smooth muscle-associated airway receptos. J Appl Physiol 101: 960–970

Berthoud H-R, Neuhuber WL (2000) Functional and chemical anatomy of the afferent vagal system. Autonom Neurosci Basic Clin 85: 1–17

Berthoud H-R, Kressel M, Neuhuber, WL (1995) Vagal afferent innervation of the rat abdominal paraganglia as revealed by anterograde DiI-tracing and confocal microscopy. Acta Anat 152: 127–132

Besedovsky HO, del Rey A (1992) Immune-neuroendocrine circuits: integrative role of cytokines. Front Neuroendocrinol 13:61–94

Bret-Dibat JL, Bluthé R-M, Kent S, Kelley K, Dantzer R (1995) Lipopolysaccharide and interleukin-1 depress food-motivated behavior in mice by a vagal-mediated mechanism. Brain Behav Immun 9: 242–246

Bluthe R-M, Walter V, Parnet P, Laye S, Lestage J, Verrier D, Poole S, Stenning BE, Kelley KW, Dantzer R (1994) Lipopolysaccharide induces sickness behavior in rats by a vagal mediated mechanism. C R Acad Sci III 317: 499–503
Bluthe R-M, Michaud B, Kelly KW, Dantzer R (1996) Vagotomy blocks behavioral effects of interleukin-1 injected via the intraperitoneal route but not by other systemic routes. Neuro Report 7: 2823–2827
Brady LS, Lynn AB, Herkenham M, Gottesfels Z (1994) Systemic interleukin-1 induces early and late patterns of c-fos mRNA expression in brain. J Neurosci 14: 4951–4964
Bronzino JD, Stern WC, Leahy JP, Morgane PJ (1976a) Sleep cycles in cats during chronic electrical stimulation of the area postrema and the anterior raphe. Brain Res Bull 1: 235–239
Bronzino JD, Stern WC, Leahy JP, Morgane PJ (1976b) Power spectral analysis of EEG activity obtained from cortical and subcortical sites during the vigilance states of the cat. Brain Res Bull 1: 285–294
Buller K, Xu Y, Dayas C, Day T (2001) Dorsal and ventral medullary catecholamine cell groups contribute differentially to systemic interleukin-1beta-induced hypothalamic pituitary adrenal axis responses. Neuroendocrinol 73: 129–138
Cano G, Sved AF, Rinaman L, Rabin BS, Card JP (2001) Characterization of the central nervous system innervation of the rat spleen using viral transneuronal tracing. J Comp Neurol 439: 1–18
Capuron L, Pagnoni G, Demetrashvili M, Wollwine BJ, Nemeroff CB, Berns GS, Miller AH (2005) Anterior cingulate activation and error processing during interferon-alpha treatment. Biol Psychiatry 58: 190–196
Clark AK, D'Aquisto F, Gentry C, Marchand F, McMahon SB, Malcangio M (2006) Rapid co-release of interleukin-1beta and caspase 1 in spinal cord inflammation. J Neurochem 99: 868–880
Clark KB, Smith DC, Hassert DC, Browning RA, Noritoku DK, Jensen RA (1998) Post-training electrical stimulation of vagal afferents with concomitant vagal efferent inactivation enhances memory storage processes in the rat. Neurobiol Learn Mem 70: 364–373
Craig AD (2002) How do you feel? The sense of the physiological condition of the body. Nat Rev Neurosci 3: 655–666
Cunha TM, Verri WA Jr, Silva JS, Poole S, Cunha FQ, Ferreira SH (2005) A cascade of cytokines mediates mechanical inflammatory hypernociception in mice. Proc Natl Acad Sci USA 102: 1755–1760
Cunningham ET Jr, Sawchenko PE (1988) Anatomical specificity of noradrenergic inputs to the paraventricular and supraoptic nuclei of the rat hypothalamus. J Comp Neurol 274: 60–76
Dogan MD, Kulchitskya, VA, Patela S, Petervari, E, Szekely M, Romanovsky AA (2003) Bilateral splanchnicotomy does not affect lipopolysaccharide-induced fever in rats. Brain Res 993: 227–229
Ek M, Kurosawa M, Lundeberg T, Ericsson A (1998) Activation of vagal afferents after intravenous injection of interleukin1b: role of endogenous prostaglandins. J Neurosci 18: 9471–9479.
Elmquist JK, Saper CB (1996) Activation of neurons projecting to the paraventricular hypothalamic nucleus by intravenous lipopolysaccharide. J Comp Neurol 374: 315–331
Elmquist JK, Scammell TE, Saper CB (1997) Mechanisms of CNS response to systemic immune challenge: the febrile response. Trends Neurosci 20: 565–570
Emch GS, Hermann G E, Rogers RC (2001) TNF-alpha induces c-Fos generation in the nucleus of the solitary tract that is blocked by NBQX and MK801. Am J Physiol 281:R1394–R1400
Ericsson A, Kovacs KJ, Sawchenko PE (1994) A functional anatomical analysis of central pathways subserving the effects of interleukin-1 on stress-related neuroendocrine neurons. J Neurosci 14: 897–913
Ericsson A, Arias C, Sawchenko PE (1997) Evidence for an intramedullary prostaglandin-dependent mechanism in the activation of stress-related neuroendocrine circuitry by intravenous interleukin-1. J Neurosci 17: 7166–7179
Ericson H, Blomqvist A, Kohler C (1989) Brainstem afferents to the tuberomammillary nucleus in the rat with special reference to monoaminergic innervation. J Comp Neurol 281: 169–192
Espana RA, Berridge CW (2006) Organization of noradrenergic efferents to arousal-related basal forebrain structures. J Comp Neurol 496: 668–683

Felton DL, Livnat S, Felton SY, Carlson SL, Bellinger DL, Yeh P (1984) Sympathetic innervation of lymph nodes in mice. Brain Res Bull 13: 693–696

Fink T, Weihe E (1988) Multiple neuropeptides in nerves supplying mammalian lymph nodes: messenger candidates for sensory and autonomic neuroimmunomodulation. Neurosci Lett 19: 39–44

Fleshner M, Goehler L, Hermann J, Relton JK, Maier SF, Watkins LR (1995) Interleukin-1b induced corticosterone elevation and hypothalamic NE depletion is vagally mediated. Brain Res Bull 37: 605–610

Fleshner M, Goehler LE, Schwartz BA, McGorry M, Martin D, Watkins LR, Maier SF (1998) Thermogenic and corticosterone responses to intravenous cytokines (IL-1b and TNF-a) are attenuated by subdiaphragmatic vagotomy. J Neuroimmunol 86: 134–141

Gamboa-Esteves FO, Lima D, Batten TFC (2001) Neurochemistry of superficial spinal neurons projecting to nucleus of the solitary tract that express c-fos on chemical and visceral nociceptive input in the rat. Metab Brain Disease 16: 151–164

Gaykema RPA, Dijkstra I, Tilders FJH (1995) Subdiaphragmatic vagotomy suppresses endotoxin-induced activation of the hypothalamic corticotropin-releasing hormones neurons and ACTH secretion. Endocrinol 136: 4717–4720

Gaykema RPA, Goehler LE, Tilders F J H, Bol JGM, McGorry MM, Maier SF, Watkins LR (1998) Bacterial endotoxin induces Fos immunoreactivity in primary afferent neurons of the vagus nerve. NeuroImmunoMod 5: 234–240

Gaykema RPA, Goehler LE, Armstrong CB, Khorsand J, Maier SF, Watkins, LR (1999) Differential FOS expression in rat brain induced by lipopolysaccharide and Staphylococcal enterotoxin B Neuro Immuno Mod 6: 220

Gaykema RPA, Goehler LE, Hansen MK, Maier SF, Watkins LR (2000) Subdiaphragmatic vagotomy blocks interleukin-1a-induced fever but does not reduce interleukin-1a levels in the circulation. Auton Neurosci Basic Clin 85: 72–77

Gaykema RPA, Park S-M, McKibbin CR, Goehler LE (2007a) Lipopolysaccharide suppresses activation of the tuberomammillary histaminergic system concomitant with behavior: a novel target of immune-sensory pathways, Neurosci, ePub ahead of print Nov 13

Gaykema RPA, Chen C-C, Goehler LE (2007b) Organization of immune-responsive medullary projections to the bed nucleus of stria terminalis, central amygdala, and paraventricular nucleus of the hypothalamus: Evidence for parallel viscerosensory pathways in the rat brain. Brain Res 1130: 130–145

Gazda LS, Milligan ED, Hansen MK, Twining CM, Poulos NM, Chacur M, O'Conner KA, Armstrong C, Maier SF, Watkins LR, Myers RR (2001) Sciatic inflammatory neuritis (SIN): behavioral allodynis in parallel with peri-sciatic proinflammatory cytokine and suproxide production. J Peripher Nerv Syst 6: 111–129

Goehler LE, Busch CR, Tartaglia N, Relton JK, Sisk D, Maier SF, Watkins LR (1995) Blockade of cytokine induced conditioned taste aversion by subdiaphramatic vagotomy: further evidence for vagal mediation of immune-brain interactions. Neurosci Lett 185: 163–166

Goehler LE, Relton JK, Dripps D, Keichle R, Tartaglia N, Maier SF, Watkins LR (1997) Vagal paraganglia bind biotinylated interleukin-1 receptor antagonist: A possible mechanism for immune-to-brain communication. Brain Res Bull 43: 357–364

Goehler LE, Gaykema RPA, Hammack SE, Maier SF, Watkins LR (1998) Interleukin-1 induces c-Fos immunoreactivity in primary afferent neurons of the vagus nerve. Brain Res 804: 306–310

Goehler LE, Gaykema RPA, Nguyen KT, Lee JL, Tilders FJH, Maier SF, Watkins LR (1999) Interleukin-1β in immune cells of the abdominal vagus nerve: an immune to nervous system link? J Neurosci 17: 2799–2806

Goehler LE, Gaykema RPA, Hansen MK, Maier SF, Watkins LR (2000) Vagally mediated fever: a visceral chemosensory modality. Auton Neurosci 85: 49–59

Goehler LE, Gaykema RPA, Opitz N, Reddaway R, Badr NA, Lyte M (2005) Activation in vagal afferents and central autonomic pathways: early responses to intestinal infection with Campylobacter jejuni. Brain Behav Immun 19: 334–344

Goehler LE, Erisir A, Gaykema RPA (2006) Neural-immune interface in the area postrema. Neurosci 140: 1415–1434

Golanov EV, Reis DJ (2001) Neurons of nucleus of the solitary tract synchronize the EEG and elevate cerebral blood flow via a novel medullary area. Brain Res 892: 1–12
Hajszan T, Zaborszky L (2002) Direct catecholaminergic-cholinergic interactions in the basal forebrain. III. Adrenergic innervation of choline acetyltransferase-containing neurons in the rat. J Comp Neurol 449: 141–157
Hansen MK, Krueger JM (1997) Subdiaphragmatic vagotomy blocks the sleep- and fever-promoting effects of interleukin-1b. Am J Physiol 273: R1246–R1253
Hansen MK, Nguyen KT, Fleshner M, Goehler LE, Gaykema RPA, Maier SF, Watkins LR (2000) Effects of vagotomy on circulating levels of endotoxin, pro-inflammatory cytokines, and corticosterone following intraperitoneal lipoplysaccharide. Am J Physiol 278: R331–R336
Hansen MK, O'Conner KA, Goehler LE, Watkins LR, Maier SF (2001) The role of the vagus nerve in interleukin-1b-induced fever is dependent on dose. Am J Physiol 280: R929–R934
Herbert H, Saper CB (1990) Cholecystokinin-, galanin-, and corticotropin-releasing factor-like immunoreactiveprojections from the nucleus of the solitary tract to the parabrachial nuclei in the rat. J Comp Neurol 293: 581–598
Herbert H, Saper CB (1992) Organization of medullary adrenergic and noradrenergic projections to the periaqueductal gray matter in the rat. J Comp Neurol 315: 34–52
Herman JP, Ostrander MM, Mueller NK, Figueidredo H (2005) Limbic system mechanisms of stress regulation: hypothalamo-pituitary-adrenocortical axis. Prog Neuro-Psychopharmacol Biol Psychiatry 29: 1201–1213
Hermann GE, Holmes GM, Rogers RC (2005) TNF(alpha) modulation of visceral and spinal sensory processing. Curr Pharm Des 11: 1391–1409
Holzer P (2006) Efferent-like roles of afferent neurons in the gut: Blood flow regulation and tissue protection. Auton Neurosci Basic Clin 125:70–75
Holzer P (2007) Role of visceral afferent neurons in mucosal inflammation and defense. Curr Opin Pharmacol 7: 563–569
Hosoi T, Okuma Y, Matsuda T, Nomura Y (2005) Novel pathway for LPS-induced afferent vagus nerve activation: possible role of nodose ganglion. Autonom Neurosci Basic Clin 120: 104–107
Hou L, Li W, Wang X (2003) Mechanism of interleukin-1b-induced calcitonin gene-related peptide production from dorsal root ganglion neurons of neonatal rats. J Neurosci Res 73: 188–197
Hwang SJ, Burette A, Valtschanoff JG (2003) VR1-positive primary afferents contact NK1-positive spinoparabrachial neurons. J Comp Neurol 460: 255–265
Kapcala LP, He JR, Gao Y, Pieper JO, DeTolla LJ (1996) Subdiaphragmatic vagotomy inhibits intra-abdominal interleukin-1β stimulation of adrenocorticotropin secretion. Brain Res 728: 247–254
Kirkup AJ, Brunsden AM, Grundy D (2001) Receptors and transmission in the brain-gut axis: potential for novel therapies. I. Receptors on visceral afferents. Am J Physiol 280:G787–G794
Konsman JP Kelley K, Dantzer R (1999) Temporal and spatial relationship between lipopolysaccharide-induced expression of Fos, interleukin-1b and inducible nitric oxide synthase in rat brain. Neurosci 89: 535–548
Krout KE, Loewy AD (2000) Parabrachial nucleus projections to the midline and intralaminar thalamic nuclei of the rat. J Comp Neurol 428: 475–494
Larson SJ, Dunn AJ (2001) Behavioral effects of cytokines. Brain Behav Immunity 15: 371–387
Lee HL, Lee KM, Son SJ, Hwang SH, Cho HJ (2004) Temporal expression of cytokines and their receptors mRNA in a neuropathic pain model. Neuroreport 15: 2807–2811
Li Y, Ji A, Weihe E, Schafer K-H (2004) Cell-specific expression and lipopolysaccharide-induced regulation of tumor necrosis factor a (TNFa) and TNF receptors in rat dorsal root ganglia. J Neurosci 24: 9623–9631
Li M, Shi J, Tang J-R, Chen D, Ai B, Chen J, Wang L-N, Cao F-Y, Li L-L, Lin C-Y, Guan X-M (2005) Effects of complete Freund's adjuvant on immunohistochemical distribution of IL-1β and IL-1R I in neurons and glia cells of dorsal root ganglion Acta Pharmacolog Sin 26: 192–198
Liu L, Yang TM, Liedtke W, Simon SA (2006) Chronic IL-1 Signaling Potentiates Voltage-Dependent Sodium Currents in Trigeminal Nociceptive Neurons. J Neurophysiol 95: 1478–1490

Lyte M, Varcoe JJ, Bailey MT (1998) Anxiogenic effect of subclinical bacterial infection in mice in the absence of overt immune activation. Physiol Behav 65: 63–69

Lyte M, Wang L, Opitz N, Gaykema RPA, Goehler LE (2006) Anxiety-like behavior during initial stage of infection with agent of colonic hyperplasia Citrobacter rodentium. Physiol Behav 89: 350–357

Marvel FA, Chen C-C, Badr NA, Gaykema RPA, Goehler LE (2004) Reversible inactivation of the dorsal vagal complex blocks lipopolysaccharide-induced social withdrawal and c-Fos expression in central autonomic nuclei. Brain Behav Immun 18: 123–143

Mascarucci P, Perego C, Terrazzino S, DeSimoni M G (1998) Glutamate release in the nucleus tractus solitarius induced by peripheral lipopolysaccharide and interleukin-1b. Neurosci 86: 1285–1290

Matsuura S (1973) Chemoreceptor properties of glomus tissue found in the carotid region of the cat. J Physiol Lond 235: 57–73

Menetry D, Basbaum AI (1987) Spinal and trigeminal projections to the nucleus of the solitary tract: a possibl substrate for somatovisceral and viscerovisceral reflex activation. J Comp Neurol 255: 439–450

Milligan E, McGorry MM, Fleshner M, Gaykema RPA, Goehler LE, Watkins LR, Maier SF (1997) Subdiaphragmatic vagotomy does not prevent fever following intracerebroventricular prostaglandin: further evidence for the importance of vagal afferents in immune-to-brain communication. Brain Res 766: 240–243

Milligan E, Zapata V, Schoeniger D, Chacur M, Green P, Poole S, Martin D, Maier SF, Watkins LR (2005) An initial investigation of spinal mechanisms underlying pain enhancement induced by fractalkine, a neuronally released chemokine. Eur J Neurosci 22: 2775–2782

Miyashita T, Williams CL (2004) Peripheral arousal-related hormones modulate norepinephrine release in the hippocampus via influences on brainstem nuclei. Behav Brain Res 153: 87–95

Nagura H, Ohtani H, Masuda T, Kimura M, Nakamura S (1991) HLA-DR expression on M cells overlaying Peyer's patches is a common feature of human small intestine. Acta Pathol Jpn 41: 818–823

Nance DM, Burns, J (1989) Innervation of the spleen in the rat: evidence for absence of afterent innervation. Brain Behav Immun 3: 281–290

Opp MR (2005) Cytokines and sleep. Sleep Med Rev 9: 355–364

Opp MR, Toth LA (1998) Somnogenic and pyrogenic effects of interleukin-1 beta and lipopolysaccharide in intact and vagotomized rats. Life Sci 62: 923–936

Ozaktay AC, Kallakuri S, Takebayashi T, Cavanaugh JM, Asik I, Deleo JA, Weinstein JN. (2006) Effects of interleukin-1 beta, interleukin-6, and tumor necrosis factor on sensitivity of dorsal root ganglion and peripheral receptive fields in rats. Eur Spine J 15: 1529–1537

Park S-M, Gaykema RPA, Goehler LE (2008) How does immune challenge inhibit ingestion of palatable food? Systemic lipopolysaccharide modulates key nodal points of feeding Neurocircuitry. Brain Behav lmmun, June 16 [epub ahead of print].

Pavlov VA, Tracey KJ (2005) The cholinergic anti-inflammatory pathway. Brain Behav Immun 19: 493–499

Peyron C, Luppi P-H, Fort P, Rampon C, Jouvet M (1996) Lower brainstem catecholamine afferents to the rat dorsal raphe nucleus. J Comp Neurol 364: 402–413

Popper P, Mantyh CR, Vigna SR, Maggio JE, Mantyh PW (1988) The localization of sensory nerve fibers and receptor binding sites for sensory neuropeptides in canine lymph nodes. Peptides 9: 257–267

Ricardo JA, Koh ET (1978) Anatomical evidence of direct projections from the nucleus of the solitary tract to the hypothalamus, amygdala, and other forebrain structures in the rat. Brain Res 153: 1–26

Rinaman L (1999) Interoceptive stress activates glucagon-like peptide-1 neurons that project to the hypothalamus. Am J Physiol 277: R582–R590

Rinaman L (2004) Hindbrain contriburions to anorexia. Am J Physiol 287:R1035–R1036

Romanovsky AA, Simons CT, Szekely M, Kulchitsky VA (1997) The vagus nerve in the thermoregulatory response to systemic inflammation. Am J Physiol 273: R407–R413

Romeo H, Tio DL, Rahman SU, Chiappelli F, Taylor AN (2001) The glossopharyngeal nerve as a novel pathway in immune-to-brain communication: relevance to neuroimmune surveillance of the oral cavity. J Neuroimmunol 115: 91–100
Roozendaal B, Williams CL, McGaugh JL (1999) Glucocorticoid receptor activation in the rat nucleus of the solitary tract facilitates memory consolidation: involvement of the basolateral amygdala. Eur J Neurosci 11: 1317–1323
Saade NE, Massaad CA, Ochoa-Chaar CI, Jabbur SJ, Safieh-Garabedian, Atweh SF (2002) Upregulation of proinflammatory cytokines and nerve growth factor by intraplantar injection of capsaicin in rats. J Physiol 343: 241–252
Sako K, Okuma Y, Hosoi T, Nomura Y (2005) STAT3 activation and c-FOS expression in the brain following peripheral administration of bacterial DNA. J Neuroimmunol 158: 40–49
Samad TA, Moore KA, Sapirstein A, Billet S, Allchorne A, Poole S, Bonventre JV, Woolf CJ (2001) Interleukin-1beta-mediated induction of Cox-2 in the CNS contributes to inflammatory pain hypersensitivity. Nature 410: 471–475
Sawchenko PE, Li HY, Ericsson A (2000) Circuits and mechanisms governing hypothalamic responses to stress: a tale of two paradigms. Prog Brain Res 122: 61–78
Scafers M, Svensson C, Sommer C, Sorkin LS (2003) Tumor necrosis factor-a induces mechanical allodynia after spinal nerve ligation by activation of p38 MAPK in primary sensory neurons. J Neurosci 23: 2517–2521
Sehic E, Blatteis CM. (1996) Blockade of lipopolysaccharide-induced fever by subdiaphragmatic vagotomy in guinea pigs. Brain Res 726: 160–166
Shapiro RE, Miselis RR (1985) The central neural connections of the area postrema. J Comp Neurol 234: 344–364
Sharkey KA, Kroese, ABA (2001) Consequenses of intestinal inflammation on the enteric nervous system: Neuronal activation induced by inflammatory mediators. Anat Rec 262: 79–90
Sharkey KA, Mawe GM (2002) Neuroimmune and epithelial interactions in intestinal inflammation. Curr Opin Pharmacols 2: 669–677
Simrin M, Axelsson J, Gillberg R, Abrahamsson H, Svedlund J, Bjornsson ES (2002) Quality of life in inflammatory bowel disease in remission: the impact of IBS-like symptoms and associated psychological factors. Am J Gastroenterol 97: 389–396
Sorkin LS, Doom CM (2000) Epineurial application of TNF elicits an acute mechanical hyperalgesia in the awake rat. J Peripher Nerv Syst 5: 96–100
Sugimoto N, Simons CT, Romanovsky AA (1999) Vagotomy does not affect thermal responsiveness to intrabrain prostaglandin E2 and cholecystokinin octapeptide. Brain Res 844: 157–163
Wan W, Janz L, Vriend CY, Sorensen CM, Greenberg AH, Nance DM (1993) Differential induction of c-Fos immunoreactivity in hypothalamus and brain stem nuclei following central and peripheral administration of endotoxin. Brain Res Bull 32: 581–587
Wan W, Wetmore L, Sorensen CM, Greenberg AH, Nance DM (1994) Neural and biochemical mediators of endotoxin and stress-induced c-fos expression in the rat brain. Brain Res Bull 34: 7–14
Wang X, Wang BR, Duan WL, Liu HL, Ju G (2000) The expression of IL-1 receptor type 1 in nodose ganglion and vagal paraganglia in the rat. Chin J Neurosci 16: 90–93
Wang X, Wang B-R, Duan X –L, Zhang P, Ding Y –Q, Jia Y, Jiao X –Y, Ju G (2002a) Strong expression of interleukin-1 receptor type 1 in the rat carotid body. J Histochem Cytochem 50: 1677–1684
Wang B, Glatzle J, Mueller MH, Kreis M, Enck P, Grundy D (2005) Lipopolsaccharide-induced changes in mesenteric afferent sensitivityof rat jejunum in vitro: role of prostaglandins. Am J Physiol Gastrointes Liver Physiol 289: G254–G260
Watkins LR, Wiertelak EP, Goehler L, Mooney-Heiberger K, Martinez J, Furness L, Smith KP, Maier SF (1994a) Neurocircuitry of illness-induced hyperalgesia. Brain Res 639: 283–299
Watkins LR, Weirtelak EP, Goehler LE, Smith KP, Martin D, Maier SF (1994b) Characterization of cytokine induced hyperalgesia. Brain Res 654: 15–26

Watkins LR, Goehler LE, Relton JK, Tartaglia N, Silbert L, Martin D, Maier SF (1995a) Blockade of interleukin-1-induced fever by subdiaphragmatic vagotomy: evidence for vagal mediation of immune-brain communication. Neurosci Lett 183: 27–31

Watkins LR, Goehler LE, Relton J, Brewer MT, Maier SF (1995b) Immune-to-brain communication: Systemic tumour necrosis factor-alpha (TNF-alpha) produces behavioral hyperalgesia via vagal afferents. Brain Res 692: 244–250

Williams CL, McGaugh JL (1993) Reversible lesions of the nucleus of the solitary tract attenuate the memory-modulatory effects of posttraining epinephrine. Behav Neurosci 107: 955–962

Wolf G, Gabay E, Tal M, Yirmiya R, Shavit Y (2006) Genetic impairment of interleukin-1 signaling attenuates neuropathic pain, autotomy, and spontaneous ectopic neuronal activity, following nerve injury in mice. Pain 120: 315–324

Zagon A (2006) Does the vagus nerve mediate the sixth sense? Trends Neurosci 24: 671–673

Zhang Y-H, Lu J, Elmquist JK, Saper CB (2000) Lipopolysaccharide activates specific populations of hypothalamic and brainstem neurons that project to the spinal cord. J Neurosci 20: 6578–6586

Zhang J-M, Li H, Liu B, Brull, SJ (2002) Acute topical application of tumor necrosis factor a evokes protein kinase A-dependent responses in rat sensory neurons. J Neurophysiol 88: 1387–1392

Molecular Basis of Cytokine Function

Pranela Rameshwar and Arlene Bardaguez

Abstract Cytokines are soluble glycoproteins that are ubiquitously produced by immune and other cells. The family of cytokines also includes a subset of small molecules, designated. Although the cytokine receptors share some subunits, they show specificity for binding and signaling. Unlike hormones that act at sites distant from the area of production, most cytokines are easily degraded. Thus, in general, cytokines mediate functions via autocrine and paracrine manner. However, cytokines can initiate functional responses at low concentrations, which could cause subtle brain effects. In contrast, cytokines have also been given attention in medicine through the `cytokine storm,' which result in general illness. Cytokine storm could arrive from bacterial and viral infections. The chapter discusses the molecular basis for behavioral dysregulation that could result from cytokine production.

Key words Cytokines · Immune system · Interleukins · Chemokines

1 Introduction

The general term for molecules belonging to chemokine, interleukin, and lymphokine families. Cytokines are soluble glycoproteins produced by several cells, but a designation of cytokine requires production by immune cells. Cytokines do not require enzymatic processing for signaling, but binds to specific receptors present on immune and neural cells, among others. In contrast to hormones that act at sites distant from the area where they are produced, most cytokines are easily degraded. The similarity between cytokines and hormones is limited to their efficacy at low levels, and high affinity binding to specific receptors. Because of their effects at low concentrations, their levels just above baseline can initiate multiple anatomical pathophysiological conditions. The recognized effects of cytokine in medicine have led to the concept of cytokine storm (Clark, 2007). This term generally refers to the

P. Rameshwar (✉)
Departments of Medicine, UMDNJ–New Jersey Medical School, Newark, NJ, USA
e-mail: rameshwa@umdnj.edu

A. Siegel, S.S. Zalcman (eds.), *The Neuroimmunological Basis of Behavior and Mental Disorders*, DOI 10.1007/978-0-387-84851-8_4,

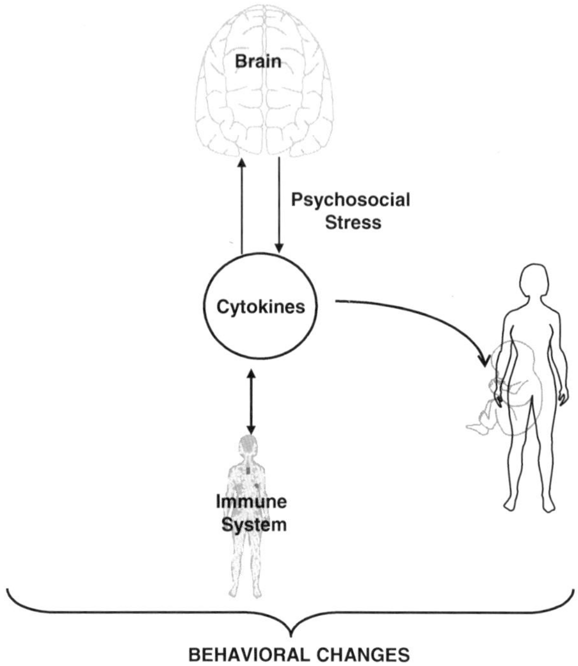

Fig 1. Cytokines are shown as central to the interactions between the brain and the immune system. Infection or psychosocial stress can change cytokine levels. In the case of pregnancy, both the mother and child can be affected.

unpleasant feeling caused by increased cytokine production following infection with the influenza virus. However, similar mechanisms are attributed to other infections, which could lead to secondary effects on other organs, including brain functions.

A review on cytokines is not completed unless credit is given to its discovery. An acknowledgment should be given to the work by Dr. Stanley Cohen and colleagues whose original designation, cytokines, has not deviated much from the current description of the growing number of cytokines (Cohen, 2004, 1986; Antonia et al., 1986).

Cytokines are subdivided into four different classes, based on their structure. The family with four alpha-helix bundles is further subdivided into interleukin (IL)-2, interferon, and IL-10 subfamilies. The other three families are grouped as IL-1, IL-17, and chemokine. The structural discrimination is important since functional redundancy is a hallmark of cytokines. In each category or subcategory, the total numbers are continually updated with the identification of new members (Uze and Monneron, 2007).

2 Interleukins

The interleukins, which currently include >30 members belong to the family of cytokines (Weinreich et al., 2006). Since the interleukins are a subgroup of cytokines, they are subjected to the rules of designation as a family of glycoproteins. The interleukins are produced by immune and other cells with varying, but redundant functions. The term interleukin was originally named because upon their discovery, it was believed that their production from immune cells resulted in communication with other immune cells. Since the immune cells were presumed to be leukocytes or white blood cells, this explains the term interleukin, which means between leukocytes.

At the early periods of discovery, the term interleukin was limited to molecules that targets white blood cells. In recent years, the number of interleukins is rapidly expanding with the current number up to 33 (Iikura et al., 2007). The stringent designation of the term interleukin to effects on leukocytes has become difficult to follow and therefore makes it difficult to define them within a narrow range of functions.

Attempts have been made among international scientists for a consensus to apply the term interleukin (WHO-IUIS, 1992, 1997). Results of multiple international meetings led to four criteria when deciding on the designation of interleukin. Firstly, the gene and amino acid sequences should be identified and also demonstrated in molecular studies to be expressed as secreted proteins. The sequence cannot be from another gene that has already been cloned with a different designation. Secondly, the molecule has to be shown to be endogenously produced by cells of the immune system and should exhibit multiple functions. Thirdly, despite the previous criteria if the gene was previously cloned and its major functions belong to tissues and/or organs other than those within the immune system, the molecule should retain its original name rather than adding to the list of interleukins. Fourthly, the discoverer is left to opt for a descriptive name rather than an interleukin. Scientists have tried to adhere to the four criteria but the difficulty arises when they are faced with the fourth. IL-24 is a typical example of an interleukin that was originally cloned in melanoma cells and has been found to impart anti-cancer effects (Chada et al., 2004). Subsequently, IL-24 was discovered to function as a cytokine and share intracellular signaling similar to several other cytokines such as those belonging to the IL-10 family (Chada et al., 2004). Since IL-24 has been shown to exhibit a potent anti-cancer effect, scientists have designated this molecule as an interleukin and also as its original name, melanoma differentiation associated gene (MDA) in combination with IL-24. Thus it is common for this gene to be designated IL-24.MDA (Oida et al., 2007; Chen et al., 2005; Fisher 2005). IL-22 represents an example that show dual roles, as an immune modulator, and in other organs (Levillayer et al., 2007; Oral et al., 2006; Zheng et al., 2007).

The IL-10 family of cytokines includes several members including IL-10, IL-22, IL-24, IL-26, IL-28, and IL-29 (Sabat et al., 2007; Kotenko and Langer, 2004). The complexity in the functions of this family of cytokines is compounded by several members sharing receptor subunits and even cross react with the receptor of each other (Kotenko and Langer, 2004). Another confounds in this large IL-10 family of cytokines is that some members (IL-19 and IL-20), although they are produced by immune cells, it is unclear if they regulate immune functions (Sabat et al., 2007). Ongoing in vivo studies on disease model suggest that the IL-10 family of cytokines might be involved in the inflammatory cascade, either as anti- and/or pro-inflammatory mediators (Sabat et al., 2007).

3 Chemokines

The chemokine group comprises a large family of small cytokines of approximately 8–10 kDa (Murphy et al., 2000). Since the chemokines belong to a subgroup of cytokines, their designations follow those linked to cytokines, as outlined above. The designation of chemokines under a separate category is mainly due to their properties in attracting immune cells to each other or to an organ of inflammation (Ruffini et al., 2007). Immune or other cells that express chemokine receptors tend to migrate towards regions of high chemokine levels, such as regions of tissue injuries. Besides the chemoattractant property, the chemokines are grouped together based on their 3D structures. Initially, the same chemokine was referred to by multiple names. This problem has been corrected after the international union of nomenclature agreed on consensus names for each member of the family (Murphy et al., 2000).

There are four cysteine residues found within conserved regions of the chemokines, which are important for their 3D structures (Fernandez and Lolis, 2002). The first two cysteines are towards the N-terminal end of the protein; the third at the centre and the fourth towards the C-terminal (Fernandez and Lolis, 2002). The first and third cyteines are connected by disulphide bonds, and the second to fourth residues are similarly connected. The 3D structures resulted in several loops (Fernandez and Lolis, 2002). The size of the loop formed between the first two cysteines, which depends on the spacing forms the N-loop. The chemokines are grouped into four categories, based on the spacing between the two cysteines: CC or β-chemokine, CXC or α-chemokine, C or γ chemokine, and CX_3C or δ-chemokines (Fernandez and Lolis, 2002). The CC chemokines comprise more than 20 members (Laing and Secombes, 2004a, b). The CXC chemokines comprise approximately 17 members (Laing and Secombes, 2004a, b). Unlike the other chemokines, the C type has only two cysteines and has few members, of which there are XCL1 and XCL2 (Laing and Secombes, 2004a, b). The fourth group, CX_3C, has only one member with three amino acids between the two cysteines, CX_3CL1 (Laing and Secombes, 2004a, b).

Chemokines are considered to exhibit mostly pro-inflammatory properties. They are produced during an immune response for the purpose of attracting additional immune cells to the site of infection (Mantovani et al., 2006). In contrast to their inducibility during infection, other members of this family, including SDF-1α/CXCL12, could be involved in homeostasis in an organ-specific method (Majka and Ratajczak, 2006). Specifically, CXCL12 is constitutively produced by bone marrow stromal cells for the purpose of retaining the hematopoietic stem cells in the region. This occurs by interactions between membrane-bound CXCL12 present on stromal cells and the receptor (CXCR4)-expressing stem cells (Majka and

Ratajczak, 2006). CXCL12 continues to maintain the movement of hematopoeitic stem cells within bone marrow through a gradient changes in its levels across bone marrow.

4 Colony Stimulating Factor (CSF) and Tumor Necrosis Factor (TNF)

These categories of cytokines require a separate subheading because of their critical roles in immune responses, but are not grouped within the interleukins and chemokine members. The CSF family is neglected when considerations are placed in behavioral and other pathophysiology caused by cytokines. The CSFs are mostly thought as growth factors mainly due to their roles as hematopoietic stimulators (Touw and van de Geijin, 2007). The three CSFs are granulocyte-colony stimulating factor (G-CSF), macrophage-colony stimulating factor (M-CSF), and granulocyte–macrophage-colony stimulating factor (GM-CSF). G-CSF and GM-CSF are commonly used in patients for neutropenia and in the case of G-CSF, for mobilizing hematopoietic stem cells from donor bone marrow for the purpose of transplantation in an allogeneic donor (Kuderer et al., 2007, Winkler and Levesque, 2006).

TNF are found in two forms, the α- (cachetin) and β- (lymphotoxin) forms and share the types 1 (TNF-R1) and 2 (TNF-R2) receptors (Locksley et al., 2001). TNF-R1 is ubiquitously expressed whereas TNF-R2 is only on immune cells (Locksley et al., 2001). As compared to studies on TNF-R1, there is limited information on the type 2 receptor. TNF induces trimerization of the receptors, which caused conformational change and activation of the receptor. The activation follows the disassociation of SODD, which suppresses the intracellular death domain. In exchange, the death domain comes in contact with the adaptor protein TRADD (Wajant et al., 2003; Chen and Goeddel, 2002). TRADD acts as an adaptor protein for other activators, which resulted in multiple intracellular pathways (Wajant et al., 2003; Chen and Goeddel, 2002). The pathways include NFκB activation, which is involved in varying inflammatory responses and the induction of anti-apoptotic mediators. In contrast to the pro-inflammatory properties by NFκB activation, TNF can activate MAPK pathways that are involved in cell differentiation, proliferation, and the activation of pro-apoptotic factors. TNF also could induce cell death, although this property is less prominent as compared to other TNF family members linked to cell death (Gaur and Aggarwal, 2003). The weak apoptotic property of TNF, combined with the pro-inflammatory property of NFκB brings up a critical point of balances between these two functions following exposures to TNF.

The properties of TNF overlaps with those of IL-1 with respect to the induction of systemic inflammation, in particular fever. Also, TNF increase the metabolism of muscle and cause loss of fat in adipocytes, which lead to cachexia. TNF has been linked to septic shock. Despite these links, trials with TNF antagonists have not shown much promise (Inanc and Direskeneli, 2006).

5 Cytokine Receptors

An understanding of cytokine receptors regardless of their secondary or tertiary structures has significance to experimental research and also to the development of therapeutic targets. Regardless of the classifications, this does not eliminate the fact that cytokines show functional pleotrophism. The cytokine receptors are classified as type 1 immunoglobulin (Ig) superfamily, type 2 interferon family, type 3 tumor necrosis factor family and those belonging to the G-protein family receptors (Zlotnik et al., 2006; Krause and Pestka, 2005; Boulay et al., 2003). The type 1 family of receptors is ubiquitously expressed and includes receptors for IL-1 and IL-2. The type 1 receptors also include those that interact with CSFs with conserved motifs in their extracellular regions. The type 2 receptors bind to the interferon family of cytokines (Zlotnik et al., 2006; Alves et al., 2007; Boraschi and Tagliabue 2006; Murphy and Young, 2006). The type 3 TNF family comprises receptors with cysteine-rich extracellular binding domains (Alves et al., 2007). The 7-transmembrane G-protein coupled receptors are limited to members of the chemokines (Zlotnik et al., 2006).

6 Cytokines in the Nervous System

The nervous and immune systems are connected anatomically by nerve fibers and by other functions such as the migration of cells into the brain as well as the movement of soluble factors into and out of the brain (Quan and Banks, 2007). The immune system includes the secondary and primary lymphoid organs of which the latter include the thymus and brain. The transport of cytokines across the blood–brain barrier has evolved from the concept where there was demarcation between normal and diseased brain to current information on the movement of cytokines via secretion of cytokines or via transporters (Quan and Banks, 2007). These series of findings have led to an understanding of the mechanisms by which cytokines are central to the peripheral diseases and brain and vice versa (Banks, 2006a).

The brain and its peripheral connections are targets of drug delivery (Banks, 2006b). During testing of drug delivery, the pharmacological dose might be different between homeostasis and in cases where cytokine levels are increased or decreased in the brain and/or peripheral organs. Thus, an understanding of the types and levels of cytokines in the brain would lead to higher efficacy in drug deliveries. During drug testing, the levels and types of cytokines are not the only consideration. Cytokines can affect neuronal functions, which might affect the efficacy of a drug (Viviani et al., 2007).

The presence of cytokines in brain could also be from endogenous sources. Microglia, which are considered as brain macrophages could produce cytokines in response to various brain insults (Kriz, 2006; Wang and Suzuki, 2007). Microglia has been reported to mediate neuroprotective functions, following acute brain injury (Simard and Rivest, 2007). Since microglial cells are sources of cytokines, the question is when these cells become protective and when they are harmful. These are relevant questions that will require robust experimental analyses at the single cell level and by electrophysiology to determine how cytokines could function as mediators of microglial effects.

In contrast to the pathogenic effects of cytokines in the brain, the interferons can also dampen the inflammatory responses in the brain such as in multiple sclerosis (Bagnato et al., 2007). An interesting role for cytokines is the effects of IL-1 on the regulation of brain volume (Oprica et al., 2007). This role is important since it might be relevant to an understanding of aging disorders associated with the central nervous system.

7 Cytokines and Neuroprotection

This section discusses the potential of cytokines for neural protection. While the experimental studies might show neural protection, genetic variation among individuals is an important point that could affect the effects of a particular cytokine in brain function (Wilson and Montgomery, 2007). Besides genetic variation, the complexity on cytokine functions is compounded by the influence of several other genes and polymorphism. Although erythropoietin (Epo) is not a cytokine, its action has been incorporated within networks of cytokines. This type of network has been studied extensively in bone marrow functions (Fliser and Haller, 2007). Epo has been shown experimentally to mediate neuroprotection and to direct cell fate toward neurogenesis while suppressing gliogenesis in neonatal stroke (Gonzalez et al., 2007).

Stem cell therapy has been proposed as a viable form of therapy for brain repair (Takahashi, 2007). Since cytokines are produced by the stem cells and endogenous cells, the microenvironment that is established by the implanted stem cells will influence the outcome by stem cells. The imperative questions will surround the types and levels of cytokine production, their receptor expression, the developmental changes in cytokines and the effects on the brain.

The protective roles of cytokines could be hindered in cases of brain tumors, which could produce cytokines that facilitate the survival of the tumor cells rather than to affect brain functions (Zisakis et al., 2007). While brain tumors exhibit acute production of cytokines, depression and dementia show chronic production of cytokines (Leonard, 2007). The chronic production of cytokines remain an unresolved issue since it is unclear if this is secondary to other underlying disorders or if this could be involved in the pathophysiology of dementia and depression. A role for cytokines in depression has been demonstrated in experimental studies in which a nonsteroidal anti-inflammatory agent, COX2 inhibitor, reversed the behavioral pattern of a rat model of depression (Myint et al., 2007). Another evidence for the effects of cytokines on depression is the role of lipopolysaccharide, which is a pro-inflammatory agent, as a depressogenic agent (Pekary et al., 2007).

8 Cytokines in Vasculogenesis: Relevance to Brain Functions

This section discusses the possibility of taking advantages of cytokines in enhancing vasculogenesis by attracting endogenous endothelial progenitors. In some cases, excess blood vessels could be undesirable since they support the pathophysiology of the disease such as cancer. In other cases, rapid increase in vasculogenesis could be a positive treatment option. Of particular relevance are disorders of the brain in which lack of blood supplies lead to brain damage. The concept proposed for enhanced vasculogenesis for wound healing can be extrapolated to brain disorders (Velazquez, 2007). At present, the research is focused on attracting endothelial progenitors from the bone marrow to the region of injuries. This source of progenitors might be delayed if rapid attraction is required in the brain. In this regard, it might be prudent to begin studies to determine how neural stem cells could be challenged to form endothelial cells and to identify which cytokines could be involved in the process. This area of investigations could lead to future therapies. Since chemokines are the prototypical chemoattractants, and they are commercially available, an injection at the site of injury could be another method to attract bone marrow-derived cells for the purpose of increasing vasculogenesis (Shireman, 2007).

9 Cytokines in Pregnancy and Behavior

Genetic causes have been the focus for mental disorders. However, in several of these studies, there is evidence that non-genetic factors might be responsible for the disease (Patterson, 2007). While most of these non-genetic causes have been associated with the environment, other endogenous changes could also be involved in behavioral changes. The changes observed during pregnancy represent prototypical cases where cytokines have been linked not only to brain functions of the mother, but to brain development of the fetus.

During infection endotoxin could enhance dopamine release, which could result in behavioral changes in the mom and possible developmental defects in the neonate's brain (Romero et al., 2006). Since endotoxin is a potent inducer of cytokines, this type of pathophysiology could be treated if the causative cytokines and the mechanisms are understood. The fact that neurons express cytokine receptors indicate that their roles as mediators by endotoxin produced during bacterial infection (Greco and Rameshwar, 2007). Animal models in which the neonates have been exposed to IL-2 showed neurobehavioral problems linked to autism (Ponzio et al., 2007). The autistic behavior of the fetus during pregnancy is not limited to IL-2. Animal studies in which IL-6 was knockout implicated this cytokine as a factor involved in predisposing the fetus to schizophrenia and autism (Smith et al., 2007).

Psychosocial stress during late stage pregnancy could lead to the production of cytokines (Coussons-Read et al., 2007). The production of cytokines during this period of pregnancy can affect the outcome such as preeclampsia and premature labor (Coussons-Read et al., 2007). In addition to the risk to pregnancy, there could be long-term risk to the fetus, based on animal models (Vanbesien-Mailliot et al., 2007).

10 Conclusion

Decades of research have identified numerous molecules that fall under the category of cytokines. Several costly clinical trials have been done, yet it has been more than 15 years since the US Food and Drug Administration has approved a new cytokine for hematological disorder (Weinreich et al., 2006). There are effective treatments with cytokines for inflammatory diseases. While TNFα has been a target in inflammation, therapies are needed to suppress inflammation while inducing anti-inflammatory mediators such as IL-10 (Tilg et al., 2007). In many cases, the benefit of cytokines appear positive, but this optimism is dampened by toxicity, partly caused by behavioral changes (Kalaaji, 2007; Minderhoud et al., 2007; Malone et al., 2007).

The question that lingers is the methods by which cytokines would fit into the general scheme of future therapy? The evolving roles of stem cells as cellular therapy, including neural diseases are of tremendous interest since the interaction between the stem cells and the diseased microenvironment would be critical to successful therapy. Since cytokines may have major roles in the cellular communication between a diseased microenvironment and the stem cells, it might be wise to continue to determine the roles of cytokines in stem cell therapy for neural disease, including brain behavior. Small molecules as targets for cytokines could benefit with current therapies of silencing RNA to inhibit specific cytokine functions (Svoboda, 2007).

In addition to cytokines and their receptors as potential drug targets, intracellular signaling molecules can also function as drug targets. STAT3, which is activated by various cytokines and is negatively regulated by the suppressor of cytokine signaling (SOCS), has been suggested as potential targets for the inflammatory lungs (Gao and Ward, 2007). AKT/PKB, which is activated by several cytokines, were also proposed as potential drug targets for type-2 diabetes and cancer (Manning and Cantley, 2007). Ultimately, we propose that specific targets of cytokines, their receptors, and/or intracellular molecules might be more efficient targets as compared to non-specific anti-inflammatory agents such as non-steroidal anti-inflammatory drugs (Rainsford, 2007). Cytokines are critical to debilitating diseases such as cancers and are involved in the related fatigue, which also involved central nervous system effects (Ryan et al, 2007). This final point underscores the need for continued dissection of molecular pathways to treat disorders since cytokines in brain will affect peripheral organs and vice versa.

Acknowledgments This work was supported by the FM Kirby Foundation.

References

Alves NL, Arosa FA, van Lier RA: Common gamma chain cytokines: dissidence in the details. *Immunol Lett* 2007, 108:113–120.

Antonia SJ, Cohen S, Cohen MC: The elaboration of a small molecular weight cytostatic factor by lymphoblastoid lines and activated lymphocytes. *Lymphokine Res* 1986, **5:**301–312.

Bagnato F, Evangelou IE, Gallo A, Gaindh D, Yao K: The effect of interferon-beta on black holes in patients with multiple sclerosis. *Expert Opin Biol Ther* 2007, **7:**1079–1091.

Banks WA: The blood-brain barrier as a regulatory interface in the gut-brain axes. *Physiol Behav* 2006a, 89:472–476.

Banks WA: The CNS as a target for peptides and peptide-based drugs. *Expert Opin Drug Deliv* 2006b, 3:707–712.

Boraschi D, Tagliabue A: The interleukin-1 receptor family. *Vitam Horm* 2006, 74:229–254.

Boulay JL, O'Shea JJ, Paul WE: Molecular phylogeny within type I cytokines and their cognate receptors. *Immunity* 2003, 19:159–163.

Chada S, Sutton RB, Ekmekcioglu S, Ellerhorst J, Mumm JB, Leitner WW, Yang HY, Sahin AA, Hunt KK, Fuson KL et al.: MDA-7/IL-24 is a unique cytokine – tumor suppressor in the IL-10 family. *Int Immunopharmacol* 2004, 4:649–667.

Chen G, Goeddel DV: TNF-R1 signaling: a beautiful pathway. *Science* 2002, 296:1634–1635.

Chen WY, Cheng YT, Lei HY, Chang CP, Wang CW, Chang MS: IL-24 inhibits the growth of hepatoma cells in vivo. *Genes Immun* 2005, 6:493–499.

Clark IA: The advent of the cytokine storm. *Immunol Cell Biol* 2007, 85:271–273.

Cohen S: Physiologic and pathologic manifestations of lymphokine action. *Hum Pathol* 1986, 17:112–121.

Cohen S: Cytokine: more than a new word, a new concept proposed by Stanley Cohen thirty years ago. *Cytokine* 2004, 28:242–247.

Coussons-Read ME, Okun ML, Nettles CD: Psychosocial stress increases inflammatory markers and alters cytokine production across pregnancy, *Brain, Behavior, and Immunity* 2007, 21:343–350.

Fernandez EJ, Lolis E: Structure, function, and inhibition of chemokines. *Annu Rev Pharmacol Toxicol* 2002, 42:469–499.

Fisher PB: Is mda-7/IL-24 a "Magic Bullet" for Cancer? *Cancer Res* 2005, 65:10128–10138.

Fliser D, Haller H: Erythropoietin and treatment of non-anemic conditions – cardiovascular protection. *Semin Hematol* 2007, 44:212–217.

Gao H, Ward PA: STAT3 and suppressor of cytokine signaling 3: potential targets in lung inflammatory responses. *Expert Opin Ther Targets* 2007, 11:869–880.

Gaur U, Aggarwal BB: Regulation of proliferation, survival and apoptosis by members of the TNF superfamily. *Biochem Pharmacol* 2003, 66:1403–1408.

Gonzalez FF, McQuillen P, Mu D, Chang Y, Wendland M, Vexler Z, Ferriero DM: Erythropoietin enhances long-term neuroprotection and neurogenesis in neonatal stroke. *Dev Neurosci* 2007, 29:321–330.

Greco SJ, Rameshwar P: Enhancing effect of IL-1{alpha} on neurogenesis from adult human mesenchymal stem cells: implication for inflammatory mediators in regenerative medicine. *J Immunol* 2007, 179:3342–3350.

Iikura M, Suto H, Kajiwara N, Oboki K, Ohno T, Okayama Y, Saito H, Galli SJ, Nakae S: IL-33 can promote survival, adhesion and cytokine production in human mast cells. *Lab Invest* 2007.

Inanc N, Direskeneli H: Serious infections under treatment with TNF-alpha antagonists compared to traditional DMARDs in patients with rheumatoid arthritis. *Rheumatol Int* 2006, 27:67–71.

Kalaaji AN: Cytokine therapy in advanced melanoma. *J Drugs Dermatol* 2007, 6:374–378.

Kotenko SV, Langer JA: Full house: 12 receptors for 27 cytokines. *Int Immunopharmacol* 2004, 4:593–608.

Krause CD, Pestka S: Evolution of the Class 2 cytokines and receptors, and discovery of new friends and relatives. *Pharmacol Ther* 2005, 106:299–346.

Kriz J: Inflammation in ischemic brain injury: timing is important. *Crit Rev Neurobiol* 2006, 18:145–157.

Kuderer NM, Dale DC, Crawford J, Lyman GH: Impact of primary prophylaxis with granulocyte colony-stimulating factor on febrile neutropenia and mortality in adult cancer patients receiving chemotherapy: a systematic review. *J Clin Oncol* 2007, 25:3158–3167.

Laing KJ, Secombes CJ: Chemokines. *Dev Comp Immunol* 2004a, 28:443–460.

Laing KJ, Secombes CJ: Trout CC chemokines: comparison of their sequences and expression patterns. *Mol Immunol* 2004b, 41:793–808.

Leonard BE: Inflammation, depression and dementia: are they connected? *Neurochem Res* 2008, 93; 52–55.

Levillayer F, Mas M, Levi-Acobas F, Brahic M, Bureau JF: Interleukin 22 Is a Candidate Gene for Tmevp3, a Locus Controlling Theiler's Virus-Induced Neurological Diseases. *Genetics* 2007, 176:1835–1844.

Locksley RM, Killeen N, Lenardo MJ: The TNF and TNF receptor superfamilies: integrating mammalian biology. *Cell* 2001, 104:487–501.

Majka M, Ratajczak MZ: Biological role of the CXCR4-SDF-1 axis in normal human hematopoietic cells. *Methods Mol Biol* 2006, 332:103–114.

Malone FR, Leisenring WM, Storer BE, Lawler R, Stern JM, Aker SN, Bouvier ME, Martin PJ, Batchelder AL, Schoch HG et al.: Prolonged anorexia and elevated plasma cytokine levels following myeloablative allogeneic hematopoietic cell transplant. *Bone Marrow Transplant* 2007.

Manning BD, Cantley LC: AKT/PKB Signaling: Navigating Downstream. *Cell* 2007, 129:1261–1274.

Mantovani A, Bonecchi R, Locati M: Tuning inflammation and immunity by chemokine sequestration: decoys and more. *Nat Rev Immunol* 2006, 6:907–918.

Minderhoud IM, Samsom M, Oldenburg B: Crohn's disease, fatigue, and infliximab: is there a role for cytokines in the pathogenesis of fatigue? *World J Gastroenterol* 2007, 13:2089–2093.

Murphy JM, Young IG: IL-3, IL-5, and GM-CSF signaling: crystal structure of the human beta-common receptor. *Vitam Horm* 2006, 74:1–30.

Murphy PM, Baggiolini M, Charo IF, Hebert CA, Horuk R, Matsushima K, Miller LH, Oppenheim JJ, Power CA: International Union of Pharmacology. XXII. Nomenclature for Chemokine Receptors. *Pharmacol Rev* 2000, 52:145–176.

Myint AM, Steinbusch HW, Goeghegan L, Luchtman D, Kim YK, Leonard BE: Effect of the COX-2 Inhibitor Celecoxib on Behavioural and Immune Changes in an Olfactory Bulbectomised Rat Model of Depression. *Neuroimmunomodulation* 2007, 14:65–71.

Oida Y, Gopalan B, Miyahara R, Branch CD, Chiao P, Chada S, Ramesh R: Inhibition of nuclear factor-{kappa}B augments antitumor activity of adenovirus-mediated melanoma differentiation-associated gene-7 against lung cancer cells via mitogen-activated protein kinase kinase kinase 1 activation. *Mol Cancer Ther* 2007, 6:1440–1449.

Oprica M, Hjorth E, Spulber S, Popescu BO, Ankarcrona M, Winblad B, Schultzberg M: Studies on brain volume, Alzheimer-related proteins and cytokines in mice with chronic overexpression of IL-1 receptor antagonist. *J Cell Mol Med* 2007, 11:810–825.

Oral HB, Kotenko SV, Yilmaz M, Mani O, Zumkehr J, Blaser K, Akdis CA, Akdis M: Regulation of T cells and cytokines by the interleukin-10 (IL-10)-family cytokines IL-19, IL-20, IL-22, IL-24 and IL-26. *Eur J Immunol* 2006, 36:380–388.

Patterson PH: Neuroscience: Maternal effects on schizophrenia risk. *Science* 2007, 318:576–577.

Pekary AE, Stevens SA, Sattin A: Lipopolysaccharide modulation of thyrotropin-releasing hormone (TRH) and TRH-like peptide levels in rat brain and endocrine organs. *J Mol Neurosci* 2007, 31:245–259.

Ponzio NM, Servatius R, Beck K, Marzouk A, Kreider TIM: Cytokine levels during pregnancy influence immunological profiles and neurobehavioral patterns of the Offspring. *Ann NY Acad Sci* 2007, 1107:118–128.

Quan N, Banks WA: Brain-immune communication pathways. *Brain, Behav Immun* 2007, 21: 727–735.

Rainsford KD: Anti-inflammatory drugs in the 21st century. *Subcell Biochem* 2007, 42:3–27.

Romero E, Ali C, Molina-Holgado E, Castellano B, Guaza C, Borrell J: Neurobehavioral and immunological consequences of prenatal immune activation in rats; influence of antipsychotics. *Neuropsychopharmacol* 2006, 32:1791–1804.

Ruffini PA, Morandi P, Cabioglu N, Altundag K, Cristofanilli M: Manipulating the chemokine-chemokine receptor network to treat cancer. *Cancer* 2007, 109:2392–2404.
Ryan JL, Carroll JK, Ryan EP, Mustian KM, Fiscella K, Morrow GR: Mechanisms of Cancer-Related Fatigue. *Oncologist* 2007, 12:22–34.
Sabat R, Wallace E, Endesfelder S, Wolk K: IL-19 and IL-20: two novel cytokines with importance in inflammatory diseases. *Expert Opin Ther Targets* 2007, 11:601–612.
Shireman PK: The chemokine system in arteriogenesis and hind limb ischemia. *J Vasc Surg* 2007, 45 Suppl A:A48–A56.
Simard AR, Rivest S: Neuroprotective effects of resident microglia following acute brain injury. *J Comp Neurol* 2007, 504:716–729.
Smith SEP, Li J, Garbett K, Mirnics K, Patterson PH: Maternal Immune Activation Alters Fetal Brain Development through Interleukin-6. J *Neurosci* 2007, 27:10695–10702.
Svoboda P: Off-targeting and other non-specific effects of RNAi experiments in mammalian cells. *Curr Opin Mol Ther* 2007, 9:248–257.
Takahashi J: Stem cell therapy for Parkinson's disease. *Expert Rev Neurotherapeutics* 2007, 7:667–675.
Tilg H, Moschen A, Kaser A: Mode of function of biological anti-TNF agents in the treatment of inflammatory bowel diseases. *Expert Opin Biol Th* 2007, 7:1051–1059.
Touw IP, van de Geijn GJ: Granulocyte colony-stimulating factor and its receptor in normal myeloid cell development, leukemia and related blood cell disorders. *Front Biosci* 2007, 12:800–815.
Uze G, Monneron D: IL-28 and IL-29: newcomers to the interferon family. *Biochimie* 2007, 89:729–734.
Vanbesien-Mailliot CCA, Wolowczuk I, Mairesse J, Viltart O, Delacre M, Khalife J, Chartier-Harlin MC, Maccari S: Prenatal stress has pro-inflammatory consequences on the immune system in adult rats. *Psychoneuroendocrinology* 2007, 32:114–124.
Velazquez OC: Angiogenesis and vasculogenesis: inducing the growth of new blood vessels and wound healing by stimulation of bone marrow-derived progenitor cell mobilization and homing. *J Vasc Surg* 2007, 45 Suppl A:A39–A47.
Viviani B, Gardoni F, Marinovich M: Cytokines and Neuronal Ion Channels in Health and Disease. *Int Rev Neurobiol Neuroinflammation Neuronal Death Repair* 2007, 82:247–263.
Wajant H, Pfizenmaier K, Scheurich P: Tumor necrosis factor signaling. *Cell Death Differ* 2003, 10:45–65.
Wang X, Suzuki Y: Microglia produce IFN-gamma independently from T cells during acute toxoplasmosis in the brain. *J Interferon Cytokine Res* 2007, 27:599–605.
Weinreich MA, Lintmaer I, Wang L, Liggitt HD, Harkey MA, Blau CA: Growth factor receptors as regulators of hematopoiesis. *Blood* 2006, 108:3713–3721.
WHO-IUIS: Nomenclature for secreted regulatory proteins of the immune system (interleukins). WHO-IUIS Nomenclature Subcommittee on Interleukin Designation. *Eur J Immunol* 1992, 22:616.
WHO-IUIS: Nomenclature for secreted regulatory proteins of the immune system (interleukins): update. IUIS/WHO Standing Committee on Interleukin Designation. *Bull World Health Organ* 1997, 75:175.
Wilson M, Montgomery H: Impact of genetic factors on outcome from brain injury. *Br J Anaesth* 2007, 99:43–48.
Winkler IG, Levesque JP: Mechanisms of hematopoietic stem cell mobilization: when innate immunity assails the cells that make blood and bone. *Exp Hematol* 2006, 34:996–1009.
Zheng Y, Danilenko DM, Valdez P, Kasman I, Eastham-Anderson J, Wu J, Ouyang W: Interleukin-22, a TH17 cytokine, mediates IL-23-induced dermal inflammation and acanthosis. *Nature* 2007, 445:648–651.
Zisakis A, Piperi C, Themistocleous MS, Korkolopoulou P, Boviatsis EI, Sakas DE, Patsouris E, Lea RW, Kalofoutis A: Comparative analysis of peripheral and localised cytokine secretion in glioblastoma patients. *Cytokine* 2007, 39: 99–105.
Zlotnik A, Yoshie O, Nomiyama H: The chemokine and chemokine receptor superfamilies and their molecular evolution. *Genome Biol* 2006, 7:243.

Interferon-α, Molecular Signaling Pathways and Behavior

Jianping Wang

Abstract Interferon-alpha (IFN-α), a prime innate immunity mediator, has been implicated in neuropsychiatric disorders in humans in addition to its other beneficial pleiotropic actions. The pathophysiology of IFN-α-induced CNS illnesses, however, remains obscure because of limited access to human brain. Demonstration of JAK/STAT pathway in IFN-α signaling provides a novel means to evaluate the functional impact of IFN-α on the brain. Recent studies in mice have indicated a direct access of systemic IFN-α to the brain showing a rapid and profound stimulation of IFN-stimulated genes in brain parenchyma cells following peripheral IFN- administration. Conflicting neurochemical data and behavioral observations of IFN-α in rodents necessitate the importance of further investigation including broad behavioral characterization and cellular/molecular dissection. Elucidation of the cerebral neuronal circuitry in mediating the behavioral dysfunctions by IFN-α will help to delineate a long-disputed etiologic and/or pathogenic link between immune activation and human mental disorders.

Keywords IFN-α · JAK/STAT signaling · IFN-stimulated gene · Brain and behavior

1 Introduction

Cytokines are important autocrine, paracrine or endocrine regulatory proteins in host response to immunological challenge and/or tissue damage. In general, they are categorized into several families that include the interferon (IFN), interleukin (IL), tumor necrosis factor (TNF), colony-stimulating factor (CSF), transforming growth factor (TGF) and chemokine (CK). As a key means of defensive response to various infections, cytokines are critical in maintaining overall homeostasis in the central nervous system (CNS), ranging from clearing infection and damaged cells to tissue repair and generation. Nevertheless, besides their beneficial actions, there

J. Wang (✉)
Division of Pharmacology and Toxicology, School of Pharmacy, University of Missouri, Kansas City, 2464 Charlotte Street, Kansas City, MO 64108, USA
e-mail: wangjp@umkc.edu

A. Siegel, S.S. Zalcman (eds.), *The Neuroimmunological Basis of Behavior and Mental Disorders*, DOI 10.1007/978-0-387-84851-8_5,

is emerging evidence that these endogenous proteins are the "double-edged" swords implicated in the development of neurological and neuropsychiatric diseases. In this chapter, we will focus on IFN signaling and its behavioral consequences. In particular, we will discuss the controversy over the studies on animals and future direction in this important field.

2 Interferons and Post-receptor Cell Signaling

Interferon (IFN) was discovered for its potent action in interfering with viral replication 50 years ago (Isaacs and Lindenmann 1957). It is a big family of inducible cytokines that are currently classified into two subfamilies: type I and type II IFNs (Samuel 2001). Type I IFN, known as viral IFN, mainly includes multiple IFN-α members, single IFN-β, IFN-ε, IFN-κ, and IFN-ω subtype, as well as IFN-δ and IFN-τ found in pig and ovine, respectively (Theofilopoulos et al. 2005). Interestingly, all the type I IFNs share a common binding site designated type I IFN receptor (IFNAR). In contrast, type II IFN, also known as immune IFN, has only one member IFN-γthat binds to a distinct receptor IFN-γreceptor (IFNGR). Beyond the biology initially identified, IFN-α has been characterized as a pleiotropic cytokine with antiviral, antiproliferative, and immune regulatory functions (Samuel 2001). As the prime member of the type I interferon family (Fung et al. 2004), the genes coding for different IFN-αs are clustered on the short arm of chromosome 9 in the human, and chromosome 4 in the mouse, respectively (Pestka 2000). However, IFN-β gene is mapped on 7p21 in human (Sehgal et al. 1986) and 4q in the mouse (Kelley et al. 1985), respectively.

While a variety of cells including astrocytes, microglia, and neurons from the CNS also express IFN-α upon stimulation with virus directly (Stewart and Sulkin 1966) or synthetic double stranded RNA poly I:C (polyriboinosinic:polyribocytidy lic acid; Cathala and Baron 1970), peripheral plasmacytoid dendritic cells (pDCs) are a major population of cell sources for IFN-α production as a whole (Theofilopoulos et al. 2005). Importantly, recent studies have also confirmed that gram negative bacterial cell wall product lipopolysaccharides (LPS) is a potent stimulator of IFN-β production through activation of the Toll-like receptor 4 (TLR 4; Toshchakov et al. 2002). This demonstrates that activation of type I IFN expression and signaling is not only induced by either viral or bacterial infection but is also essential for LPS-induced lethality (Karaghiosoff et al. 2003).

Primarily released from virus-infected cells, IFN-α binds to type I IFNAR on the membrane of its target cells, leading to the activation of the receptor. The molecular basis for cell signaling by IFN-α has been demonstrated recently and involves phosphorylation and activation of the members of the Janus kinase (JAK)/ signal transducer and activator of transcription (STAT) protein kinase pathway (Stark et al. 1998). Binding of IFN-α to its receptor triggers activation of receptor associated tyrosine kinase 2 (Tyk2) and JAK1, which phosphorylate STAT2 and STAT1 sequentially to form heterodimers. Subsequently, the STAT1/STAT2 heterodimer migrates into the nucleus and associates with another transcription factor IFN regulatory factor (IRF)-9 (p48 or ISGF-3γ) forming IFN-stimulated gene factor 3 (ISGF-3). Finally, ISGF3 interacts with a specific DNA motif (termed the interferon-stimulated response element or ISRE) present in the 5′-regulatory region of IFN-regulated target genes thus modulating their transcriptional activity.

The two key transcription factors for IFN-α signaling, STAT1 and STAT2, are not only activated, but also their expression is upregulated dramatically by IFN-α (Marie et al. 1998). This positive feedback mechanism for IFN-α production helps to mount a robust first line response of host defense system against microbial infection. In the CNS, the genes encoding IFNs and their receptors are expressed both under normal physiological (Brandt et al. 1993) and pathological conditions (Sandberg et al. 1994) in the brain. A recent global gene profiling recorded a profound transcriptional activation of IFN-regulated genes by IFN-α in primary cultured neurons (Wang and Campbell 2005) similar to those identified from non-neural peripheral cells (Der et al. 1998). Among the most highly expressed genes are IFN-induced 15 kDa protein (ISG15), ubiquitin-specific proteinase 18 (USP18), IFN-induced 10 kDa protein (IP-10 or CXCL10), STAT1, and IFN-induced guanylate-binding protein 3 (GBP3). In consideration of the expressional amplitude of these genes and the concentration of IFN-α used in the study, these findings indicate that neurons are a very sensitive population of cells to this cytokine. This lays the basis for neurons in brain physiology and pathophysiology when IFN-α is involved.

Consistent with the importance of this pathway in mediating the actions of IFN-α, mice that lack either STAT1 (Durbin et al. 1996, Meraz et al. 1996) or STAT2 (Park et al. 2000) have impaired IFN-regulated ISGF3-dependent gene expression and are highly sensitive to viral infection. Nevertheless, in addition to the well-characterized JAK/STAT pathway for IFN-α signaling, recent studies in elucidating the cellular responses to IFN-α have identified other signaling pathways in the post receptor signal cascade for IFN-α *in vitro* (Li et al. 2004). The two major alternative pathways include STAT1-indepdendent phosphoinositide 3 kinase (PI3K; Uddin et al. 1997) and mitogen-activated protein (p38 MAP) kinase (Uddin et al. 1999) pathways. It should be pointed out that these JAK/STAT-independent pathways for IFN-α signaling were identified based solely on studies using transformed (neoplastic) or immortalized cell lines (Li et al. 2004) that may already bear abnormality of cellular machinery. In fact, a recent study *in vivo* did not detect significant difference of either phospho-p38 MAPK or total p38 MAPK at the protein level by Western blot in the brains of GFAP-IFNα transgenic mice compared with wildtype controls despite of the highly increased STAT1 expression (total STAT1) and activation (phosphorylated STAT1) by chronic IFN-α stimulation (Wang et al. 2002). Furthermore, the transcriptional stimulation of highly activated IFN-regulated genes (PKR, USP18, STAT1, ISG15 and GBP3) in response to IFN-α appears to be STAT1-dependent both in *in vitro* primary cultured neurons (Wang and Campbell 2005) as well as in STAT1 knockout mice (Wang et al. 2008). Depletion of STAT1 gene not only decreased basal expression of those ISGs, but also completely abolished transcriptional response to IFN-α, which suggests little, if any, role for alternative pathway signaling in response to IFN-α. Hence, the physiological and pathophysiological significance of the alternative pathways proposed in IFN-α signaling remains to be further defined *in vivo*.

3 Interferon-α and Its CNS Actions

3.1 Interferon-α and Human Neuropsychiatric Disorders

As a prime innate immune mediator produced following viral infection and autoimmune development, IFN-α is also found in association with a number of human psychiatric illnesses including CNS lupus (Shiozawa et al. 1992), neuroAIDS (Krivine et al. 1999) and Schizophrenia (Waltrip et al. 1990). For example, elevated serum IFN-α is not only correlated with disease activity in patients with active systemic lupus erythematosus (SLE; Hooks et al. 1979), but also detected in the cerebrospinal fluid (CSF) of SLE patients with psychiatric manifestations in which its level is directly correlated with the psychotic attack (Lebon et al. 1983). In addition, immunohistochemical studies have detected an intense staining of IFN-α and its receptor in senile plaques of the brain tissue from Alzheimer's patients (Yamada et al. 1994).

Direct evidence for IFN-α and its relationship with human neuropsychiatric disorder comes from clinical observations (Adams et al. 1984). As an FDA-approved therapeutic agent, IFN-α has been prescribed to treat patients with chronic type C or B virus-induced hepatitis (Dieperink et al. 2000) and a host of different malignancies (Jonasch and Haluska 2001) including hairy cell leukemia, chronic myelogenous leukemia (CML), metastatic melanoma, AIDS-related Kaposi's sarcoma, and follicular lymphoma. Despite its well-documented antiviral, immunoregulatory, and antitumor functions, chronic treatment with IFN-α has been frequently reported to result in severe side effects in different systems (Jonasch and Haluska 2001,Vial et al. 2000), in particular the CNS toxicities leading to neurological and behavioral dysfunctions. The CNS side effects are among the most problematic ones and often lead to discontinuation of therapy.

Over the last two decades, human studies have identified numerous neuropsychiatric signs and symptoms ranging from depression, anxiety, personality changes, memory loss, psychosis to suicide in as high as 50% of the patients who undergo chronic IFN-α therapy (Caraceni et al. 1998). The psychopathological manifestation occurs as early as one to two weeks following IFN-α treatment (Beratis et al. 2005, Capuron et al. 2002) and appears to be highly dose-related and cumulative, worsening over the time. In general, these neurobehavioral disturbances dissipate gradually upon the discontinuation of IFN-α treatment. However, some changes can be persistent and lasts for a long period of time (Meyers et al. 1991). Mood and anxiety disorders are among the most prevalent one in patients including irritability, depression, and anxiety. But, mania and hypomania are frequently reported in patients including those who undergo significant dose reduction or treatment breaks (Greenberg et al. 2000). Renalt et al. (1987) proposed to classify these neuropsychiatric complications into three main categories.

(i) Organic personality syndrome: irritability and personality changes, which quickly reverse upon dose reduction or discontinuation of therapy; (ii) Organic affective syndrome: depression and emotional liability, which occurred in up to 45% of patients depending on the duration of treatment; and iii) Psychotic manifestations: involving a small proportion of patients, who developed agitation, delirium, paranoia, and even suicide.

Despite frequent clinical reports and increasing awareness, the molecular basis for IFN-α-induced psychiatric complication remains largely unknown. Functional imaging studies by [^{18}F]-deoxyglucose positron emission tomography (FDG-PET) or functional magnetic resonance imaging (fMRI) found that chronic IFN-α treatment increases and decreases the glucose metabolism in basal ganglia and prefrontal cortex, respectively in association with the neuropsychiatric symptoms (Capuron et al. 2005, Juengling et al. 2000). Chronic IFN-α therapy also significantly changes the absolute powers of slow waves of the resting electroencephalogram (EEG; Kamei et al. 2002). Together with the decreased slow waves by serial quantitative EEG measurement, these alterations in humans indicate an encephalopathy caused by peripheral IFN-α and suggest the possible brain circuitry responsible for development of neuropsychiatric symptoms during IFN-α treatment. While relatively descriptive, human studies from different laboratories also observed that the depressive symptoms are associated with decreased tryptophan in serum (Capuron et al. 2002), decreased serotonin (5-HT; Bonaccorso et al. 2002), increased HPA (hypothalamic-pituitary-adrenal) activation (Capuron et al. 2003) and increased indoleamine 2,3-dioxygenase (IDO) activity and its pathway product kynurenine (Wichers et al. 2005) in the CSF of IFN-α-treated patients. The relationship among these changes and relative contribution of individual biochemical pathway to the behavioral dysfunction observed remains unclear. But these findings suggest potential candidate molecule(s) or biochemical pathway in mediating a psychopathology of IFN-α. Nevertheless, antidepressant medication, especially the selective serotonin reuptake inhibitors (SSRIs) Paroxetine and Fluoxetine have shown some effectiveness in the management of the psychiatric manifestations in these patients (Musselman et al. 2001). Taken together, a dysregulation of serotonergic neurotransmission is suggested for the IFN-α-induced behavioral dysfunction and thereby hypothesized as a neurochemical basis for the depression and anxiety caused by chronic IFN-α therapy.

3.2 *Neurochemical and Behavioral Impact of Interferon-α in Animals*

Besides its antiviral action, IFN-α was the first cytokine identified possessing neuromodulatory actions (Calvet and Gresser 1979) that mediate the immune-to-brain signaling. In early studies, acute treatment of IFN-α by single injection in rabbits, rats, guinea pigs, and mice were found to induce fever (Kimura et al. 1994, Krueger et al. 1987), anorexia (Plata-Salaman 1992), sleepiness (Kimura et al. 1994, Krueger et al. 1987), and decreased activity (Crnic and Segall MA 1992) collectively comprising the so-called sickness behavior. Although controversial, some neuroregulatory effects especially analgesic effect of IFN-α can be attenuated by the opioid receptor antagonist naloxone suggesting an involvement of the opioid receptor (Hori et al. 1998,Yamano et al. 2000). Nevertheless, such result still awaits validation by approaches other than pharmacological ones. On the other hand, electrophysiological studies by intracellular recordings have shown that IFN-α inhibits glutamate-induced excitatory postsynaptic potentials (EPSPs) and blocks long-term potentiation (LTP) by high-frequency titanic stimulation of CA1 hippocampal neurons in rats (Mendoza-Fernandez et al. 2000), but decreases evoked inhibitory postsynaptic potential (IPSP) amplitude in CA3 pyramidal cells leading to epileptiform bursts (Muller et al. 1993). The latter finding is consistent with the seizures reported frequently in humans during IFN-α therapy (Shakil et al. 1996) as well as in GFAP-IFNα transgenic mice (Campbell et al. 1999). Also, the relationship between IFN-α and HPA axis from animal work is also controversial (Gisslinger et al. 1993, Raber et al. 1997), as those in humans (Gisslinger et al. 1993, Shimizu et al. 1995). *Ex vivo* studies recorded a direct stimulation of CRH (corticotropin-releasing hormone) release from the hypothalamus (Gisslinger et al. 1993, Raber et al. 1997) as well as amygdala in the presence of IFN-α (Raber et al. 1997). But, whether the HPA activation by IFN-α mediates the behavioral dysfunction such as depression and anxiety is far from defined since chronic treatment of IFN-α did not have any stimulatory effect on HPA axis in both humans (Kauppila et al. 1997) and animals (De La Garza et al. 2005).

In the effort to elucidate cellular and molecular mechanisms for the behavioral dysfunctions after IFN-α treatment in humans, rodents are among the choice of model animal. This is particularly true with mice because of the availability of numerous lines of genetically manipulated mice generated in recent years for functional validation. Although an initial study reported increased anxiety profile in IFN-α-treated mice (Schrott and Crnic 1996), majority of the animal studies have focused on examining depression-like behaviors by using either Porsolt swim-test or tail suspension test (TST; Schaefer et al. 2002), the two commonly used paradigms in screening for potential antidepressants (Cryan et al. 2002). A number of early studies from one laboratory reported that at a dose of 6×10^4 IU/kg single or repeated (one injection daily for 7 days) systemic administration of human IFN-α, but not IFN-β or IFN-γ, by intravenous injection (i.v.) resulted in increased immobility on both Porsolt swim and TST in mice (1,800 units for a 30-gram mouse; Makino et al. 1998) or rats (1.8×10^4 units for a 300-gram rat) (Makino et al. 2000), respectively. Since the same treatment showed no change on total locomotor activity, a depression-like behavior was indicated in these animals by IFN-α. In support of identified depressive behavior, antidepressant treatment started before the IFN-α injection reversed decreased immobility (Makino et al. 1998). However, when the same dosage of homologous IFN-α was given to a group of the same strain of mice, the immobility profile was not altered (Makino et al. 2000). Furthermore, recent studies on rodents by other laboratories have not observed such behavioral changes even when much higher doses of human IFN-α are administered (De La Garza et al. 2005, Loftis et al. 2006).

In separate studies that have examined neurochemical influences of IFN-α, the results again differ from laboratory to laboratory. In rats, chronic peripheral and bolus intracerebroventricular (i.c.v) application of IFN-α increased binding sites and affinity for the low-affinity binding sites for 5-HT in the brain (Abe et al. 1999), but decreased 5-HT and its metabolite 5-hydroxyindoleacetic acid (5-HIAA) in different brain regions (Kamata et al. 2000). In contrast, Shuto et al. (1997) reported that repeated but not a bolus intraperitoneal (i.p.) injection of human IFN-α at 1.5×10^7 IU/kg significantly inhibited dopaminergic neurotransmission, decreased dopamine and its metabolite 3,4-dihydroxyphenylacetic acid (DOPAC), but not 5-HT or its metabolite 5-HIAA levels in mouse brain. In another similar study in mice, IFN-α treatment caused no detectable changes on cerebral neurotransmission at all (Dunn 1992).

In summary, the neurochemical and behavioral action of IFN-α in rodents have been highly inconsistent, which have raised a lot of questions with regard to the cause for opposite observations. The differences between studies, including the type (homologous or heterologous) and source of IFN-α, route of administration, and dose and duration of the treatment may be responsible for the conflicting results.

3.3 Homologous Interferon-α in Mice: From Gene Expression to Behavioral Evaluations

Careful examination of the studies in rodents revealed that most of the previous animal studies used human IFN-α rather the homologous IFN-α besides differences in IFN-α treatment regimen (dose and duration) between different studies (De La Garza et al. 2005, Makino et al. 1998, 2000). Such practice of using human IFN-α in rodents may not be appropriate since earlier studies demonstrated a clear species restriction for its activity by *in vitro* bioassay except for a hybrid IFN-αA/D constructed from human IFN-αA and IFN-αD that crosses the species barrier and is considered a "universal interferon-α" (Trown et al. 1986, Zoon et al. 1992). On the other hand, several afferent pathways can be used to gain access to the brain by a cytokine protein from the periphery (Maier and Watkins 1998, Quan and Banks 2007), and these include the neural route, circumventricular organs, blood–brain barrier (BBB) transport of cytokines, and secretions from barrier structural cells. However, pharmacokinetic studies estimated that only a tiny proportion (less than 0.2%) of IFN-α reaches the CNS after peripheral systemic administration because of the blood–brain barrier (Billiau et al. 1981, Collins et al. 1985), which poses a critical question of whether systemic IFN-α can enter the brain for a direct biological activity. Thus, despite neuromodulatory actions by IFN-α given via peripheral injection, the site of action of IFN-α for its CNS effects (centrally or peripherally) is still not clear. A major obstacle has been the lack of detectable and specific biomarker after binding of cytokines to their receptors. In addition, cellular localization of IFN-α receptor in the brain is not well understood due to its low expression under physiological conditions measured by receptor-binding assay (Janicki 1992).

With recently characterized cellular cascades for IFN-α signaling (Darnell et al. 1994, Stark et al. 1998), we took a new approach by measuring transcription factors for IFN-α signaling as well as IFN-α-regulated genes as the marker for direct and specific actions of systemic IFN-α in the CNS, and to evaluate the species-restricted activity of IFN-α in mice *in vivo* following peripheral administration of homologous (mouse) or heterologous (human) IFN-α. Expression of several prominent IFN-stimulated genes (ISGs) identified from recent gene-profiling studies of IFN-α-treated cell cultures (Der et al. 1998, Wang and Campbell 2005) was analyzed and we found that ip administration of mouse IFN-α induced a robust expression of several prototypic ISGs, in particular STAT1, ISG15 and USP18 and GBP3 in the brain and liver (Wang et al. 2008). However, in contrast to mouse IFN-α, application of human IFN-α at the same dose or 10–15 times that of the mouse IFN-α, had no such stimulatory effect on the expression of these IFN-regulated genes, confirming the species-restricted activity of IFN-α *in vivo* (Fig. 1). Time course study showed that expression of IFN-α-regulated genes in the brain was rapidly upregulated at 2 h, peaked at 8 h and returned to baseline after 24 h following IFN-α administration similar to the temporal profile of the same set of the genes in liver and spleen

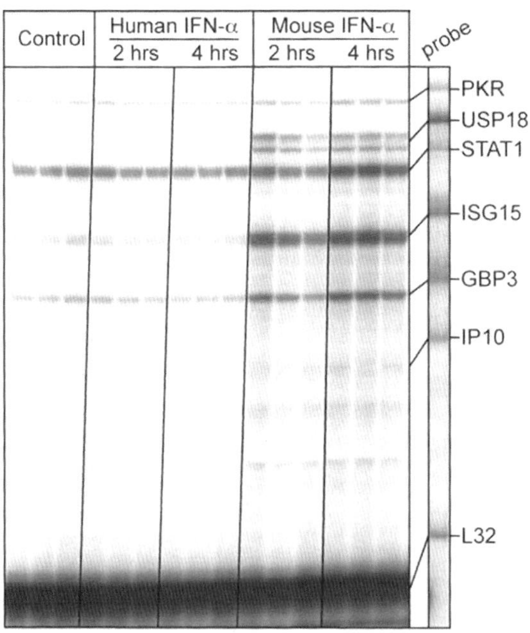

Fig. 1 Activation of the cerebral expression of IFN-stimulated genes after peripheral IFN-α challenge in mice. Mice were injected with a single dose of 10,000 IU of mouse IFN-α, human IFN-α or PBS by ip and the brain were removed following injections for RNA extraction. One μg of poly (A+) RNA was analyzed by RPA and each lane represents an individual animal

induced by IFN-α. Moreover, chronic treatment of IFN-α every other day for 2 weeks resulted in a comparable level of activation of the same IFN-stimulated genes in the brain, as did a single IFN-α injection. Additionally, the overall transcriptional response to IFN-α challenge is STAT1-dependent. More importantly, *in situ* hybridization revealed that STAT1 transcripts activated by IFN-α were distributed widely in the brain parenchyma with highest expression in cerebellum and hippocampus. Dual labeling *in situ* hybridization combined with immunocytochemical staining demonstrated a wide distribution of the STAT1 transcripts in different parenchyma cells of the brain, such as chorodial/pendymal epithelia, capillary endothelia, neurons, and astrocytes (Fig. 2).

Following the confirmation of a direct CNS action of homologous IFN-α after peripheral administration, it prompted us to evaluate the behavioral impact of this IFN-α in mice. In this study, male adult C57BL/6J mice at 8–10 weeks of age were treated with carrier-free mouse IFN-α by ip injection and tested on a battery of behavioral paradigms including elevated plus-maze (EPM), light dark box (LDB),

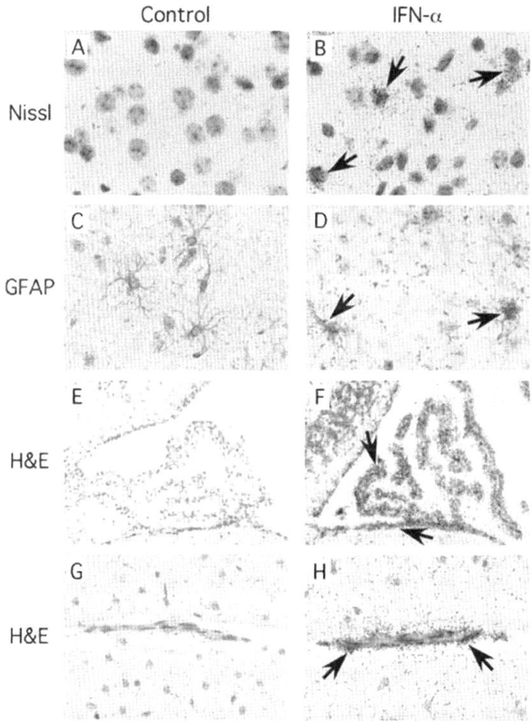

Fig. 2 Cellular localization of the STAT1 gene transcripts in the brains of mice following systemic IFN-α treatment. Dual *in situ* hybridization of STAT1 RNA transcripts and immunohistochemical localization of Nissl staining (*A–B; arrows*), GFAP (*C–D; arrows*), and H&E staining of choroid plexi (*E–F; arrows*) and vascular endothelia (*G–H; arrows*). The region shown in A–D is the cortex and E–H is the cerebellum respectively. *Published in Molecular Psychiatry (*2008 13. 293–301*)*

forced-swimming test (FST) and TST; Wang and Zhang 2007). In contrast to previous observations, we found that mouse IFN-α at a dose of 50,000 IU/mouse equivalent to that previously used (Schrott and Crnic 1996, Wang et al. 2008), decreased immobile time on FST by either acute or repeated ip injection while general activity (locomotion) was not significantly affected. However, EPM test revealed a decreased percentage of time in open arms in IFN-α-treated mice, indicating an increased anxiety profile in these mice. Also, the body weight gain was significantly slower in IFN-α-treated mice over a period of five to seven daily IFN-α injections than in corresponding controls. Nevertheless, no significant differences were observed for the measurements on either light dark box or TST after IFN-α treatment. To explore the neurochemical basis for the altered behaviors, HPLC analysis indicated that IFN-α treatment increased tryptophan concentration and altered the turnover of 5-HT in a number of brain regions including hypothalamus, brain stem, and cerebellum.

In summary, these studies show that peripheral administration of IFN-α acts in the brain directly leading to a significant expression of IFN-regulated genes. With changed cerebral neurotransmission and behavioral alterations, systemic IFN-α treatment of mouse may provide a model to elucidate the pathogenesis of the neuropsychiatric disorders caused by chronic IFN-α therapy in humans. Despite the opposite results on FST, increased anxiety profile after IFN-α challenge is consistent with more recent observation in rats (Myint et al. 2007) and primates (Felger et al. 2007). Altered tryptophan and 5-HT neurotransmission observed may be responsible for the behavioral dysfunction induced by this cytokine. Also, our studies have revealed the importance of homologous IFN-α in animal studies (Zoon et al. 1992). This may, at least in part, explain the negative results reported frequently in rodents when human IFN-α preparations were used (De La Garza and Asnis 2003, De La Garza et al. 2005).

4 Future Directions

Emerging evidence has shown that immune-related molecules including interferons and MHC (major histocompatibility complex) play critical role in brain physiology and pathophysiology, from synapse formation, neurodevelopment, and cognition (Boulanger and Shatz 2004, Wang et al. 2004). As a prime innate immune mediator triggered by viral and bacterial infections, type I IFN may directly serve as an etiopathogenic factor for human neurological and psychiatric disorders. Nevertheless, more studies are necessary in determining the behavioral impact of this cytokine and elucidating the molecular basis for behavioral dysfunction in animal models because of inaccessibility of human brain for mechanistic studies. However, in order to collect reproducible data in animal studies, a number of confounding factors must be kept in mind and these include: (i) the species specificity of IFN-α preparation in related animal investigation; (ii) the carrier protein added to IFN-α preparation, such as BSA (bovine serum albumin) or HSA (human serum albumin) which stimulates cytokine expression by contaminated molecules (Shacter et al. 1993, Steere et al.

1978); (iii) the regimen of IFN-α treatment in which direct dosage extrapolation from human studies may not reflect sensitivity differences to the cytokine in different species; and finally (iv) the bioactivity of IFN-α from different sources and/or even different batches from same vendor because of the nature of recombinant protein which change its activity due to repeated freeze-thawing cycle and long-term storage.

With the emergence of similar imaging equipment for small animals, functional imaging studies on humans as well as rodents will help to map the neuronal circuitry in mediating the CNS actions and behavioral alterations by cytokines. To dissect the molecular entity for neurochemical and neurobehavioral action of IFN-α, integrated molecular and pharmacological approach using animal studies will address whether decreased serotonergic neurotransmission by IFN-α resulted from increased reuptake (e.g., increased expression of serotonin transporter (SERT)), decreased release (such as 5-HT1A receptor activation), or increased 5-HT catabolism (enzymatic activation). Meanwhile, comprehensive behavioral studies including development of new testing paradigms for animals remain essential in defining the mental impact of IFN-α. In general, further studies will not only help to delineate long-disputed etiologic and/or pathogenic links between immune activation and human mental disorder; but may also provide insights into novel therapeutic strategies for designing new treatment for anxiety, depression, Schizophrenia, and autism.

Acknowledgments The author thanks Dr. Orisa J. Igwe (Division of Pharmacology & Toxicology, University of Missouri-Kansas City) for his helpful comments on the manuscript. This work was supported by NIH grant MH 69524.

References

Abe S, Hori T, Suzuki T, Baba A, Shiraishi H, Yamamoto T (1999) Effects of chronic administration of interferon α A/D on serotonergic receptors in rat brain. Neurochem Res 24: 359–363

Adams F, Quesada JR, Gutterman JU (1984) Neuropsychiatric manifestations of human leukocyte interferon therapy in patients with cancer. JAMA 252: 938–941

Beratis S, Katrivanou A, Georgiou S, Monastirli A, Pasmatzi E, Gourzis P, Tsambaos D (2005) Major depression and risk of depressive symptomatology associated with short-term and low-dose interferon-alpha treatment. J Psychosom Res 58: 15–18

Billiau A, Heremans H, Ververken D, van Damme J, Carton H, de Somer P (1981) Tissue distribution of human interferons after exogenous administration in rabbits, monkeys, and mice. Arch Virol 68: 19–25

Bonaccorso S, Marino V, Puzella A, Pasquini M, Biondi M, Artini M, Almerighi C, Verkerk R, Meltzer H, Maes M (2002) Increased depressive ratings in patients with hepatitis C receiving interferon-α-based immunotherapy are related to interferon-α-induced changes in the serotonergic system. J Clin Psychopharmacol 22: 86–90

Boulanger LM, Shatz CJ (2004) Immune signalling in neural development, synaptic plasticity and disease. Nat Rev Neurosci 5: 521–531

Brandt ER, Mackay IR, Hertzog PJ, Cheetham BF, Sherritt M, Bernard CC (1993) Molecular detection of interferon-α expression in multiple sclerosis brain. J Neuroimmunol 44: 1–6

Calvet MC, Gresser I (1979) Interferon enhances the excitability of cultured neurones. Nature 278: 558–560

Campbell IL, Krucker T, Steffensen S, Akwa Y, Powell HC, Lane T, Carr DJ, Gold LH, Henriksen SJ, Siggins GR (1999) Structural and functional neuropathology in transgenic mice with CNS expression of IFN-α. Brain Res 835: 46–61

Capuron L, Pagnoni G, Demetrashvili M, Woolwine BJ, Nemeroff CB, Berns GS, Miller AH (2005) Anterior cingulate activation and error processing during interferon-alpha treatment. Biol Psychiatry 58: 190–196

Capuron L, Raison CL, Musselman DL, Lawson DH, Nemeroff CB, Miller AH (2003) Association of exaggerated HPA axis response to the initial injection of interferon-α with development of depression during interferon-α therapy. Am J Psychiatry 160: 1342–1345

Capuron L, Ravaud A, Neveu PJ, Miller AH, Maes M, Dantzer R (2002) Association between decreased serum tryptophan concentrations and depressive symptoms in cancer patients undergoing cytokine therapy. Mol Psychiatry 7: 468–473

Caraceni A, Gangeri L, Martini C, Belli F, Brunelli C, Baldini M, Mascheroni L, Lenisa L, Cascinelli N (1998) Neurotoxicity of interferon-alpha in melanoma therapy: results from a randomized controlled trial. Cancer 83: 482–489

Cathala F, Baron S (1970) Interferon in rabbit brain, cerebrospinal fluid and serum following administration of polyinosinic-polycytidylic acid. J Immunol 104: 1355–1358

Collins JM, Riccardi R, Trown P, O'Neill D, Poplack DG (1985) Plasma and cerebrospinal fluid pharmacokinetics of recombinant interferon alpha A in monkeys: comparison of intravenous, intramuscular, and intraventricular delivery. Cancer Drug Deliv 2: 247–253

Crnic LS, Segall MA (1992) Behavioral effects of mouse interferons-α and -γ and human interferon-α in mice. Brain Res 590: 277–284

Cryan JF, Markou A, Lucki I (2002) Assessing antidepressant activity in rodents: recent developments and future needs. Trends Pharmacol Sci 23: 238–245

Darnell JE, Jr., Kerr IM, Stark GR (1994) Jak-STAT pathways and transcriptional activation in response to IFNs and other extracellular signaling proteins. Science 264: 1415–1421

De La Garza R, 2nd, Asnis GM (2003) The non-steroidal anti-inflammatory drug diclofenac sodium attenuates IFN-α induced alterations to monoamine turnover in prefrontal cortex and hippocampus. Brain Res 977: 70–79

De La Garza R, 2nd, Asnis GM, Pedrosa E, Stearns C, Migdal AL, Reinus JF, Paladugu R, Vemulapalli S (2005) Recombinant human interferon-α does not alter reward behavior, or neuroimmune and neuroendocrine activation in rats. Prog Neuropsychopharmacol Biol Psychiatry 29: 781–792

Der SD, Zhou A, Williams BRG, Silverman RH (1998) Identification of genes differentially regulated by interferon α, β, or γ using oligonucleotide arrays. Proc Natl Acad Sci USA 95: 15623–15628

Dieperink E, Willenbring M, Ho SB (2000) Neuropsychiatric symptoms associated with hepatitis C and interferon alpha: A review. Am J Psychiatry 157: 867–876

Dunn AJ (1992) The role of interleukin-1 and tumor necrosis factor α in the neurochemical and neuroendocrine responses to endotoxin. Brain Res Bul 29: 807–812

Durbin JE, Hackenmiller R, Simon MC, Levy DE (1996) Targeted disruption of the mouse *Stat1* gene results in compromised innate immunity to viral disease. Cell 84: 443–450

Felger JC, Alagbe O, Hu F, Mook D, Freeman AA, Sanchez MM, Kalin NH, Ratti E, Nemeroff CB, Miller AH (2007) Effects of interferon-alpha on rhesus monkeys: a nonhuman primate model of cytokine-induced depression. Biol Psychiatry 62: 1324–1333

Fung MC, Sia SF, Leung KN, Mak NK (2004) Detection of differential expression of mouse interferon-alpha subtypes by polymerase chain reaction using specific primers. J Immunol Methods 284: 177–186

Gisslinger H, Svoboda T, Clodi M, Gilly B, Ludwig H, Havelec L, Luger A (1993) Interferon-α stimulates the hypothalamic-pituitary-adrenal axis *in vivo* and *in vitro*. Neuroendocrinology 57: 489–495

Greenberg DB, Jonasch E, Gadd MA, Ryan BF, Everett JR, Sober AJ, Mihm MA, Tanabe KK, Ott M, Haluska FG (2000) Adjuvant therapy of melanoma with interferon-alpha-2b is associated with mania and bipolar syndromes. Cancer 89: 356–362

Hooks JJ, Moutsopoulos HM, Geis SA, Stahl NI, Decker JL, Notkins AL (1979) Immune interferon in the circulation of patients with autoimmune disease. N Engl J Med 301: 5–8
Hori T, Katafuchi T, Take S, Shimizu N (1998) Neuroimmunomodulatory actions of hypothalamic interferon-α. Neuroimmunomodulation 5: 172–177
Isaacs A, Lindenmann J (1957) Virus interference. 1. The interferon. Proc R Soc Lond B 147: 258–267
Janicki PK (1992) Binding of human α-interferon in the brain tissue membranes of rat. Res Commun Chem Pathol Pharmacol 75: 117–120
Jonasch E, Haluska FG (2001) Interferon in oncological practice: review of interferon biology, clinical applications, and toxicities. Oncologist 6: 34–55
Juengling FD, Ebert D, Gut O, Engelbrecht MA, Rasenack J, Nitzsche EU, Bauer J, Lieb K (2000) Prefrontal cortical hypometabolism during low-dose interferon alpha treatment. Psychopharmacology 152: 383–389
Kamata M, Higuchi H, Yoshimoto M, Yoshida K, Shimizu T (2000) Effect of single intracerebroventricular injection of alpha-interferon on monoamine concentrations in the rat brain. Eur Neuropsychopharmacol 10: 129–132
Kamei S, Sakai T, Matsuura M, Tanaka N, Kojima T, Arakawa Y, Matsukawa Y, Mizutani T, Oga K, Ohkubo H, Matsumura H, Hirayanagi K (2002) Alterations of quantitative EEG and minimental state examination in interferon-α-treated hepatitis C. Eur Neurol 48: 102–107
Karaghiosoff M, Steinborn R, Kovarik P, Kriegshauser G, Baccarini M, Donabauer B, Reichart U, Kolbe T, Bogdan C, Leanderson T, Levy D, Decker T, Muller M (2003) Central role for type I interferons and Tyk2 in lipopolysaccharide-induced endotoxin shock. Nat Immunol 4: 471–477
Kauppila M, Koskinen P, Remes K, Viikari J, Irjala K (1997) Hypothalamic-pituitary axis remains intact after interferon-alpha treatment in hematologic diseases. J Interferon Cytokine Res 17: 543–550
Kelley KA, Kozak CA, Pitha PM (1985) Localization of the mouse interferon-β 1 gene to chromosome 4. J Interf Res 5: 409–413
Kimura M, Majde JA, Toth LA, Opp MR, Krueger JM (1994) Somnogenic effects of rabbit and recombinant human interferons in rabbits. Am J Physiol 267: R53–61
Krivine A, Force G, Servan J, Cabee A, Rozenberg F, Dighiero L, Marguet F, Lebon P (1999) Measuring HIV-1 RNA and interferon-alpha in the cerebrospinal fluid of AIDS patients: insights into the pathogenesis of AIDS Dementia Complex. J Neurovirol 5: 500–506
Krueger JM, Dinarello CA, Shoham S, Davenne D, Walter J, Kubillus S (1987) Interferon alpha-2 enhances slow-wave sleep in rabbits. Int J Immunopharmacol 9: 23–30
Lebon P, Lenoir GR, Fischer A, Lagrue A (1983) Synthesis of intrathecal interferon in systemic lupus erythematosus with neurological complications. Br Med J 287: 1165–1167
Li Y, Srivastava KK, Platanias LC (2004) Mechanisms of type I interferon signaling in normal and malignant cells. Arch Immunol Ther Exp (Warsz) 52: 156–163
Loftis JM, Wall JM, Pagel RL, Hauser P (2006) Administration of pegylated interferon-α-2a or -2b does not induce sickness behavior in Lewis rats. Psychoneuroendocrinology 31: 1289–1294
Maier SF, Watkins LR (1998) Cytokines for psychologists: implications of bidirectional immune-to-brain communication for understanding behavior, mood, and cognition. Psychol Rev 105: 83–107
Makino M, Kitano Y, Hirohashi M, Takasuna K (1998) Enhancement of immobility in mouse forced swimming test by treatment with human interferon. Eur J Pharmacol 356: 1–7
Makino M, Kitano Y, Komiyama C, Takasuna K (2000) Human interferon-α increases immobility in the forced swimming test in rats. Psychopharmacology 148: 106–110
Marie I, Durbin JE, Levy DE (1998) Differential viral induction of distinct interferon-alpha genes by positive feedback through interferon regulatory factor-7. EMBO J 17: 6660–6669
Mendoza-Fernandez V, Andrew RD, Barajas-Lopez C (2000) Interferon-α inhibits long-term potentiation and unmasks a long-term depression in the rat hippocampus. Brain Res 885: 14–24
Meraz MA, White JM, Sheehan KC, Bach EA, Rodig SJ, Dighe AS, Kaplan DH, Riley JK, Greenlund AC, Campbell D, Carver-Moore K, DuBois RN, Clark R, Aguet M, Schreiber RD (1996)

Targeted disruption of the *Stat1* gene in mice reveals unexpected physiologic specificity in the JAK-STAT signaling pathway. Cell 84: 431–442

Meyers CA, Scheibel RS, Forman AD (1991) Persistent neurotoxicity of systemically administered interferon-α. Neurology 41: 672–676

Muller M, Fontana A, Zbinden G, Gahwiler BH (1993) Effects of interferons and hydrogen peroxide on CA3 pyramidal cells in rat hippocampal slice cultures. Brain Res 619: 157–162

Musselman DL, Lawson DH, Gumnick JF, Manatunga AK, Penna S, Goodkin RS, Greiner K, Nemeroff CB, Miller AH (2001) Paroxetine for the prevention of depression induced by high-dose interferon alfa. N Engl J Med 344: 961–966

Myint AM, O'Mahony S, Kubera M, Kim YK, Kenny C, Kaim-Basta A, Steinbusch HW, Leonard BE (2007) Role of paroxetine in interferon-α-induced immune and behavioural changes in male Wistar rats. J Psychopharmacol 21: 843–850

Park C, Li S, Cha E, Schindler C (2000) Immune response in Stat2 knockout mice. Immunity 13: 795–804

Pestka S (2000) The human interferon alpha species and receptors. Biopolymers 55:254–287

Plata-Salaman CR (1992) Interferons and central regulation of feeding. Am J Physiol 263: R1222–1227

Quan N, Banks WA (2007) Brain-immune communication pathways. Brain Behav Immun 21: 727–735

Raber J, Koob GF, Bloom FE (1997) Interferon-α and transforming growth factor-β 1 regulate corticotropin-releasing factor release from the amygdala: comparison with the hypothalamic response. Neurochem Int 30: 455–463

Renault PF, Hoofnagle JH, Park Y, Mullen KD, Peters M, Jones DB, Rustgi V, Jones EA (1987) Psychiatric complications of long-term interferon alfa therapy. Arch Intern Med 147: 1577–1580

Samuel CE (2001) Antiviral actions of interferons. Clin Microbiol Rev 14:778–809

Sandberg K, Eloranta ML, Campbell IL (1994) Expression of alpha/beta interferons (IFN-α/β) and their relationship to IFN-α/β-induced genes in lymphocytic choriomeningitis. J Virol 68: 7358–7366

Schaefer M, Engelbrecht MA, Gut O, Fiebich BL, Bauer J, Schmidt F, Grunze H, Lieb K (2002) Interferon alpha (IFNα) and psychiatric syndromes: a review. Prog Neuropsychopharmacol Biol Psychiatry 26: 731–746

Schrott LM, Crnic LS (1996) Increased anxiety behaviors in autoimmune mice. Behav Neurosci 110: 492–502

Sehgal PB, Zilberstein A, Ruggieri RM, May LT, Ferguson-Smith A, Slate DL, Revel M, Ruddle FH (1986) Human chromosome 7 carries the beta 2 interferon gene. Proc Natl Acad Sci USA 83: 5219–5222

Shacter E, Arzadon GK, Williams JA (1993) Stimulation of interleukin-6 and prostaglandin E2 secretion from peritoneal macrophages by polymers of albumin. Blood 82: 2853–2864

Shakil AO, Di Bisceglie AM, Hoofnagle JH (1996) Seizures during alpha interferon therapy. J Hepatol 24: 48–51

Shimizu H, Ohtani K, Sato N, Nagamine T, Mori M (1995) Increase in serum interleukin-6, plasma ACTH and serum cortisol levels after systemic interferon-α administration. Endocr J 42: 551–556

Shiozawa S, Kuroki Y, Kim M, Hirohata S, Ogino T (1992) Interferon-alpha in lupus psychosis. Arthritis Rheum 35: 417–422

Shuto H, Kataoka Y, Horikawa T, Fujihara N, Oishi R (1997) Repeated interferon-α administration inhibits dopaminergic neural activity in the mouse brain. Brain Res 747: 348–351

Stark GR, Kerr IM, Williams BRG, Silverman RH, Schreiber RD (1998) How cells respond to interferons. Ann Rev Biochem 67: 227–264

Steere AC, Rifaat MK, Seligmann EB, Jr., Hochstein HD, Friedland G, Dasse P, Wustrack KO, Axnick KJ, Barker LF (1978) Pyrogen reactions associated with the infusion of normal serum albumin (human). Transfusion 18: 102–107

Stewart WE, 2nd, Sulkin SE (1966) Interferon production in hamsters experimentally infected with rabies virus. Proc Soc Exp Biol Med 123: 650–654

Theofilopoulos AN, Baccala R, Beutler B, Kono DH (2005) Type I interferons (α/β) in immunity and autoimmunity. Annu Rev Immunol 23: 307–336

Toshchakov V, Jones BW, Perera PY, Thomas K, Cody MJ, Zhang S, Williams BR, Major J, Hamilton TA, Fenton MJ, Vogel SN (2002) TLR4, but not TLR2, mediates IFN-β-induced STAT1α/β-dependent gene expression in macrophages. Nat Immunol 3: 392–398

Trown PW, Wills RJ, Kamm JJ (1986) The preclinical development of Roferon-A. Cancer 57: 1648–1656

Uddin S, Fish EN, Sher DA, Gardziola C, White MF, Platanias LC (1997) Activation of the phosphatidylinositol 3-kinase serine kinase by IFN-α. J Immunol 158: 2390–2397

Uddin S, Majchrzak B, Woodson J, Arunkumar P, Alsayed Y, Pine R, Young PR, Fish EN, Platanias LC (1999) Activation of the p38 mitogen-activated protein kinase by type I interferons. J Biol Chem 274: 30127–30131

Vial T, Choquet-Kastylevsky G, Liautard C, Descotes J (2000) Endocrine and neurological adverse effects of the therapeutic interferons. Toxicology 142: 161–172

Waltrip RW, 2nd, Carrigan DR, Carpenter WT, Jr. (1990) Immunopathology and viral reactivation. A general theory of schizophrenia. J Nerv Ment Dis 178: 729–738

Wang J, Campbell IL (2005) Innate STAT1-dependent genomic response of neurons to the antiviral cytokine alpha interferon. J Virol 79: 8295–8302

Wang J, Campbell IL, Zhang H (2008) Systemic interferon-α regulates interferon-stimulated genes in the central nervous system. Mol Psychiatry:13: 293–301

Wang J, Lin W, Popko B, Campbell IL (2004) Inducible production of interferon-γin the developing brain causes cerebellar dysplasia with activation of the Sonic hedgehog pathway. Mol Cell Neurosci 27: 489–496

Wang J, Schreiber RD, Campbell IL (2002) STAT1 deficiency unexpectedly and markedly exacerbates the pathophysiological actions of IFN-α in the central nervous system. Proc Natl Acad Sci USA 99: 16209–16214

Wang J, Zhang H (2007) Behavioral changes by peripheral interferon-α treatment in mice: a revisit with homologous interferon-α. In: 37th annual meeting of the Society for Neuroscience (1711.9/FF10), San Diego, California (November 3–7, 2007)

Wichers MC, Koek GH, Robaeys G, Verkerk R, Scharpe S, Maes M (2005) IDO and interferon-α-induced depressive symptoms: a shift in hypothesis from tryptophan depletion to neurotoxicity. Mol Psychiatry 10: 538–544

Yamada T, Horisberger MA, Kawaguchi N, Moroo I, Toyoda T (1994) Immunohistochemistry using antibodies to α-interferon and its induced protein, MxA, in Alzheimer's and Parkinson's disease brain tissues. Neurosci Lett 181: 61–64

Yamano M, Yuki H, Yasuda S, Miyata K (2000) Corticotropin-releasing hormone receptors mediate consensus interferon-α YM643-induced depression-like behavior in mice. J Pharmacol Exp Ther 292: 181–187

Zoon KC, Miller D, Bekisz J, zur Nedden D, Enterline JC, Nguyen NY, Hu RQ (1992) Purification and characterization of multiple components of human lymphoblastoid interferon-α. J Biol Chem 267: 15210–15216

Exercise and Stress Resistance: Neural-Immune Mechanisms

Monika Fleshner, Sarah L. Kennedy, John D. Johnson, Heidi E. W. Day, and Benjamin N. Greenwood

Abstract Stimulation of the sympathetic nervous system (SNS) and the release of norepinephrine (NE) is a powerful feature of the acute stress response that is adaptive when the response is acute and constrained. If SNS activation is frequent or excessive, however, it can contribute to negative health consequences including "metabolic syndrome" and immunosuppression. We recently reported that sedentary rats exposed to a well-characterized acute stressor (uncontrollable tailshock) excessively activate the SNS leading to depletion of NE below basal levels in some peripheral tissues. NE depletion specifically in the spleen suppresses the in vivo immunoglobulin response to an antigenic protein challenge (keyhole limpet hemocyanin, αKLH immunoglobulin (Ig)). Regular moderate physical activity (voluntary wheel running) buffers a wide array of negative consequences of acute stressor exposure including splenic NE depletion and αKLH Ig suppression. *In the current chapter we will present the hypothesis that adaptations in the central sympathetic neurocircuit produced by physical activity constrain excessive stress-induced SNS responses, thereby preventing splenic NE depletion and αKLH Ig and anti-tetanus toxoid Ig suppression in physically active stressed rats.*

Keywords Sympathetic nervous system Antibody · Enkephalin · Bednucleus of the Stria terminalis · Wheel running

Abbreviations ADR, adrenergic receptor; BAR, barrington's nucleus; BSTov, oval region of bed nucleus of the stria terminalis; CRH, corticotropin releasing hormone; dpPVN, dorsoparvicellular cap hypothalamus; ENK, enkephalin; Ig, immunoglobulin; ip, intraperitoneal; IFNγ, interferon gamma; KLH, keyhole limpet hemocyanin; LC, locus coeruleus; NE, norepinephrine; PVNenk, enkephalinergic subdivision paraventricular hypothalamus; SNS, sympathetic nervous system; RVLM, rostral ventral lateral medulla; TH, tyrosine hydroxylase; TTX, tetanus toxoid; Sed, sedentary (locked wheel); Run, housed with mobile running wheel.

M. Fleshner (✉)
Department of Integrative Physiology, Center for Neuroscience, University of Colorado-Boulder, Clare Small Building, Boulder, CO 80309–0354, USA
e-mail: fleshner@colorado.edu

A. Siegel, S.S. Zalcman (eds.), *The Neuroimmunological Basis of Behavior and Mental Disorders*, DOI 10.1007/978-0-387-84851-8_6,

1 Excessive Sympathetic Nervous System Output Is Detrimental to Health

Stimulation of the SNS is a hallmark of the acute stress response (Goldstein 1996). SNS activation has many physiological consequences that work in concert to promote the "fight/flight" response (Jansen et al. 1995; Goldstein 1996). SNS activation is a powerful feature of the acute stress response that is adaptive when the response is acute and constrained. If, however, SNS activation is frequent or excessive, it can produce negative health consequences (Seals and Bell 2004). For example, chronically elevated SNS responses are believed to mechanistically contribute to the etiology of "metabolic syndrome," a key antecedent to clinical atherosclerotic diseases that includes visceral adiposity, glucose intolerance, insulin resistance, dyslipidemia, and hypertension (Baron 1990; Julius et al. 1992; Lind and Lithell 1993). In addition, it has been reported in both the human and animal literatures that chronic or excessive SNS activation can lead to arterial wall thickening (Pauletto et al. 1991; Chen et al. 1995; Xin et al. 1997), hypertension (Lind and Lithell 1993), α- and β- adrenergic receptor desensitization (Dinenno et al., 2000), and immunosuppression (Irwin 1993; Kennedy et al. 2005). The negative consequences of frequent and/or excessive SNS activity have been convincingly demonstrated in transgenic mice lacking functional α_{2A}ADR autoinhibition in the midbrain. Due to the lack of normal α_{2A}ADR central nervous system constraint on SNS drive, these mice have chronically activated peripheral SNS responses and rapidly develop cardiac dysfunction and reduced exercise capacity (Baum et al. 1992). Thus discovery of interventions that can promote SNS constraint could have important health implications. Regular, moderate physical activity is one such intervention.

2 Regular, Moderate Exercise Is Associated with Improved Overall Health

Regular, moderate physical activity positively influences many aspects of health. For example, a physically active lifestyle is associated with decreased risks of developing metabolic syndrome, hypertension, and coronary heart disease (Berlin 1990). Regular physical activity also improves the maintenance of autonomic tone across the lifespan (Seals et al. 1994) and can prevent excessive SNS responses to intense acute stressor exposure (Fleshner 2000; Kennedy et al. 2005). Also, a physically active lifestyle is associated with decreases in bacterial and viral illness (Cannon and Kluger 1984; Cannon 1993; Eichner 1993). It is possible that the reported reduction in infectious disease associated with exercise is due to an indirect health benefit of exercise, i.e., stress reduction (Antoni et al. 1990; LaPerriere et al. 1994). Exposure to physical and/or psychological stress modulates the immune response (Plotnikoff et al. 1991; Laudenslager 1994; Maier et al. 1994). Stress is neither globally immunosuppressive nor immunopotentiating. Factors that impact the effect of stress on the immune response include the following: the duration of stressor exposure (Monjan 1976); the perceived controllability of the stressor (Laudenslager et al. 1983); the measure of the immune response (Fleshner et al. 1998; Deak et al. 1999); and the physiological state of the organism (e.g., young vs. old, anxious vs. calm, healthy vs. ill, and physically active vs. sedentary (Brown and Siegal 1988; Ader 1991; Dishman et al. 1995; Bonneau 1997; Moraska and Fleshner 2001; Fleshner et al. 2002). *Physical activity may constrain the SNS response to stress, thereby contributing to improved health and resistance to immunosuppression.*

3 Exercise, Stress, Disease and Immune Function

There is early evidence to support the stress-buffering effect of regular, moderate exercise on disease and immunity. Brown and Siegal (Brown and Siegal 1988) assessed teenage girls (364 subjects) for their levels of self-reported exercise schedules, stress levels, and disease incidence. Although there are limitations to the conclusions that can be drawn from this study due to the failings of self-report and lack of disease verification, the primary findings are clear. Girls who were sedentary and under high levels of stress had elevated disease incidence. In contrast, girls who were moderately physically active and under high stress were protected against the stress-induced increases in disease incidence. Physical activity had no effect on disease incidence in the low stress group. Thus, the hypothesis that physical activity may improve health by preventing the deleterious consequences of stress has support in the human literature.

Similar support can be found in the animal literature using immunological measures as an endpoint. Dishman et al. (1995) examined the protective effects of moderate wheel running on footshock-induced suppression in NK cytotoxicity and found that 6 weeks of wheel running prior to footshock prevented the stress-induced suppression in NK cell function. Fleshner (2000), and Moraska and Fleshner (2001) reported a similar stress-buffering effect of wheel running on stress-induced suppression in antibody or immunoglobulin generated against KLH (αKLH Ig). Rats were housed with running wheels that were either locked or mobile. After 6 weeks, rats were immunized with KLH and exposed to a single session of stress (100, 1.6 mA, 5 s uncontrollable and unpredictable tailshocks). The concentration of antibody response generated across weeks after immunization with KLH was measured in blood using ELISA. Stress reduced the αKLH Ig response in sedentary (Sed, housed with locked wheels), but not in physically active (Run, housed with mobile wheels) rats (Moraska and Fleshner 2001). Voluntary wheel running in the absence of stress did not affect αKLH Ig (Moraska and Fleshner 2001).

We have recently replicated and extended this previous study using tetanus toxoid (TTX) instead of KLH and subcutaneous instead of intraperitoneal immunizations. This was done because subcutaneous TTX is a more human relevant vaccination protocol, and would make the research findings more readily translatable to human studies. Adult, male, Fischer 344 rats were either allowed access to running wheels or remained sedentary. Following 8 weeks of voluntary activity, animals were immunized subcutaneously with 12 LF TTX, then either exposed to tailshock stress (100, 1.6 mA, 5 s) or remained as home cage controls. Blood samples were taken from the tail vein on day zero and every 7 days for the next 6 weeks and levels

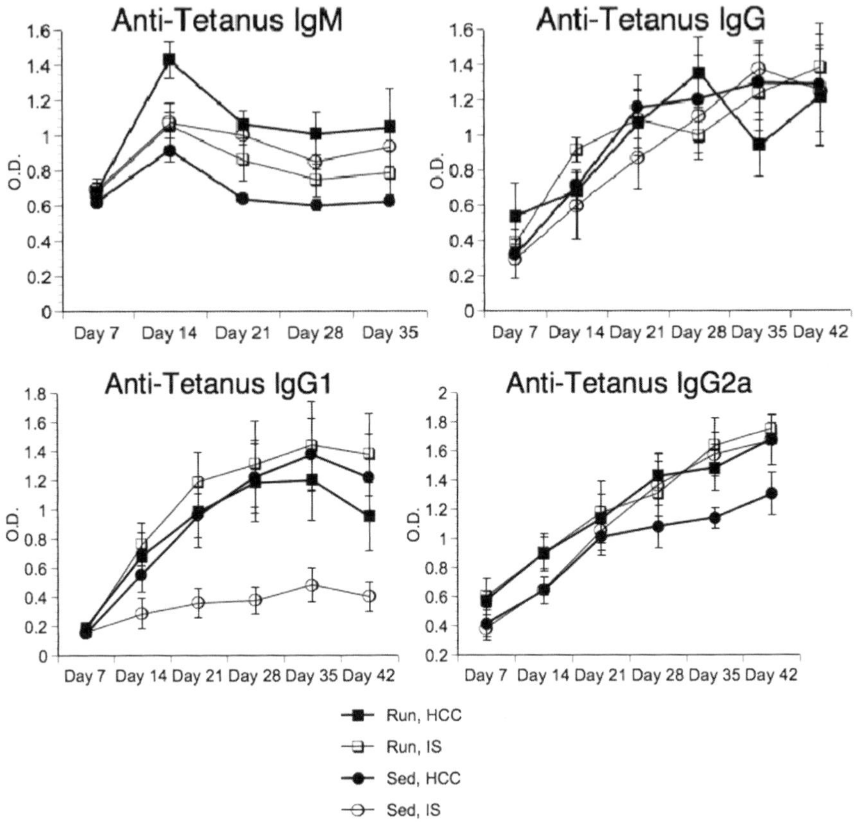

Fig. 1 Adult male F344 rats (10 per group) were either housed with a running wheel (Run) or no wheel (Sed) for 6 weeks. All animals were immunized subcutaneously with TTX and exposed to tailshock stress (IS) or no stress (HCC). Exercise per se increased both α TTX IgM and IgG2a ($p<0.05$). IS suppressed αTTX IgG1 ($p<0.001$), and this effect was prevented in Run animals

of αTTX Ig were measured using ELISA. The results are shown in Fig. 1. Physical activity alone enhanced αTTX IgM and IgG2a response compared to sedentary animals after subcutaneous immunization. Exposure to tailshock suppressed production of αTTX IgG1 (63% reduction), but not IgG2a or IgM compared to sedentary non-stressed rats. Neither stress nor activity affected αTTX IgG. The pattern of changes in αTTX antibody isotypes following stressor exposure in the sedentary animals was similar to our previous findings, testing subcutaneous KLH (Gazda et al. 2003). Importantly, physical activity prevented the stress-induced reduction in antibody, in this instance specifically prevented αTTX IgG1 suppression. These results replicate and extend the findings of Moraska and Fleshner (2001).

Regular, moderate physical activity can prevent the negative consequences of stress on immune function. One intriguing alternative interpretation of the current work is that sedentary organisms may be more stress sensitive, or more negatively impacted by activation of the stress response. *Regardless of one's interpretation it is clear that regular, moderate physical activity in the form of daily wheel running changes stress physiology, and allows animals to better buffer the negative impact of stress on acquired immunity.*

4 The Spleen Is Site of Stress-Induced KLH Antibody Suppression

The generation of an antibody response to a T cell-dependent soluble protein, such as KLH, involves the interaction of antigen presenting cells (APC; B cells, and/or dendritic cells), T helper cells (Th) and B cells. Following ip injection of KLH, antigen is transported to the draining lymph nodes and spleen. B cells expressing the B cell receptor (BCR) that bind KLH must receive T cell help from the KLH-specific T helper cells in the form of costimulation and cytokines. The Th "help" facilitates B cell proliferation, B cell differentiation into antibody secreting cells, and Ig isotype switching (IgM to IgG or IgG2a, (Janeway 1997)). The proliferation of KLH-specific Th and B cells is greatest in the draining lymph nodes and spleen 4–7 days after KLH (Fleshner et al. 1995, 1998; Gazda et al. 2003) immunization. Rats that are immunized with KLH and exposed to a single session of inescapable tailshock have a long-term reduction in serum levels of αKLH IgM, IgG, and IgG2a (Laudenslager et al., 1988; Fleshner et al. 1995, 1998; Gazda et al., 2003). We know that the final site of stress-induced immunomodulation is the spleen because if we remove the spleen from adult male rats prior to ip immunization with KLH and stressor exposure, we eliminate the stress-induced reduction in αKLH Ig (Fleshner and Laudenslager 2004).

Importantly, the stress-associated suppressive effect is specific to the generation of antibody to the antigen. Total serum IgM and IgG is not reduced (Fleshner 1992; Smith et al. 2004). Using flow cytometric analysis (Fleshner 1995; Fleshner et al. 1998), ELISPOT (Laudenslager and Fleshner 1994), and antigen-specific proliferation assays (Gazda et al. 2003), we have determined that the suppression in αKLH Ig is likely due to a failure of the stressed rats to increase KLH-specific T helper cell numbers (Fleshner et al. 1995, 1998). With fewer αKLH T helper cells, there is less T cell help and fewer KLH-specific B cells in the spleen (Laudenslager and Fleshner 1994). Fewer KLH-specific B cells lead to a reduction in serum αKLH Ig.

5 Splenic Norepinephrine Depletion Is Necessary and Sufficient for KLH Antibody Suppression

Although the signal(s) responsible for stress-induced suppression of αKLH Ig remain unclear, changes in the SNS response likely play a role. Most primary and secondary lymphoid tissues (including the spleen) receive dense SNS innervation (Felten 1987) and Th cells (Sanders 1997; Kohm et al., 2000; Kohm and Sanders 2001; Swanson et al. 2001), B cells (Kasprowicz et al. 2000; Kohm et al. 2002; Podojil and Sanders 2003; Podojil et al. 2004) and monocytes–macrophages–dendritic cells (Takahashi et al. 2004) express α_2ADRs. If we focus on the role of the SNS in stress-induced immunomodulation, there is evidence that SNS contributes to stress-induced suppression of specifically the αKLH Ig response (Irwin 1993). Although earlier work suggested that high concentrations of NE could suppress various aspects of immunity, more recent data support the hypothesis that splenic NE depletion, not circulating or splenic NE elevation, may be responsible for stress-induced suppression of *in vivo* αKLH Ig responses.

5.1 Dogmatic Change on the Role of NE

There are several lines of evidence to support this shift in dogma from "too much NE" to "too little NE." First, examination of the previous literature demonstrating that high levels of NE are immunosuppressive, many studies were done *in vitro*, examined mitogen-stimulated proliferative or cytokine responses, and tested pharmacological concentrations of NE (Ramer-Quinn 1997; Malarkey et al. 2002). Under these circumstances, NE suppresses immune function and can be fatal to immune cells (Del Rey et al. 2003). Second, activation status of the immune cells was rarely considered in these earlier studies. For example, modulation of dendritic cell function following NE exposure occurred only in the early phases of dendritic cell activation (Maestroni 2002), and β_2ADR are differentially expressed on naïve versus stimulated B cells (Sanders et al. 2003). Thus previous research supporting a simple view that too much NE is responsible for stress-induced suppression of in vivo immune responses has limitations.

Recent evidence is consistent with the dogmatic shift that too little NE may be responsible for stress-induced suppression of *in vivo* antibody responses and that dynamic interactions between SNS and immune cells occur to produce optimal Ig responses. For example, during the generation of an in vivo antibody response to KLH, NE is released from peripheral nerves innervating the spleen (Kohm and Sanders 2000). NE binding to the B cell β_2ADR stimulates the expression of costimulatory molecules (Kohm et al. 2002), Ig production (Kasprowicz et al., 2000), and splenic germinal center formation (Kohm and Sanders1999). Depletion of splenic NE content by cutting the splenic nerve (Fleshner 2006), pharmacological lesion (6-OHDA, (Kohm and Sanders 1999)) or pharmacological competition (α-methyl-*p*-tyrosine) prior to in vivo KLH immunization reduces αKLH Ig. Thus splenic NE depletion in the absence of stress is sufficient to suppress αKLH Ig. In addition, we have evidence that stress-induced suppression of αKLH Ig requires splenic NE depletion and not circulating NE elevation (Kennedy et al. 2005). Rats treated with a substrate for NE synthesis (tyrosine) prior to stressor exposure are protected from stress-induced splenic NE depletion and αKLH Ig suppression. Importantly, blood concentrations of NE in the tyrosine-treated stressed rats were equal to saline injected stressed rats, yet tyrosine completely prevented the suppression in αKLH Ig. Tyrosine is a precursor for the synthesis of NE (and DA), and during times of intense SNS drive can be rate limiting (Gibson and Wurtman 1977; Milner and Wurtman 1987; Acworth et al. 1988). Furthermore, activation of the

SNS in the absence of stressor exposure with an α_{2A}ADR antagonist (Mirtazapine, Mirt) that acts in the brain to release the SNS from α_{2A}ADR-mediated inhibition (Dazzi et al. 2002), elevates blood NE equal to, or greater than, that produced by stress. Yet, in spite of very high blood concentrations of NE at the time of immunization, Mirt at this dose and injected ip, produced neither splenic NE nor αKLH Ig suppression (Kennedy et al. 2005).

6 "The Central Sympathetic Circuit": Pathways of Splenic Sympathetic Innervation

The central control of the autonomic nervous system involves highly complex reciprocal interactions between many areas in the brain. Although simplification of these interactions is difficult, utilization of cell labeling techniques has identified a central autonomic pathway. Specifically, if we focus on the innervation pathway of the spleen, a more detailed description of the central sympathetic circuits responsible for splenic innervation can be described using retrograde pseudorabies virus staining and cFos immunohistochemistry. Pseudorabies virus is retrogradely transported from nerve terminals to cell bodies crossing synapses, thus allowing the tracing of multisynaptic pathways (Cano et al. 2000, 2001, 2004). Increased expression of the transcription factor cFos is indicative of neural activation or increased neural metabolism (Dragunow and Faull 1989). Using these techniques, Cano et al. (2000, 2001) have described specific areas in the brain that are both active during stress (cFos) and that innervate the spleen (pseudorabies virus+). They include the dorsal parvicellular cap of the paraventricular hypothalamic nucleus (dpPVN), enkephaglinergic subdivisions of the paraventricular hypothalamic nucleus (PVNenk), amygdala, bed nucleus of the stria terminalis (BST), A5 cell group (A5), and rostral ventrolateral medulla (RVLM). Additionally Barrington's nucleus (BAR) and locus coeruleus (LC), regions not traditionally thought to be directly involved in modulation of the peripheral SNS, also contain pseudorabies virus rapidly following splenic injection (Cano et al. 2000, 2001). We will refer to these brain regions with synaptic connections to peripheral sympathetically innervated targets as the "central sympathetic circuit." Modulation of nuclei within this circuit affect both splenic nerve activity (Katafuchi 1993; Hori et al. 1995; Stauss et al., 1997) and splenic immune function (Katafuchi 1993; Nistico et al., 1994; Rassnick et al., 1994; Hori et al. 1995), supporting a functional regulatory role for the central sympathetic circuit in splenic sympathetic and immune modulation.

It is important to note, however, that although these specific regions of the brain have been shown to send projections to the spleen, these areas also innervate other peripheral tissues. Thus alterations in stress-reactivity of these brain areas are indicative of overall SNS constraint, of which the spleen is one example. Thus the impact that exercise has on SNS constraint could have far reaching implications for other negative health consequences of excessive SNS responses.

7 Physical Activity Modulates Peripheral Sympathetic Nervous System Output

Can physical activity improve constraint of stress-induced sympathetic output and prevent splenic NE depletion? Historically it is clear that endurance training reduces plasma catecholamine levels in response to submaximal exercise (Galbo et al. 1977; Winder 1982; Brooks et al. 1999) and differentially impacts tissue sympathetic content (Mazzeo 1984, 1986; Mazzeo and Grantham 1989; Mazzeo 1991, 1996). The majority of the animal literature on this topic, however, is not directly relevant to our model because rats were trained using forced treadmill running, and tested after an additional treadmill exercise challenge. This is problematic because treadmill running produces immunological and physiological changes indicative of chronic stress (Hoffman-Goetz et al. 1988, 1994; Lin et al. 1995; Blank et al. 1997; Moraska et al., 2000). Nonetheless, there is evidence that exercise training changes SNS responses to a subsequent exercise challenge. Mazzeo and Grantham (1989) reported increased NE synthesis/turnover in the liver after an exercise challenge, while Overton et al. (Overton 1991) reported a reduction in the autonomic response (arterial pressure and heart rate) to a psychological stressor (noise) in trained rats. Treadmill training, however, was once again used.

We reported that wheel running also blunts activation of the SNS and prevents tissue NE depletion. Although the precise mechanism(s) for this effect remains unknown, there is evidence that wheel running produces adaptations in both peripheral sympathetic nerve NE synthesis/release and central sympathetic circuit activation that function to prevent/delay tissue NE depletion. Interestingly, the adaptations produced by physical activity are not equal across all tissues. For example, 6weeks of voluntary wheel running prevents stress-induced NE depletion in the liver and spleen (Fleshner 2000; Greenwood et al. 2003), but not the adrenals (Greenwood et al. 2003). Based on recent series of studies, we can conclude that wheel running produces an increase in the rate of turnover (synthesis/reuptake) of NE in the liver and not in the spleen. Thus for liver, physical activity may prevent NE depletion via peripheral adaptations in synthesis/reuptake, whereas for spleen a different adaptation, possibly in the central sympathetic circuit, may be responsible.

8 Physical Activity Produces Central Sympathetic Circuit Adaptations

The majority of the literature examining the impact of physical activity on the central nervous system deals with central regulation of peripheral physiological systems responding directly to the demands of exercise. From this literature, it appears that central noradrenergic pathways are activated during exercise (Meeusen et al. 1997). The majority of these studies, however, once again employed forced treadmill exercise regimens. Although there is some work investigating the changes in brain activity associated with voluntary wheel running (Rhodes et al. 2003), there are few studies that test the effect of voluntary wheel running on the brain's response to a subsequent stressor or environmental challenge. Lambert and Jonsdottier (1998) reported that 5–6weeks of voluntary wheel running reduced hypothalamic concentrations of NE in spontaneously hypertensive (physiologically stressed) rats. The reduction in hypothalamic NE was responsible for reducing SNS activity and improving cardiovascular function (Friberg et al. 1988; Jonsdottir et al. 1996). In addition, Dishman (1997) reported that voluntary wheel running prevents foot-shock induced NE depletion in LC and reduces foot-shock induced pre-frontal cortex NE elevations (in vivo microdialysis).

We have reported that voluntary wheel running produces changes in stress-induced activation of the central sympathetic circuit (Greenwood et al. 2003). If we focus on those sites that most directly provide splenic sympathetic innervation (Sved et al., 2001), we find that rats that voluntarily run on a running wheel for 6weeks prior to exposure to tailshock stress have a reduction in neural activation or the number of cFos cells in BAR, LC, A5 cell group (A5), and RVLM. In contrast, tailshock stress produced an increase in neuronal activity or a greater number of cFos cells in the basal lateral or oval BST (BSTov; Day et al. 2004) and PVNenk (Greenwood et al. 2003) in physically active vs. Sed rats. Not all regions involved with the central autonomic pathway, however, were affected. For example, tailshock stress equally stimulated cFos expression in neurons of the amygdala, A7, dpPVN and caudal VLM (CVLM) in wheel run compared to Sed rats (Greenwood et al. 2003).

Those regions of the central sympathetic circuit that are most closely connected to the spleen (based on tracer studies) and are less active (i.e., less cFos) during uncontrollable stress in physically active vs. Sed rats are BAR, LC, A5, and RVLM. These brain stem regions comprise the primary sources of noradrenergic innervation in the brain and tightly regulate descending peripheral SNS output/activation to tissues including the spleen (Romagnano et al. 1991). BAR likely does not contribute to the splenic effect because it is primarily but not exclusively (Cano et al. 2000), involved in parasympathetic outflow, and the rat spleen does not receive parasympathetic innervation (Bellinger et al. 1993).

8.1 Inhibitory Projections from BST and/or PVNenk to Sympathetic Nuclei (LC, A5, RVLM).

Wheel running produces adaptations in the central sympathetic circuit that could involve structures that are not necessarily directly linked to peripheral organs, but can *modulate* the activity of the brainstem sympathetic nuclei during times of stress. Stimulation of the amygdala, BST or PVN, for example, can *modulate* peripheral sympathetic activity as measured by changes in heart rate and blood pressure (Ciriello and Janssen 1993; Gelsema et al. 1993; Dunn and Williams 1995, 1998; Ciriello and Roder 1999). The sympathetic effects of amygdala, BST and PVN stimulation, however, are not due to direct connections with sympathetic preganglionic neurons. Instead, neurons in these regions synapse on other sympathetic nuclei, such as the LC and RVLM, which do have direct connections with the peripheral SNS (Holets et al. 1988; Hosoya et al. 1991; Loewy 1991; Romagnano et al. 1991; Giancola et al. 1993).

Based on anterograde tracing studies and immunoelectron microscopy, the central amygdala and BSTov project to the peri-LC, a region just dorsal to the LC that contains many NE dendrites of LC neurons (Van Bockstaele et al. 1999b; Dong et al. 2001). This is particularly important because the LC is uniquely capable of activating the central sympathetic circuit during times of stress (Ashton-Jones et al. 1995; Valentino and Van Bockstaele 2001). The majority of neurons projecting from the amygdala contain corticotropin-releasing hormone (CRH) that increases peripheral sympathetic drive (Nijsen et al. 2000, 2001; Goncharuk et al. 2002; Arlt et al. 2003). In contrast, only ~13% of BSTov projections to the peri-LC are CRH (Van Bockstaele et al. 1999). The remaining neurons are unidentified, however, given the high prevalence of dense core vesicles in these specific BSTov afferents and the fact that the majority of neurons are not CRH immunoreactive, it is possible that other neuropeptides, such as ENK, may be involved (Van Bockstaele et al. 1999).

δ and μ opioid receptors are on presynaptic axon terminals and dendrites of the LC (Van Bockstaele et al. 1996a, 1997). Acute administration of opiate agonists inhibit LC neuronal discharge and inhibit LC output (Williams et al. 1982; North and Williams 1983, 1985; van Bockstaele et al. 1996). The effect of opioids on LC function is somewhat equivocal, however, as opioids can activate as well as inhibit LC neurons (Pan et al. 2004). These different responses may depend on the precise location of opiate agonist administration (i.e., peri-LC vs. central LC (Travagli et al. 1996; Ennis et al. 1998)). Nonetheless, it has been proposed that during stress, opioids (potentially ENK) may function as endogenous inhibitory factors opposing the excitation of CRH in the LC (Valentino and Van Bockstaele 2001).

Besides a potential inhibitory projection from the BSTov to peri-LC, the PVNenk may also send inhibitory projections to regions of the central sympathetic circuit that are directly connected to sympathetic preganglionic neurons. For example, PVNenk is an origin of opioidergic projections to the RVLM (Saper et al. 1976) and enkephalinergic terminals have been found in RVLM (Boone and Corry 1996). Stressor exposure, including tailshock stress, increases enkephalin mRNA within the PVNenk (Helmreich et al. 1999; Dumont et al. 2000). Neurons within RVLM express opioid receptors, and opioid administration into the RVLM (but not the caudal VLM) inhibits sympathetic output and cardiovascular responses to an exercise challenge (Ishide et al. 2000). Selective activation of δ and μ opioid receptors in the RVLM also inhibits NE release and cardiovascular responses after an exercise challenge (Nauli et al. 2001). Wheel running may be a sufficient exercise stimulus to activate endogenous brain opioid systems. For example, spontaneously hypertensive rats allowed to run on running wheels had 6–12 times greater dynorphin-converting enzyme activity in cerebrospinal fluid (Persson et al., 1993), suggesting increased activity of endogenous opioid systems in the brain.

Consistent with previous studies we reported that sedentary rats exposed to tailshock stress robustly activate BSTov (Greenwood et al. 2003) and PVNenk (Greenwood et al. 2003). Physically active rats had greater numbers of stress-induced cFos neurons in BSTov (projects to the peri-LC) and PVNenk (projects to the RVLM). In a serious of recent preliminary studies, we have evidence that the BSTov neurons that are more active in physically active rats after tailshock stress are not CRH positive but are enkephalinergic. As depicted in Fig. 2, 6 weeks, but not 3 weeks of wheel running produces an increase in cFos in CRH negative BSTov cells. Because virtually no CRH cells contain enkephalin and we did not find increased cFos expression in CRH cells after tailshock stress, the cells that are cFos positive could be enkephalinergic.

Using dual-probe in situ hybridization (Fig. 3), we directly tested this and found that the cfos mRNA expressed after tailshock was found in enkephalinergic cells

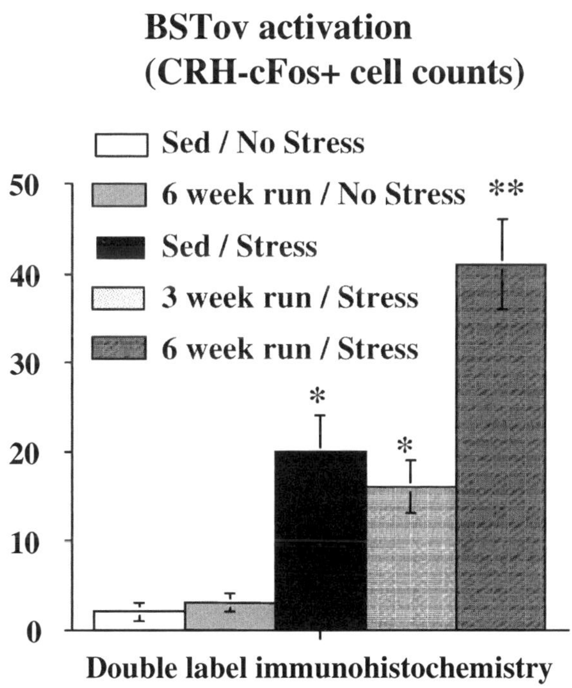

Fig. 2 Adult male F344 rats (8 per group) were housed with either running wheels (*Run*) or with no wheels (*Sed*). After 3 or 6 weeks, rats were exposed to tailshock stress (*Stress*) or remained in their home cages (*No Stress*) and were sacrificed and brains processed for double labeled immunohistochemistry. Data are presented as cell counts per region that expressed cFos and did not express corticotropin-releasing hormone (CRH). The results were that stress increased the number of cFos+ cells, and 6 weeks, but not 3 weeks, of wheel running increased the number ($p<0.001$)

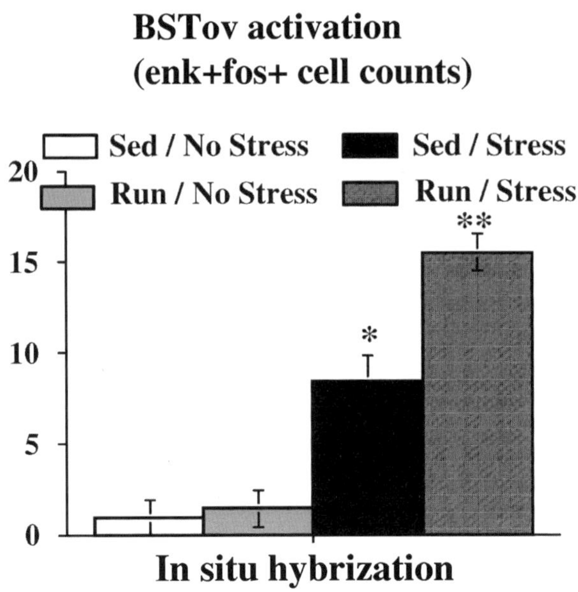

Fig. 3 Adult male F344 rats (8 per group) were housed with either running wheels (*Run*) or with no wheels (*Sed*). After 6 weeks, rats were exposed to tailshock stress (*Stress*) or remained in their home cages (*No Stress*) and were sacrificed and brains processed for in situ hybridization. Data are presented as cell counts per region that expressed cfos mRNA and enkephalin mRNA. The results were that stress increased the number of double positive cells, and 6 weeks of wheel running increased the number ($p<0.001$)

in BSTov and in the enkephalinergic regions of the PVNenk (not shown). In the absence of stress we found that 6 weeks of running produced an increase in preproenkephalin hnRNA in BSTov (Fig. 4) and met enkephalin in LC and RVLM (Fig. 5) compared to sedentary animals.

Thus wheel running (1) *reduces* tailshock-induced activity of tyrosine hydroxylase+ LC, A5, and RVLM (and not caudal VLM) neurons (Greenwood et al., 2003); (2) *increases* tailshock-induced activation of CRH negative BSTov neurons and PVNenk neurons (Fig. 2), (3) *increases* tailshock-induced activation of enkephalinergic cells in the BSTov and PVNenk (Fig. 3); (4) in the absence of stress, increases preproenkephalin heteronuclear RNA in BSTov (Fig. 4); and (5) in the absence of stress, increases met enkephalin content of the LC and RVLM (Fig. 5). *We hypothesize, therefore, that wheel running increases the inhibition of LC via enkephalinergic projections from BSTov to peri-LC and increases inhibition of RVLM via enkephalinergic projections from PVNenk to RVLM. Constraint of stress-induced activation of LC and/or RVLM could contribute to the reduction in A5 (Romagnano et al. 1991).*

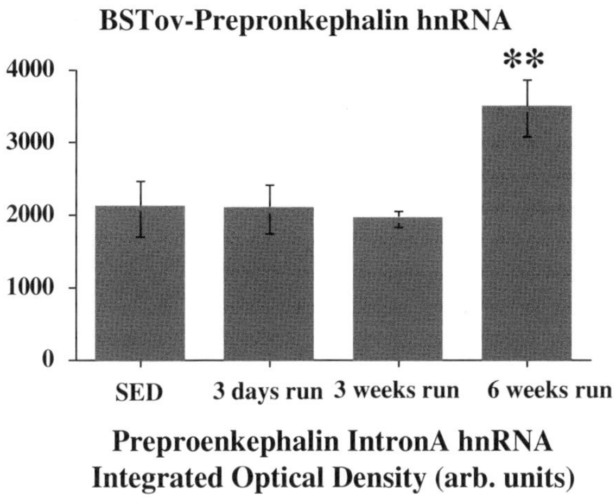

Fig. 4 Adult male F344 rats (8 per group) were housed with either running wheels (*Run*) or with no wheels (*Sed*). After 3 days, 3 weeks, or 6 weeks of running, animals were sacrificed. Using in situ hybridization, preproenkephalin hn RNA levels were assessed. Only 6 weeks of wheel running increased enkephalin ($p<0.001$)

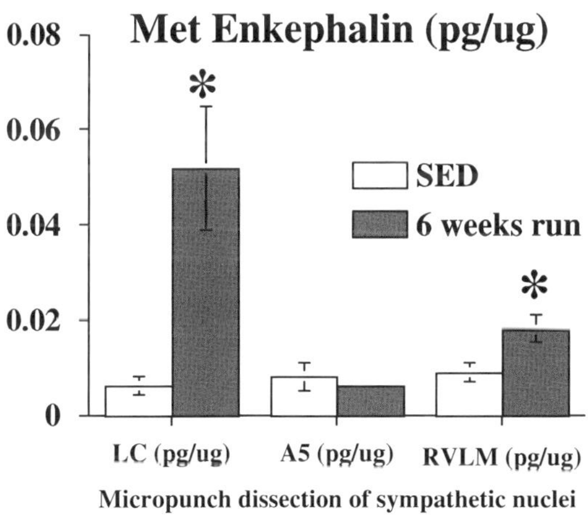

Fig. 5 Adult, male F344 rats (6 per group) were allowed to run or remained Sed. After 6 weeks, rats were sacrificed ~5 h after running had ceased during their inactive (light) cycle. Brains were micropunch dissected and enkephalin was measured using RIA. Results were that 6 weeks of wheel running increased basal content of enkephalin in the locus coeruleus (LC) and rostral ventral lateral medulla (RVLM), but not A5. In contrast to the LC and RVLM that are known to receive enkephalin projections from oval region of bed nucleus of the stria terminalis (BSTov) and enkephalinergic subdivision paraventricular hypothalamus (PVNenk) respectively, evidence for direct enkephalin innervation of A5 is limited. This may explain the lack of enkephalin increase in this site

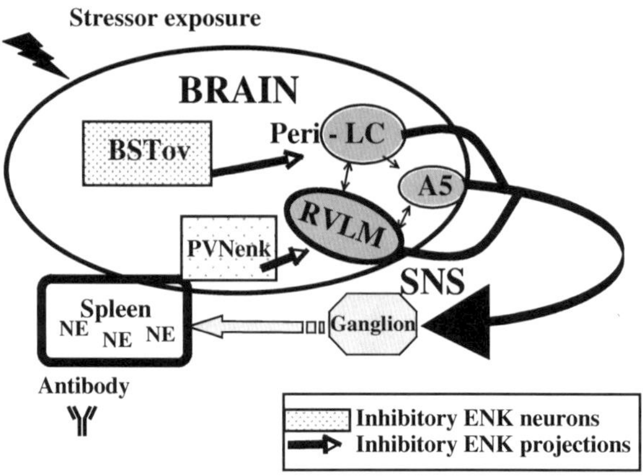

Fig. 6 We hypothesize that habitual voluntary physical activity increases the activity of enkephalinergic BSTov and paraventricular hypothalamic (PVN) neurons. These neurons modulate stress-responsive autonomic output circuits. In physically active organisms compared to sedentary organisms, exposure to an intense stressor activates more inhibitory enkephalin neurons that better constrain central sympathetic nervous system (SNS) drive. This prevents splenic norepinephrine (NE) depletion and consequent antibody suppression and hence promotes an adaptive versus maladaptive SNS response

10 Conclusions

Our current hypothesis is depicted in Fig. 6. Exposure to acute stress stimulates the central sympathetic circuit. Physically active rats, compared to sedentary rats, have increased inhibitory BSTov or PVNenk (enkephalin) projections to sympathetic nuclei (LC, A5, RVLM). These adaptations in the central sympathetic circuit constrain stress-induced SNS drive, thereby preventing splenic NE depletion and Ig and suppression in physically active stressed rats.

Additional studies will continue to test the validity of this hypothesis. Such experiments are certain to produce important results that could both add to our basic understanding of central autonomic/sympathetic regulation of SNS drive and potentially lead to development of future effective behavioral and pharmacological interventions that can be used to prevent the detrimental consequences of stressor exposure and SNS dysregulation on health.

Acknowledgments This work was supported by grants from the National Institutes of Health to M. Fleshner RO1-AI48555 and RO1 MH068283.

References

Acworth, I. N., M. J. During and R. J. Wurtman (1988). Tyrosine: effects on catecholamine release. *Brain Res Bull* 21(3): 473–477.

Ader, R., D. L. Felten and N. Cohen (1991). *Psychoneuroimmunology*, New York: Academic Press.

Antoni, M., N. Schneiderman, M. A. Fletcher, D. A. Goldstein, G. Ironson and A. LaPerriere (1990). Psychoneuroimmunology and HIV-1. *J Consult Clin Psycholo* 58(1): 38–49.
Arlt, J., H. Jahn, M. Kellner, A. Strohle, A. Yassouridis and K. Wiedemann (2003). Modulation of sympathetic activity by corticotropin-releasing hormone and atrial natriuretic peptide. *Neuropeptides* 37(6): 362–368.
Ashton-Jones, G., M. T. Shipley and R. Grzanna (1995). The locus coeruleus, A5 and A7 noradrenergic cell groups. *The Rat Nervous System* Second edition: 183–213.
Baron, A., M. Laakso, G. Brechtel, B. Hoit, C, Watt and S. Edelman, (1990). Reduced postprandial skeltal muscle blood flow contributes to glucose intolerance in human obesity. *J Clin Endocr Metab* 70: 1525–1533.
Baum, K., D. Essfeld, D. Leyk and J. Stegemann (1992). Blood pressure and heart rate during rest-exercise and exercise-rest transitions. *Eur J Appl Physiol Occup Physiol* 64(2): 134–138.
Bellinger, D. L., D. Lorton, R. W. Hamill, S. Y. Felten and D. L. Felten (1993). Acetylcholinesterase staining and choline acetyltransferase activity in the young adult rat spleen: lack of evidence for cholinergic innervation. *Brain Behav Immun* 7(3): 191–204.
Berlin, J. A. and G. A. Colditz (1990). A meta-analysis of physical activity in the prevention of coronary heart disease. *Am J Epidemiol* 132: 612–628.
Blank, S. E., T. B. Jones, E. G. Lee, J. Brahler, R. M. Gallucci, M. L. Fox and G. G. Meadows (1997). Modulation of NK cell cytolytic activity by macrophages in chronically exercise-stressed mice. *J Appl Physiol* 83(3): 845–850.
Bonneau, R. H., M. A. Brehm, and A. M. Kern (1997). The impact of psychological stress on the efficacy of anti-viral adoptive immunotherapy in an immunocomprimised host. *J Neuroimmunol* 78: 19–33.
Boone, J. B., Jr. and J. M. Corry (1996). Proenkephalin gene expression in the brainstem regulates post-exercise hypotension. *Brain Res Mol Brain Res* 42(1): 31–38.
Brooks, G. A., T. D. Fahey, T. P. White and K. M. Baldwin (1999). *Exercise Physiology: Human Bioenergetics and Its Applications*. Mountain View, Mayfield Publishing Company.
Brown, J. D. and J. M. Siegel (1988). Exercise as a buffer of life stress: A prospective study of adolescent health. *Health Psychol* 7(4): 341–353.
Cannon, J. G. (1993). Exercise and resistance to infection. *J Appl Physiol* 73(4): 973–981.
Cannon, J. G., and Kluger, M. (1984). Exercise enhances survival rates in mice infected with *Salmonella typhimurium*. *Soc Exp Biol Med* 175: 518–521.
Cano, G., J. P. Card, L. Rinaman and A. F. Sved (2000). Connections of Barrington's nucleus to the sympathetic nervous system in rats. *J Auton Nerv Syst* 79(2–3): 117–128.
Cano, G., J. P. Card and A. F. Sved (2004). Dual viral transneuronal tracing of central autonomic circuits involved in the innervation of the two kidneys in rat. *J Comp Neurol* 471(4): 462–481.
Cano, G., A. F. Sved, L. Rinaman, B. S. Rabin and J. P. Card (2001). Characterization of the central nervous system innervation of the rat spleen using viral transneuronal tracing. *J Comp Neurol* 439(1): 1–18.
Chen, C. Y., S. E. DiCarlo and T. J. Scislo (1995). Daily spontaneous running attenuated the central gain of the arterial baroreflex. *Am J Physiol* 268(2 Pt 2): H662–H669.
Ciriello, J. and S. A. Janssen (1993). Effect of glutamate stimulation of bed nucleus of the stria terminalis on arterial pressure and heart rate. *Am J Physiol* 265(5 Pt 2): H1516–H1522.
Ciriello, J. and S. Roder (1999). GABAergic effects on the depressor responses elicited by stimulation of central nucleus of the amygdala. *Am J Physiol* 276(1 Pt 2): H242–H247.
Day, H. E., B. N. Greenwood, S. E. Hammack, L. R. Watkins, M. Fleshner, S. F. Maier and S. Campeau (2004). Differential expression of 5HT-1A, alpha 1b adrenergic, CRF-R1, and CRF-R2 receptor mRNA in serotonergic, gamma-aminobutyric acidergic, and catecholaminergic cells of the rat dorsal raphe nucleus. *J Comp Neurol* 474(3): 364–378.
Dazzi, L., S. Ladu, F. Spiga, G. Vacca, A. Rivano, L. Pira and G. Biggio (2002). Chronic treatment with imipramine or mirtazapine antagonizes stress- and FG7142-induced increase in cortical norepinephrine output in freely moving rats. *Synapse* 43(1): 70–77.
Deak, T., K. T. Nguyen, M. Fleshner, L. R. Watkins and S. F. Maier (1999). Acute stress may facilitate recovery from a subcutaneous bacterial challenge. *Neuroimmunomodulation* 6(5): 344–354.

Del Rey, A., A. Kabiersch, S. Petzoldt and H. O. Besedovsky (2003). Sympathetic abnormalities during autoimmune processes: potential relevance of noradrenaline-induced apoptosis. *Ann N Y Acad Sci* 992: 158–167.

Dinenno, F. A., P. P. Jones, D. R. Seals and H. Tanaka (2000). Age-associated arterial wall thickening is related to elevations in sympathetic activity in healthy humans. *Am J Physiol Heart Circ Physiol* 278(4): H1205–10.

Dishman, R. K. (1997). Brain monamines, exercise, and behavioral stress: animal models. *Medicine and Science in Sports and Exercise* 29: 63–74.

Dishman, R. K., J. M. Warren, S. D. Youngstedt, H. Yoo, B. N. Bunnell, E. H. Mougey, J. L. Meyerhoff, L. Jaso-Friedmann and D. L. Evans (1995). Activity-wheel running attenuates suppression of natural killer cell activity after footshock. *J Appl Physiol* 78(4): 1547–1554.

Dong, H. W., G. D. Petrovich, A. G. Watts and L. W. Swanson (2001). Basic organization of projections from the oval and fusiform nuclei of the bed nuclei of the stria terminalis in adult rat brain. *J Comp Neurol* 436(4): 430–455.

Dragunow, M. and R. Faull (1989). The use of c-fos as a metabolic marker in neuronal pathway tracing. *J Neurosci Methods* 29(3): 261–265.

Dumont, E. C., R. Kinkead, J. F. Trottier, I. Gosselin and G. Drolet (2000). Effect of chronic psychogenic stress exposure on enkephalin neuronal activity and expression in the rat hypothalamic paraventricular nucleus. *J Neurochem* 75(5): 2200–2211.

Dunn, J. D. and T. J. Williams (1995). Cardiovascular responses to electrical stimulation of the bed nucleus of the stria terminalis. *J Comp Neurol* 352(2): 227–234.

Dunn, J. D. and T. J. Williams (1998). Effect of sinoaortic denervation on arterial pressure changes evoked by bed nucleus stimulation. *Brain Res Bull* 46(4): 361–365.

Eichner, E. R. (1993). Infection, immunity, and exercise. *Phys Sport Med* 21: 125–135.

Ennis, M., M. T. Shipley, G. Ashton-Jones and J. T. Williams (1998). Afferent control of nuclues ceruleus: Differntial regulation by "shell" and "core" inputs. *Adv Pharmacol* 42: 767–771.

Felten, D. L., K. D. Ackerman, S. J.Wiegand and S. Y. Felten (1987). Noradrenergic sympathetic innervation of the spleen I. Nerve fibers associate with lymphocytes and macrophages in specific compartments of the splenic white pulp. *J Neurosci Res* 18: 28–36.

Fleshner, M. (2000). Exercise and neuroendocrine regulation of antibody production: protective effect of physical activity on stress-induced suppression of the specific antibody response. *Int J Sports Med* 21 Suppl 1: S14–S19.

Fleshner, M., (2006). *Stress-induced sympathetic nervous system activation contributes to both suppressed acquired immunity and potentiated innate immunity: The role of splenic NE depletion and extracellular Hsp72.* In: Neural and Neuroendocrine Mechanisms in Host Defense and Autoimmunity. New York, NY, Springer Publishing.

Fleshner, M., J. Campisi, T. Deak, B. N. Greenwood, J. A. Kintzel, T. H. Leem, T. P. Smith and B. Sorensen (2002). Acute stressor exposure facilitates innate immunity more in physically active than in sedentary rats. *Am J Physiol Regul Integr Comp Physiol* 282(6): R1680–R1686.

Fleshner, M., J. Hermann, L. L. Lockwood, L. R. Watkins, M. L. Laudenslager and S. F. Maier. (1995). Stressed rats fail to expand the CD45RC+CD4+ (Th1-like) T cell subset in response to KLH: Possible involvement of IFN-gamma. *Brain, Behav, Immun* 9: 101–112.

Fleshner, M. and M. L. Laudenslager (2004). Psychoneuroimmunology: then and now. *Behav Cogn Neurosci Rev* 3(2): 114–130.

Fleshner, M., K. T. Nguyen, C. S. Cotter, L. R. Watkins and S. F. Maier (1998). Acute stressor exposure both suppresses acquired immunity and potentiates innate immunity. *Am J Physiol* 275(3 Pt 2): R870–R878.

Fleshner, M., L. R. Watkins, D. Bellgrau, M. L. Laudenslager and S. F. Maier (1992). Specific changes in lymphocyte subpopulations: a potential mechanism for stress-induced immunosuppression. *J Neuroimmunol* 41: 131–142.

Friberg, P., P. Hoffmann, M. Nordlander and P. Thoren (1988). Effects of voluntary physical exercise on cardiac function and energetics in the spontaneously hypertensive rat. *Acta Physiol Scand* 133: 495–500.

Galbo, H., E. A. Richter, J. J. Holst and N. J. Christensen (1977). Diminished hormonal responses to exercise in trained rats. *J Appl Physiol* 43: 953–958.

Gazda, L. S., T. Smith, L. R. Watkins, S. F. Maier and M. Fleshner (2003). Stressor exposure produces long-term reductions in antigen-specific T and B cell responses. *Stress* 6(4): 259–267.

Gelsema, A. J., N. E. Copeland, G. Drolet and H. Bachelard (1993). Cardiovascular effects of neuronal activation of the extended amygdala in rats. *Brain Res* 626(1–2): 156–166.

Giancola, S. B., S. Roder and J. Ciriello (1993). Contribution of caudal ventrolateral medulla to the cardiovascular responses elicited by activation of bed nucleus of the stria terminalis. *Brain Res* 606(1): 162–166.

Gibson, C. J. and R. J. Wurtman (1977). Physiological control of brain catechol synthesis by brain tyrosine concentration. *Biochem Pharmacol* 26(12): 1137–1142.

Goldstein, D. S. (1996). The sympathetic nervous system and the 'fight-or-flight' response: outmoded ideas? *Mol Psychiatry* 1(2): 95–97.

Goncharuk, V. D., J. Van Heerikhuize, D. F. Swaab and R. M. Buijs (2002). Paraventricular nucleus of the human hypothalamus in primary hypertension: activation of corticotropin-releasing hormone neurons. *J Comp Neurol* 443(4): 321–331.

Greenwood, B. N., S. Kennedy, T. P. Smith, S. Campeau, H. E. Day and M. Fleshner (2003). Voluntary freewheel running selectively modulates catecholamine content in peripheral tissue and c-Fos expression in the central sympathetic circuit following exposure to uncontrollable stress in rats. *Neuroscience* 120(1): 269–281.

Helmreich, D. L., L. R. Watkins, T. Deak, S. F. Maier, H. Akil and S. J. Watson (1999). The effect of stressor controllability on stress-induced neuropeptide mRNA expression within the paraventricular nucleus of the hypothalamus. *J Neuroendocrinol* 11(2): 121–128.

Hoffman-Goetz, L. and Pedersen, B. K. (1994). Exercise and the immune system: a model of the stress response. *Immunol Today* 15: 382–387.

Hoffman-Goetz, L., Thorne, R. J. and Houston, M. E. (1988). Splenic immune response following treadmill exercise in mice. *Can J Physiol Pharmacol* 66: 1415–1419.

Holets, V. R., T. Hokfelt, A. Rokaeus, L. Terenius and M. Goldstein (1988). Locus coeruleus neurons in the rat containing neuropeptide Y, tyrosine hydroxylase or galanin and their efferent projections to the spinal cord, cerebral cortex and hypothalamus. *Neuroscience* 24(3): 893–906.

Hori, T., T. Katafuchi, S. Take, N. SHimizu and A. Niijima (1995). The autonomic nervous system as a communication channel between the brain and the immune system. *Neuroimmunomodulation* 2: 203–215.

Hosoya, Y., Y. Sugiura, N. Okado, A. D. Loewy and K. Kohno (1991). Descending input from the hypothalamic paraventricular nucleus to sympathetic preganglionic neurons in the rat. *Exp Brain Res* 85(1): 10–20.

Irwin, M. (1993). Brain corticotropin-releasing hormone and interleukin-1 beta-induced suppression of specific antibody production. *Endocrinology* 133: 1352–1360.

Ishide, T., M. Mancini, T. J. Maher, P. Chayaikul and A. Ally (2000). Rostral ventrolateral medulla opioid receptor activation modulates glutamate release and attenuates the exercise pressor reflex. *Brain Res* 865(2): 177–185.

Janeway, C. A. a. P. T. (1997). *ImmunoBiology: The immune system in health and disease*, Current Biology/Garland.

Jansen, A. S., X. V. Nguyen, V. Karpitskiy, T. C. Mettenleiter and A. D. Loewy (1995). Central command neurons of the sympathetic nervous system: basis of the fight-or-flight response. *Science* 270(5236): 644–646.

Jonsdottir, I. H., A. Asea, P. Hoffman, K. Hellstrand and P. Thoren (1996). Voluntary chronic exercise in the spontaneously hypertensive rat. *Life Sci* 53: 643–652.

Julius, S., T. Gudbrandsson, K. Jamerson and O. Andersson (1992). The interconnection between sympathetics, microcirculation, and insulin resistance in hypertension. *Blood Pressure* 1: 9–19.

Kasprowicz, D. J., A. P. Kohm, M. T. Berton, A. J. Chruscinski, A. Sharpe and V. M. Sanders (2000). Stimulation of the B cell receptor, CD86 (B7–2), and the beta 2-adrenergic receptor

intrinsically modulates the level of IgG1 and IgE produced per B cell. *J Immunol* 165(2): 680–690.
Katafuchi, T., S. Take, and T. Hori (1993). Roles of sympathetic nervous system in the suppression of cytotoxic splenic natural killer cells in the rat. *J Physiol* 465: 343–357.
Kennedy, S. L., M. Nickerson, J. Campisi, J. D. Johnson, T. P. Smith, C. Sharkey and M. Fleshner (2005). Splenic norepinephrine depletion following acute stress suppresses in vivo antibody response. *J Neuroimmunol* 165(1–2): 150–160.
Kohm, A. P., A. Mozaffarian and V. M. Sanders (2002). B cell receptor- and beta 2-adrenergic receptor-induced regulation of B7–2 (CD86) expression in B cells. *J Immunol* 168(12): 6314–6322.
Kohm, A. P. and V. M. Sanders (1999). Suppression of antigen-specific Th2 cell-dependent IgM and IgG1 production following norepinephrine depletion in vivo. *J Immunol* 162: 5299–5308.
Kohm, A. P. and V. M. Sanders (2000). Norepinephrine: a messenger from the brain to the immune system. *Immunol Today* 21(11): 539–542.
Kohm, A. P. and V. M. Sanders (2001). Norepinephrine and beta 2-adrenergic receptor stimulation regulate CD4+ T and B lymphocyte function in vitro and in vivo. *Pharmacol Rev* 53(4): 487–525.
Kohm, A. P., Y. Tang, V. M. Sanders and S. B. Jones (2000). Activation of antigen-specific CD4+ Th2 cells and B cells in vivo increases norepinephrine release in the spleen and bone marrow. *J Immunol* 165(2): 725–733.
Lambert, G. W., and Jonsdottir, I. H. (1998). Influence of voluntary exercise on hypothalamic norepinephrine. *J Appl Physiol* 85(3): 962–966.
LaPerriere, A., Ironside, G., Antoni, M., Schneiderman, N., Klimas, N. and Fletcher, M.A. (1994). Exercise and Psychoneuroimmunology. *Med Sci Sports Exercise* 26(2): 182–190.
Laudenslager, M. L. and M. Fleshner (1994). Stress and Immunity: Of mice, monkeys, models, and mechanisms. *The Handbook of Human Stress and Immunity.* a. J. K.-G. R. Glaser, Academic Press: 161–181.
Laudenslager, M. L., M. Fleshner, P. Hofstadter, P. E. Held, L. Simons and S. F. Maier (1988). Suppression of specific antibody production by inescapable shock: stability under varying conditions. *Brain Behav Immun* 2(2): 92–101.
Laudenslager, M. L., S. M. Ryan, R. L. Drugen, R. L. Hyson and S. F. Maier (1983). Coping and immunosuppression: inescapable but not escapable shock suppresses lymphocyte proliferation. *Science* 221: 568–570.
Lin, Y.-S., M-S. Jan, T-J., Tsai and H-I. Chen (1995). Immunomudulatory effects of acute exercise bout in sedentary and trained rats. *Med Sci Sports Exercise* 27(1): 73–78.
Lind, L. and H. Lithell (1993). Decreased peripheral blood flow in the pathogenisis of the metabolic syndrome comprising hypertension, hyperlimidemia and hyperinsulinemia. *Am Heart J* 125: 1474–1497.
Loewy, A. D. (1991). Forebrain nuclei involved in autonomic control. *Prog Brain Res* 87: 253–268.
Maestroni, G. J. (2002). Short exposure of maturing, bone marrow-derived dendritic cells to norepinephrine: impact on kinetics of cytokine production and Th development. *J Neuroimmunol* 129(1–2): 106–114.
Maier, S. F., L. R. Watkins and M. Fleshner (1994). Psychoneuroimmunology. The interface between behavior, brain, and immunity. *Am Psychol* 49(12): 1004–1017.
Malarkey, W. B., J. Wang, C. Cheney, R. Glaser and H. Nagaraja (2002). Human lymphocyte growth hormone stimulates interferon gamma production and is inhibited by cortisol and norepinephrine. *J Neuroimmunol* 123(1–2): 180–187.
Mazzeo, R. S. (1991). Catecholamine responses to acute and chronic exercise. *Med Sci Sports Exercise* 23(7): 839–845.
Mazzeo, R. S., G. A. Brooks and S. A. Horvath (1984). Effects of age on metabolic responses to endurance training in rats. *J Appl Physiol* 57(5): 1369–1374.
Mazzeo, R. S. and A. Brownlow (1996). Influence of microgravity on heart and liver catecholamine content in rats. *FASEB J* 10: A573.

Mazzeo, R. S., R. W.Colburn and S. M.Horvath (1986). Effect of aging and endurance training on tissue catecholamine response to strenuous exercise in Fischer 344 rats. *Metabolism* 35: 602–607.

Mazzeo, R. S. and P. A. Grantham (1989). Norepinephrine turnover in various tissues at rest and during exercise: evidence for a training effect. *Metabolism* 38(5): 479–483.

Meeusen, R., I. Smolders, S. Sarre, K. de Meirleir, H. Keizer, M. Serneels, G. Ebinger and Y. Michotte (1997). Endurance training effects on neurotransmitter release in rat striatum: an in vivo microdialysis study. *Acta Physiol Scand* 159(4): 335–341.

Milner, J. D. and R. J. Wurtman (1987). Tyrosine availability: a presynaptic factor controlling catecholamine release. *Adv Exp Med Biol* 221: 211–221.

Monjan, A., and Collector (1976). Stress-induced modulation of the immune response. *Science* 196: 307–308.

Moraska, A., T. Deak, R. L. Spencer, D. Roth and M. Fleshner (2000). Treadmill running produces both positive and negative physiological adaptations in Sprague-Dawley rats. *Am J Physiol Regul Integr Comp Physiol* 279(4): R1321–R1329.

Moraska, A. and M. Fleshner (2001). Voluntary physical activity prevents stress-induced behavioral depression and anti-KLH antibody suppression. *Am J Physiol Regul Integr Comp Physiol* 281(2): R484–R489.

Nauli, S. M., T. J. Maher, W. J. Pearce and A. Ally (2001). Effects of opioid receptor activation on cardiovascular responses and extracellular monoamines within the rostral ventrolateral medulla during static contraction of skeletal muscle. *Neurosci Res* 41(4): 373–383.

Nijsen, M. J., G. Croiset, M. Diamant, D. De Wied and V. M. Wiegant (2001). CRH signalling in the bed nucleus of the stria terminalis is involved in stress-induced cardiac vagal activation in conscious rats. *Neuropsychopharmacology* 24(1): 1–10.

Nijsen, M. J., G. Croiset, R. Stam, A. Bruijnzeel, M. Diamant, D. de Wied and V. M. Wiegant (2000). The role of the CRH type 1 receptor in autonomic responses to corticotropin- releasing hormone in the rat. *Neuropsychopharmacology* 22(4): 388–399.

Nistico, G., M. C. Caroleo, M. Arbitrio and L. Pulvirenti (1994). Corticotropin-releasing factor microinfused into the locus coeruleus produces electrocortical desynchronization and immunosuppression. *Neuroimmunomodulation* 1(2): 135–140.

North, R. A. and J. T. Williams (1983). Opiate activation of potassium conductance inhibits calcium action potentials in rat locus coeruleus neurones. *Br J Pharmacol* 80(2): 225–228.

North, R. A. and J. T. Williams (1985). On the potassium conductance increased by opioids in rat locus coeruleus neurones. *J Physiol* 364: 265–280.

Overton, J. M., K. C. Kregel, G. Davis-Gorman, D. R. Seals, C. M. Tipton and L. A. Fisher (1991). Effects of exercise training on responses to central injection of CRF and noise stress. Physiol Behav 49: 93–98.

Pan, Y. Z., D. P. Li, S. R. Chen and H. L. Pan (2004). Activation of mu-opioid receptors excites a population of locus coeruleus-spinal neurons through presynaptic disinhibition. *Brain Res* 997(1): 67–78.

Pauletto, P., G. Scannapieco and A. C. Pessina (1991). Sympathetic drive and vascular damage in hypertension and atherosclerosis. *Hypertension* 17(4 Suppl): III75–III81.

Persson, S., I. Jonsdottir, P. Thoren, C. Post, F. Nyberg and P. Hoffmann (1993). Cerebrospinal fluid dynorphin-converting enzyme activity is increased by voluntary exercise in the spontaneously hypertensive rat. *Life Sci* 53(8): 643–652.

Plotnikoff, N., A. Murgo, R. Faith and J. Wybran, Ed. (1991). *Stress and Immunity.* Boca Raton, CRC Press, Inc.

Podojil, J. R., N. W. Kin and V. M. Sanders (2004). CD86 and beta2-adrenergic receptor signaling pathways, respectively, increase Oct-2 and OCA-B Expression and binding to the 3'-IgH enhancer in B cells. *J Biol Chem* 279(22): 23394–23404.

Podojil, J. R. and V. M. Sanders (2003). Selective regulation of mature IgG1 transcription by CD86 and beta 2-adrenergic receptor stimulation. *J Immunol* 170(10): 5143–5151.

Ramer-Quinn, D. S., R. A. Baker and V. M. Sanders (1997). Activated T helper 1 and T helper 2 cells differentially express the beta-2-adrenergic receptor. *J Immunol* 159: 4857–4867.

Rassnick, S., A. F. Sved and B. S. Rabin (1994). Locus coeruleus stimulation by corticotropin-releasing hormone suppresses in vitro cellular immune responses. *J Neurosci* 14(10): 6033–6040.
Rhodes, J. S., T. Garland, Jr. and S. C. Gammie (2003). Patterns of brain activity associated with variation in voluntary wheel-running behavior. *Behav Neurosci* 117(6): 1243–1256.
Romagnano, M. A., R. J. Harshbarger and R. W. Hamill (1991). Brainstem enkephalinergic projections to spinal autonomic nuclei. *J Neurosci* 11(11): 3539–3555.
Sanders, V. M., R. A. Baker, D. S. Ramer-Quinn, D. J. Kasprowicz, B. A. Fuchs and N. E. Street (1997). Differential expression of the beta2-adrenergic receptor by Th1 and Th2 clones. *J Immunol* 158: 4200–4210.
Sanders, V. M., D. J. Kasprowicz, M. A. Swanson-Mungerson, J. R. Podojil and A. P. Kohm (2003). Adaptive immunity in mice lacking the beta(2)-adrenergic receptor. *Brain Behav Immun* 17(1): 55–67.
Saper, C. B., L. W. Swanson and W. M. Cowan (1976). The efferent connections of the ventromedial nucleus of the hypothalamus of the rat. *J Comp Neurol* 169(4): 409–442.
Seals, D. R. and C. Bell (2004). Chronic sympathetic activation: consequence and cause of age-associated obesity? *Diabetes* 53(2): 276–284.
Seals, D. R., J. A. Taylor, A. V. Ng and M. D. Esler (1994). Exercise and aging: autonomic control of the circulation. *Med Sci Sports Exerc* 26(5): 568–576.
Smith, T. P., S. L. Kennedy and M. Fleshner (2004). Influence of age and physical activity on the primary in vivo antibody and T cell-mediated responses in men. *J Appl Physiol* 97(2): 491–498.
Stauss, H. M., P. B. Persson, A. K. Johnson and K. C. Kregel (1997). Frequency-response characteristics of autonomic nervous system function in conscious rats. *Am J Physiol* 273(2 Pt 2): H786–H795.
Sved, A. F., G. Cano and J. P. Card (2001). Neuroanatomical specificity of the circuits controlling sympathetic outflow to different targets. *Clin Exp Pharmacol Physiol* 28(1–2): 115–119.
Swanson, M. A., W. T. Lee and V. M. Sanders (2001). IFN-gamma production by Th1 cells generated from naive CD4+ T cells exposed to norepinephrine. *J Immunol* 166(1): 232–240.
Takahashi, H. K., H. Iwagaki, S. Mori, T. Yoshino, N. Tanaka and M. Nishibori (2004). Beta 2-adrenergic receptor agonist induces IL-18 production without IL-12 production. *J Neuroimmunol* 151(1–2): 137–147.
Travagli, R. A., M. Wessendorf and J. T. Williams (1996). Dendritic arbor of locus coeruleus neurons contributes to opioid inhibition. *J Neurophysiol* 75(5): 2029–2035.
Valentino, R. J. and E. Van Bockstaele (2001). Opposing regulation of the locus coeruleus by corticotropin-releasing factor and opioids. Potential for reciprocal interactions between stress and opioid sensitivity. *Psychopharmacology (Berl)* 158(4): 331–342.
Van Bockstaele, E. J., E. E. Colago, A. Moriwaki and G. R. Uhl (1996a). Mu-opioid receptor is located on the plasma membrane of dendrites that receive asymmetric synapses from axon terminals containing leucine-enkephalin in the rat nucleus locus coeruleus. *J Comp Neurol* 376(1): 65–74.
van Bockstaele, E. J., E. E. Colago and V. M. Pickel (1996b). Enkephalin terminals form inhibitory-type synapses on neurons in the rat nucleus locus coeruleus that project to the medial prefrontal cortex. *Neuroscience* 71(2): 429–442.
van Bockstaele, E. J., K. Commons and V. M. Pickel (1997). Delta-opioid receptor is present in presynaptic axon terminals in the rat nucleus locus coeruleus: relationships with methionine5-enkephalin. *J Comp Neurol* 388(4): 575–586.
Van Bockstaele, E. J., J. Peoples and P. Telegan (1999a). Efferent projections of the nucleus of the solitary tract to peri-locus coeruleus dendrites in rat brain: evidence for a monosynaptic pathway. *J Comp Neurol* 412(3): 410–428.
Van Bockstaele, E. J., J. Peoples and R. J. Valentino (1999b). A.E. Bennett Research Award. Anatomic basis for differential regulation of the rostrolateral peri-locus coeruleus region by limbic afferents. *Biol Psychiatry* 46(10): 1352–1363.

Williams, J. T., T. M. Egan and R. A. North (1982). Enkephalin opens potassium channels on mammalian central neurones. *Nature* 299(5878): 74–77.

Winder, W. W., M. A. Bettie and R. T. Holman (1982). Endurance training attenuates stress hormone responses to exercise in fasted rats. *Am J Physiol* 243: R179–R184.

Xin, X., N. Yang, A. D. Eckhart and J. E. Faber (1997). alpha1D-adrenergic receptors and mitogen-activated protein kinase mediate increased protein synthesis by arterial smooth muscle. *Mol Pharmacol* 51: 764–775.

Part II
Neuroimmunological Basis of Behavior

Alteration of Neurodevelopment and Behavior by Maternal Immune Activation

Stephen E.P. Smith and Paul H. Patterson

Abstract The immune system rapidly responds to pathogens by releasing a variety of signaling molecules that trigger a number of infection-fighting cellular programs. These same signaling pathways (e.g., NF-κB, JAX/STAT, ERK) are used by the developing brain to orchestrate programs of cell proliferation, differentiation, and migration. Thus, when a pregnant woman falls ill, there is the potential for cross-talk between the maternal immune response and the developing fetal brain. In fact, maternal infections are significant environmental risk factors for schizophrenia and autism. There are several animal models in which infection-induced maternal immune activation causes behavioral, histological, and gene expression changes in the offspring that are reminiscent of human mental disorders. We review both human and animal data that demonstrate these effects of maternal immune activation, and discuss potential mechanisms through which the maternal immune system may alter brain development.

Keywords Autism · Schizophrenia · Cytokines · IL-6 · Maternal infection · LPS · Poly(I:C) · Influenza

1 Genes vs. Environment in Mental Disease

It is well known that genes play a major role in several mental disorders. However, in our view, the genetic contributions to schizophrenia and autism, in particular, can be overemphasized. While genes undeniably play a major role, only 5–10% of autism cases can be attributed to known chromosomal abnormalities or single gene mutations. Furthermore, while early estimates of monozygotic twin concordance rates in autism were as high as 90%, recent reports put that number closer to 60% (Lemery-Chalfant et al., 2006). Similarly, with notable exceptions (e.g., DISC1), very few cases of schizophrenia can be traced to a known genetic cause. While this discrepancy may be

P.H. Patterson (✉)
Biology Division, California Institute of Technology, Pasadena, CA 91125, USA
e-mail: php@caltech.edu

A. Siegel, S.S. Zalcman (eds.), *The Neuroimmunological Basis of Behavior and Mental Disorders*, DOI 10.1007/978-0-387-84851-8_7,

attributed to the complexity of studying diseases where multiple genes each contribute small amounts to risk, the concordance rate for monozygotic twins in schizophrenia (50%) leaves much room for environmental factors. Moreover, indirect evidence reveals an important finding: monozygotic twins that share a placenta have a high concordance rate (60%) for schizophrenia, while those with separate placentas have a concordance rate similar to that of dizygotic twins (Davis et al., 1995; Phelps et al., 1997). In addition, the concordance rate of dizygotic twins (~17%) is almost twice as high as that of siblings (~9%), even though these groups have identical genetic relatedness. These findings highlight the importance of the intrauterine environment, which is further emphasized by human and animal studies of maternal infection.

2 Prenatal Infections and Mental Disorders

Maternal infection by several different organisms during early- to mid-pregnancy has been linked to both schizophrenia and autism. The strongest evidence for maternal infection increasing risk for a mental disorder in the offspring is the connection between schizophrenia and maternal respiratory infection. In a pioneering study, Mednick et al. (1988) found a higher rate of schizophrenia among a cohort of Swedish adults who were *in utero* during the 1957 influenza epidemic. Since then, over 25 epidemiological studies have assessed the rate of schizophrenia in people who were *in utero* during influenza epidemics, and the majority have found an increased incidence of disease among the exposed offspring (reviewed by Bagalkote et al., 2001). However, this epidemiological data is population-based and therefore is unable to document a direct relationship between respiratory infection in individual mothers and later development of schizophrenia in the offspring. While this caveat should decrease the probability of finding an association, it nevertheless creates uncertainty in the conclusions. Brown, Susser, and colleagues were able to overcome this limitation by using a large pool of banked maternal serum samples that were linked to detailed medical records of both mothers and offspring (Brown et al., 2004a). They found that in cases where they were able to confirm maternal influenza infection by antibody assays of the banked serum, the resulting offspring were 3–7 times more likely to develop schizophrenia as adults. Due to the high prevalence of influenza, they estimate that 14–21% of the schizophrenia cases would not have occurred if maternal infection had been prevented. The same group has also found associations between schizophrenia in the offspring and maternal toxoplasmosis (Brown et al., 2005), genital/reproductive viral infection (Babulas et al., 2006) or elevated levels of maternal interleukin(IL)-8 (Brown et al., 2004b). Remarkably, this association was detected despite the inability to screen for a susceptibility genotype. If one assumes that only genetically susceptible individuals will develop schizophrenia after maternal infection, the increased risk due to infection will be much higher than 3–7-fold in this group.

Although there is much less evidence available, a link has also been found between maternal infection and autism in the offspring. Rubella epidemics in the 1960s were associated with greatly increased risk for autism in children that were exposed *in utero,* as well as physical abnormalities and mental retardation (Desmond et al., 1967; Chess, 1977). Since the development of the rubella vaccine, many fewer cases of maternal rubella infection are seen. Small studies of other maternal infections, such as toxoplasma, syphilis, varicella, and rubeola also support the idea that maternal infection can be a risk factor for autism (Ciaranello and Ciaranello, 1995; Hyman et al., 2005). While the phenotypic heterogeneity and complex genetics of schizophrenia and autism makes it difficult to establish maternal infection as a definitive cause of the disorders, there is considerable evidence implicating it as an important risk factor.

3 Animal Models of Immune Activation

The diversity of infections that have been implicated as risk factors for mental disorders, as well the fact that many of these infections do not have direct access to the fetal compartment, has led to the hypothesis that the maternal immune system, rather than a specific pathogen, is responsible for the increased incidence of mental disease in the offspring (Patterson, 2002, 2005). While this hypothesis is not testable in humans, animal models have shown that maternal immune activation is able to cause a variety of behavioral, histological, and transcriptional changes in the adult offspring. The models use one of three methods to induce maternal immune activation in pregnant rodents: influenza infection, injection of the synthetic double stranded RNA, poly(I:C), or injection of bacterial lipopolysaccharide (LPS). Neuropathology, gene expression profiling, electrophysiology, behavioral assays, and antipsychotic drug treatment demonstrate similarities between the "exposed" offspring and humans suffering from mental disease. Unfortunately, it is often difficult to directly compare the results from different research groups, as variables such as the species used, the compound administered, dosing, timing and method of treatment, and tests of outcome are often different among laboratories. Taken together, however, these models strongly support the hypothesis that maternal immune activation can have deleterious effects on the offspring *in utero* (Table 1).

3.1 Assaying Features of Mental Illness in Mice

Before discussing the specific rodent models, it will be useful to briefly review some of the more common behavioral tests that are used to model schizophrenic- and autistic-like symptoms in mice. Schizophrenia and autism are both assessed using the diagnostics and statistical manual, now in its fourth volume (DSMIV). The diagnosis involves an interview, and is based upon behaviors, some of which are difficult to model in animals (e.g., disordered thoughts or language delay). Fortunately, there are several endophenotypes, traits that are not diagnostic but are found at a greater frequency in populations of affected individuals, which can be measured in experimental animals.

Table 1 Behavior and Histology Outcomes Following Various Types of MIA

References	Species and Treatment	Behavioral Findings	Histological Findings
Cai et al. (2000)	4 mg/kg LPS I.P. E18,19 rat P8 histology	N.R.	Increased GFAP, decreased MBP, altered microglial staining
Ling et al. (2004)	10,000EU/kg LPS I.P. E10.5 rat Adult (>1yr) histology	N.R.	Fewer TH+ neurons and increased microglial staining in substantia nigra
Borrell et al. (2002)	1 mg/kg LPS S.C. Alternate days throughout rat pregnancy; Adult histology	PPI deficits corrected by antipsychotic drugs	Increased GFAP, MHCII staining of microglia, TH increase in nucl accumb
Bakos et al. (2004)	20–80 μg/kg LPS S.C. E15–19 rat (increasing dose schedule)	Increased entries into all arms of plus maze, slips in beam walking test.	N.R.
Fortier et al. (2004)	50μg/kg LPS I.P. E18 & 19 rat	Increased amph.-induced locomotion, acoustic startle response	N.R.
Paintlia et al. (2004)	1 mg/kg LPS I.P. E18 rat E20 or P9–30 Histology	N.R.	Less MBP, PLP and myelin staining at P9–30, more microglia at E20
Golan et al., (2005, 2006)	0.12 mg/kg LPS I.P. E17 mouse Adult histology	Normal exploration and motor function, mostly normal learning/memory but specific deficits	Smaller, denser neurons in hippocampus, more pyknotic cells in cortex
Fatemi et al. (2002); Shi et al. (2003); Shi et al. (2008)	Intranasal influenza E9 mouse Adult histology	PPI, open field, novel object, social interaction deficits	Large adult brain, pyramidal cell atrophy, Purkinje cell deficit
Smith et al. (2007) Shi et al. (2008)	20mg/kg poly(I:C) I.P. E12 mouse Adult histology	PPI, LI,open field, social interaction deficits	Purkinje cell deficit
Zuckerman et al. (2003); Zuckerman and Weiner (2005)	4mg/kg poly(I:C) I.V. E15 rat Adult histology	LI deficit, enhanced reversal learning, normal water maze, increased amph. and MK-801 locomotion	Pyknotic cells in hippocampus, increased KCl-stimulated dopamine release in striatum
Meyer et al. (2006a); Nyffeler et al. (2006)	5 mg/kg poly(I:C) I.V. E9 mouse Adult histology	PPI, LI, open field, working memory deficits; increased amph-induced locomotion	GABAA receptor increase, no increase in pyknotic cells in hippocampus

Table 1 (continued)

References	Species and Treatment	Behavioral Findings	Histological Findings
Ozawa et al. (2006)	5 mg/kg poly(I:C) I.P. Daily E12–17 mouse Adult histology	PPI, open field, working memory deficit, increased amph locomotion	Altered dopamine metabolism in striatum

Different rows represent different research groups. Amph, amphetamine; GABA, g-aminobutyric acid; GFAP, glial fibrillary acidic protein; I.P, intraperitoneal; I.V., intravenous; MBP, myelin basic protein; N.R., not reported; PLP, proteolipid protein; PPI, prepulse inhibition; LI, latent inhibition; LPS, lipopolysaccharide; nucl accumb, shell of nucleus accumbens; S.C., subcutaneous; TH, tyrosine hydroxylase.

Prepulse inhibition (PPI) is a measure of sensory-motor gating that is disrupted in schizophrenic (Turetsky et al., 2007) and autistic (Perry et al., 2007) individuals, as well as in people who have other mental health problems. High functioning autistics describe being bombarded with an overwhelming amount of sensory information, and deficits in PPI may reflect an underlying inability to quickly classify sensory input as relevant or irrelevant. PPI refers to the inhibition of a startle response to an aversive stimulus (a "pulse") when the startling stimulus is preceded by a smaller, nonstartling stimulus (a "prepulse"). The interval between the pulse and prepulse ranges from 50 to 500 ms, which does not allow time for higher-level processing of the two stimuli. In rodents, PPI is assayed by placing the animal in a small enclosure and measuring its startle response to a loud pulse of white noise; in humans, the stimuli are usually airpuffs to the eye.

Latent inhibition (LI) is another behavioral test that measures the ability of a subject to ignore or "filter out" irrelevant information, and is highly pertinent to schizophrenia (Weiner, 2003). The neural circuitry for LI lies in the hippocampus and nucleus accumbens, and relies heavily on dopaminergic transmission between these areas. LI is disrupted in schizophrenic subjects and in amphetamine-treated humans and rats, is restored to normal levels in schizophrenics by antipsychotic drugs, and is enhanced in normal humans and rats by these drugs (Weiner, 2003). Measuring LI involves repeatedly exposing the subject to a conditioned stimulus (CS), and then pairing that stimulus with a different, unconditioned stimulus (US). Pre-exposed (PE) animals will not associate the CS and the US as strongly as non-pre-exposed (NPE) animals; the difference between the responses of PE and NPE animals is termed LI. Both PPI and LI tests are easily administered to both humans and experimental animals, making them ideal tools for the validation of animal models.

Several other tests have been developed in animals to model features of human mental disorders. Heightened fear and anxiety are features of many mental diseases, and the open field test, which involves placing a mouse in a brightly lit enclosure and monitoring its movement, measures anxiety levels in rodents. Fewer entries into the center of the field and shorter distances moved in exploration are relevant measures. Interestingly, studies of the exploratory behavior of autistic children have validated the rodent paradigm by showing reduced exploration of a novel object-filled room by affected children (Pierce and Courchesne, 2001; Akshoomoff et al., 2004).

Finally, social interaction is a central deficit in autism, and several tests have been developed to monitor the interaction between two mice placed in a shared enclosure (Crawley, 2007).

3.2 Maternal Influenza Infection

The influenza infection model is based directly on human data showing a higher incidence of schizophrenia in offspring of mothers who developed influenza infections during the second trimester of pregnancy (Fatemi et al., 2002; Shi et al., 2003). Mice are inoculated intranasally with a mouse-adapted human influenza virus on day 9.5 of pregnancy. Over about 7 days, the mice develop fluid in the lungs, show noticeable sickness behavior, and have elevated serum levels of several cytokines. As mouse pregnancy lasts 18–19 days, the sickness persists for the second half of mouse pregnancy. Due to differences in mouse vs. human fetal development, namely that mice are born with their brains in a less mature state than humans, this time period corresponds to the human second trimester in terms of brain development milestones (for an excellent review of inter-species developmental stages, see Clancy et al., 2001). Thus, this model closely recapitulates the human risk factor of a second trimester influenza infection.

The adult offspring of influenza-infected mice appear superficially normal, but display several behavior abnormalities that are highly relevant to schizophrenia and autism. They have lower PPI than controls, and this deficit is rescued by acute treatment with antipsychotic drugs. They display heightened anxiety in the open field, as measured by a reduced total distance moved, less rearing, and fewer entries into the center of the field. Finally, they show less social interaction with an unfamiliar, same-sex conspecific (Shi et al., 2003). Offspring of flu-infected mothers also display several histological abnormalities that are reminiscent of those found in mental disorders. For example, the most commonly reported histological finding in postmortem autistic brains is a selective loss of Purkinje cells in lobules VI and VII of the cerebellum, and structural MRI studies have also found reduced volume of autistic cerebella (Palmen et al., 2004). Remarkably, offspring of the influenza-exposed mice also show a highly selective reduction in the linear density of PCs in lobules VI and VII of the cerebellum, a deficit that seems to be of developmental origin (Shi et al., 2008). Other relevant histological findings include altered levels of synaptosome-associated protein-25 (SNAP-25), reduced reelin immunoreactivity in the cortex, and smaller, more densely packed pyramidal cells in the hippocampus (Fatemi et al., 1998, 1999, 2002).

3.3 Maternal Immune Activation

Several lines of evidence suggest that maternal immune activation (MIA), rather than direct infection of the fetus, is responsible for the behavioral and histological changes seen in the offspring of infected mothers. First, in human studies, the fact that a wide variety of pathogens have similar effects suggests that they act via a similar mechanism. Furthermore, many of the implicated infections are confined to specific areas of the body, and do not involve the fetus. For example, influenza infection is typically confined to the respiratory system. This was confirmed in the influenza mouse model: using a sensitive RT-PCR assay that can detect as little as one plaque-forming unit of virus, no virus was detected in the exposed fetuses (Shi et al., 2005). Finally, and most convincingly, two rodent models have been developed in which behavioral deficits are induced in adult offspring by directly activating the maternal immune system in the absence of pathogens.

3.4 Maternal Poly(I:C) Administration

Poly(I:C) is a synthetic double-stranded RNA that is a potent agonist of the toll-like receptor(TLR)-3. Double-stranded RNA is a sign of viral infection for the innate immune system; activation of TLR3 induces an inflammatory cascade that results in the production of antiviral cytokines and chemokines. Injection of poly(I:C) in a pregnant rodent at mid-gestation produces offspring that are remarkably similar to the offspring of mice given a flu infection. These offspring display deficits in PPI, LI, open field exploration and social interaction (Shi et al., 2003; Zuckerman et al., 2003; Zuckerman and Weiner, 2005; Smith et al., 2007). Many of these behavioral deficits respond to antipsychotic drugs (Zuckerman et al., 2003; Zuckerman and Weiner, 2005; Ozawa et al., 2006). Furthermore, the PPI and LI deficits only occur after puberty, mimicking the adult onset of schizophrenia (Zuckerman et al., 2003; Ozawa et al., 2006). Poly(I:C) also causes enlarged ventricals (Piontkewitz et al., 2007) and altered dopamine metabolism in the adult offspring (Ozawa et al., 2006), which are relevant to schizophrenia, and increased $GABA_A$ receptor expression in the adult offspring (Nyffeler et al., 2006), which is also relevant for autism. Also in common with the influenza model, maternal poly(I:C) treatment causes a deficit of Purkinje cells in lobule VII of the cerebellum (Shi et al., 2008).

3.5 Maternal LPS Administration

Another method of inducing MIA is the injection of bacterial LPS, a natural ligand for the TLR4. Intrauterine bacterial infection is commonly associated with preterm birth, and neurological disorders such as cerebral palsy, and mental retardation (Saliba and Henrot, 2001; Dammann et al., 2002). Intrauterine infection leads to pathology, such as white matter damage, which is more severe than that found in schizophrenia and autism. Very limited evidence links maternal bacterial infection to the later development of schizophrenia (Watson et al., 1984; O'Callaghan et al., 1994). However, activation of TLR4 activates many of the same signaling pathways as TLR3, and elevates levels of many of the same cytokines in the maternal circulation, notably IL-6. The specific combination of cytokines and chemokines is different than for TLR3, but like poly(I:C), LPS produces a very strong, but transient, immune activation. Many of the behavioral abnormalities seen in the offspring of poly(I:C)-treated mothers are also seen in the offspring of LPS-treated mothers. Using a very severe protocol of LPS injections daily throughout pregnancy, one group has reported PPI deficits (Borrell et al., 2002) that are corrected by administration of antipsychotic drugs (Romero et al., 2007) in adult offspring. However, two injections of 50 μg/kg LPS in late pregnancy (E18 and 19) does not yield PPI deficits (Fortier et al., 2004). Increased anxiety-like behavior and abnormal social behavior (Hava et al., 2006), as well as enhanced amphetamine-induced locomotion (Fortier et al., 2004) and abnormal learning and memory (Golan et al., 2005) have been reported in the offspring of mice given LPS injections on E17, 18, or 19. Histological findings include fewer, more densely packed neurons in the hippocampus (Golan et al., 2005), increased microglial staining (Cai et al., 2000; Borrell et al., 2002; Ling et al., 2004; Paintlia et al., 2004), increased glial fibrillary acidic protein (GFAP) staining (Cai et al., 2000; Borrell et al., 2002), altered tyrosine hydroxylase (TH) staining (Borrell et al., 2002; Ling et al., 2004) and decreased myelin basic protein (MBP) staining (Cai et al., 2000; Paintlia et al., 2004), all potentially relevant for mental illness.

3.6 Other Relevant Animal Models of Environmental Risk Factors

Several other protocols for inducing brain pathology and behavioral abnormalities in pre or early postnatal animals exist, including intrauterine infection with periodontal bacteria (Lin et al., 2003a, b; Han et al., 2004; Bobetsis et al., 2006), injection of LPS directly into the fetus (reviewed by Wang et al., 2006), or injection of cytokines in early postnatal animals (reviewed by Nawa and Takei, 2006). However, as these methods are not meant to model a second trimester maternal infection, and are not known to be directly relevant to schizophrenia and autism, they are not reviewed here. There is also a significant body of research on maternal stress as a risk factor for schizophrenia (reviewed by Relier, 2001), and in rodents, maternal behavioral stress leads to abnormal behavior in the adult offspring, although the assays used are different (reviewed by Weinstock, 2001).

4 Mechanisms of Behavioral Abnormalities Caused by MIA

The mechanisms and pathways by which MIA alters behavior and fetal brain development is beginning to be explored, with particular focus on cytokines. Cytokines are small (8–30 kDa) signaling molecules that are released in response to a wide range of immune challenges. They are attractive candidates for causing the observed changes in fetal brain development for several reasons. First, accessibility: they are released into the maternal serum in response to infection/MIA; thus, even though an influenza infection is confined to the lungs, cytokines produced at the infection site will have access to the fetus. Moreover, evidence indicates that cytokines can cross the placenta and access the fetus (Dahlgren et al., 2006; Ponzio et al., 2007). Second, activity: Cytokines signal through several key developmental pathways, including the STAT, NF-κB, and ERK cascades, allowing for the possibility of interference with those signaling pathways. Moreover, many cytokine receptors are expressed in both developing and mature neurons and glia, and when activated, can cause morphological and functional changes in those cells (Jankowsky and Patterson, 1999; Gilmore et al., 2004; Bauer et al., 2007). Thus, cytokines are logical candidates to perturb fetal brain development.

Altered serum cytokine levels in response to MIA have been documented in several of the animal models discussed above (Table 2). Consistently, regardless of the method employed, the proinflammatory cytokines IL-6, IL-1β, and tumor necrosis factor (TNF)α are elevated in the maternal serum and placenta. Our group has identified at least 10 more cytokines and chemokines that are upregulated in maternal serum after poly(I:C) administration. At least some of these cytokines are able to cross the placenta and gain access to the fetal brain. Radiolabeled IL-6 and IL-2 have both been found to cross the placenta; when injected iv in pregnant rodents, radioactivity levels in the fetuses are 15–20% of those in maternal serum (Dahlgren et al., 2006; Ponzio et al., 2007). Whether cytokines actually cross into the fetal brain during experimentally induced MIA is a matter of contention. Some groups have reported that IL-6 protein is significantly elevated in the fetal brain following

Table 2 Maternal Immune Activation Increases Cytokine Levels in the Fetal Brain

Reference	Treatment	Findings
Cai et al. (2000)*	4 mg/kg LPS I.P. E18 rat	TNFα, IL1β increased
Urakubo et al. (2001)	2.5 mg/kg LPS I.P. E16 rat	TNFα increased
Liverman et al. (2006)	50μg LPS I.P. E18 mouse	IL1β, IL6, MCP-1, VEGF increased
Paintlia et al. (2004)*	1 mg/kg LPS I.P. E18 rat	TNFα, IL1β, iNOS increased
Golan et al. (2005)	0.12 mg/kg LPS I.P. E17 mouse	IL6 increased
Ashdown et al. (2006)	0.05 mg/kg LPS I.P. E18 rat	No change in TNFα, IL1β, IL6
Meyer et al. (2006a)	5 mg/kg poly(I:C) I.V. E9 mouse	IL1β, IL6 increased
Meyer et al. (2006b)	5 mg/kg poly(I:C) I.V. E17 mouse	IL1β, IL6, IL10 increased
Meyer et al. (2008)	2 mg/kg poly(I:C) I.V. E9 mouse	TNFα, IL1β, IL6, IL10 increased
Gilmore et al. (2005)	20 mg/kg poly(I:C) I.P. E16 rat	No change in TNFα

Assays were for cytokine protein, except where noted (*mRNA assayed). While some authors report no changes in cytokine levels, the majority of studies show significant increases. The studies that report no changes use less severe methods of immune activation (lower dose of lipopolysaccharide (LPS) or I.P. administration of poly(I:C)) which may not produce detectable changes.

MIA (Meyer et al, 2006b), although negative or inconclusive results have also been reported (Meyer et al., 2008, Ashdown et al., 2006). Similar mixed results have been reported for TNFα; reports have shown a small increase (Urakubo et al., 2001), a significant increase (Bell et al., 2004) or undetectable (Ashdown et al., 2006) levels of TNAα protein or mRNA in fetal brains after MIA. The different cytokine responses in the fetal brain are likely due to the different severity of the treatment in different laboratories. For example, induction of severe inflammation by injecting LPS directly into the uterine horn produces large increases in the levels of cytokine mRNAs in the fetal brain (Elovitz et al., 2006). Cytokine mRNA increases in fetal brain have also been reported following iv injection of poly(I:C); Meyer et al., 2006b, 2008). Our group has administered a relatively mild dose of poly(I:C); single ip injection), and we observe only small, nonsignificant changes in the levels of cytokine proteins in the fetal brain (W. Xu, B. Deverman, S. Smith, unpublished data). However, since the cytokine levels under discussion approach the lower detection limit of the assays, some of the negative results may reflect limitations of the assays rather than a lack of cytokine access to the brain. The radiolabeled cytokine experiments, as well as preliminary data from our group showing increased mRNA of downstream genes in fetal brains of poly(I:C)-treated mothers (E. Hsaio, S. Smith, unpublished data), suggest that cytokines are able to cross the placenta and gain access to the fetal brain, despite remaining undetectable by standard ELISA assays. Since there are also reports of cytokine mRNA induction in the fetal brain, it is also possible that other signaling mechanisms (fever, ischemia, nutritional changes) could induce cytokines in the fetus directly, or indirectly, by altering the placenta.

Recent work has shown that the cytokine IL-6 plays a critical role in the manifestation of behavioral deficits caused by MIA. Samuelsson et al. (2006) administered three ip injections of IL-6 to pregnant rats over the course of 6 days. They found a working memory deficit in the adult offspring, as well as elevated IL-6 levels in the adult hippocampus, indicating an ongoing inflammation triggered by early events. This inflammatory state is reminiscent of the profound inflammation in autistic brains (Vargas et al., 2005), in which IL-6 was among the most prominently upregulated cytokines in subjects ranging in age from 5 to 44. Moreover, the IL-6-exposed adult rat offspring had elevated GFAP and $GABA_A$ receptor levels, similar to those reported in some MIA offspring (Nyffeler et al., 2006).

We have also studied the effects of both raising IL-6 levels in pregnant mice and blocking endogenous IL-6 in the poly(I:C) MIA model (Smith et al., 2007). In pregnant mice injected once with 5 μg of IL-6, we found both PPI and LI deficits in the adult offspring. Other injected cytokines (IL-1α, TNFα, and interferon (IFN)γ) had no effect on the behavior of the adult offspring. We used neutralizing antibodies to selectively block cytokines during poly(I:C)-induced MIA; co-administration of an anti-IL6 antibody completely prevented all of the behavioral deficits induced by poly(I:C); deficits in PPI, LI, and social interaction, as well as increased open field anxiety). In contrast, neutralization of IL-1β or IFNγ did not rescue the poly(I:C)-induced behavioral deficits. Moreover, IL-6 knockout (KO) mice are insensitive to the effects of MIA; offspring of pregnant IL-6 KO mice that were treated with poly(I:C) do not display PPI or LI deficits. We also used a microarray analysis to monitor changes in the adult brain transcriptome caused by MIA. Of 61 genes whose expression was altered by MIA, 55 were normalized in the offspring of pregnant mice that were co-injected with poly(I:C) and anti-IL-6. Thus, blocking IL-6 prevents the changes in behavior and gene expression caused by MIA (Smith et al., 2007).

Based on these results, an attractive, but preliminary, hypothesis can be proposed for the mechanism of MIA-induced behavioral deficits (Fig. 1). Maternal immune activation induces production of cytokines, particularly IL-6, which enter the maternal circulation. In mid-, but not late gestation, IL-6 crosses into the fetal circulation (Dahlgren et al., 2006), which correlates with human epidemiological data suggesting that early, not late, infections increase risk for mental disorders (Brown et al., 2004a). IL-6 can have variety of direct effects on the developing brain (reviewed by Bauer et al., 2007). Recent studies in our group indicate that offspring of influenza-infected mice have abnormal neuron migration to cortical layers II/III (Limin Shi, personal communication) as well as fewer Purkinje cells in lobules VI and VII of the cerebellum (Shi et al., 2008). In addition, IL-6 causes neurons in tissue culture to retract their processes, which suggests that early morphological changes in the developing brain could precipitate future behavioral abnormalities (Gilmore et al., 2004). Finally, the STAT-3 pathway, through which IL-6 signals, regulates developmental processes such as the switch between neurogenesis and gliogenesis (Bauer et al., 2007). The potential for IL-6 to alter fetal development

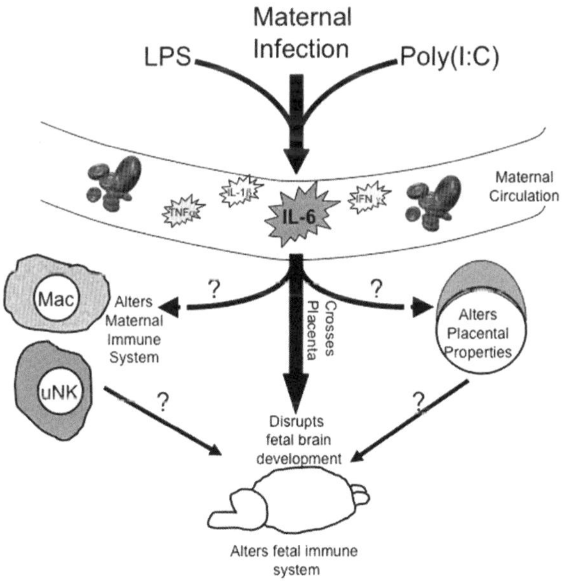

Fig. 1 Proposed mechanism through which MIA leads to behavior abnormalities. Maternal infection, lipopolysaccharide (LPS), or poly(I:C) all lead to increased levels of cytokines in the maternal circulation. The cytokine interleukin (IL)-6 disrupts fetal brain development, either by crossing the placenta and directly interfering with signaling pathways in the developing brain, or indirectly via alterations to the placenta or the maternal immune system

indirectly by changing placental properties such as vascularization (Paul et al., 2003), or by breaking down maternal immune tolerance of the fetus (Sargent et al., 2006), should also be considered.

5 Implications and Potential Therapies

One important implication of the demonstration of the key role of IL-6 in the effects of MIA is that it suggests prevention or treatments based on anti-cytokine or anti-inflammatory therapies. There are two potential time-points for intervention: at the time of the maternal infection, and postpartum, when the behavioral deficits have already manifested. Potential interventions at the time of infection are complicated by the need to fight the infection. Although blocking IL-6 prevents the deficits caused by poly(I:C), if IL-6 is blocked during an influenza infection in a pregnant mouse, the mouse suffers miscarriage and will often succumb to the illness (S. Smith, unpublished). A similar effect is observed in IL-6 KO mice, indicating that IL-6 is necessary to fight infection. Thus, direct neutralization of IL-6 is not a viable clinical option. Other treatments that reduce the inflammatory response, but still allow effective control of infection may be possible. Anti-inflammatory treatment with N-acetylcysteine suppresses the fetal inflammatory response after maternal LPS administration (Beloosesky et al., 2006), but it remains to be seen if this treatment would be viable in an infection model. The anti-inflammatory cytokine IL-10 is naturally increased during normal pregnancy, and endogenous IL-10 is essential for resistance to LPS-induced pregnancy loss and preterm labor in mice (Robertson et al., 2006). Recently, Meyer et al. (2008) showed that macrophage-specific overexpression of IL-10 prevents behavioral deficits in the offspring that are caused by maternal poly(I:C) administration. The behavioral deficits may be prevented by the IL-10-induced reduction in the concentration of IL-6 and TNFα in maternal serum after poly(I:C) injection. One caveat of this work is that the genotype of the offspring was not addressed, so the behavioral results may have stemmed from a postnatal, anti-inflammatory action of IL-10 overexpression, which would be the equivalent of treatments discussed below. Further, the observation that IL-10 overexpression in the absence of poly(I:C) treatment induces behavior abnormalities in the offspring (Meyer et al., 2008) highlights the inherent dangers of prenatal interventions. Without a way to predict which offspring might be susceptible to develop schizophrenia or autism when exposed to MIA, it is unlikely that the FDA would approve clinical trials for these types of early interventions.

It is also possible that MIA sets in motion an ongoing immune activation or dysregulation in the brain that may be responsible for some of the behavioral abnormalities observed in the adult offspring. Both schizophrenic (Garver et al., 2003; Zhang et al., 2005) and autistic (Singh et al., 1991; Croonenberghs et al., 2002; Zimmerman et al., 2005) subjects show signs of abnormal peripheral immune systems, with reports of elevated cytokines in blood. Recent microarray data show dysregulation of immune-related transcripts in both schizophrenic (Arion et al., 2007; Saetre et al., 2007) and autistic (Garbett et al., 2007) brains. A 50-fold increase in TNFα levels was found in a study of cerebrospinal fluid (CSF) from ten autistic patients (Chez et al., 2007). Moreover, there is severe inflammation in the brains of autistic patients, from a broad range of ages (5–44 years) and heterogeneity in diagnoses (regressive vs. nonregressive, epilepsy, retardation) (Vargas et al., 2005). Many cytokines and chemokines are elevated in tissue from both the cortex and the cerebellum, and a high density of activated microglia and astrocytes are present, indicating an active cellular inflammatory process. A replication of this work was recently presented, showing increased Iba-1 microglial immunoreactivity in six autistic subjects compared to age-matched controls (Morgan et al., 2007). It is likely that this chronic elevation of cytokines and associated cellular inflammation would have an adverse, acute effect on the behavior of the patients, perhaps even causing some of the core features of the disorders. It is clear, for instance, that exogenous as well as endogenous IL-6 and IL-1 regulate neuronal excitability, long-term potentiation, and learning (Jankowsky and Patterson, 1999; Balschun et al., 2004; Bauer et al., 2007). IL-6 and related cytokines also regulate the stress response, feeding, sleep, and depressive behaviors in the adult brain, and injections of certain cytokines can induce psychiatric symptoms in adult humans (Capuron and Dantzer, 2003; Theoharides et al., 2004; Schiepers et al., 2005; Bauer et al., 2007). These considerations raise the possibility that treating the inflammation and lowering the level of inflammatory cytokines in the adult or young brain might be able to improve symptoms. In fact, a recent pilot study of 25 autistic children suggested that behavioral symptoms improve after treatment with Pioglitazone, an anti-inflammatory drug that is especially active against microglia (Boris et al., 2007). It should be noted, however, that this trial was not placebo-controlled and outcome was based on parental reporting. A clinical trail of the anti-inflammatory drug minocycline is ongoing at the NIH.

MIA has the potential to cause an ongoing inflammatory process such as is seen in autism. The developing immune system may need to find an appropriate balance between pro and anti-inflammatory signaling, and MIA may permanently alter this setpoint. Such an alteration in development is seen in the offspring of pregnant rats that have been exposed to restraint stress throughout pregnancy. Offspring of stressed females have an altered behavioral response to stress as adults, which correlates with a significantly longer time for serum corticosterone to return to baseline following hypothalamo-pituitary-adrenocortical (HPA) axis activation. This alteration is due to early exposure to elevated glucocorticoid, as adrenalectomy of the mothers prevents the changes in the offspring, and administration of a synthetic glucocorticoid, dexamethasone, induces the changes (Barbazanges et al., 1996). Importantly, maternal stress alters corticosteroid receptor expression in the hippocampus of the adult offspring, an area important for terminating the stress response (Barbazanges et al., 1996; Levitt et al., 1996; Francis et al., 1999). A mechanism whereby early exposure to a biological signal causes an alteration in the later response to that signal could be applicable to the immune system.

In fact, an early postnatal inflammatory challenge can cause the organism to respond differently to subsequent challenges in adulthood. Rats exposed to LPS on

P14 show a blunted febrile response to LPS as adults, and abnormally high COX-2 levels under baseline conditions, which are reduced after LPS challenge (normal animals show the opposite: low COX-2 levels under baseline conditions, which increase upon LPS stimulation) (Boisse et al., 2004). In a similar study, rats given a neonatal *E. coli* infection show impaired memory and increased brain inflammation after a subsequent challenge with LPS as adults (Bilbo et al., 2005a, b), compared to control rats given only an adult LPS injection. Maternal immune challenge may have a similar programming effect on the fetus as these early postnatal examples. The offspring of LPS-treated mothers display blunted or absent responses to a preweaning LPS injection (Hodyl et al., 2007; Lasala and Zhou, 2007), and fewer circulating monocytes are present in the adult offspring (Hodyl et al., 2007). MIA can also affect the way that the brain responds to subsequent nonimmunologic challenges. Wang et al. (2007) injected LPS into the uterus of E15 mice, and then induced hypoxia-ischemia in neonatal or adult offspring. They report that in neonatal (P5 or 9) mice, ischemia was more severe in LPS-exposed mice than in controls, whereas in adult ischemia, LPS exposure was protective (Wang et al., 2007).

Perhaps alteration of the fetal immune system, such that it is hyper-responsive to later challenges, could account for the frequent anecdotal connections between regressive autism and illness at the time of regression. The sudden onset of autistic symptoms in children and adults has been reported following encephalitis or infection with herpes simplex, varicella, cytomegalovirus (Libbey et al., 2005), and malaria (Mankoski et al., 2006). Central nervous system (CNS) infections of this type are known to rapidly induce proinflammatory cytokine expression. In contrast, infections in autistic children are associated with acute amelioration of behavioral symptoms, consistent with ongoing regulation of behavior by cytokines (Curran et al., 2007).

As with the behavioral data, cytokines probably play a large role in mediating the long-term immunologic effects of MIA. In a recent study involving the response of mice to maternal IL-2, daily injections of IL-2 from E12–17 resulted in elevated B and T cell counts in response to antigenic stimulation in preweaning pups. The results were interpreted as an acceleration of T cell development and a skewing of TH responses towards TH-1 (Ponzio et al., 2007). Moreover, a recent preliminary report showed that IL-1 administration to neonatal rats triggers microglial activation in the brain that persists into adulthood, and which is accompanied by a PPI deficit. Remarkably, treatment with the anti-inflammatory drug minocycline normalized the levels of microglia in the brain as well as the PPI (Tsuda et al., 2007).

It would be informative to develop an MIA model in mice with inflammation similar to that observed in autism, namely increased inflammatory cytokines in CSF, and microglial and astrocyte activation (Vargas et al., 2005). To date, most groups have reported only mild increases or no changes in inflammatory parameters in the adult rodent brain. The exception is a recent report (Romero et al., 2007) using a severe protocol of daily LPS administration throughout rat pregnancy. This protocol yields strikingly elevated serum levels of IL-1β, IL-2, IL-6, and TNF-α in the adult offspring. The high serum levels of cytokines would be expected to have dramatic consequences on the behavior of these animals. The only behavior that the authors assayed, PPI, is abnormal in the exposed offspring. Remarkably, both serum levels of inflammatory cytokines, as well as PPI are normalized by treatment of the adult offspring with the antipsychotic drug haloperidol. In addition, even 2 weeks after ending haloperidol treatment, the levels of IL-6, IL-2 and IL-12 remain lower than untreated, prenatally exposed animals. This study demonstrates not only that MIA can alter both behavior and immunological parameters in the adult offspring, but also that behavior and immunological parameters are tightly correlated. Coupled with the neonatal IL-1 data showing behavioral improvements in adult animals after anti-inflammatory treatment (Tsuda et al., 2007), these studies highlight the potential for treatment of abnormal behavior through normalization of inflammatory cytokines.

Finally, it is also of interest that Piontkewitz et al. (2007) have recently reported that treatment of adult offspring of poly(I:C)-treated pregnant rats with the antipsychotic drug clozapine normalizes both behavioral abnormalities and ventricular dilation as observed by structural magnetic resonance imaging. Moreover, these positive effects can also be achieved during the prodromal period, before the postpubertal onset of pathology and behavior. This result suggests the possibility of preventative treatment, highlighting the potential clinical relevance of the MIA model.

6 Conclusions and Perspectives

Schizophrenia and autism are thought to result from an interaction of a susceptibility genotype and environmental risk factors. The recent trend to focus on susceptibility genes has yielded some interesting candidates, and study of environmental factors that interact with those genes could reveal much about pathogenesis, prevention, and potential treatments. Maternal infection is an environmental risk factor for both schizophrenia and autism, and a preponderance of evidence suggests that the maternal immune system, rather than pathogen invasion of the fetus, is detrimental. Several animal models of MIA are now well characterized, and these models mimic diverse behavioral and pathological symptoms of the disorders. Coupled with their etiological relevance, these models offer attractive research opportunities.

Although full description of the MIA models continue, work has recently begun on the second phase of study: dissecting the mechanisms of how MIA alters fetal brain development, and the long-term changes in immune status that are set in motion by MIA. Regarding the former issue, maternal IL-6 produced in response to infection likely crosses the placenta and interacts with the fetal brain, although it may also alter the placenta itself. Ongoing research is examining the site of IL-6 action, with the eventual goal of characterizing the cellular and molecular changes caused by MIA. Regarding the permanent immune dysregulation in the postnatal brain, recent reports have shown that antipsychotic drugs can suppress an overactive immune system caused by MIA, and treatment with anti-inflammatory (or antipsychotic) drugs can normalize behavioral abnormalities produced by neonatal or MIA-induced inflammatory cytokine exposure. A more thorough understanding of the mechanisms relating MIA with later psychiatric disease will hopefully lead to better prevention and treatment of these devastating disorders.

Acknowledgements We wish to thank Kathleen Hamilton for administrative assistance, Limin Shi and Ben Deverman for assistance with the experiments described from this laboratory, and the Autism Speaks foundation, the McKnight Neuroscience of Brain Disorders Award, and the National Institute of Mental Health for financial support.

References

Akshoomoff N, Lord C, Lincoln AJ, Courchesne RY, Carper RA, Townsend J, Courchesne E (2004) Outcome classification of preschool children with autism spectrum disorders using MRI brain measures. J Am Acad Child Adolesc Psychiatry 43:349–357.

Arion D, Unger T, Lewis DA, Levitt P, Mirnics K (2007) Molecular evidence for increased expression of genes related to immune and chaperone function in the prefrontal cortex in schizophrenia. Biol Psychiatry 62:711–721.

Ashdown H, Dumont Y, Ng M, Poole S, Boksa P, Luheshi GN (2006) The role of cytokines in mediating effects of prenatal infection on the fetus: implications for schizophrenia. Mol Psychiatry 11:47–55.

Babulas V, Factor-Litvak P, Goetz R, Schaefer CA, Brown AS (2006) Prenatal exposure to maternal genital and reproductive infections and adult schizophrenia. Am J Psychiatry 163:927–929.

Bagalkote H, Pang D, Jones P (2001) Maternal influenza and schizophrenia in the offspring. Intl J Ment Health 39:3–21.

Bakos J, Duncko R, Makatsori A, Pirnik Z, Kiss A, Jezova D (2004) Prenatal immune challenge affects growth, behavior, and brain dopamine in offspring. Ann N Y Acad Sci 1018:281–287.

Balschun D, Wetzel W, Del Rey A, Pitossi F, Schneider H, Zuschratter W, Besedovsky HO (2004) Interleukin-6: a cytokine to forget. Faseb J 18:1788–1790.

Barbazanges A, Piazza PV, Le Moal M, Maccari S (1996) Maternal glucocorticoid secretion mediates long-term effects of prenatal stress. J Neurosci 16:3943–3949.

Bauer S, Kerr BJ, Patterson PH (2007) The neuropoietic cytokine family in development, plasticity, disease and injury. Nat Rev Neurosci 8:221–232.

Bell MJ, Hallenbeck JM, Gallo V (2004) Determining the fetal inflammatory response in an experimental model of intrauterine inflammation in rats. Pediatr Res 56:541–546.

Beloosesky R, Gayle DA, Amidi F, Nunez SE, Babu J, Desai M, Ross MG (2006) N-acetyl-cysteine suppresses amniotic fluid and placenta inflammatory cytokine responses to lipopolysaccharide in rats. Am J Obstet Gynecol 194:268–273.

Bilbo SD, Biedenkapp JC, Der-Avakian A, Watkins LR, Rudy JW, Maier SF (2005a) Neonatal infection-induced memory impairment after lipopolysaccharide in adulthood is prevented via caspase-1 inhibition. J Neurosci 25:8000–8009.

Bilbo SD, Levkoff LH, Mahoney JH, Watkins LR, Rudy JW, Maier SF (2005b) Neonatal infection induces memory impairments following an immune challenge in adulthood. Behav Neurosci 119:293–301.

Bobetsis YA, Barros SP, Offenbacher S (2006) Exploring the relationship between periodontal disease and pregnancy complications. J Am Dental Assoc 137:7s–13s.

Boisse L, Mouihate A, Ellis S, Pittman QJ (2004) Long-term alterations in neuroimmune responses after neonatal exposure to lipopolysaccharide. J Neurosci 24:4928–4934.

Boris M, Kaiser CC, Goldblatt A, Elice MW, Edelson SM, Adams JB, Feinstein DL (2007) Effect of pioglitazone treatment on behavioral symptoms in autistic children. J Neuroinflammation 4:3.

Borrell J, Vela JM, Arevalo-Martin A, Molina-Holgado E, Guaza C (2002) Prenatal immune challenge disrupts sensorimotor gating in adult rats. Implications for the etiopathogenesis of schizophrenia. Neuropsychopharmacology 26:204–215.

Brown AS, Begg MD, Gravenstein S, Schaefer CA, Wyatt RJ, Bresnahan M, Babulas VP, Susser ES (2004a) Serologic evidence of prenatal influenza in the etiology of schizophrenia. Arch Gen Psychiatry 61:774–780.

Brown AS, Hooton J, Schaefer CA, Zhang H, Petkova E, Babulas V, Perrin M, Gorman JM, Susser ES (2004b) Elevated maternal interleukin-8 levels and risk of schizophrenia in adult offspring. Am J Psychiatry 161:889–895.

Brown AS, Schaefer CA, Quesenberry CP, Jr., Liu L, Babulas VP, Susser ES (2005) Maternal exposure to toxoplasmosis and risk of schizophrenia in adult offspring. Am J Psychiatry 162:767–773.

Cai Z, Pan ZL, Pang Y, Evans OB, Rhodes PG (2000) Cytokine induction in fetal rat brains and brain injury in neonatal rats after maternal lipopolysaccharide administration. Pediatr Res 47:64–72.

Capuron L, Dantzer R (2003) Cytokines and depression: the need for a new paradigm. Brain Behav Immun 17 Suppl 1:S119–124.

Chess S (1977) Follow-up report on autism in congenital rubella. J Autism Child Schizophr 7:69–81.

Chez MG, Dowling T, Patel PB, Khanna P, Kominsky M (2007) Elevation of tumor necrosis factor-alpha in cerebrospinal fluid of autistic children. Pediatr Neurol 36:361–365.

Ciaranello AL, Ciaranello RD (1995) The neurobiology of infantile autism. Annu Rev Neurosci 18:101–128.

Clancy B, Darlington RB, Finlay BL (2001) Translating developmental time across mammalian species. Neuroscience 105:7–17.

Crawley JN (2007) Mouse behavioral assays relevant to the symptoms of autism. Brain Pathol 17:448–459.

Croonenberghs J, Bosmans E, Deboutte D, Kenis G, Maes M (2002) Activation of the inflammatory response system in autism. Neuropsychobiology 45:1–6.

Curran LK, Newschaffer CJ, Lee LC, Crawford SO, Johnston MV, Zimmerman AW (2007) Behaviors associated with fever in children with autism spectrum disorders. Pediatrics 120: e1386–1392.

Dahlgren J, Samuelsson AM, Jansson T, Holmang A (2006) Interleukin-6 in the maternal circulation reaches the rat fetus in mid-gestation. Pediatr Res 60:147–151.

Dammann O, Kuban KC, Leviton A (2002) Perinatal infection, fetal inflammatory response, white matter damage, and cognitive limitations in children born preterm. Ment Retard Dev Disabil Res Rev 8:46–50.

Davis JO, Phelps JA, Bracha HS (1995) Prenatal development of monozygotic twins and concordance for schizophrenia. Schizophr Bull 21:357–366.

Desmond MM, Wilson GS, Melnick JL, Singer DB, Zion TE, Rudolph AJ, Pineda RG, Ziai MH, Blattner RJ (1967) Congenital rubella encephalitis. Course and early sequelae. J Pediatr 71:311–331.

Elovitz MA, Mrinalini C, Sammel MD (2006) Elucidating the early signal transduction pathways leading to fetal brain injury in preterm birth. Pediatr Res 59:50–55.

Fatemi SH, Sidwell R, Kist D, Akhter P, Meltzer HY, Bailey K, Thuras P, Sedgwick J (1998) Differential expression of synaptosome-associated protein 25 kDa [SNAP-25] in hippocampi of neonatal mice following exposure to human influenza virus in utero. Brain Res 800:1–9.

Fatemi SH, Emamian ES, Kist D, Sidwell RW, Nakajima K, Akhter P, Shier A, Sheikh S, Bailey K (1999) Defective corticogenesis and reduction in Reelin immunoreactivity in cortex and hippocampus of prenatally infected neonatal mice. Mol Psychiatry 4:145–154.

Fatemi SH, Earle J, Kanodia R, Kist D, Emamian ES, Patterson PH, Shi L, Sidwell R (2002) Prenatal viral infection leads to pyramidal cell atrophy and macrocephaly in adulthood: implications for genesis of autism and schizophrenia. Cell Mol Neurobiol 22:25–33.

Fortier ME, Joober R, Luheshi GN, Boksa P (2004) Maternal exposure to bacterial endotoxin during pregnancy enhances amphetamine-induced locomotion and startle responses in adult rat offspring. J Psychiatr Res 38:335–345.

Francis D, Diorio J, Liu D, Meaney MJ (1999) Nongenomic transmission across generations of maternal behavior and stress responses in the rat. Science 286:1155–1158.

Garbett K, Ebert P, Lintas C, Mirnics K, Persico A (2007) Immune transcript disturbances in temporal cortex of autistic brains. 2007 Society for Neuroscience poster presentation.

Garver DL, Tamas RL, Holcomb JA (2003) Elevated interleukin-6 in the cerebrospinal fluid of a previously delineated schizophrenia subtype. Neuropsychopharmacology 28:1515–1520.

Gilmore JH, Jarskog LF, Vadlamudi S, Lauder J (2004) Prenatal infection and risk for schizophrenia: IL-I beta, IL-6, and TNF alpha inhibit cortical neuron dendrite development. Neuropsychopharmacology 29:1221–1229.

Gilmore JH, Jarskog LF, Vadlamudi S (2005) Maternal poly I:C exposure during pregnancy regulates TNF alpha, BDNF, and NGF expression in neonatal brain and the maternal-fetal unit of the rat. J Neuroimmunol 159:106–112.

Golan HM, Lev V, Hallak M, Sorokin Y, Huleihel M (2005) Specific neurodevelopmental damage in mice offspring following maternal inflammation during pregnancy. Neuropharmacology 48:903–917.

Golan H, Stilman M, Lev V, Huleihel M (2006) Normal aging of offspring mice of mothers with induced inflammation during pregnancy. Neuroscience 141:1909–1918.

Han YW, Redline RW, Li M, Yin L, Hill GB, McCormick TS (2004) Fusobacterium nucleatum induces premature and term stillbirths in pregnant mice: implication of oral bacteria in preterm birth. Infect Immun 72:2272–2279.

Hava G, Vered L, Yael M, Mordechai H, Mahoud H (2006) Alterations in behavior in adult offspring mice following maternal inflammation during pregnancy. Dev Psychobiol 48:162–168.

Hodyl NA, Krivanek KM, Lawrence E, Clifton VL, Hodgson DM (2007) Prenatal exposure to a pro-inflammatory stimulus causes delays in the development of the innate immune response to LPS in the offspring. J Neuroimmunol 190:61–71.

Hyman SL, Arndt TL, Rodier PM (2005) Environmental Agents and Autism: Once and Future Associations. Int Rev Res Ment Ret 30:171–194.

Jankowsky JL, Patterson PH (1999) Cytokine and growth factor involvement in long-term potentiation. Mol Cell Neurosci 14:273–286.

Lasala N, Zhou H (2007) Effects of maternal exposure to LPS on the inflammatory response in the offspring. J Neuroimmunol 189:95–101.

Lemery-Chalfant K, Goldsmith HH, Schmidt NL, Arneson CL, Van Hulle CA (2006) Wisconsin Twin Panel: current directions and findings. Twin Res Hum Genet 9:1030–1037.

Levitt NS, Lindsay RS, Holmes MC, Seckl JR (1996) Dexamethasone in the last week of pregnancy attenuates hippocampal glucocorticoid receptor gene expression and elevates blood pressure in the adult offspring in the rat. Neuroendocrinology 64:412–418.

Libbey JE, Sweeten TL, McMahon WM, Fujinami RS (2005) Autistic disorder and viral infections. J Neurovirol 11:1–10.

Lin D, Smith MA, Champagne C, Elter J, Beck J, Offenbacher S (2003a) Porphyromonas gingivalis infection during pregnancy increases maternal tumor necrosis factor alpha, suppresses maternal interleukin-10, and enhances fetal growth restriction and resorption in mice. Infect Immun 71:5156–5162.

Lin D, Smith MA, Elter J, Champagne C, Downey CL, Beck J, Offenbacher S (2003b) Porphyromonas gingivalis infection in pregnant mice is associated with placental dissemination, an increase in the placental Th1/Th2 cytokine ratio, and fetal growth restriction. Infect Immun 71:5163–5168.

Ling Z, Chang QA, Tong CW, Leurgans SE, Lipton JW, Carvey PM (2004) Rotenone potentiates dopamine neuron loss in animals exposed to lipopolysaccharide prenatally. Exp Neurol 190:373–383.

Liverman CS, Kaftan HA, Cui L, Hersperger SG, Taboada E, Klein RM, Berman NE (2006) Altered expression of pro-inflammatory and developmental genes in the fetal brain in a mouse model of maternal infection. Neurosci Lett 399:220–225.

Mankoski RE, Collins M, Ndosi NK, Mgalla EH, Sarwatt VV, Folstein SE (2006) Etiologies of autism in a case-series from Tanzania. J Autism Dev Disord 36:1039–1051.

Mednick SA, Machon RA, Huttunen MO, Bonett D (1988) Adult schizophrenia following prenatal exposure to an influenza epidemic. Arch Gen Psychiatry 45:189–192.

Meyer U, Feldon J, Schedlowski M, Yee BK (2006a) Immunological stress at the maternal-foetal interface: a link between neurodevelopment and adult psychopathology. Brain Behav Immun 20:378–388.

Meyer U, Nyffeler M, Engler A, Urwyler A, Schedlowski M, Knuesel I, Yee BK, Feldon J (2006b) The time of prenatal immune challenge determines the specificity of inflammation-mediated brain and behavioral pathology. J Neurosci 26:4752–4762.

Meyer U, Murray PJ, Urwyler A, Yee BK, Schedlowski M, Feldon J (2008) Adult behavioral and pharmacological dysfunctions following disruption of the fetal brain balance between pro-inflammatory and IL-10-mediated anti-inflammatory signaling. Mol Psychiatr 13(2): 208–221.

Morgan JT, Chana G, Buckwalter J, Courchesne E, Everall IP (2007) Increased Iba-1 positive microglial cell denisty in the autistic brain. 2007 Society for Neuroscience poster presentation.

Nawa H, Takei N (2006) Recent progress in animal modeling of immune inflammatory processes in schizophrenia: implication of specific cytokines. Neurosci Res 56:2–13.

Nyffeler M, Meyer U, Yee BK, Feldon J, Knuesel I (2006) Maternal immune activation during pregnancy increases limbic GABAA receptor immunoreactivity in the adult offspring: implications for schizophrenia. Neuroscience 143:51–62.

O'Callaghan E, Sham PC, Takei N, Murray G, Glover G, Hare EH, Murray RM (1994) The relationship of schizophrenic births to 16 infectious diseases. Br J Psychiatry 165:353–356.

Ozawa K, Hashimoto K, Kishimoto T, Shimizu E, Ishikura H, Iyo M (2006) Immune activation during pregnancy in mice leads to dopaminergic hyperfunction and cognitive impairment in the offspring: a neurodevelopmental animal model of schizophrenia. Biol Psychiatry 59:546–554.

Paintlia MK, Paintlia AS, Barbosa E, Singh I, Singh AK (2004) N-acetylcysteine prevents endotoxin-induced degeneration of oligodendrocyte progenitors and hypomyelination in developing rat brain. J Neurosci Res 78:347–361.

Palmen SJ, van Engeland H, Hof PR, Schmitz C (2004) Neuropathological findings in autism. Brain 127:2572–2583.

Patterson PH (2002) Maternal infection: window on neuroimmune interactions in fetal brain development and mental illness. Curr Opin Neurobiol 12:115–118.

Patterson PH (2005) Maternal influenza infection leads to neuropathology and behavioral abnormalities in adult offspring. Neuropsychopharmacology 30:S9–S9.

Paul R, Koedel U, Winkler F, Kieseier BC, Fontana A, Kopf M, Hartung HP, Pfister HW (2003) Lack of IL-6 augments inflammatory response but decreases vascular permeability in bacterial meningitis. Brain 126:1873–1882.

Perry W, Minassian A, Lopez B, Maron L, Lincoln A (2007) Sensorimotor Gating Deficits in Adults with Autism. Biol Psychiatry 61(4): 482–486.

Phelps JA, Davis JO, Schartz KM (1997) Nature, Nurture, and Twin Research Strategies. Curr Dir Psychol Sci 6:117–121.

Pierce K, Courchesne E (2001) Evidence for a cerebellar role in reduced exploration and stereotyped behavior in autism. Biol Psychiatry 49:655–664.

Piontkewitz Y, Weiner I, Assaf Y (2007) Post-pubertal emergence of schizophrenia-like abnormalaties following prenatal immune system activation and thier prevention: Modeling the disorder and its prodrome. In: 7th IBRO World Congress of Neuroscience, p 45. Melbourne, Australia.

Ponzio NM, Servatius R, Beck K, Marzouk A, Kreider T (2007) Cytokine levels during pregnancy influence immunological profiles and neurobehavioral patterns of the offspring. Ann N Y Acad Sci 1107:118–128.

Relier JP (2001) Influence of maternal stress on fetal behavior and brain development. Biol Neonate 79:168–171.

Robertson SA, Skinner RJ, Care AS (2006) Essential role for IL-10 in resistance to lipopolysaccharide-induced preterm labor in mice. J Immunol 177:4888–4896.

Romero E, Ali C, Molina-Holgado E, Castellano B, Guaza C, Borrell J (2007) Neurobehavioral and immunological consequences of prenatal immune activation in rats. Influence of antipsychotics. Neuropsychopharmacology 32:1791–1804.

Saetre P, Emilsson L, Axelsson E, Kreuger J, Lindholm E, Jazin E (2007) Inflammation-related genes up-regulated in schizophrenia brains. BMC Psychiatry 7:46.

Saliba E, Henrot A (2001) Inflammatory mediators and neonatal brain damage. Biol Neonate 79:224–227.

Samuelsson AM, Jennische E, Hansson HA, Holmang A (2006) Prenatal exposure to interleukin-6 results in inflammatory neurodegeneration in hippocampus with NMDA/GABA(A) dysregulation and impaired spatial learning. Am J Physiol Regul Integr Comp Physiol 290:R1345–1356.

Sargent IL, Borzychowski AM, Redman CW (2006) NK cells and human pregnancy – an inflammatory view. Trends Immunol 27:399–404.

Schiepers OJ, Wichers MC, Maes M (2005) Cytokines and major depression. Prog Neuropsychopharmacol Biol Psychiatry 29:201–217.

Shi L, Fatemi SH, Sidwell RW, Patterson PH (2003) Maternal influenza infection causes marked behavioral and pharmacological changes in the offspring. J Neurosci 23:297–302.

Shi L, Tu N, Patterson PH (2005) Maternal influenza infection is likely to alter fetal brain development indirectly: the virus is not detected in the fetus. Int J Dev Neurosci 23:299–305.

Shi L, Smith SE, Malkova N, Tse D, Su Y, Patterson PH (2008) Activation of the maternal immune system alters cerebellar development in the offspring Brain Behav Immun. Aug 9 [Epub ahead of print].

Singh VK, Warren RP, Odell JD, Cole P (1991) Changes of soluble interleukin-2, interleukin-2 receptor, T8 antigen, and interleukin-1 in the serum of autistic children. Clin Immunol Immunopathol 61:448–455.

Smith SE, Li J, Garbett K, Mirnics K, Patterson PH (2007) Maternal immune activation alters fetal brain development through interleukin-6. J Neurosci 27:10695–10702.

Theoharides TC, Weinkauf C, Conti P (2004) Brain cytokines and neuropsychiatric disorders. J Clin Psychopharmacol 24:577–581.

Tsuda N, Eda T, Mizuno M, Sotoyama H, Nawa H (2007) Minocycline improves cognitive and behavioral impairments resulted from neonatal exposure to interleukin-1. 2007 Society for Neuroscience poster presentation.

Turetsky BI, Calkins ME, Light GA, Olincy A, Radant AD, Swerdlow NR (2007) Neurophysiological Endophenotypes of Schizophrenia: The Viability of Selected Candidate Measures. Schizophr Bull 33:69–94.

Urakubo A, Jarskog LF, Lieberman JA, Gilmore JH (2001) Prenatal exposure to maternal infection alters cytokine expression in the placenta, amniotic fluid, and fetal brain. Schizophr Res 47:27–36.

Vargas DL, Nascimbene C, Krishnan C, Zimmerman AW, Pardo CA (2005) Neuroglial activation and neuroinflammation in the brain of patients with autism. Ann Neurol 57:67–81.

Wang X, Rousset CI, Hagberg H, Mallard C (2006) Lipopolysaccharide-induced inflammation and perinatal brain injury. Semin Fetal Neonatal Med 11:343–353.

Wang X, Hagberg H, Nie C, Zhu C, Ikeda T, Mallard C (2007) Dual role of intrauterine immune challenge on neonatal and adult brain vulnerability to hypoxia-ischemia. J Neuropathol Exp Neurol 66:552–561.

Watson CG, Kucala T, Tilleskjor C, Jacobs L (1984) Schizophrenic birth seasonality in relation to the incidence of infectious diseases and temperature extremes. Arch Gen Psychiatry 41:85–90.

Weiner I (2003) The "two-headed" latent inhibition model of schizophrenia: modeling positive and negative symptoms and their treatment. Psychopharmacology (Berl) 169:257–297.

Weinstock M (2001) Alterations induced by gestational stress in brain morphology and behaviour of the offspring. Prog Neurobiol 65:427–451.

Zhang XY, Zhou DF, Cao LY, Wu GY, Shen YC (2005) Cortisol and cytokines in chronic and treatment-resistant patients with schizophrenia: association with psychopathology and response to antipsychotics. Neuropsychopharmacology 30:1532–1538.

Zimmerman AW, Jyonouchi H, Comi AM, Connors SL, Milstien S, Varsou A, Heyes MP (2005) Cerebrospinal fluid and serum markers of inflammation in autism. Pediatr Neurol 33:195–201.

Zuckerman L, Weiner I (2005) Maternal immune activation leads to behavioral and pharmacological changes in the adult offspring. J Psychiatr Res 39:311–323.

Zuckerman L, Rehavi M, Nachman R, Weiner I (2003) Immune activation during pregnancy in rats leads to a postpubertal emergence of disrupted latent inhibition, dopaminergic hyperfunction, and altered limbic morphology in the offspring: a novel neurodevelopmental model of schizophrenia. Neuropsychopharmacology 28:1778–1789.

Interleukin-2 and Septohippocampal Neurons: Neurodevelopment and Autoimmunity

John M. Petitto, Zhi Huang, Grace K. Ha, and Daniel Dauer

Abstract As is the case for IL-2's biology in the peripheral immune system, the effects of IL-2 on brain development, function and disease also appear to be bidirectional and highly complex. Determining whether, and under what circumstances (e.g., development, acute injury) these different actions of IL-2 are operative in the brain is essential to make significant advances in understanding the multifaceted affects of IL-2 on CNS function and disease. Good animal models such as our IL-2 knockout mouse model could provide valuable new insight into how this cytokine may have simultaneous, dynamics effects on multiple systems (e.g., regulating homeostasis in the brain and immune system, autoimmunity that can affect both systems). This chapter presents some of our research and synthesizes the relevant literature in an attempt to understand better the complex actions IL-2 on septohippocampal neurons, and how this cytokine may influence brain neurodevelopment and autoimmunity. Such information may provide new insight into the role of brain cytokines and autoimmunity in prominent neurological and neuropsychiatric diseases (e.g., multiple sclerosis, Alzheimer's disease, schizophrenia).

Keywords Cytokines · IL-2 · Autoimmunity · Neurodevelopment · Neurodegeneration · Genetics

1 Introduction

Groundbreaking studies such as those showing that peripheral immunization could activate hypothalamic neurons and that lymphocytes had the capacity to produce neuropeptides (for reviews see Besedovsky et al., 1991 and Blalock, 1994) stimulated a wealth of research establishing that cytokines modulate brain cell function. Cytokines and their receptors, once believed to be derived solely from lymphoid cells, have been

---

J.M. Petitto (✉)
Departments of Psychiatry, Neuroscience, and Pharmacology & Therapeutics, McKnight Brain Institute, FL, USA
e-mail: jpetitto@ufl.edu

A. Siegel, S.S. Zalcman (eds.), *The Neuroimmunological Basis of Behavior and Mental Disorders*, DOI 10.1007/978-0-387-84851-8_8,

identified in normal brain where they function as neuromodulators, neurotrophic factors, and mediators of immune-like responses involved in central nervous system (CNS) pathology. Whereas some of the earliest ideas about the actions of cytokines in the brain posited that they had redundant functional properties on brain cell, mirroring their putative effects in the peripheral immune system, it was later found that brain cytokine had selective effects on neurons. A landmark study by Zalcman and colleagues advance that view by demonstrating that interleukin (IL)-1, IL-2, and IL-6 exhibited cytokine-specific changes in monoamine activity in hypothalamus, hippocampus, and prefrontal cortex (Zalcman et al., 1994). Such studies have laid the foundation for the growing body of research that has sought to dissect the potential mechanisms whereby different cytokines can influence complex neurobiological processes involved in understanding complex behaviors (e.g., learning and memory) and neuropsychiatric diseases affecting such domains of behavior. Basic research in neuroimmunology has given more attention to the examination of IL-1 and other proinflammatory cytokines including tumor necrosis factor (TNF)-α and IL-6. In this section a few highlights related to limbic function and learning and memory are presented pertaining to those cytokines, and in the subsequent sections we will focus on IL-2's actions in the hippocampus and cognition, and our laboratory's research that attempts to dissect the complex actions of peripheral and central IL-2 on brain development and autoimmunity.

Neuroimmunological research has laid a foundation to begin to understand how cytokines can influence complex neurobiological processes involved in understanding learning and memory, and neuropsychiatric diseases involving this and other domains of behavior. As described elsewhere in this text, cytokines derived from peripheral immune cells (and in some cases perhaps other tissues), do not readily cross the blood–brain barrier (BBB). The mechanisms of cytokine transport (e.g., active vs. passive transport) and the degree to which they enter the CNS differ for different cytokines (Pan and Kastin, 1999). Goehler et al. (2006) have described how the area of postrema acts as an anatomical "interface" between the peripheral immune system and the brain. Though less well studied, afferent sensory fibers of the vagus can carry signals initiated by IL-1 to brain stem areas (e.g., nucleus tractus solitarius), and vagal sensory activation may occur during infection and provide input to the brain that modifies behavior (Maier and Watkins, 1998; Goehler et al., 2007). It remains to be determined the degree to which other cytokines can signal the CNS via the vagus or other afferent nerves in the periphery, and how such events might lead to changes in behaviors (e.g., learning and memory, emotional behaviors) that influence the well being of the organism.

A number of cytokines and cytokine receptors are synthesized by endogenous brain cells. In the brain, some classical immune cytokines (e.g., IL-1) have been found to have neuromodulatory and neurotrophic effects that are limited to very specific neural pathways, while at the same time exerting more general effects on the brain's endogenous immune cells (e.g., microglia, astrocytes). Cytokine receptors are generally more readily detectable in the brain than cytokines themselves. Although gene expression studies have been successful in detecting cytokine receptors, and under some conditions cytokines as well, it has been more challenging to unequivocally detect cytokines and cytokine receptors using immunohistocytochemistry (or to reliably detect cytokine receptors using radioligand receptor binding or autoradiography; Petitto and Huang, 2001). Receptors for IL-1R (Cunningham et al., 1992), IL-1R antagonist (Licinio et al, 1991), and IL-6R (Schobitz et al., 1992), for example, are also detectable in the rodent dentate gyrus (DG) by in situ hybridization. As noted later in this chapter, IL-2 receptor genes are expressed throughout CA1–CA4 of the hippocampus and DG (Lapchak et al., 1991; Petitto and Huang, 1994; Petitto et al., 1998; Petitto and Huang, 2001). Thus, receptors for these cytokines in the hippocampus place them in a position to influence learning and memory and other related behaviors, and some of these same cytokines that have been found to target receptors in the hippocampus have the capacity to modulate neurobiological processes known to mediate these behaviors.

Animal studies provide compelling evidence that IL-1 can modify learning and memory performance. Using contextual fear condition as a test of hippocampal-dependent learning, for example, Pugh et al. (1998) showed that intracerebroventricular (ICV) administration IL-1β impaired contextual fear conditioning but did not change auditory-cue fear conditioning (a form of conditioning that is not dependent on the hippocampus). Systemic administration of LPS can disrupt learning and memory performance as well, and this disruption was antagonized by antibody to the IL-1β (Gibertini, et al., 1995). Along these lines, pyrogens which have reported to induce sickness behavior (e.g., fever, decreased activity and exploration, reduced social interaction, depressive-like symptoms), sometimes elicit reduced cognitive performance (Aubert, et al., 1995; Dantzer et al., 1998). IL-1 mRNA and protein have been found to be increased in the hippocampus following some forms of peripheral immune system activation such as occurs following LPS administration (Laye et al., 1994). Hippocampal long-term potentiation (LTP) is believed to be an important neurobiological mechanism involved in learning and memory storage at the cellular and molecular level. There is substantial literature that has established that cytokines can modulate LTP (for review see Jankowsky and Patterson, 1999).

2 IL-2 and the Septohippocampal System

One of the earliest observations suggesting that cytokines could influence cognitive function in humans came from cancer studies where IL-2 immunotherapy was used to treat some types of neoplasias (Denicoff et al., 1987, West et al., 1987). There is now considerable evidence indicating that IL-2 may be involved in CNS development, homeostatic repair mechanisms in response to brain injury, and neurodegenerative processes. IL-2 mRNA transcripts and IL-2-like immunoreactivity have been identified in rodent and human brain (Lapchak, 1992; Eizenberg et al., 1995). Some evidence suggests that neurons, microglia, and astrocytes can produce IL-2 (Eizenberg et al., 1995; Hanisch and Quirion, 1996; Hanisch et al., 1997; Labuzek et al., 2005). IL-2 immunoreactivity has been mapped in discrete areas of perfused normal rat forebrain including the septohippocampal system and related limbic regions (Lapchak et al., 1991; Villemain et al., 1991), and is present in hippocampal tissue (Araujo and Lapchak, 1994).

IL-2 has been implicated in the pathogenesis of several CNS disorders, including those that exhibit neuropathological alterations of the septohippocampal system (Hanisch and Quirion, 1995). In humans receiving IL-2 treatment for cancer therapy, prolonged exposure to IL-2 was found to induce cognitive dysfunction and other untoward neuropsychiatric side effects. Many effects of IL-2 in the brain occur in the hippocampal formation where receptors for this cytokine are enriched (Araujo et al., 1989b; Hanisch and Quirion, 1995; Lapchak et al., 1991; Petitto and Huang, 1994; Petitto et al., 1998). IL-2 may, for example, modify cellular and molecular substrates of learning and memory such as LTP (Tancredi et al., 1990, Tancredi et al., 1992), and affect multiple parameters of cognitive behavioral performance in animals (Bianchi and Panerai, 1993; Mennicken and Quirion, 1997; Hanisch et al., 1997; Lacosta et al., 1999; Nemni et al., 1992). IL-2 can provide trophic support to primary cultured neurons from multiple regions of the rat brain, including the hippocampus and medial septum (Awatsuji et al., 1993; Sarder et al., 1993), and positively affects the morphology of neurite branching in hippocampal cultures (Sarder et al., 1993, 1996). Furthermore, IL-2 has been shown to be one of the most potent modulators of acetylcholine (ACh) release from rat hippocampal slices (Hanisch et al., 1993; Seto et al., 1997).

3 IL-2 and Cognition

As noted earlier, IL-2 immunotherapy has been found to impair cognition in humans. Doses of cytokine used to treat cancer are significantly above the physiological range, and repeated dosing appears to account for the untoward side effects (Capuron et al., 1998). In postmortem hippocampi of Alzheimer's disease patients, IL-2 levels were found to be elevated compared to controls (Araujo and Lapchak, 1994). When aging mice were dosed chronically with IL-2 it produced memory deficits and neuronal damage, which was selective to hippocampus (Nemni et al., 1992). It is unknown if older patients receiving IL-2 immunotherapy for cancer treatment are more vulnerable to IL-2 related neurotoxicity. Other animal studies have confirmed clinical observations that exogenous IL-2 administration can alter parameters of cognitive functioning. IL-2 has been found in several studies to enhance the effects of scopolamine-induced amnesia in a passive paradigm (Bianchi and Panerai, 1993, 1998). This effect was antagonized completely by acetylcarnitine, a drug that enhances cholinergic activity (the actions of IL-2 on cholinergic neurotransmission in the hippocampus are described in the section that follows). In hippocampal slices, IL-2 modulates Ach in a dose-dependent biphasic manner, potentiating release at very low (fM) concentrations and inhibiting release at higher (nM) concentrations (Seto et al., 1997). In contrast, cholinergic interneurons in the striatum do not respond to IL-2. Since the hippocampus is critical in spatial learning and memory consolidation, it has been postulated that IL-2 may alter memory processing via interactions with septohippocampal cholinergic nerve terminals in hippocampus (Hanisch and Quirion, 1996).

In rats chronic ICV infusion of human IL-2 produced transient changes in behavior in the Morris water maze test, which was accompanied by changes in hippocampal muscarinic cholinergic receptors (Hanisch et al, 1997). In a well-designed set of experiments, Lacosta et al. (1999) assessed the effect of exogenous IL-2 on learning and memory performance in mice in the Morris water maze. Using standard submerged platform testing procedures where animals must locate a submerged platform that is always in the same position in the pool, they found that neither acute nor chronic IL-2 had an effect. They subsequently used a modification of the standard submerged platform testing regimen where they changed the location of the submerged platform on each testing day (the platform remained in the same location however throughout each testing day). Using this modification, they found that the between-trial improvement seen in saline treated mice was impaired in the IL-2 treated mice. This impairment was found in animals that received chronic dosing of IL-2, but not in animals receiving the acute dosing regimen. Thus, these data indicate that chronic IL-2 treatment interfered with the working component of spatial memory.

4 IL-2 Knockout Mice: IL-2s Intrinsic Actions, Neurotrophins, and Hippocampal Neurons

Despite numerous studies documenting various actions of IL-2 in the brain including trophic actions on cultured neurons and release of several major neuromodulatory transmitters, virtually all of these studies have used the strategy of administering exogenous IL-2. Most of the available data comes from in vitro studies, and to lesser degree, from studies in animals where IL-2's effects on various target behaviors or functional neurobiological outcomes (e.g., LTP in vivo) are used to make inferences about the action of the endogenous cytokine. Thus, one of the goals of our research has been to study IL-2 knockout mice to better understand the role of endogenous IL-2 on brain function.

In our initial studies in IL-2 knockout mice, we found that gene deletion impaired learning and memory performance, sensorimotor gating, and reductions in hippocampal infrapyramidal (IP) mossy fiber length (Petitto et al., 1999). Mossy fiber length has been shown to correlate positively with spatial learning ability in a number of studies (Schopke et al., 1991; Schwegler and Crusio, 1995; Schwegler et al., 1988). We have shown that IL-2 knockout mice also have fewer IP granule cells (Beck et al., 2005a). Given the various neurotrophic and neuromodulatory effects on hippocampal neurons in vitro noted above, our data indicates that hippocampal IL-2 may provide trophic support for hippocampal neurons. Absence or impairment of brain IL-2 function may play a key role, for example, in the ongoing increase in dentate granule cells during the first year of life (Altman and Bayer, 1990; Cameron and McKay, 2001) and effect the integrity of axons in the DG (Schwegler et al., 1991).

Our experiments have also shown that, compared to wild-type mice, IL-2 knockout mice had significantly reduced concentrations of brain-derived neurotrophic factor (BDNF) protein and increased concentrations of nerve growth factor (NGF) in the hippocampus. Though the receptors for IL-2 are enriched in the hippocampus, including the granule cell layer (GCL) of the DG (Petitto and Huang, 1994; Petitto et al., 1998), it is not clear whether IL-2s trophic actions observed on neurons in vitro (Awatsuji et al., 1993; Sarder et al., 1996, 1993) operate in vivo, or whether IL-2 regulates the expression of neurotrophic factors. The observed differences in the level of BDNF were consistent with our hypothesis that we would find reductions in trophic factors important to hippocampal development and maintenance. BDNF plays a role in the maintenance and repair of septal cholinergic neurons (Alderson et al., 1990; Morse et al., 1993; Ward and Hagg, 2000), can implement a positive feedback mechanism with these neurons to enhance the release of Ach (Knipper et al., 1994), and can also modulate postnatal neurogenesis (Larsson et al., 2002; Lee et al., 2002), thus potentially impacting granule cell number. The mechanism of the interaction between IL-2 gene deletion and the reduction of BDNF levels remains unclear. Though BDNF is expressed in the peripheral immune system by lymphocytes, IL-2 does not stimulate its production or release in these cells. IL-2 can, however, upregulate the expression of TrkB, the receptor for BDNF, in lymphocytes (Besser and Wank, 1999). Furthermore, some evidence suggests that BDNF can stimulate a positive feedback mechanism of its own via the TrkB receptor in hippocampal neurons (Canossa et al., 1997; Saarelainen et al., 2001). In IL-2 knockout mice, one yet untested possibility is that the absence of IL-2 may therefore potentially lead to a downregulation of the TrkB receptor, thereby partially inhibiting the positive feedback production of BDNF. By contrast, NGF protein levels were actually increased in the IL-2 knockout mice. Interestingly, such an imbalance between BDNF and NGF levels (decreased BDNF and increased NGF concentrations) has been observed in the hippocampus of Alzheimer's disease brains (Hock et al., 2000). Given the reduction in cholinergic survival in the MS/vDB of IL-2 knockout mice that we have observed (Beck et al., 2002, 2005a), it is possible that the hippocampal target neurons in these animals may produce higher protein levels of NGF as a compensatory response. Similarly, moderate lesions of rat septohippocampal projections led to increased expression of mRNA for NGF, but not BDNF in hippocampal target cells (Hellweg et al., 1997). Together, these data in IL-2 knockout mice suggest that IL-2 has direct and/or indirect effects on BDNF.

5 IL-2 Knockout Mice: Peripheral Autoimmunity and Loss of Cholinergic Neurons

We have been testing the hypothesis that there is a loss of neuronal survival and maintenance occuring primarily due to a unique form of autoimmunity that targets medial septal cholinergic neurons. Our data show that neither striatal cholinergic neurons nor medial septal GABAergic neurons are affected in IL-2 knockout mice, indicating that the autoimmunity is selective for medial septal cholinergic neurons in the brain (Beck et al., 2002, 2005). IL-2 immunoregulatory cytokine is indispensable for maintaining immunological homeostasis (e.g., self-tolerance, T regulatory cell development and function). Schorle et al. (1991) performed the targeted deletion of the IL-2 gene (Schorle et al., 1991). They found that these mice had marked alterations in peripheral immune homeostasis and autoimmune disease that depends in part on genetic background (e.g., autoimmune bowel disease on B6 x 129, hemolytic anemia on Balb/c). This and related research has led to new understanding of IL-2 biology. It is now appreciated that IL-2 is critical for T regulatory cell development and self-tolerance. Loss of peripheral IL-2 impairs negative regulatory processes, the immune system becomes dysregulated, and mice eventually develop peripheral autoimmune disease (Horak, 1995; Horak et al., 1995; Kundig et al., 1993), the rate and degree to which this occurs depends on the genetic background of the animal. Balb/c-IL-2–/– mice develop autoimmune disease rapidly, whereas C57BL/6-IL-2–/– mice are healthy and develop clinical signs of autoimmune disease at a much slower rate; C3H and 129 x C57BL/6 IL-2 knockout mice are intermediate (Grassl et al., 1997; Ma et al., 1995; Sadlack et al., 1993, 1995). It is possible that subclinical autoimmune processes (e.g., certain cytokines crossing the BBB) as well as frank clinical autoimmune disease (e.g., antibody or T cell mediated destruction of neurons) could lead to alterations in the septohippocampal system of IL-2 knockout mice.

Autoimmune-mediated damage of medial septal cholinergic neurons has been described previously in experimental animals by Kalman et al. (1997). They found evidence for immune-mediated (e.g., Immunoglobulin (Ig)G autoantibodies) neurodegeneration that was selective for medial septal and diagonal band of Broca (MS/vDB) ChAT-positive neurons. We have examined the number of MS/vDB ChAT-positive neurons of IL-2 knockout and wild-type littermates at 3 weeks of age and 8–12 weeks of age (Beck et al., 2002, 2005). At 3 weeks of age, IL-2 knockout mice on the C57BL/6 background have not yet developed autoimmunity (e.g., spleen size is the same as wild-type littermates, no bowel involvement), whereas by 8–12 weeks of age IL-2 knockout mice show considerable evidence of peripheral autoimmunity (e.g., marked splenomegaly). We found that compared to wild-type littermates, adult IL-2 deficient mice have a marked reduction of MS/vDB ChAT-positive cell bodies. Generation of the medial septal cholinergic neurons is essentially complete by e17 (Semba and Fibiger, 1988), thus the reduction of these neurons in IL-2 knockout mice is due to decreased survival (or an unlikely loss of phenotype). Moreover, adult wild-type and knockout mice do not differ in ChAT-positive neurons in the striatum or in GABAergic neurons in the MS/vDB, indicating that the effect is selective for MS/vDB cholinergic neurons. We are currently conducting studies to examine the hypothesis that IL-2 knockout mice will exhibit a decline in medial septal neurons that will coincide (vary inversely) with indices of peripheral autoimmunity (e.g., proliferation of activated T cells in the spleen and splenomegaly).

Since IL-2 is an important factor in immune physiology, it is possible that immune dysregulation caused by the absence of IL-2 is a key mechanism that may contribute to neurodegenerative processes in IL-2 knockout mice. IL-2-deficiency leads to generalized systemic autoimmune disease in adult mice that affects several organs in the periphery including the colon and kidney (Horak, 1995). These peripheral autoimmune effects in IL-2 knockout mice are mediated largely by infiltrating T cells. In the colon, for example, adult IL-2 knockout mice develop chronic inflammatory bowel disease with features common to inflammatory ulcerative colitis in humans, where the lamina propria is infiltrated with activated T cells responsible for the development of this inflammatory disease (Ma et al., 1995). In addition, there is a disruption of immune homeostasis that is evidenced by changes in the gene expression of several Th1, Th2, and various proinflammatory cytokines in this organ (Autenrieth et al., 1997; Meijssen et al., 1998). Moreover, these cytokine changes and the onset of inflammatory bowel disease are preceded by increased gene expression of IL-15, which shares the same signal transducing receptor subunits with IL-2 (Meijssen et al., 1998). Thus, it is possible that immune dysregulation in the brain of IL-2 knockout mice may be induced by activated T cells and/or proinflammatory cytokines (e.g., IL-1, TNFα, IL-6) from the periphery crossing the BBB (Hickey et al., 1991). We postulate further that the loss of neurons in the medial septum will be preceded in time by autoantibody deposition (likely IgG) and/or infiltrating T cells, and will likely be accompanied by increased concentration of several proinflammatory cytokines and chemokines.

6 Significance for Understanding the Neuroimmunology of Neurological and Neuropsychiatric Diseases

As is the case for IL-2s biology in the peripheral immune system, the effects of IL-2 on brain development, function, and disease also appear to be bidirectional and highly complex. There has been considerable interest in evidence suggesting that early neurodevelopmental alterations may account for key pathophysiological abnormalities seen decades later in the mature brain of individuals with schizophrenia. Epidemiological findings have contributed, for example, to theory that prenatal viral infection and/or the maternal immune response to such a putative agent may serve as an environmental trigger for the expression of schizophrenia in individuals with genetic loading for the disorder. During early development an event like viral infection or birth trauma in a genetically susceptible individual could disrupt the normal timing at which IL-2 may stimulate neuronal growth and migration. If this occurred in a region like the hippocampus where IL-2 receptors are enriched, this could contribute to the alterations in this region (e.g., abnormal orientation of subsets of hippocampal neurons) that are seen in postmortem brains of individuals with schizophrenia. Immunological disturbances in the peripheral immune system during development may also lead to abnormalities in neurodevelopment. Determining when and under what conditions (e.g., development, injury) these different actions of IL-2 are operative in the brain is essential to make significant advances in understanding of the multifaceted affects of IL-2 on CNS function and disease. For several decades there has been a great deal of speculation about the role of autoimmunity in brain disease. More recently, in the field of neuroimmunology we have learned a great deal about the role of cytokines on neurobiological processes, and there have been many studies that have found peripheral immune alterations in patients with neurological and neuropsychiatric diseases. Despite a plethora of published studies, almost all of this data in humans is correlative and much of the basic research has understandably relied on simpler models (e.g., in vitro models). Thus, informative animal models such as our IL-2 knockout mouse model could provide valuable new insight in understanding how the complex biology of a cytokine such as IL-2 can have simultaneous, dynamic effects on multiple systems (e.g., regulating homeostasis in the brain and immune system, autoimmunity that can affect both systems). We are currently breeding novel congenic mice with and without the IL-2 gene and/or the Rag2 gene (leading to immunodeficiency) and combining these with various leukocyte adoptive transfer manipulations to dissect the relative contributions of IL-2 in the brain versus the peripheral immune system on brain development and neurodegeneration. These experiments should provide essential new information into brain IL-2s intrinsic actions in the septohippocampal system in vivo. Moreover, this unique mouse model will also provide much needed new data elucidating neuroimmunological and autoimmune processes involved in brain development and disease. Such information may ultimately provide critical new insight into the role of brain cytokines and autoimmunity in prominent neurological and neuropsychiatric diseases (e.g., multiple sclerosis, Alzheimer's disease, schizophrenia).

References

Alderson, R.F., Alterman, A.L., Barde, Y.A., Lindsay, R.M., 1990. Brain-derived neurotrophic factor increases survival and differentiated functions of rat septal cholinergic neurons in culture. Neuron 5, 297–306.

Altman, J., Bayer, S.A., 1990. Migration and distribution of two populations of hippocampal granule cell precursors during the perinatal and postnatal periods. J Comp Neurol 301, 365–381.

Araujo, D.M., Lapchak, P.A., 1994. Induction of immune system mediators in the hippocampal formation in Alzheimer's and Parkinson's diseases: selective effects on specific interleukins and interleukin receptors. Neuroscience 61, 745–754.

Araujo, D.M., Lapchak, P.A., Collier, B., Quirion, R., 1989b. Localization of interleukin-2 immunoreactivity and interleukin-2 receptors in the rat brain: interaction with the cholinergic system. Brain Res 489, 257–266.

Aubert A., Vega C., Dantzer R., Goodall G., 1995. Pyrogens specifically disrupt the acquisition of a task involving cognitive processing in the rat. Brain Behav Immun, Jun 9(2),129–148.

Autenrieth, I.B., Bucheler, N., Bohn, E., Heinze, G., Horak, I., 1997. Cytokine mRNA expression in intestinal tissue of interleukin-2 deficient mice with bowel inflammation. Gut 41, 793–800.

Awatsuji H., Furukawa Y., Nakajima M., Furukawa S., Hayashi K., 1993. Interleukin-2 as a neurotrophic factor for supporting the survival of neurons cultured from various regions of fetal rat brain. J Neurosci Res 35, 305–311.

Beck, R.D., Jr., King, M.A., Ha, G.K., Cushman, J.D., Huang, Z., Petitto, J.M., 2005a. IL-2 deficiency results in altered septal and hippocampal cytoarchitecture: relation to development and neurotrophins. J Neuroimmunol 160, 146–153.

Beck, R.D., Jr., King, M.A., Huang, Z., Petitto, J.M., 2002. Alterations in septohippocampal cholinergic neurons resulting from interleukin-2 gene knockout. Brain Res 955, 16–23.

Beck, R.D., Jr., Wasserfall, C., Ha, G.K., Cushman, J.D., Huang, Z., Atkinson, M.A., Petitto, J.M., 2005b. Changes in hippocampal IL-15, related cytokines, and neurogenesis in IL-2 deficient mice. Brain Res 1041, 223–230.
Besedovsky H.O., del Rey A., Klusman I., Furukawa H., Monge Arditi G., Kabiersch A., 1991. Cytokines as modulators of the hypothalamus-pituitary-adrenal axis. J Steroid Biochem Mol Biol 40, 613–618.
Besser M., Wank R. 1999. Cutting edge: clonally restricted production of the neurotrophins brain-derived neurotrophic factor and neurotrophin-3 mRNA by human immune cells and Th1/Th2-polarized expression of their receptors. J Immunol 162, 6303–6306.
Bianchi M., Panerai A.E., 1993. Interleukin-2 enhances scopolamine-induced amnesia and hyperactivity in the mouse. Neuroreport 4, 1046–1048.
Bianchi, M., Sacerdote, P., Panerai A.E., 1998. Cytokines and Cognitive Function in Mice. Biol Signals Recept 7, 45–54.
Blalock J.E., 1994. The syntax of immune-neuroendocrine communications. Immunol Today 15, 504–511.
Cameron, H.A., McKay, R.D., 2001. Adult neurogenesis produces a large pool of new granule cells in the dentate gyrus. J Comp Neurol 435, 406–417.
Canossa, M., Griesbeck, O., Berninger, B., Campana, G., Kolbeck, R., Thoenen, H., 1997. Neurotrophin release by neurotrophins: implications for activity-dependent neuronal plasticity. Proc Natl Acad Sci USA 94, 13279–13286.
Capuron, L., Ravaud, A., Radat, F., Dantzer, R., Goodall, G., 1998. Effects of interleukin-2 and alpha interferon cytokine immunotherapy on the mood and cognitive performance of cancer patients. Neuroimmunomodulation 22, 9.
Cunningham E.T.J., Wada E., Carter D.B., Tracey D.E., Battey J.F., 1992. De Souza E.B. In situ histochemical localization of type I interleukin-1 receptor messenger RNA in the central nervous system, pituitary, and adrenal gland of the mouse. J Neurosci 12, 1101–1114.
Dantzer R., Bluthe R.M., Gheusi G., Cremona S., Laye, S., Parnet P., Kelley K.W., 1998. Molecular basis of sickness behavior. Ann NY Acad Sci 856, 132–138.
Denicoff, K.D., Rubinow, D.R., Papa, M.Z., Simpson, C., Seipp, C.A., Lotze, M.T., Chang, A.E., Rosenstein, D., Rosenberg, S.A., 1987. The neuropsychiatric effects of treatment with interleukin-2 and lymphokine-activated killer cells. Ann Intern Med 107, 293–300.
Elizenberg, O., Faber-Elman, A., Loton, M., Cchwartz, M., 1995. Inteleukin-2 transcripts in human and rodent brains: possible expression by astrocytes. J Neurochem 64(5), 1928–36.
Gibertini M., Newton C., Klein T.W., Friedman H., 1995. Legionella pneumophila-induced visual learning impairment reversed by anti-interleukin-1 beta. Proceedings of the Society for Experimental Biology and Medicine. Soc Exp Biol Med, Oct 210(1), 7–11.
Goehler, L.E., Erisir A., Gaykema R.P. 2006. Neural-immune interface in the rat area postrema. Neuroscience 140, 1415–1434.
Goehler, L.E., Lyte M., Gaykema R.P. 2007. Infection-induced viscerosensory signals from the gut enhance anxiety: implications for psychoneuroimmunology. Brain Behav Immun 21, 721–726.
Grassl, G., Pummerer, C.L., Horak, I., Neu, N., 1997. Induction of autoimmune myocarditis in interleukin-2-deficient mice. Circulation 95, 1773–1776.
Hanisch U., Neuhaus J., Rowe W., Van-Rossum D., Möller T., Kettenmann H., Quirion R., 1997. Neurotoxic consequences of central long-term administration of interleukin-2 in rats. Neuroscience 79, 799–818.
Hanisch U.K., Quirion R., 1995. Interleukin-2 as a neuroregulatory cytokine. Brain Res Rev 21, 246–284.
Hanisch U.K., Seto D., Quirion R., 1993. Modulation of hippocampal acetylcholine release: a potent central action of interleukin-2. J Neurosci 13(8), 3368–3374.
Hellweg, R., Humpel, C., Lowe, A., Hortnagl, H., 1997. Moderate lesion of the rat cholinergic septohippocampal pathway increases hippocampal nerve growth factor synthesis: evidence for long-term compensatory changes? Brain Res Mol Brain Res 45, 177–181.
Hock, C., Heese, K., Hulette, C., Rosenberg, C., Otten, U., 2000. Region-specific neurotrophin imbalances in Alzheimer disease: decreased levels of brain-derived neurotrophic factor and

increased levels of nerve growth factor in hippocampus and cortical areas. Arch Neurol 57, 846–851.
Horak, I., 1995. Immunodeficiency in IL-2-knockout mice. Clin Immunol Immunopathol 76, S172–173.
Horak, I., Lohler, J., Ma, A., Smith, K.A., 1995. Interleukin-2 deficient mice: a new model to study autoimmunity and self-tolerance. Immunol Rev 148, 35–44.
Jankowsky J., Patterson P., 1999. Cytokine and growth factor involvement in long-term potentiation. Mol Cell Neurosci 14, 273–286.
Kalman, J., Engelhardt, J.I., Le, W.D., Xie, W., Kovacs, I., Kasa, P., Appel, S.H., 1997. Experimental immune-mediated damage of septal cholinergic neurons. J Neuroimmunol 77, 63–74.
Hickey, W.F., Hsu, B.L., Kimura, H., 1991. T-lymphocyte entry into the central nervous system. J Neurosci. Res. 1991; 28, 254–260.
Knipper, M., da Penha Berzaghi, M., Blochl, A., Breer, H., Thoenen, H., Lindholm, D., 1994. Positive feedback between acetylcholine and the neurotrophins nerve growth factor and brain-derived neurotrophic factor in the rat hippocampus. Eur J Neurosci 6, 668–671.
Kundig, T.M., Schorle, H., Bachmann, M.F., Hengartner, H., Zinkernagel, R.M., Horak, I., 1993. Immune responses in interleukin-2-deficient mice. Science 262, 1059–1061.
Lacosta, S., Merali, Z., Anisman, H., 1999. Influence of acute and repeated interleukin-2 administration on spatial learning, locomotor activity, exploratory behaviors, and anxiety. Behav Neurosci 113, 1030–1041.
Lapchak, P.A., 1992. A role for interleukin-2 in the regulation of striatal dopaminergic function. Neuroreport 3, 165–168.
Lapchak P.A., Araujo D.M., Quirion R., Beaudet A., 1991. Immunoautoradiographic localization of interleukin 2-like immunoreactivity and interleukin 2 receptors (Tac antigen-like immunoreactivity) in the rat brain. Neuroscience 44, 173–184.
Larsson, E., Mandel, R.J., Klein, R.L., Muzyczka, N., Lindvall, O., Kokaia, Z., 2002. Suppression of insult-induced neurogenesis in adult rat brain by brain-derived neurotrophic factor. Exp Neurol 177, 1–8.
Lee, J., Duan, W., Mattson, M.P., 2002. Evidence that brain-derived neurotrophic factor is required for basal neurogenesis and mediates, in part, the enhancement of neurogenesis by dietary restriction in the hippocampus of adult mice. J Neurochem 82, 1367–1375.
Laye S., Parnet P., Goujon E., Dantzer R., 1994. Peripheral administration of lipopolysaccharide induces the expression of cytokine transcripts in the brain and pituitary of mice. Brain research. Mol Brain Res, Nov 27(1), 157–162.
Licinio J., Wong M.L., Gold P.W., 1991. Localization of interleukin-1 receptor antagonist mRNA in rat brain. Endocrinology 129, 562–564.
Labuzek, K., Kowalski, J., Gabryel, B., Herman, Z.S., 2005. Chlorpromazine and loxapine reduce interleukin-1beta and interleukin-2 release by rat mixed glial and microglial cell cultures. Eur Neuropsychopharmacol 15, 23–30.
Ma, A., Datta, M., Margosian, E., Chen, J., Horak, I., 1995. T cells, but not B cells, are required for bowel inflammation in interleukin 2-deficient mice. J Exp Med 182, 1567–1572.
Maier S.F., Watkins L.R. Cytokines for psychologists: Implications of bidirectional immune-to-brain communication for understanding behavior, mood and cognition. Psychological Review 1998; 83–107.
Meijssen, M.A., Brandwein, S.L., Reinecker, H.C., Bhan, A.K., Podolsky, D.K., 1998. Alteration of gene expression by intestinal epithelial cells precedes colitis in interleukin-2-deficient mice. Am J Physiol 274, G472–479.
Mennicken F., Quirion R., 1997. Interleukin-2 increases choline acetyltransferase activity in septal-cell cultures. Synapse 26, 175–183.
Nemni R., Iannaccone S., Quattrini A., Smirne S., Sessa M., Lodi M., Erminio C., Canal N., 1992. Effect of chronic treatment with recombinant interleukin-2 on the central nervous system of adult and old mice. Brain Res 591, 248–252.
Morse, J.K., Wiegand, S.J., Anderson, K., You, Y., Cai, N., Carnahan, J., Miller, J., DiStefano, P.S., Altar, C.A., Lindsay, R.M., et al., 1993. Brain-derived neurotrophic factor (BDNF) prevents

the degeneration of medial septal cholinergic neurons following fimbria transection. J Neurosci 13, 4146–4156.
Pan W., Kastin A.J., 1999. Penetration of neurotrophins and cytokines across the blood-brain/blood-spinal cord barrier. Adv Drug Deliver Rev, Apr 5; 36(2–3), 291–298
Petitto J.M., Huang Z., 1994. Molecular cloning of a partial cDNA of the interleukin-2 receptor-β in normal mouse brain: in situ localization in the hippocampus and expression by neuroblastoma cells. Brain Res 650, 140–145.
Petitto J.M., Huang Z., 2001. Cloning the full-length IL-2/15 receptor-β cDNA sequence from mouse brain: evidence of enrichment in hippocampal formation neurons. Regul Peptides 98, 77–87.
Petitto J.M., Huang Z., Raizada M., Rinker C.M., McCarthy D.B., 1998. Molecular cloning of the cDNA coding sequence of IL-2 receptor-$\gamma(\gamma_c)$ from human and murine forebrain: expression in the hippocampus in situ and by brain cells in vitro. Mol Brain Res 53, 152–162.
Petitto J.M., McNamara R., Gendreau P.L., Huang Z., Jackson A., 1999. Impaired learning and memory and altered hippocampal neurodevelopment resulting from IL-2 gene deletion. J Neurosci Res 56, 441–446.
Pugh C.R., Kumagawa K., Fleshner M., Watkins L.R., Maier S.F., Rudy J.W., 1998,. Selective effects of peripheral lipopolysaccharide administration on contextual and auditory-cue fear conditioning. Brain Behav Immun, Sep 12(3), 212–229.
Saarelainen, T., Vaittinen, S., Castren, E., 2001. trkB-receptor activation contributes to the kainate-induced increase in BDNF mRNA synthesis. Cell Mol Neurobiol 21, 429–435.
Sadlack, B., Merz, H., Schorle, H., Schimpl, A., Fellar, A.C., Horak, I., 1993. Ulcerative colitis-like disease in mice with a disrupted interleukin-2 gene. Cell 72(2), 253–61.
Sadlack, B., Lohler, J., Schorle, H., Klebb, G., Haber, H., Sickel, E., Noelle, R.J., Horak, I., 1995. Generalized autoimmune disease in interleukin-2-deficient mice is triggered by an uncontrolled activation and proliferation of CD4+ T cells. Eur J Immunol 25(11), 3053–9.
Sarder M., Abe K., Saito H., Nishiyama N., 1996. Comparative effect of IL-2 and IL-6 on morphology of cultured hippocampal neurons from fetal rat brain. Brain Res 715, 9–16.
Sarder M., Saito H., Abe K., 1993. Interleukin-2 promotes survival and neurite extension of cultured neurons from fetal rat brain. Brain Res 625, 347–350.
Schopke, R., Wolfer, D.P., Lipp, H.P., Leisinger-Trigona, M.C., 1991. Swimming navigation and structural variations of the infrapyramidal mossy fibers in the hippocampus of the mosue. Hippocampus 3, 315–328.
Schorle, H., Holrscke, T., Hunig, T., Schimpl, A., Horak, I., 1991. Development and function of T cells in mice rendered interleukin-2 deficient by gene targeting. Nature 352(6336), 621–4.
Schwegler, H., Crusio, W.E., 1995. Correlations between radial-maze learning and structural variations of septum and hippocampus in rodents. Behav Brain Res 67, 29–41.
Schwegler, H., Crusio, W.E., Lipp, H.P., Brust, I., Mueller, G.G., 1991. Early postnatal hyperthyroidism alters hippocampal circuitry and improves radial-maze learning in adult mice. J Neurosci 11, 2102–2106.
Schwegler, H., Crusio, W.E., Lipp, H.P., Heimrich, B., 1988. Water-maze learning in the mouse correlates with variation in hippocampal morphology. Behav Genet 18, 153–165.
Schobitz, B., Voorhuis, D.A., De Kloet, E.R., 1992. Localization of interleukin-6 mRNA and interleukin-6 receptor mRNA in rat brain. Neurosci Lett 136, 189–192.
Semba, K., Fibiger, H.C., 1988. Time of origin of cholinergic neurons in the rat basal forebrain. J Comp Neurol 269, 87–95.
Seto D., Satyabrata K., Quirion R., 1997. Evidence for direct and indirect mechanisms in the potent modulatory action of interleukin-2 on the release of acetylcholine in rat hippocampal slices. Brit J Pharmcol 120, 1151–1157.
Tancredi V., D'Archangelo G., Grassi F., Tarroni P., Palmieri G., Santoni A., Eusebi F., 1992. Tumor necrosis factor alters synaptic transmission in rat hippocampal slices. Neurosci Lett 146, 176–178.
Tancredi V., Zona C., Velotti F., Eusebi F., Santoni, A., 1990. Interleukin-2 suppresses established long-term potentiation and inhibits its induction in the rat hippocampus. Brain Res 525, 149–151.

Villemain, F., Owens, T., Renno, T., Beaudet, A., 1991. Localization of mRNA for interleukin-2 (IL-2) in mouse brain by in situ hybridization. Soc Neurosci Abstr 17, 1199.

Ward, N.L., Hagg, T., 2000. BDNF is needed for postnatal maturation of basal forebrain and neostriatum cholinergic neurons in vivo. Exp Neurol 162, 297–310.

West, W.H., Tauer, K.W., Yannelli, J.R., Marshall, G.D., Orr, D.W., Thurman, G.B., Oldham, R.K., 1987. Constant-infusion recombinant interleukin-2 in adoptive immunotherapy of advanced cancer. New Engl J Med 316, 898–905.

Zalcman, S., Green-Johnson-J.M., Murray L., Nance D.M., Dyck D., Anisman H., Greenberg A.H., 1994. Cytokine-specific central monoamine alterations induced by interleukin-1, -2, and -6. Brain Res 643, 40–49.

Cytokine-Induced Sickness Behavior and Depression

Q. Chang, S.S. Szegedi, J.C. O'Connor, R. Dantzer, and K.W. Kelley

Abstract Sickness behavior refers to a set of non-specific responses that develop in humans and animals during the course of an infection. Sickness symptoms possess motivational features and evolutionary values that favor survival of organisms during infection. The discovery that proinflammatory cytokines induce sickness behaviors forms a cornerstone for elucidating immune-to-brain communication systems. Cytokines produced in the periphery by leukocytes during infection and chronic inflammatory diseases serve as messengers that are sent to the brain via neural and humoral routes to activate a diffuse cytokine system that mirrors that in the periphery. The brain initiates a series of events that induce behavioral changes collectively known as sickness behaviors. Sickness behavior is now recognized to be part of a highly-organized host response to infection. However, prolonged activation of the innate immune system, as occurs during chronic infectious diseases and non-infectious diseases with inflammatory components, can lead to symptoms of depression in vulnerable individuals. Recent clinical findings have implicated the tryptophan-degrading enzyme, indoleamine 2, 3 dioxygenase (IDO), as a potential mediator of inflammation-associated depression. Experimental data obtained in animal studies have provided molecular support for such a relationship. Here we discuss the current evidence that favors the view that acute inflammation induces sickness behavior whereas chronic inflammation can lead to depressive-like behaviors that are mediated by IDO.

Keywords Sickness behavior · Depression · Indoleamine 2, 3-dioxygenase · Inflammation · Cytokines · Kynurenine · Tryptophan

---

K.W. Kelley (✉)
Integrative Immunology and Behavior Program, Department of Animal Sciences, College of ACES, Department of Pathology, College of Medicine, University of Illinois at Urbana-Champaign, Urbana, IL 61801, USA
e-mail: kwkelley@illinois.edu

Supported by grants from NIH to KWK (R01 MH 51569 and R01 AG 029573) and RD (R01 MH 079829 and R01 MH 71349).

A. Siegel, S.S. Zalcman (eds.), *The Neuroimmunological Basis of Behavior and Mental Disorders*, DOI 10.1007/978-0-387-84851-8_9,

1 Introduction

In the mid-1930s, Hans Selye recognized that a wide variety of aversive stimuli cause similar physiological changes, such as adrenal hypertrophy, shrinkage of the thymus gland and an increase in the number of circulating neutrophils. He termed this phenomenon, "the syndrome of just being sick," and these symptoms became part of the characteristics that framed his theory of the General Adaptation Syndrome (Selye 1936). Selye recognized early on that this sickness syndrome is somehow related to inflammation. As a better understanding of inflammation has developed during the past 75 years, it is now recognized that acute and chronic inflammation affects the brain as well as the rest of the body. Symptoms of acute infection caused by a wide array of pathogens are known to all of us: malaise, fatigue, listlessness, reduced appetite, fever, sleepiness, and social withdrawal. These and other symptoms characterize nonspecific symptoms of infection and acute inflammation, and they change both animal and human behaviors. These symptoms are known as sickness behaviors (Hart 1988; Kent et al. 1992b). If Hans Selye were alive today, he would likely include sickness behavior as another essential component of his General Adaptation Syndrome. Since sickness behaviors are nonspecific responses to a wide diversity of pathogenic microbes as well as noxious stimuli, symptoms of sickness are not typically regarded as important by clinicians. Rather, they are considered as uncomfortable but unavoidable elements of pathogen-induced pathological processes. This chapter will highlight how the discovery of inflammation-induced sickness behavior has formed a solid foundation to better understand communication systems between the immune system and the brain that regulate behavior.

2 Cytokine-Induced Sickness Behavior

2.1 Sickness Behavior Is an Adaptive Response to Infection

Sickness behavior is an adaptive response that reflects an organized strategy used by organisms to cope with infection. Sick individuals are able to recognize the internal and external constraints to which they are exposed and make appropriate behavioral adjustments. In other words, sickness behavior has motivational properties that increase the probability of survival and facilitate recovery from illness. When considered in this way, the behavioral response to sickness is similar to most aversive stimuli. For example, sickness behavior helps animals resist infections just as fear helps them manifest appropriate defensive behaviors when confronted by a threat (fight-or-flight response; Dantzer 2001; Kelley et al. 2003; Dantzer and Kelley, 2007). Psychologist Neal Miller was the first scientist to point out that sickness symptoms are the expression of a motivational state rather than a consequence of physical illness (Aubert 1999; Dantzer 2004). To demonstrate this principle, Miller found that after injection with endotoxin lipopolysaccharide (LPS), thirsty rats trained to press a bar to obtain water reduced, rather than increased, their bar pressing activities (Holmes and Miller 1963). In other experiments, Miller used rats that had been trained to press a lever to obtain rest periods while they were in a continuous rotating drum.

These experiments demonstrated that LPS administration increased the rate of lever presses in rats, indicating that the rats would work more to obtain rest following exposure to LPS. The fact that LPS treatment could decrease or increase behavioral responses was an important finding because these data provided strong support for a motivational interpretation of sickness behaviors (Miller 1964).

More than 30 years after these initial experiments, the motivational interpretation of sickness behavior was reexamined from the perspective of the emerging field of psychoneuroimmunology (Aubert et al. 1995a, 1997a, b). Experiments conducted by Aubert et al. (1997a) used LPS to evaluate the choices of sick animals under environmental conditions that lead to motivational conflicts. Lactating mice normally build nests to provide warmth and a safe environment for their pups and retrieve their pups when they wander out of the nest. Aubert et al. induced sickness in lactating mice by injection of LPS. Despite their sickness, lactating mice remained able to efficiently retrieve their pups when the pups were dispersed throughout the cage. The mothers did not engage in nest-building when the nest was removed and replaced by cotton wool. However, when the ambient temperature was lowered from 22 to 6°C the mothers actively engaged in nest-building for their young pups. These data were interpreted as direct proof that the motivation of sickness competes with the motivation of maternal behavior, and this competition ultimately determines the behavioral performance of sick animals.

Another component of sickness that confers adaptive value is the febrile response, which is classically defined as an increase in the hypothalamic thermoregulatory set-point. Mounting a febrile response is metabolically expensive. In humans, a rise in core body temperature of 1°C causes metabolic rate to increase by about 13%. Fever has been demonstrated to promote survival in a variety of experimental infections (Kluger 1979, 1986, 1991). A number of clinical observations and experimental studies have demonstrated that elevated body temperature activates leukocytes and depress as growth of bacterial and viral pathogens. Animals infected with these pathogens have a lower survival rate if fevers are prevented by cold ambient temperatures or treatment with antipyretic drugs (Kluger 1986, Hasday and Garrison 2000, Hasday et al. 2000). Due to such a high energetic cost, febrile animals must minimize their thermal losses. Although the febrile response is not a behavior per se, success in mounting a fever often requires behavioral changes that reduce heat loss, such as huddling, postural changes, and seeking shelter.

The anorexia that accompanies sickness and fever can seem paradoxical from an adaptive point of view: i.e., given the metabolic expense of fever, how can restriction of energy intake be adaptive? This conundrum is probably explained by viewing sickness behavior from an ecological perspective. Anorexia decreases the interest of animals in searching for food, a process known as foraging. A substantial number of calories are required for animals to search for food, so a reduction in foraging translates into reduced energy expenditure. Secondly, a reduction in foraging minimizes the risk of sick animals becoming exposed to a predator during the search for food. If this perspective is correct, sick animals should demonstrate more suppression in foraging than in consummatory components of food intake. This appears to be the case since in experimental circumstances, sick animals stopped pressing a lever for food (Crestani et al. 1991; Aubert et al. 1995b) but still consumed some food pellets that were provided directly, although consumption of the pellets was less than normal (Aubert et al. 1995b)

Recent evidence has established that inflammation is associated with an increase in a variety of forms of clinical pain (Watkins et al. 2007). From an ecological perspective, Watkins and Maier pointed out that hyperalgesia should be included as a behavioral component of sickness behavior (Watkins et al. 1995b). For example, saliva contains anti-microbial compounds (e.g., lactoperoxidase) and growth factors (e.g., epidermal growth factor) that promote wound healing. A common behavior of animals is to lick wounded or infected sites. Since these wounded sites are often painful, recuperative behaviors such as licking are likely to promote animal survival. Hyperalgesia would also discourage animals from moving, which would promote energy conservation.

2.2 Proinflammatory Cytokines and Sickness Behavior

Sickness symptoms are the outward expression of an innate immune response that has been activated by pathogenic microorganisms or noxious stimuli (e.g., trauma, toxins). The immediate defense against these insults is the so-called acute phase response. Although often considered to consist of only changes in blood proteins, the acute phase response comprises immune, physiological, metabolic, and behavioral responses. Several cell types, such as monocytes and macrophages, as well as their CNS-differentiated microglial relatives, play a pivotal role in innate immunity. These innate immune cells express toll-like receptors (TLRs) that constitute a family of pattern recognition receptors. There are presently 13 recognized TLR genes in mammals, all of which recognize molecules that are expressed by a variety of pathogens (Carpentier et al. 2008). Once these receptors are bound by a particular pathogen-associated molecular pattern, a critical intracellular protein known as MyD88 is recruited and activated (with the exception of TLR3). A variety of downstream kinases and transcription factors, such as members of the mitogen-activated protein kinase (MAPKs) and nuclear factor-kappaB (NFκB) families, are recruited three subsequently induce the synthesis of a number of mediators of the acute phase response, including proinflammatory cytokines (van der Bruggen et al. 1999; Abraham 2000; Barton and Medzhitov 2003; Kim et al. 2004).

Cytokines are soluble messenger proteins produced and released primarily by myeloid and lymphoid cells in order to regulate the immune response. Although cytokines are heterogeneous and their biological functions are both pleiotropic and redundant, two general categories of cytokines that are often distinguished have been classified as proinflammatory and anti-inflammatory cytokines. To date, dozens of cytokines have been recognized, and 33 of them are named interleukins (IL; Dinarello 2007). The major proinflammatory cytokines synthesized that have been studied in psychoneuroimmunology are some of the first ones that were identified and cloned, including IL-1 (existing in two molecular forms IL-1α and IL-1β), interleukin-6 (IL-6), and tumor necrosis factor α (existing in two molecular forms

TNFα and TNFβ). IL-1 is produced by macrophages, monocytes and dendritic cells in the periphery and glial cells in the central nervous system (CNS). IL-6 is secreted by glia, macrophages, and T cells, but other cells such as muscle tissues produce it as well. IL-6 is an important mediator of fever and is responsible for inducing synthesis of hepatic acute phase proteins in response to inflammation. TNFα is produced by several types of cells, including macrophages, lymphocytes, endothelial cells and fibroblasts during inflammatory responses and regulates a wide spectrum of cellular processes, including apoptosis, proliferation, and differentiation. Another important cytokine that can augment synthesis of proinflammatory cytokines is interferon gamma (IFN-γ), which is synthesized mainly by T cells and natural killer (NK) cells. IFN-γ possesses antiviral, anti-tumor, and immunoregulatory properties, being well-known to prime macrophages for a number of activities.

2.3 Experimental Studies of Sickness Behavior in Animal Models

The study of sickness behavior merged with immunology soon after cytokines were recognized as immune modulators that could induce the complete repertoire of sickness behaviors, independent of a pathogenic infection. Following the pioneering studies that showed that IL-1 activates the hypothalamie-pituitary–adrenocortical (HPA) axis (reviewed by Besedovsky and Rey 2007), several research groups embarked on determining the effects of cytokines on behavior and the brain and their relation to stress (Anisman and Merali 2003; Larson and Dunn 2001; Watkins and Maier 2000; Yirmiya et al. 2000). In this section, we will mainly concentrate on results of our own group.

Soon after IL-1 was isolated and cloned in the early 1980s, it was tested in the conditioned taste aversion paradigm. Rats injected with IL-1 formed an aversion to the taste of a saccharin solution (Tazi et al. 1988). The sickness-inducing property of IL-1 was further confirmed by showing that IL-1-injected rats were less prone to press a lever in order to obtain a food reward in a Skinner box and to explore a novel juvenile introduced as a social stimulus into the home cage (Crestani et al. 1991; Kent et al. 1992a, b). By that time, several reports had established that peripheral and central administration of IL-1α, IL-1β and TNFα into healthy laboratory animals induced fever, thereby more precisely defining the term "endogenous pyrogen" (Kluger 1991). The question naturally arose as to whether elevated body temperature was responsible for symptoms of sickness behavior. To test this idea, the IL-1 receptor antagonist (IL-1ra) was injected into the ventricles of the brain just prior to administration of IL-1β (Fig. 1). Intracerebroventricular (i.c.v.) administration of IL-1ra significantly impaired the ability of IL-1β given intraperitoneally (i.p.) to reduce lever presses for food and investigation of a novel juvenile. However, IL-1ra given i.c.v. did not impair the elevated body temperature or increased metabolic rate caused by injection of IL-1β (Kent et al. 1992c). These were some of the first data to establish dissociation between the pyrogenic and behavioral properties of IL-1β. During the

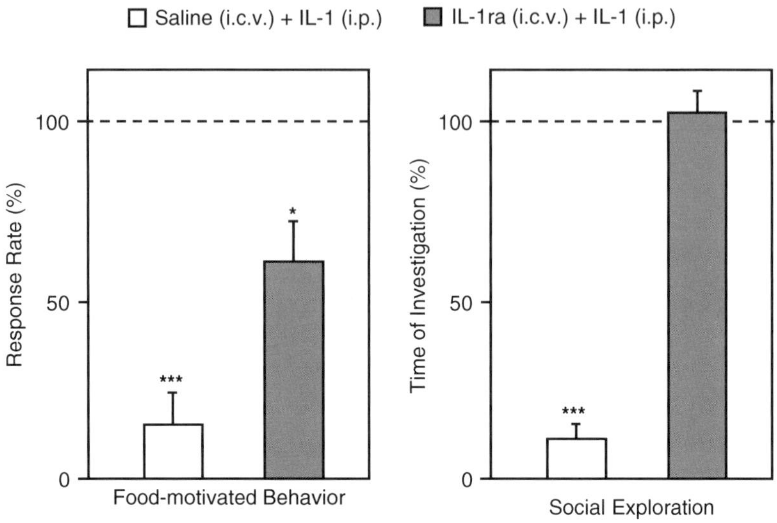

Fig. 1 Sickness behavior is caused by an increase in brain proinflammatory cytokines. Central administration of an IL-1ra (24 μg/rat) into the ventricles of the brain significantly impaired the reduction in food motivated behavior caused by peripheral administration (i.p.) of IL-1 (4 μg/rat) and entirely reversed the inhibition of this cytokine on social exploratory behavior. These early findings provided one of first demonstrations that sickness behavior induced in the periphery is caused by proinflammatory cytokines acting in the brain. Although the mechanism was unknown at that time, the data suggested that peripherally administrated proinflammatory cytokine(s) somehow communicated with the brain, perhaps by inducing the brain to produce endogenous cytokines. The figure represents the percentage change in relation to baseline, shown as the horizontal dashed line ($^{*}p < .05$; $^{***}p < .001$). Data represent means ± SEM. (From Kent et al. 1992c)

past 15 years, cytokine-induced sickness behavior has been intensively studied. The findings in this field, especially the mechanisms through which peripheral cytokines influence brain function and behavior, now form a cornerstone in psychoneuroimmunology (Dantzer 2001; Kelley et al. 2003; Dantzer 2004, 2006; De La Garza II 2005; Dantzer and Kelley 2007; Quan and Banks 2007; Hopkins 2007; Dantzer et al. 2008).

2.4 Measurement of Sickness Behavior in Animal Models

A variety of techniques have been used to quantify sickness behaviors in response to either pathogen-associated molecular patterns such as LPS or recombinant cytokines. Pathogenic microorganisms such as influenza virus (Swiergiel et al. 1997) or mycoplasma (Yirmiya et al. 1997) have been used more rarely. The three most common assays used to measure sickness behavior are described here.

Food intake and food-motivated behavior: Disruption of food-motivated behavior and food consumption reflect the anorexia of sickness. As an example,

IL-1 given i.p. (5 μg/mouse) suppresses the number of operant responses of rats trained to press a lever for food by ~70–90% at 1 h and ~80–95% at 4 h following injection (Crestani et al. 1991 and Fig. 1). Administration of LPS also disrupts nose-pokes in mice trained to work for food by poking their noses into the hole of a cage, a task that measures food-motivated behavior (Johnson et al. 1997; Bret-Dibat et al. 1997). Measurements of daily changes in food intake and body weight are less time-consuming indices of sickness behavior. LPS administration results in a significant decrease in body weight as well as consumption of food and water (Kent et al. 1994; Johnson et al. 1997; Neveu et al. 1998; Laye et al. 2000; Castanon et al. 2001). For instance, LPS given i.p. (5 or 10 μg/mouse) caused more than an 85% decrease in food intake during the 10 h period after injection (~0.4 g vs. ~3.2 g in controls). When IL-1ra was administered i.c.v. (4 μg/mouse) immediately following LPS injection, the 10 h food intake recovered to ~1.9 g. These results suggest that upregulation of IL-1 in the brain is at least partially responsible for the LPS induced decrease in food consumption (Laye et al. 2000).

Locomotor Activity: Locomotor activity is a validated readout of sickness behavior. A sick animal usually displays a reduction in locomotor activity. Several reports have shown that both LPS and proinflammatory cytokines, given either peripherally or i.c.v., reduces locomotor activity within 1–4 h which gradually recovers by 24 h (Laye et al. 1995; Lenczowski et al. 1999; Godbout et al. 2005b). In these kinds of experiments, rodents are usually placed in a novel environment so as to stimulate locomotor activity. Locomotor activity is intermittently videotaped for a few minutes at several time points and the data are subsequently analyzed manually or with appropriate computer software. This activity can be quantified by the number of movements between artificial compartments in an open field (line-crossing). For example, Fig. 3a shows that line crossing is reduced around 50% within 6 h after injection with LPS (0.83 mg/kg), and this activity returns to nearly baseline 24 h later. Computer programs are also available to measure the distance moved. Finally, mice normally explore and investigate objects by assuming an upright posture known as rearing. The reduction in the number of rearings in the open field is another sensitive index of sickness behavior (Palin et al. 2007; Palin et al. 2008)

Social Exploration: Exploration of an unfamiliar young rodent is widely used as a measure of sickness behavior, and it resembles social withdrawal that occurs during many human illnesses. To assess motivation to engage in social exploration, a novel conspecific juvenile is introduced into the test subject's home cage. The juvenile presents a strong stimulus for investigatory behaviors of adult mice. This social exploratory behavior is video taped and the amount of time the treated mouse engages in investigating and sniffing the juvenile is recorded (Dantzer et al. 1991; Bluthe et al. 1991, 1995, 1997, 2000, 2006; Kent et al. 1992b). Some strains of mice, such as C57/BL6 mice, are very aggressive when investigating novel juveniles. To protect the juvenile from aggressiveness of the test animal, a protective version of social exploration test can be used in which the juveniles are protected in a small wire-mesh cage (Berton et al. 2006).

2.5 Endogenous Antagonists of Proinflammatory Cytokines

Sickness behavior initiates following an acute episode of inflammation, so it is important that the organism is endowed with endogenous mechanisms to recover. Chronic activation of cytokine signaling in the CNS can result in irreversible damage, including neuronal death, astrogliosis, and demyelination (Allan and Rothwell 2003). Fortunately, the acute inflammation that leads to sickness behavior most often resolves within a day or two, and no permanent damage can be detected. Several types of molecules may play a role in restricting the potentially damaging effects of proinflammatory cytokines in CNS. These include anti-inflammatory cytokines, cytokine antagonists, and inhibitors of cytokine synthesis.

Anti-inflammatory cytokines belong to a very heterogeneous family. These cytokines are produced and secreted by leukocytes and posses the property of inhibiting expression of proinflammatory cytokines and their receptors, thereby shifting the balance of cytokine profiles. Anti-inflammatory cytokines include IL-1ra, IL-4, IL-10, IL-13, and transforming growth factor beta (TGF-β). During the course of an immune response, these proteins downregulate the synthesis and action of proinflammatory cytokines and their receptors and induce the synthesis of proinflammatory cytokine receptor antagonists. They can also directly interfere with proinflammatory cytokine receptor signaling (Strle et al. 2008). Anti-inflammatory cytokines play a physiological role in the regulation of neuronal inflammation, and IL-10 has been investigated most extensively (Strle et al. 2001; Milligan et al 2006). Administration of IL-10 i.c.v abrogates LPS-induced suppression of social exploration in rats (Bluthe et al. 1999). Similarly, LPS-induced fever is exaggerated and prolonged in IL-10-deficient mice (Leon et al. 1999). IL-10 inhibits proinflammatory cytokine production by microglial cells through inhibition of NFκB (Heyen et al. 2000). The effects of other anti-inflammatory cytokines, including IL-4 and IL-13, appear more complex. For example, IL-4 and IL-13 can be protective and synergistic, depending on the time of administration in relation to the induction of sickness behavior (Bluthe et al. 2001, 2002). Central administration of IL-1ra following peripheral administration of IL-1 completely blocks the suppression of both social exploration and partially restores food-motivated behavior caused by IL-1 (Fig. 1; Kent et al. 1992c). Some peptides, such as IGF-1 (McCusker et al. 2006; Palin et al. 2008) and α-melanocyte stimulating hormone (Getting 2006), antagonize certain aspects of inflammation. Although the generality of anti-inflammatory action and its mechanisms remain to be elucidated, IGF-1 behaves as an anti-inflammatory cytokine in the brain. When administrated centrally, IGF-1 attenuates sickness behavior induced by i.c.v. injection of LPS (Dantzer et al. 1999), TNFα (Bluthe et al. 2006; Palin et al. 2007) and IL-1 (Bluthe et al. 2006). Chronic administration of IGF-1 into the lateral ventricles of the brain also attenuates the impairment of spatial memory caused by kainate, an agonist of excitatory amino acid receptors (Bluthe et al. 2005). This effect is probably mediated by interfering with TNFα

since the effects of IGF-1 can be mimicked by pentoxifylline, an inhibitor of TNFα synthesis.

2.6 Propagation of Peripheral Cytokine Signals into the Brain

Although far from being completely understood, it was not until recently that the intricate details by which peripherally produced cytokines induce changes in the brain have been partially elucidated (Quan and Banks 2007). The evidence available to date points to parallel channels through which immune signals propagate from periphery to the brain (Dantzer 2001; Kelley et al. 2003; Dantzer and Kelley 2007; Fig. 2).

Afferent nerves, especially the afferent branch of the vagus, have been demonstrated to play an important role in relaying the status of the peripheral immune system to the brain (Watkins et al. 1995a; Goehler et al. 1997, 1998 and 1999; Gaykema et al. 2000). Neurons in the nodose ganglia monitor inflammation in abdomen. After i.p. administration of LPS in rats, IL-1β is induced not only in macrophages and dendritic cells within connective tissues surrounding afferent vagal fibers, but also within the fibers themselves (Watkins et al. 1995c; Goehler et al.

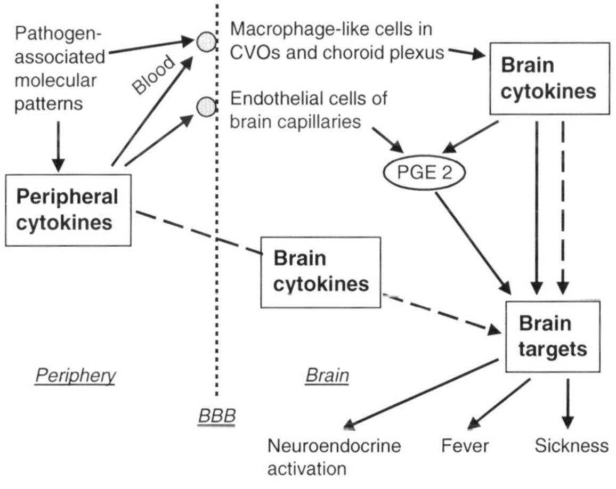

Fig. 2 Mechanisms involved in cytokine-induced sickness. Activation of pathogen-associated molecular pattern receptors induce the production of proinflammatory cytokines in the periphery and in macrophage-like cells of circumventricular organs (CVOs) and choroid plexuses. Peripheral cytokines induce the expression of cytokines in the brain via vagal neural afferents or a relay at the level of CVOs and brain vasculature endothelial cells. PGE 2 series are produced locally and diffuse to brain targets to alter the set point for various regulatory processes. Alternatively, brain cytokines propagate by volume diffusion to their neural targets. The dotted lines with arrows represent neutrally transmitted actions of cytokines to distant targets. (Modified from Konsman et al. 2002)

2000). Subdiaphragmatic vagotomy blocks the LPS-induced reduction in social exploration, even though it has no effect on peripheral macrophages (Bluthe et al. 1994). Subdiaphragmatic vagotomy also blocks LPS-induction of IL-1β mRNA expression in the hypothalamus (Laye et al. 1995).

Besides afferent vagal fibers, other afferent neural pathways may play roles in relaying inflammatory signals to the brain. For instance, vagotomy did not, but local anesthesia did, block fever induced by local subcutaneous administration of LPS (Gourine et al. 2001; Roth and De Souza 2001). Also, afferent terminals of the glossopharyngeal nerve are activated when orolingual infections occur, and these afferent nerves transmit a signal to the brain (Romeo et al. 2001). Collectively, these findings indicate that the neural pathway that is involved in relaying peripheral inflammation to the brain depends on the location in the body where inflammation occurs.

Another immune-to-brain communication pathway is represented by circumventricular organs (CVOs). CVOs are brain regions positioned around the margin of the ventricular system of the brain. In mammals, eight CVOs are divided into two groups, secretory and sensory. The secretory CVOs include the median eminence, neurohypophysis, intermediate lobe of the pituitary gland and pineal gland. The subcommisural organ, although poorly understood, is also a secretory CVO. Some also consider the choroid plexuses to be CVOs. The sensory CVOs include the subfornical organ (SFO), the organum vasculosum of the lamina terminalis (OVLT) and area postrema (AP). A distinct feature of the CVOs is that the usually highly packed endothelial cells which compose the blood–brain barrier (BBB) are fenestrated. As a result, circulating substances can pass through the "leaky" vessels and reach the brain parenchyma (Quan and Banks 2007). Despite its attractiveness, the leakage hypothesis of cytokines through CVOs has been questioned. The diffusion of blood-borne IL-1α entering the choroid plexus and the SFO is greatly restricted so that this cytokine does not freely penetrate into surrounding regions (Maness et al. 1998). The tanycytic barrier surrounding CVOs (Rodriguez et al. 2005) can prevent large molecules from gaining access to other brain structures, even after these large molecules have entered CVOs (Peruzzo et al. 2000). Because cytokines entering CVOs do not necessarily leak into adjacent brain regions, an alternative "relay hypothesis" is more likely. This already ancient hypothesis (Blatteis 1992) proposes that cytokines such as IL-1 activate their receptors in CVOs, and the latter relays soluble signals such as prostaglandin E2 (PGE2) to other brain areas (Katsuura et al. 1990; Komaki et al. 1992; Quan and Banks 2007)

Perivascular cells, including brain macrophages and pericytes, are present in CVOs, the choroid plexus, and leptomeninges (Thomas 1999; Rivest 2003). TLR 4 is expressed in these macrophage-like cells and recognizes LPS together with CD14 (Quan et al. 1998; Rivest 2003). These pericytes respond to circulating pathogen-associated molecular patterns and cytokines by producing cytokines. Locally produced cytokines are likely to act directly or indirectly on neurons that project to other brain structures or propagate via the CSF into adjacent brain areas by volume diffusion. This process could recruit other cytokine-producing cells, such as parenchymal microglial cells, to activate the brain cytokine system (Konsman et al. 2000; Vitkovic et al. 2000).

Receptor-mediated active transport is another mechanism by which peripheral cytokines can be transported both into and out of the brain. Cytokine transport is a complex event because not all cytokines are transported with the same efficiency. Even for those that are transportable, the rates of movement differ among cytokines and brain regions. Saturable transport systems that carry IL-1α, IL-1β, IL-1ra, IL-6, TNFα, and IFN-γ from blood to brain have been described. The transport of IL-6, TNFα, and IL-1 all occur via different systems, and IFN-γ is only saturable in the brain but not in all the spinal regions (Banks et al. 1995, 2002; Watkins et al. 1995a; Pan and Kastin 2001; Banks 2006). Comparison of saturable transport to extracellular pathways (CVO "leakage") showed that the major mechanism of brain entry of IL-1 after i.v. injection is the saturable transport system, with extracellular pathways accounting for only a small fraction of that entering the brain. However, after i.c.v. injection, IL-1 entering the brain largely relies on diffusion and leakage through extracellular pathways (Plotkin et al. 1996).

The way these different pathways of immune-to-brain communication can interact with each other has hardly been studied (Dantzer et al. 2000). Rapid activation of afferent neural pathways probably sensitize target brain structures to the subsequent production and action of cytokines that slowly diffuse by volume diffusion from the CVOs to recruit parenchymal microglia and perivascular macrophages-like cells.

3 Inflammation-Associated Depression

3.1 From Sickness Behavior to Depression

In the United States, the National Institutes of Mental Health was the first public funding agency to support research in communication pathways between the immune system and brain. The fruit of those investments has now been documented with the discoveries that cytokines are involved in a number of mental health issues, covering topics as diverse as depression, emotion, pain, autism and learning, and memory. The use of recombinant cytokines for treatment of cancer, as originally described in 1989 (Dantzer and Kelley 1989), has prompted a surge of interest in the potential role of cytokines in sickness behavior and now depression. The theory for a psychopathological role of proinflammatory cytokines in induction of depression was first independently proposed by American scholar R.S. Smith (1991) and the Belgian psychiatrist M. Maes et al. (1991, 1992 and 1995). This theory has been vigorously advanced at the clinical level by Maes, and it is mainly based on the observation of a positive correlation between increased severity of symptoms in patients suffering from depression and the elevation in inflammatory biomarkers in their blood, as well as hyperactivity of HPA axis.

A role for inflammation in the pathophysiology of depression has also been proposed based on the increased prevalence of major depressive disorders in somatic patients with a chronic inflammatory condition and the iatrogenic effects of cytokine immunotherapy in cancer patients and subjects infected with hepatitis C virus (Bonaccorso and Dantzer. 2001; Capuron and Dantzer 2003; Raison et al. 2006; Dantzer and Kelley 2007; Dantzer et al. 2008).

3.2 Immunotherapy Induces Depression

Since cytokines were first discovered, cloned, and expressed beginning in the 1980s, they have been tested as treatments for a number of diseases, including cancers, viral infections, and autoimmune diseases. IFNα was the first and is now the most widely used cytokine in treating the aforementioned diseases. The proinflammatory cytokines IL-2, IL-1, IFNβ, and TNFα are also used in immune therapies (Meyers 1999; Tayal and Kalra 2008). However, the deleterious side effects of proinflammatory cytokines on the CNS continue to be a limiting complication, and it was these side effects that provided the first clue that cytokines somehow acted in the brain (Dantzer and Kelley 1989). Clinically-relevant side-effects include fever, anorexia, fatigue and malaise, and life-threatening effects can be caused by the capillary leak syndrome. Although IFNα treatment is better tolerated, repeated IFNα administration to hepatitis C patients induces depressive symptoms in 40% of patients undergoing treatment (Irwin and Miller 2007) and increases the risk of suicide. Similarly, administration of either IL-2 or IFNα to patients with metastatic melanoma or kidney cancer induces a number of psychiatric side effects (Renault et al. 1987; Denicoff, et al. 1987; Irwin and Miller 2007). The cytokine therapy-associated depression sometimes reaches a level of severity that leads to discontinuation of the treatment (Meyers and Valentine 1995). The side effects of cytokine therapy usually fall into two categories. One set includes early onset flu-like symptoms such as malaise, fever, and headache, as well as neurovegetative symptoms, such as fatigue and loss of appetite. These symptoms usually appear within a few days after cytokine treatment and are very common in cytokine-treated patients. For instance, fatigue was reported to appear in 80% of patients treated with IFNα (Capuron et al. 2002). The second category comprises psychological symptoms that usually occur within a few weeks after initiation of cytokine therapy. These symptoms, which are experienced by about one-third to one-half of the patients, include depression, anxiety, and cognitive dysfunction (Capuron et al. 2000, 2002; Schiepers et al.2005; Dantzer et al. 2008).

The fact that psychological symptoms of depression are less frequent than neurovegetative symptoms in response to cytokine immunotherapy indicates the existence of vulnerability factors. Studies using a variety of psychological depression rating scales have demonstrated that pretreatment scores predict occurrence of depressive symptoms following cytokine therapies (Capuron and Ravaud 1999; Miyaoka et al. 1999; Musselman et al. 2001). Cognitive disturbances, sleep difficulties, and poor social support are among the risk factors that associated with cytokine-induced depression (Capuron and Dantzer 2003). The exact nature of these susceptibility factors remain to be defined and their biological basis is yet to be elucidated.

3.3 Depressive Disorders in Chronic Inflammatory Disorders

A higher than normal prevalence of depression is observed in several chronic inflammatory disorders. These medical conditions include cardiovascular diseases, multiple forms of cancer (and the chemotherapy and radiation treatments that accompany cancer treatment), multiple sclerosis, Alzheimer's disease, stroke, obesity, metabolic syndrome, and rheumatoid arthritis. Depressive symptoms are often attributed to emotional reactions, such as hopelessness, distress, fear, and frustration, and these emotions can ultimately progress to illness, disability and lower quality of life. Psychopathological responses that occur in these conditions are also associated with activation of the innate and adaptive immune responses and cytokine release (Schiepers et al. 2005; Spalletta et al. 2006; McCaffery et al. 2006). These findings support the concept that there exist potentially important relationships between mood disorders and cytokines

Cardiovascular disease: It is well-established that behavioral and psychosocial factors are associated with cardiovascular diseases, such as events that occur in acute coronary syndromes (Rozanski et al. 2005). Prevalence of major depression has been reported to occur in 17–27% of hospitalized patients with coronary artery disease (CAD; Rudisch and Nemeroff, 2003), and elevated inflammatory biomarkers have been proposed as risk factors for coronary heart disease (CHD; Ridker et al. 2000a, b). Current evidence suggests that depression and inflammation in CHD are related (Shimbo et al. 2005), but little is known about the mechanisms that mediate this relationship. It has been reported that the cholesterol-lowering drug statin reduces depression scores in CHD patients independently of its cholesterol-lowering effects (Young-Xu et al. 2003), and this activity is probably associated with its anti-inflammatory properties (Blake and Ridker 2000; Forrester and Libby 2007; Sola et al. 2006). Taken together, these findings indicate that inflammation not only influences the course of cardiovascular diseases, but may also be a causal factor in the development of depressive symptoms in patients with this disease.

Stroke: Incidence of depression is very common following stroke. It was reported that 40% or more of ischemic stroke patients are diagnosed with post-stroke depression (Pohjasvaara et al. 1998; Hachinski 1999; Nys et al. 2005). Brain inflammation induced by stroke may be related to clinical depression. Proinflammatory cytokines, such as IL-6 and TNFα, increase markedly in the brain as a result of stroke in human patients (Vila et al. 2000; Zaremba and Losy 2001; Zaremba et al. 2001; Intiso et al. 2004). Experimental stroke models also show upregulation of a number of cytokines at both the mRNA and protein levels (Allan and Rothwell 2003). In a study that used the middle cerebral artery occlusion (MCAO) model, MCAO-treated mice displayed a reduction in sucrose consumption. Central administration of IL-1ra but not corticosteroid antagonists restored sucrose consumption in MCAO-treated mice (Craft and DeVries 2006). These are in accordance with the proposed role of proinflammatory cytokines in post-stroke depression (Spalletta et al. 2006).

Multiple sclerosis (MS) and experimental autoimmune encephalomyelitis: Multiple sclerosis is the most common demyelinating disease of the CNS in humans. It is an autoimmune disease caused by immune attack on oligodendrocytes, the resident cells of the CNS. These cells ensheath myelinated neurons, and the loss of myelin impairs neurotransmission in these neurons. Depression is common in MS patients, with annual prevalence rates as high as 20% and lifetime prevalence rates of about 50% (Siegert and Abernethy 2005). In addition, immunotherapy (such as administration of IFNβ) can promote depression in MS patients (Weinstock-Guttman et al. 1995; Goeb et al. 2006). Experimental autoimmune encephalomyelitis (EAE) is a generally accepted animal model of MS. It is induced by vaccination of animals with myelin proteins in adjuvant. EAE is a T cell-mediated autoimmune disease in which CD4+T helper cells are activated, which then attack myelin sheaths in the CNS. EAE and MS share many clinical and pathogenetic similarities, and these similarities are also shared at the level of psychopathology. For example, sickness and depressive-like behaviors have been documented in animals with EAE, including decreased food intake, reduced consumption of a palatable sucrose solution, and decreased social interest. Development of depressive-like behavior appears to be associated with brain inflammatory processes, in particular the production of proinflammatory cytokines (Pollak and Yirmiya 2002).

Obesity, Metabolic Syndrome, and Depression: Epidemiological surveys have established significant comorbidities between depression and obesity (Faith et al. 2002; Stunkard et al. 2003; Bornstein et al. 2006; Strine et al. 2008). Major depression before adulthood is a risk factor to develop overweight; abdominal obesity is associated with depression (McElroy et al. 2004). In a broader view beyond obesity, there has been a growing interest into the association between depression and the metabolic syndrome (MetS). MetS is a clustering of metabolic abnormalities in one person, including obesity, atherogenic dyslipidemia, hypertension, insulin resistance or glucose intolerance, and a proinflammatory state. Individuals with MetS have an elevated risk of cardiovascular diseases and type 2 diabetes (Mensah et al. 2004). Subjects with the MetS are at increased risk for developing depression, and reciprocally, depression facilitates the development of MetS (Kinder et al. 2004; Lowe et al. 2006; Skilton et al. 2007; see for review Soczynska et al. 2007). The possible role of inflammation in association between depression and obesity or MetS is still unclear despite the well-known occurrence of low-grade inflammation in obesity and MetS (e.g., Ferroni et al. 2004; Sutherland et al. 2004; Dandona et al. 2005; Cancello and Clement 2006; Greenberg and Obin 2006; Ferrante 2007; Yudkin 2007; Lee and Pratley 2005).

3.4 Effects of Anti-depressants and Anti-inflammatory Drugs on Cytokine-Induced Depression

The fact that depressive disorders occur in patients afflicted with chronic infectious diseases, in subjects with autoimmune diseases such as rheumatoid arthritis, and in patients undergoing cytokine therapy raise a number of important questions: Are cytokines somehow involved in the etiology and severity of clinical depression? Would traditional anti-depressants be effective in treating cytokine-induced depressive symptoms? Do anti-depressants affect cytokine production or action? Would treatment with cytokine antagonists improve some symptoms of clinical depression?

There is a wealth of data on the anti-inflammatory properties of anti-depressant drugs (see e.g., Kenis and Maes, 2002 for a review). Several clinical observations now suggest that anti-depressants can be of therapeutic value in treating cytokine-induced depressive disorders. For instance, paroxetine is a selective serotonin reuptake inhibitor (SSRI). It was reported in a prospective, double blind experiment that 45% of malignant melanoma patients developed clinical depression during the course of IFNα therapy, but only 11% of the paroxetine-treated patients became depressed. Severe depression caused 35% of patients to be removed from the IFNα-treatment as compared to only 5% of the patients treated with paroxetine and IFNα (Musselman et al. 2001). A case report established that the SSRI anti-depressant fluoxetine (Prozac) initiated 2 weeks before restarting IFNα treatment was associated with that patient effectively continuing IFNα therapy. This patient, who had a history of clinical depression, had been discontinued twice due to a severe depressive response following IFNα treatment to prevent recurrence of malignant melanoma (Hauser et al. 2000). Even in hepatitis C patients who are not treated with IFNα, plasma TNFα and IL-1β concentrations were significantly and positively associated with depression scores on the Beck Depression Inventory II questionnaire (Loftis et al. 2008).

In experimental studies conducted in rats, depressive-like behavior induced by LPS was ameliorated by chronic anti-depressant treatment (Yirmiya 1996; Shen et al. 1999; Yirmiya et al. 2001). The anti-depressants used in those experiments included the tricyclic anti-depressants imipramine (Yirmiya 1996) and desipramine (Shen et al. 1999) and the SSRI anti-depressant fluoxetine (Yirmiya et al. 2001). Results of these experiments showed a broad, general effectiveness of chronic administration of anti-depressants, regardless of their pharmacological class (i.e., monoamine oxidase inhibitors, tricyclics, or SSRIs; Capuron and Dantzer 2003). There is evidence that tricyclic anti-depressants are more effective in treating cytokine-induced depressive-like behavior than those in other categories (Shen et al. 1999). A possible mechanism of action for the tricyclic anti-depressants' specific effects in reducing proinflammatory cytokine-induced depressive-like behavior is their anti-inflammatory properties, which appear to be more potent than that of other classes of anti-depressants. For example, in vitro experiments showed that incubation of human blood monocytes with the tricyclic anti-depressants imipramine and clomipramine inhibited the release of proinflammatory cytokines IL-1β, TNFα, and IL-6. However, the SSRI anti-depressant citalopram only showed minimal effects on proinflammatory cytokine secretion (Xia et al. 1996). The immunosuppressive effects of tricyclic anti-depressants were also supported by in vivo studies in rats. Chronic treatment with the tricyclic anti-depressant desipramine suppressed IL-1β and TNFα secretion following LPS injection (Connor et al. 2000). Moreover, only the tricyclic anti-depressant desipramine, but not paroxetine and venlafaxine which belong to other pharmacological classes, suppressed LPS-stimulated TNFα synthesis (Shen, et al. 1999).

If inflammation somehow contributes to development or maintenance of depressive disorders, anti-inflammatory treatments might be beneficial for patients with depressive disorders. In a recent double-blind controlled study, a specific antagonist of COX2 was found to add significantly to the effect of the anti-depressant reboxetin (Müller et al. 2006). We recently showed that pretreatment with the tetracycline derivative, minocycline that has central anti-inflammatory properties, blocks depressive-like behavior in the forced swim test (FST) and tail suspension test (TST). This reduction in depressive-like behavior was associated with attenuated expression of TNFα, IFNγ, IL-1β and IDO steady-state mRNAs in the brain (Fig. 4; O'Connor et al. 2008).

3.5 Experimental Studies on Cytokine-Induced Depression

Like other fields of biomedical studies, animal models of depression have been used in both industry and academia (Willner 1995; Dalvi and Lucki 1999; Rupniak 2003; McArthur and Borsini 2006). Since major depression in human encompasses a wide variety of symptoms, it is unlikely that one animal model can mimic all aspects of this complex multifaceted disorder. During the past few decades, a number of rodent models of depression have been developed to address this issue. Common models include olfactory bulbectomy (Jesberger and Richardson 1988; Song and Leonard 2005), neurotransmitter manipulations (Carlsson 1976; Dilsaver 1986a, b; Nakamura 1991), learned helplessness paradigms (Maier 1984), and exposure to chronic mild stress (CMS; Willner 2005). Some genetic models of depression have also been developed (El Yacoubi and Vaugeois 2007). Here, we focus on only one of these models of depression developed using cytokines.

In the domain of behavioral evaluation, a number of tests are recommended when measuring "depressive-like" behavior in animals. Since sickness and depressive-like behaviors can overlap with each other, behavioral tests of sickness behavior (e.g., food consumption, locomotor activity, social exploration) are sometimes used as readouts of depressive-like behavior. While these dependent variables of sickness mimic neurovegetative symptoms, such as fatigue and reduced appetite, they are often not good measures of psychopathological symptoms, especially when the depressive-like behavior is dissociated from sickness behavior (Frenois et al. 2007; O'Connor et al. 2008). Therefore, in order to separate these two similar but yet distinct forms of behavior, tests have been developed that specifically point to psychological dimensions of depression. These tests are especially useful when studying cytokine-induced depressive-like behavior, and the four major ones are described below.

Sucrose preference test: This was one of the first tests used to validate animal models of depression, especially CMS. When given a choice, rodents prefer to consume sweetened solutions, and this response has been interpreted by psychologists as a "pleasurable" experience. Decreased consumption of a palatable sucrose solution has been assumed to reflect anhedonia (Katz 1982; Monleon et al. 1995). In general, this test is valid, not only because animals consume less sucrose following chronic stress, but also because chronic treatment with anti-depressants from a variety of pharmacological classes restores sucrose preference (Katz 1982; Willner et al. 1987; Muscat et al. 1992; Monleon et al. 1995). In practice, saccharin is sometimes used to replace sucrose because of the caloric contribution of sucrose, which can lead to satiation.

The two-bottle paradigm in which experimental animals can choose between a bottle containing water and a juxtaposed bottle containing sucrose or saccharin is commonly used in this test. The sucrose/saccharin preference test can be repeated daily in animals during chronic experiments. In recent years, the sucrose/saccharin preference test has been used in psychoneuroimmunology research on animal models. A number of studies have shown the validity of this test in reflecting cytokine-induced sickness behavior and depressive-like behavior (Yirmiya 1996; Pollak et al. 2000; Mormède et al. 2002; Frenois et al. 2007). Also, anti-depressants are effective in reversing the putative anhedonic condition that occurs following administration of proinflammatory cytokines or LPS (Yirmiya 1996; Pollak and Yirmiya 2002).

Forced swim and tail suspension tests: The FST and TST are the two most commonly used animal behavioral tests for anti-depressant screening in the pharmaceutical industry (Cryan et al. 2002; Crowley et al. 2004; Bourin et al. 2005). Both the FST and TST are based on the measurement of time that mice and rats spend in an immobile position. Increased immobility in these two tests is claimed to reflect a helpless or resignation-like state. Although the constructs are similar, the two tests are slightly different in terms of the biological substrates that underlie the behaviors (Cryan and Mombereau 2004).

The FST was developed by the French neuropharmacologist Roger Porsolt et al. (1977 a, b). In this test, a mouse or rat is usually placed in a cylinder of water for a single session lasting 6 min. After swimming in this inescapable environment for a short while, animals that display depressive-like behavior usually show increased immobility and passive floating. The TST was developed by the Belgian neuropharmacologist Lucien Steru and colleagues and is regarded as a derivative of FST (Steru et al. 1985). In the TST, adhesive tape is used to suspend a mouse or rat by the tail to a hook for 6 or 10 min. Usually, after escape-oriented struggling during the first 1–2 min, the animals show increasing bouts of immobility. Both FST and TST have been shown to be sensitive to all major class of anti-depressants because these drugs significantly shorten the duration of immobility if given to mice or rats prior to the FST or TST (Borsini and Meli 1988; Teste et al. 1993;

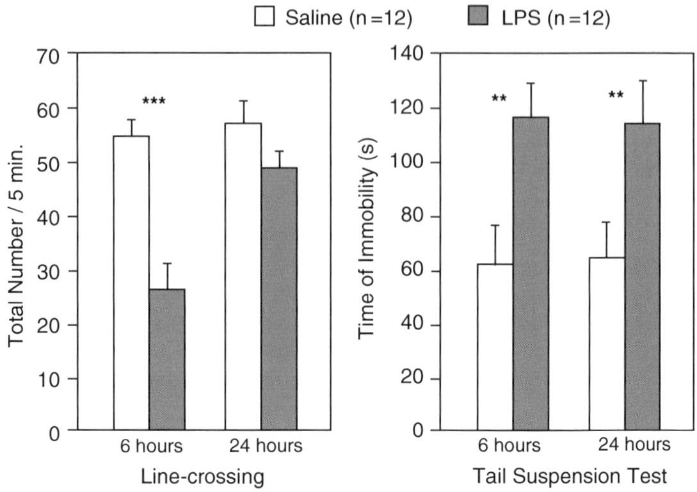

Fig. 3 Depressive-like behavior can be separated from sickness behavior. Mice were injected i.p. with saline or LPS (0.83 mg/kg). Sickness behavior was measured as locomotor activity (line-crossing number) and depressive-like behavior was determined with the tail suspension test (TST). Both sickness and depressive-like behaviors were obvious 6 h following LPS injection, as shown by the reduction in line crossing and increase in time spent motionless in the TST. By 24 h, sickness behavior had retuned to nearly baseline, but depressive-like behavior remained (**$p<.01$, ***$p<.001$). Data represent means ± SEM. These findings provide strong evidence for a dissociation between LPS-induced sickness behavior and LPS-induced depressive-like behavior. (From Frenois et al. 2007)

Crowley et al. 2004). Nowadays, motor tracking and analysis software for FST and computer-controlled commercially available TST equipment have made both tests more efficient and reliable.

The validity of these tests in screening anti-depressant drugs does not necessarily mean that they are also validated in testing cytokine-induced depressive-like behavior. We recently used both these behavioral tests to measure cytokine-induced depressive-like behavior in mice. The data indicated that FST and TST are useful in measuring cytokine-induced depressive-like behavior, which is reflected as an increase in immobility of duration in both tests (Godbout et al. 2007; Frenois et al. 2007; O'Connor et al. 2008). An example of results we recently obtained using the TST is shown in Fig. 3b. These data established that at both 6 and 24 h following an i.p. injection of LPS (0.83 mg/kg), immobility was significantly increased compared to saline-injected mice. It is important to note that sickness behavior, as measured by line-crossing, had returned to baseline at 24 h (Fig. 3b). Comparing FST and TST with the tests of sickness behavior, these experiments also established that in some experimental conditions, depressive-like behavior can be separated from sickness behavior, i.e., depressive-like behavior remains after sickness behavior disappears (Frenois et al. 2007; O'Connor et al. 2008; Fig. 4)

Voluntary wheel running: Voluntary wheel running (VWR) is one of the most intensively studied behaviors performed by laboratory rodents (Sherwin 1998). In

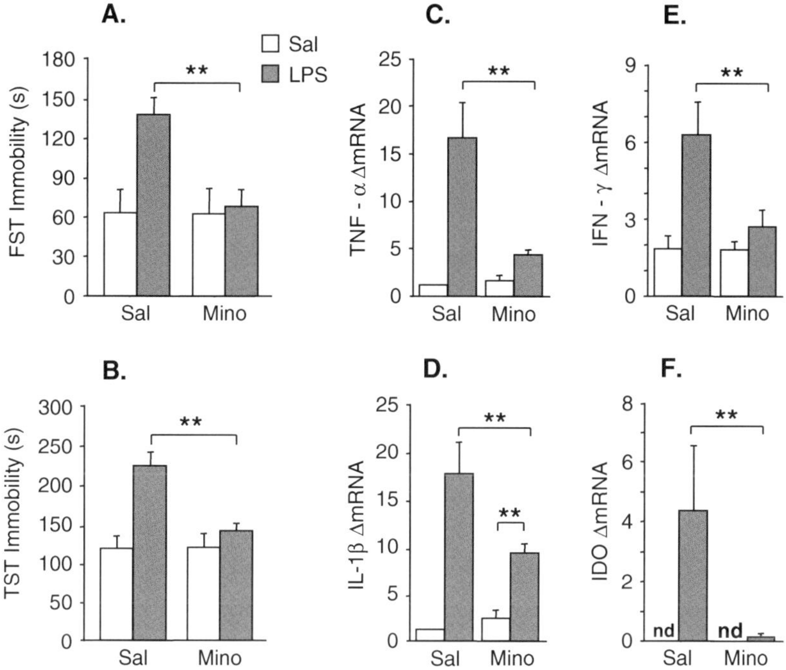

Fig. 4 Minocycline inhibits LPS-induced depressive-like behavior that is associated with blockade of brain proinflammatory cytokine and IDO expression. Mice were injected i.p. with either saline or minocycline (50 mg/kg) for 3 successive days. Immediately following the final injection, mice received either i.p. injection of non-pyrogenic saline or LPS (0.83 mg/kg). (A) and (B). Duration of immobility during a 6 min forced swim test (FST) and 10 min tail suspension test (TST) 24 h (FST) and 28 h (TST) following administration of LPS. Immediately following the TST at 28 h, mice were sacrificed and steady-state expression of mRNA transcripts in the brain were measured by real-time RT-PCR for (C). TNF-α, (D). IL-1β, (E). IFN γ and (F). IDO. Data represent means ± SEM. (* $P<0.05$, ** $P<0.01$). (From O'Connor et al. 2008)

this test, a rat or mouse is individually housed in a cage that is attached to a running wheel. The number of wheel turns is registered by a mechanical counter or a magnetic sensor, and the distance run is calculated either manually or with computer software by counting the number of revolutions.

Because of the obvious face validity and its linkage to the brain motivation system, VWR has been used as a model of human physical exercise, which has been reported to counteract depression (Martinsen 1994; Dunn et al. 2001, 2005; Hunsberger et al. 2007). Indeed, VWR has been shown to possess therapeutic effects on depressive-like disorders in animal models. For instance, consistent with effect of human exercise reducing development of stress-related depression, wheel running has been shown to reduce stress-induced depressive-like behavior, including reversing disability in a shuttle box escape paradigm (Moraska and Fleshner 2001; Greenwood et al. 2003, 2007a, b). These effects of VWR on shuttle box escape were similar to the effects of anti-depressants in the same paradigm (Telner and Singhal 1981; Telner et al. 1981; Massol et al. 1989). There is no consensus to explain why rodents engage in VWR. Several factors that affect VWR, and could be responsible for it, include the curiosity of mice to explore, general activity, stereotypic activity, escape, play, food or water deprivation, social rank, hormonal status, adrenal activity, and body weight maintenance (Sherwin 1998). Brain dopamine and opioid reward systems are necessary for VWR. Using transgenic mice and high VWR strains, key regions of the brain reward circuit have been identified in VWR, including the prefrontal cortex, NAc, caudate-putamen, and the lateral hippocampus (Vargas-perez et al. 2004; Rhodes et al. 2005).

In recent years, VWR has been applied in psychoneuroimmunology experiments. Proinflammatory cytokines, such as IL-6, IL-1β, and TNFα, caused sickness symptoms, including a reduction in VWR (Harden et al. 2006, 2008). In one of our recent studies (Moreau et al. 2008, in press), mice were inoculated by *Bacillus Calmette-Guerin* (BCG), a vaccine developed for protection against tuberculosis. BCG-treated mice showed a dramatic acute reduction in VWR (to ~30% of baseline level) at 24 h, followed by a modest longer-term decline (to ~60–70% of baseline level) in VWR for up to 10 days after BCG injection. The marked decrease in VWR during the first day following BCG injection can be interpreted as physical sickness and fatigue. However, the chronic reduction of VWR is unlikely to be attributed to sickness because the locomotor activity largely recovered within two days following BCG administration. Decreased VWR 2–10 days following BCG treatment is more likely to represent a depressive-like symptom.

3.6 Mechanisms of Inflammation-Associated Depression

The molecular mechanisms that mediate cytokine-induced depression are only beginning to be understood. One major player is indoleamine 2, 3 dioxygenase (IDO; EC 1.13.11.17; Dantzer et al. 2008). IDO was first characterized for its key role in the regulation of the immune response in the periphery (Mellor and Munn 1999). In the brain, it is responsible for the production of tryptophan metabolites that have potent activities (Schwarcz and Pellicciari 2002).

IDO and tryptophan indoleamine 2.3 dioxygenase (TDO; EC 1.13.11.11) are the two major enzymes involved in tryptophan (Trp) catabolism. IDO degrades Trp along the kynurenine (Kyn) pathway. IDO is present in macrophages, monocytes, and microglia in the brain, as well as in extrahepatic organs such as the lungs. TDO expression is restricted to the liver (Taylor and Feng 1991). IDO is also involved in the catabolism of related indoleamine-containing compounds such as serotonin (5-hydroxytryptamine, 5-HT), melatonin, and tryptamine (Taylor and Feng 1991; Grohmann et al. 2003), whereas TDO only utilizes Trp.

Tryptophan is an essential amino acid that is actively transported into brain, being utilized for the synthesis of 5-HT, a key neurotransmitter involved in the regulation of number of brain functions including emotions. It was first reported that HIV-1-positive patients display a reduction in serum Trp concentrations. This decline in Trp correlated with the severity of psychiatric symptoms, along with increased Kyn concentrations, which was used as an indicator of heightened Trp catabolism through IDO (Fuchs et al. 1990). Depressive symptoms that develop following cytokine immunotherapy are accompanied by a reduction in plasma levels of Trp (Capuron et al. 2002). The Kyn to Trp ratio (Kyn/Trp) in blood can also be used as a general indicator of Trp degradation by IDO (Schrocksnadel et al. 2006). An increase of this ratio is associated with an elevation in plasma levels of neopterin, a marker of macrophage activation, suggesting the upregulation of IDO (Widner et al. 2002). Evidence obtained to date indicates that the connection between proinflammatory cytokines, Trp and depression is at least in part due to the increased activity of IDO (reviewed in Schiepers et al. 2005; Dantzer et al, 2008).

At the molecular level, IDO is a cytosolic 45 kD monomeric heme-containing oxygenase-type metalloenzyme. Human IDO is encoded by the Indo gene on chromosome 8 (King and Thomas 2007). The Indo gene contains 10 exons spread over ~15 kbp of DNA. The Indo promoter contains interferon-stimulated response elements (ISREs) and an IFNα-activated site (GAS) that are recognized by the transcription factor interferon regulatory factor 1 (IRF-1) and signal transducer and activator of transcription 1α (STAT-1α), respectively (King and Thomas 2007). The promoter also contains nonspecific IFN-stimulated response elements which can respond to IFNα and IFNβ as well. The expression of IDO can also be regulated by other proinflammatory cytokines, including TNFα, IL-6, and IL-1β, through the NFκB and p38MAPK pathways (Fujigaki et al. 2006). Anti-inflammatory cytokines, including IL-4, IL-10, and TGFβ, have been found to downregulate IDO expression (Musso et al. 1994; Yuan et al. 1998; Wichers and Maes 2004). It therefore appears that inflammation plays an important role in the regulation of IDO expression.

To switch Trp metabolism to the Kyn pathway, IDO (as well as TDO in the liver) catalyzes the oxidative cleavage of the 2,3 double bond in the pyrrole ring of Trp and indoleamine derivatives, forming N-formylkynurenine, which then deformylates to form Kyn (Taniguchi et al. 1979; Sugimoto et al. 2006). Kyn undergoes further conversions generating a number of metabolites, including the neurotoxin quinolinic acid (QA), 3-hydroxyanthranilic acid (3-HAA), 3-hydroxykynurenine (3-HK), neuroprotective kynurenic acid (KA), and nicotinamide adenine dinucleotide (NAD). In humans, over 90% of Trp is catabolized along the Kyn pathway (Littlejohn et al. 2003). Analysis of tissue homogenates from LPS-treated mice versus nontreated mice found that the enzymatic activity of IDO is highly induced not only in the lung, stomach, cecum, colon, testes, seminal vesicle, and epididymis (Takikawa et al. 1986) but also in the brain (Lestage et al. 2002).

IDO is the rate-limiting, regulatory enzyme in the Kyn pathway (Schwarcz R, Pellicciari R, 2002). Kyn catabolism leads to the generation of metabolites that can be either neurotoxic or neuroprotective and may contribute to the pathophysiology of depression. For instance, 3-HK, QA, and 3-HAA participate in the formation of reactive oxygen species leading to neuronal damage. QA activates

N-methyl-D-aspartate (NMDA) receptors, resulting in the depolarization of neurons and possibly excitotoxicity (Wichers et al. 2005; Stone et al. 2007). KA, on the other hand, is an antagonist of several glutamate receptor subtypes and is neuroprotective. Peripherally generated Kyn is able to cross the BBB, and human microglia are capable of metabolizing this substance into its toxic metabolites (Schwarcz and Pellicciari 2002). Importantly, new data have recently established that human endothelial cells and pericytes that form the BBB constitutively express IDO and downstream enzymes that degrade Kyn along the Trp pathway (Owe-Young et al. 2008). This pathway is likely to generate a flux of Kyn into the human brain and could lead to immune tolerance at the level of the BBB itself.

Besides the Kyn pathway-associated neurotoxicity, neurodegeneration and negative influence on glutamatergic neurotransmission, another principle route of cytokine-induced IDO activation that could contribute to depression is the reduced synthesis and accelerated metabolism of 5-HT. Although the etiology of major depression is still elusive, the link between major depression and dysfunction of brain serotonergic neurotransmission has been strongly suggested (Byerley and Risch 1985; Eison 1990; Nutt 2002; Muller and Schwarz 2007). It is unlikely that proinflammatory cytokine-induced depressive disorders spare this association. Indeed, IFNγ and TNFα have been shown to increase 5-HT transporter mRNA expression and 5-HT uptake, respectively (Morikawa et al. 1998; Mossner et al. 2000). The serotonergic system could also be affected by cytokine-induced upregulation of IDO activity. Reduction in the concentrations of Trp and 5-HT via IDO enzymatic activity as well as decreased 5-HT synthesis due to IDO activation can result in reduced levels of 5-HT.

We recently obtained direct evidence that depressive-like behavior in mice induced by systemic inflammation is mediated by upregulation of IDO activity. In these experiments, LPS treatment induced prolonged depressive-like behavior (measured with FST and TST) in aged mice when compared to young adult mice. The enhanced psychopathological response in aged mice was associated with a more pronounced induction of peripheral and brain IDO (represented by higher Kyn/Trp ratio) and a markedly higher turnover rate of brain serotonin (as measured by the ratio of 5-hydroxyindoleacetic acid (5-HIAA) to 5-HT) compared to young adult mice at 24 h post-LPS injection (Godbout et al. 2007). In another study, we found that peripheral administration of LPS activates IDO and results in distinct depressive-like behaviors in mice, also measured by FST and TST. Blockade of IDO activation either indirectly with the anti-inflammatory tetracycline derivative minocycline, which attenuates LPS-induced expression of proinflammatory cytokines, or directly with 1-methyltryptophan (1-MT), a competitive inhibitor of IDO, prevented the development of depressive-like behaviors. Both minocycline and 1-MT normalized the Kyn/Trp ratio in plasma and brain of LPS-treated mice without changing the LPS-induced increase in turnover of brain 5-HT. Administration of L-Kyn to naïve mice dose-dependently induced depressive-like behavior. These new results implicate IDO as a critical molecular mediator of inflammation-induced depressive disorders, probably mainly through the catabolism of Trp along the Kyn pathway (O'Connor et al. 2008; Fig. 5).

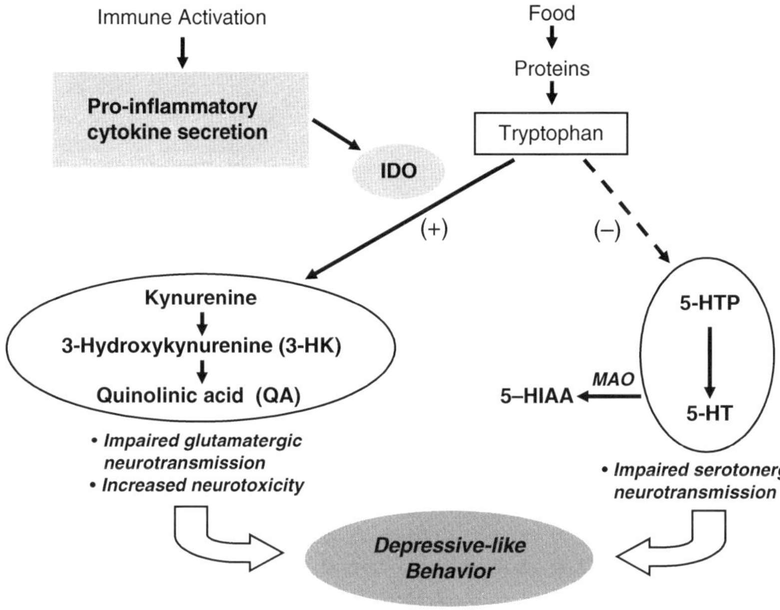

Fig. 5 Mechanisms that regulate proinflammatory cytokine-induced depressive-like behavior. Peripheral immune signals, which can be caused by infectious microbes, non-infectious inflammatory conditions, trauma or immunotherapy propagate to the brain. This activates the diffuse brain cytokine system, resulting in an exaggerated production of proinflammatory cytokines, including IL-1β, IL-6, TNFα, and IFNγ. These inflammatory cytokines, especially IFN$\gamma$and TNFα, induce the transcription and enzymatic activity of indoleamine 2,3-dioxygenase (IDO). IDO diverts tryptophan from the synthesis of 5-hydroxytryptophan (5-HTP) and serotonin (5-hydroxytryptamine; 5-HT) to the kynurenine pathway, generating kynurenine and its downstream end products 3-hydroxykynurenine (3-HK) and quinolinic acid (QA). As a result, glutamatergic neurotransmission is impaired and neurotoxicity increases since QA acts as an NMDA receptor agonist, and 3-HK and QA generate free radicals. Activation of the kynurenine pathway may also impact serotonergic neurotransmission via decreased synthesis of 5-HT and increased turnover of 5-HT by upregulation of monoamine oxidase (MAO). The IDO-mediated neurotoxicity and dysfunction of serotonergic and glutamatergic neurotransmission may ultimately cause depressive symptoms (Modified from Godbout et al. 2007)

4 Conclusions and Perspectives

Based on the collective findings of many investigators over the past 20 years, it now appears that systemic inflammation is an important biological event that causes sickness behavior and increases the risk for occurrence of depression. The explosion in knowledge about immune-to-brain communication offers new approaches in clinical practice. Sickness symptoms undoubtedly have a negative influence on quality of life, but most current medications alleviate only some of the symptoms (e.g., fever). Depressive disorders in patients with prolonged infection or undergoing immunotherapy now appears to be linked to neuroinflammation and upregulation of brain IDO activity. These findings call for new understanding of depressive disorders in patients that are afflicted with disorders characterized by excessive inflammation. These include some of the major health issues of our time such as obesity, CHD, and type II diabetes. The compounds targeting inflammatory mediators and IDO, especially those that act directly in the brain to alleviate neuroinflammation and downregulate brain IDO expression, may create new opportunities for drug development (Dantzer et al. 2008).

Developing a clear and precise understanding of body and mind relationships is a challenging scientific endeavor. Psychoneuroimmunology is a field that tries to answer this question by exploring the scientific basis for communication systems between systemic organs and the brain, but the field remains in its infancy. Although the elucidation of immune–brain communication pathways is far from complete, it is quite reasonable to achieve useful clinical progress along the way. An important basic area of research that needs to be more intensively studied is the fundamental neuroanatomy that underlies the behavioral effects of cytokines. Since microglia and astrocytes can be activated in nearly all parts of the brain, the mechanisms may be similar throughout the brain. In contrast to the excellent neuroscience that has been reported, such as the identification of specific brain regions that are responsible for the pyrogenic effects of proinflammatory cytokines and the neuroendocrine effects of cytokines on CRH release (Rivest 2001; Ferri et al. 2005), only a few studies have linked specific alterations in inflammatory mediators in well-defined brain regions with sickness behavior and/or depressive-like behavior (Quan et al. 1997, 1998, 1999 and 2000; Konsman et al. 1999, 2000; Sparkman et al. 2006; Goehler et al. 2000, 2006 and 2007; Frenois et al. 2007). Much more work is needed to provide a clearer picture of the brain circuitry involved in both sickness behavior and depressive-like behavior. A number of techniques can facilitate the definition of cause–effect relationships during neuroinflammation. For example, in vivo brain microdialysis has been widely used to better understand the neurobiology of learning and memory (Chang and Gold 2003, 2004 and 2008; Chang et al. 2006). This technique has the potential to answer some important questions, especially the changes in neurotransmission that occur in the extracellular space of different brain regions and their temporal relation with the development of cytokine-induced sickness behavior and depression (Merali et al. 1997; Linthorst et al. 1995). Live brain imaging techniques, coupled with genetically encoded reporters of some of the relevant genes induced by proinflammatory cytokines such as COX2 and IDO (Sheng et al. 2000; Babcock and Carlin 2000), can enlighten our understanding of functional physiology of cytokine-induced sickness and depression. Similarly, routine techniques that have long been used in the neurosciences, such as in situ hybridization (Quan and Herkenham 2002) and organotypic brain slice cultures (Bernardino et al. 2005), offer important useful approaches in psychoneuroimmunology.

Experiments using models of autoimmune diseases have already broadened our understanding in brain inflammation and depressive disorders, but much more needs to be done. Recent studies using the MS model of EAE show that immune cells infiltrate and traffic to the brain after onset of the disease (McMahon et al. 2005; Bailey et al. 2007). These immune cells, and the inflammatory mediators released by them, are likely to contribute to both sickness and depressive-like behaviors in this disease. The temporal dynamics of immune cell recruitment into the EAE brain has been established in multiple variants of the disease (McMahon et al. 2005; Miller et al. 2007). The time course of infiltration of immune cells provides a distinct opportunity for future studies on cellular aspects involved in the development of sickness and depressive-like behaviors. Other neurodegenerative diseases, such as Alzheimer's disease, possess obvious features of neuroinflammation (Tuppo and Arias 2005; Rosenberg 2005; Weisman et al. 2006). Indeed, many patients with Alzheimer's disease manifest with depressive disorders (Tune 1998; Lyketsos and Olin 2002; Lyketsos and Lee 2004; Starkstein and Mizrahi 2006). However, few studies have been conducted on neuroinflammation-associated depression in patients with Alzheimer's disease or depressive-like behavior in animal models of this disease. Expanding research in inflammation on topics such as these will provide important insights into psychopathological aspects of depression, an intensively studied but still elusive malady.

References

Abraham E (2000) NF-kappaB activation. Crit Care Med 28:N100–104.

Allan SM, Rothwell NJ (2003) Inflammation in central nervous system injury. Philos Trans R Soc Lond B Biol Sci 358:1669–1677.

Anisman H, Merali Z (2003) Cytokines, stress and depressive illness: brain-immune interactions. Ann Med 35:2–11.

Aubert A (1999) Sickness and behaviour in animals: a motivational perspective. Neurosci Biobehav Rev 23: 1029–1036.

Aubert A, Goodall G, Dantzer R (1995a) Compared effects of cold ambient temperature and cytokines on macronutrient intake in rats. Physiol Behav 57:869–873.

Aubert A, Kelley KW, Dantzer R (1997a) Differential effect of lipopolysaccharide on food hoarding behavior and food consumption in rats. Brain Behav Immun 11:229–238.

Aubert A, Vega C, Dantzer R, Goodall G (1995b) Pyrogens specifically disrupt the acquisition of a task involving cognitive processing in the rat. Brain Behav Immun 9:129–148.

Aubert A, Goodall G, Dantzer R, Gheusi G (1997b) Differential effects of lipopolysaccharide on pup retrieving and nest building in lactating mice. Brain Behav Immun 11:107–118.

Babcock TA, Carlin JM (2000) Transcriptional activation of indoleamine dioxygenase by interleukin 1 and tumor necrosis factor alpha in interferon-treated epithelial cells. Cytokine 12:588–594.

Bailey SL, Schreiner B, McMahon EJ, Miller SD (2007) CNS myeloid DCs presenting endogenous myelin peptides 'preferentially' polarize CD4+ T(H)-17 cells in relapsing EAE. Nat Immunol 8:172–180.

Banks WA (2006) The blood-brain barrier in psychoneuroimmunology. Neurol Clin 24:413–419.

Banks WA, Plotkin SR, Kastin AJ (1995) Permeability of the blood-brain barrier to soluble cytokine receptors. Neuroimmunomodulation 2:161–165.

Banks WA, Farr SA, Morley JE (2002) Entry of blood-borne cytokines into the central nervous system: effects on cognitive processes. Neuroimmunomodulation 10:319–327.

Barton GM, Medzhitov R (2003) Toll-like receptor signaling pathways. Science 300:1524–1525.

Bernardino L, Xapelli S, Silva AP, Jakobsen B, Poulsen FR, Oliveira CR, Vezzani A, Malva JO, Zimmer J (2005) Modulator effects of interleukin-1beta and tumor necrosis factor-alpha on AMPA-induced excitotoxicity in mouse organotypic hippocampal slice cultures. J Neurosci 25:6734–6744.

Berton O, McClung CA, Dileone RJ, Krishnan V, Renthal W, Russo SJ, Graham D, Tsankova NM, Bolanos CA, Rios M, Monteggia LM, Self DW, Nestler EJ (2006) Essential role of BDNF in the mesolimbic dopamine pathway in social defeat stress. Science 311:864–868.

Besedovsky HO, Rey AD (2007) Physiology of psychoneuroimmunology: a personal view. Brain Behav Immun 21:34–44.

Blake GJ, Ridker PM (2000) Are statins anti-inflammatory? Curr Control Trials Cardiovasc Med 1:161–165.

Blatteis CM (1992) Role of the OVLT in the febrile response to circulating pyrogens. Prog Brain Res 91:409–412.

Bluthe RM, Dantzer R, Kelley KW (1991) Interleukin-1 mediates behavioural but not metabolic effects of tumor necrosis factor alpha in mice. Eur J Pharmacol 209:281–283.

Bluthe RM, Dantzer R, Kelley KW (1997) Central mediation of the effects of interleukin-1 on social exploration and body weight in mice. Psychoneuroendocrinology 22:1–11.

Bluthe RM, Kelley KW, Dantzer R (2006) Effects of insulin-like growth factor-I on cytokine-induced sickness behavior in mice. Brain Behav Immun 20:57–63.

Bluthe RM, Beaudu C, Kelley KW, Dantzer R (1995) Differential effects of IL-1ra on sickness behavior and weight loss induced by IL-1 in rats. Brain Res 677:171–176.

Bluthe RM, Frenois F, Kelley KW, Dantzer R (2005) Pentoxifylline and insulin-like growth factor-I (IGF-I) abrogate kainic acid-induced cognitive impairment in mice. J Neuroimmunol 169:50–58.

Bluthe RM, Bristow A, Lestage J, Imbs C, Dantzer R (2001) Central injection of interleukin-13 potentiates LPS-induced sickness behavior in rats. Neuroreport 12:3979–3983.

Bluthe RM, Lestage J, Rees G, Bristow A, Dantzer R (2002) Dual effect of central injection of recombinant rat interleukin-4 on lipopolysaccharide-induced sickness behavior in rats. Neuropsychopharmacology 26:86–93.

Bluthe RM, Castanon N, Pousset F, Bristow A, Ball C, Lestage J, Michaud B, Kelley KW, Dantzer R (1999) Central injection of IL-10 antagonizes the behavioural effects of lipopolysaccharide in rats. Psychoneuroendocrinology 24:301–311.

Bluthe RM, Walter V, Parnet P, Laye S, Lestage J, Verrier D, Poole S, Stenning BE, Kelley KW, Dantzer R (1994) Lipopolysaccharide induces sickness behaviour in rats by a vagal mediated mechanism. C R Acad Sci III 317:499–503.

Bonaccorso S, Puzella A, Marino V, Pasquini M, Biondi M, Artini M, Almerighi C, Levrero M, Egyed B, Bosmans E, Meltzer HY, Maes M (2001) Immunotherapy with interferon-alpha in patients affected by chronic hepatitis C induces an intercorrelated stimulation of the cytokine network and an increase in depressive and anxiety symptoms. Psychiatry Res 105:45–55.

Bornstein SR, Schuppenies A, Wong ML, Licinio J (2006) Approaching the shared biology of obesity and depression: the stress axis as the locus of gene-environment interactions. Mol Psychiatry 11:892–902.

Borsini F, Meli A (1988) Is the forced swimming test a suitable model for revealing antidepressant activity? Psychopharmacology (Berl) 94:147–160.

Bourin M, Chenu F, Ripoll N, David DJ (2005) A proposal of decision tree to screen putative antidepressants using forced swim and tail suspension tests. Behav Brain Res 164:266–269.

Bret-Dibat JL, Creminon C, Couraud JY, Kelley KW, Dantzer R, Kent S (1997) Systemic capsaicin pretreatment fails to block the decrease in food-motivated behavior induced by lipopolysaccharide and interleukin-1beta. Brain Res Bull 42:443–449.

Byerley WF, Risch SC (1985) Depression and serotonin metabolism: rationale for neurotransmitter precursor treatment. J Clin Psychopharmacol 5:191–206.

Cancello R, Clement K (2006) Is obesity an inflammatory illness? Role of low-grade inflammation and macrophage infiltration in human white adipose tissue. Bjog 113:1141–1147.

Capuron L, Ravaud A (1999) Prediction of the depressive effects of interferon alfa therapy by the patient's initial affective state. N Engl J Med 340:1370.

Capuron L, Dantzer R (2003) Cytokines and depression: the need for a new paradigm. Brain Behav Immun 17 Suppl 1:S119–124.

Capuron L, Ravaud A, Dantzer R (2000) Early depressive symptoms in cancer patients receiving interleukin 2 and/or interferon alfa-2b therapy. J Clin Oncol 18:2143–2151.

Capuron L, Gumnick JF, Musselman DL, Lawson DH, Reemsnyder A, Nemeroff CB, Miller AH (2002) Neurobehavioral effects of interferon-alpha in cancer patients: phenomenology and paroxetine responsiveness of symptom dimensions. Neuropsychopharmacology 26:643–652.

Carlsson A (1976) The contribution of drug research to investigating the nature of endogenous depression. Pharmakopsychiatr Neuropsychopharmakol 9:2–10.

Carpentier PA, Duncan DS, Miller SD (2008) Glial toll-like receptor signaling in central nervous system infection and autoimmunity. Brain Behav Immun 22:140–147.

Castanon N, Bluthe RM, Dantzer R (2001) Chronic treatment with the atypical antidepressant tianeptine attenuates sickness behavior induced by peripheral but not central lipopolysaccharide and interleukin-1beta in the rat. Psychopharmacology (Berl) 154:50–60.

Chang Q, Gold PE (2003) Switching memory systems during learning: changes in patterns of brain acetylcholine release in the hippocampus and striatum in rats. J Neurosci 23:3001–3005.

Chang Q, Gold PE (2004) Impaired and spared cholinergic functions in the hippocampus after lesions of the medial septum/vertical limb of the diagonal band with 192 IgG-saporin. Hippocampus 14:170–179.

Chang Q, Gold PE (2008) Age-related changes in memory and in acetylcholine functions in the hippocampus in the Ts65Dn mouse, a model of Down syndrome. Neurobiol Learn Mem 89:167–177.

Chang Q, Savage LM, Gold PE (2006) Microdialysis measures of functional increases in ACh release in the hippocampus with and without inclusion of acetylcholinesterase inhibitors in the perfusate. J Neurochem 97:697–706.

Connor TJ, Harkin A, Kelly JP, Leonard BE (2000) Olfactory bulbectomy provokes a suppression of interleukin-1beta and tumour necrosis factor-alpha production in response to an in vivo challenge with lipopolysaccharide: effect of chronic desipramine treatment. Neuroimmunomodulation 7:27–35.

Craft TK, DeVries AC (2006) Role of IL-1 in poststroke depressive-like behavior in mice. Biol Psychiatry 60:812–818.

Crestani F, Seguy F, Dantzer R (1991) Behavioural effects of peripherally injected interleukin-1: role of prostaglandins. Brain Res 542:330–335.

Crowley JJ, Jones MD, O'Leary OF, Lucki I (2004) Automated tests for measuring the effects of antidepressants in mice. Pharmacol Biochem Behav 78:269–274.

Cryan JF, Mombereau C (2004) In search of a depressed mouse: utility of models for studying depression-related behavior in genetically modified mice. Mol Psychiatry 9:326–357.

Cryan JF, Markou A, Lucki I (2002) Assessing antidepressant activity in rodents: recent developments and future needs. Trends Pharmacol Sci 23:238–245.

Dalvi A, Lucki I (1999) Murine models of depression. Psychopharmacology (Berl) 147:14–16.

Dandona P, Aljada A, Chaudhuri A, Mohanty P, Garg R (2005) Metabolic syndrome: a comprehensive perspective based on interactions between obesity, diabetes, and inflammation. Circulation 111:1448–1454.

Dantzer R (2001) Cytokine-induced sickness behavior: where do we stand? Brain Behav Immun 15:7–24.

Dantzer R (2004) Cytokine-induced sickness behaviour: a neuroimmune response to activation of innate immunity. Eur J Pharmacol 500:399–411.

Dantzer R (2006) Cytokine, sickness behavior, and depression. Neurol Clin 24:441–460.

Dantzer R, Kelley KW (1989) Stress and immunity: an integrated view of relationships between the brain and the immune system. Life Sci 44:1995–2008.

Dantzer R, Kelley KW (2007) Twenty years of research on cytokine-induced sickness behavior. Brain Behav Immun 21:153–160.

Dantzer R, Bluthe RM, Kelley KW (1991) Androgen-dependent vasopressinergic neurotransmission attenuates interleukin-1-induced sickness behavior. Brain Res 557:115–120.

Dantzer R, Gheusi G, Johnson RW, Kelley KW (1999) Central administration of insulin-like growth factor-1 inhibits lipopolysaccharide-induced sickness behavior in mice. Neuroreport 10:289–292.

Dantzer R, Konsman JP, Bluthe RM, Kelley KW (2000) Neural and humoral pathways of communication from the immune system to the brain: parallel or convergent? Auton Neurosci 85:60–65.
Dantzer R, O'Connor JC, Freund GG, Johnson RW, Kelley KW (2008) From inflammation to sickness and depression: when the immune system subjugates the brain. Nat Rev Neurosci 9:46–56.
De La Garza R, 2nd (2005) Endotoxin- or pro-inflammatory cytokine-induced sickness behavior as an animal model of depression: focus on anhedonia. Neurosci Biobehav Rev 29:761–770.
Denicoff KD, Rubinow DR, Papa MZ, Simpson C, Seipp CA, Lotze MT, Chang AE, Rosenstein D, Rosenberg SA (1987) The neuropsychiatric effects of treatment with interleukin-2 and lymphokine-activated killer cells. Ann Intern Med 107:293–300.
Dilsaver SC (1986a) Pharmacologic induction of cholinergic system up-regulation and supersensitivity in affective disorders research. J Clin Psychopharmacol 6:65–74.
Dilsaver SC (1986b) Cholinergic mechanisms in depression. Brain Res 396:285–316.
Dinarello CA (2007) Historical insights into cytokines. Eur J Immunol 37 Suppl 1:S34–45.
Dunn AL, Trivedi MH, O'Neal HA (2001) Physical activity dose-response effects on outcomes of depression and anxiety. Med Sci Sports Exerc 33:S587–597; discussion 609–510.
Dunn AL, Trivedi MH, Kampert JB, Clark CG, Chambliss HO (2005) Exercise treatment for depression: efficacy and dose response. Am J Prev Med 28:1–8.
Eison MS (1990) Serotonin: a common neurobiologic substrate in anxiety and depression. J Clin Psychopharmacol 10:26S–30S.
El Yacoubi M, Vaugeois JM (2007) Genetic rodent models of depression. Curr Opin Pharmacol 7:3–7.
Faith MS, Matz PE, Jorge MA (2002) Obesity-depression associations in the population. J Psychosom Res 53:935–942.
Ferrante AW, Jr. (2007) Obesity-induced inflammation: a metabolic dialogue in the language of inflammation. J Intern Med 262:408–414.
Ferri CC, Yuill EA, Ferguson AV (2005) Interleukin-1beta depolarizes magnocellular neurons in the paraventricular nucleus of the hypothalamus through prostaglandin-mediated activation of a non selective cationic conductance. Regul Pept 129:63–71.
Ferroni P, Basili S, Falco A, Davi G (2004) Inflammation, insulin resistance, and obesity. Curr Atheroscler Rep 6:424–431.
Forrester JS, Libby P (2007) The inflammation hypothesis and its potential relevance to statin therapy. Am J Cardiol 99:732–738.
Frenois F, Moreau M, O'Connor J, Lawson M, Micon C, Lestage J, Kelley KW, Dantzer R, Castanon N (2007) Lipopolysaccharide induces delayed FosB/DeltaFosB immunostaining within the mouse extended amygdala, hippocampus and hypothalamus, that parallel the expression of depressive-like behavior. Psychoneuroendocrinology 32:516–531.
Fuchs D, Moller AA, Reibnegger G, Stockle E, Werner ER, Wachter H (1990) Decreased serum tryptophan in patients with HIV-1 infection correlates with increased serum neopterin and with neurologic/psychiatric symptoms. J Acquir Immune Defic Syndr 3:873–876.
Fujigaki H, Saito K, Fujigaki S, Takemura M, Sudo K, Ishiguro H, Seishima M (2006) The signal transducer and activator of transcription 1alpha and interferon regulatory factor 1 are not essential for the induction of indoleamine 2,3-dioxygenase by lipopolysaccharide: involvement of p38 mitogen-activated protein kinase and nuclear factor-kappaB pathways, and synergistic effect of several proinflammatory cytokines. J Biochem 139:655–662.
Gaykema RP, Goehler LE, Hansen MK, Maier SF, Watkins LR (2000) Subdiaphragmatic vagotomy blocks interleukin-1beta-induced fever but does not reduce IL-1beta levels in the circulation. Auton Neurosci 85:72–77.
Getting SJ (2006) Targeting melanocortin receptors as potential novel therapeutics. Pharmacol Ther 111:1–15.
Godbout JP, Chen J, Abraham J, Richwine AF, Berg BM, Kelley KW, Johnson RW (2005b) Exaggerated neuroinflammation and sickness behavior in aged mice following activation of the peripheral innate immune system. Faseb J 19:1329–1331.

Godbout JP, Moreau M, Lestage J, Chen J, Sparkman NL, O'Connor JC, Castanon N, Kelley KW, Dantzer R, Johnson RW (2007, Sep) Aging Exacerbates Depressive-like Behavior in Mice in Response to Activation of the Peripheral Innate Immune System. Neuropsychopharmacology 33(10):2341–51. Epub 2007 Dec 12.

Goeb JL, Even C, Nicolas G, Gohier B, Dubas F, Garre JB (2006) Psychiatric side effects of interferon-beta in multiple sclerosis. Eur Psychiatry 21:186–193.

Goehler LE, Erisir A, Gaykema RP. (2006, Jul 21) Neural-immune interface in the rate area postrema. Neuroscience 140(4):1415–34. Epub 2006 May 2.

Goehler LE, Lyte M, Gaykema RP (2007) Infection-induced viscerosensory signals from the gut enhance anxiety: implications for psychoneuroimmunology. Brain Behav Immun 21:721–726.

Goehler LE, Gaykema RP, Hammack SE, Maier SF, Watkins LR (1998) Interleukin-1 induces c-Fos immunoreactivity in primary afferent neurons of the vagus nerve. Brain Res 804:306–310.

Goehler LE, Gaykema RP, Hansen MK, Anderson K, Maier SF, Watkins LR (2000) Vagal immune-to-brain communication: a visceral chemosensory pathway. Auton Neurosci 85:49–59.

Goehler LE, Relton JK, Dripps D, Kiechle R, Tartaglia N, Maier SF, Watkins LR (1997) Vagal paraganglia bind biotinylated interleukin-1 receptor antagonist: a possible mechanism for immune-to-brain communication. Brain Res Bull 43:357–364.

Goehler LE, Gaykema RP, Nguyen KT, Lee JE, Tilders FJ, Maier SF, Watkins LR (1999) Interleukin-1beta in immune cells of the abdominal vagus nerve: a link between the immune and nervous systems? J Neurosci 19:2799–2806.

Gourine AV, Rudolph K, Korsak AS, Kubatko J, Tesfaigzi J, Kozak W, Kluger MJ (2001) Role of capsaicin-sensitive afferents in fever and cytokine responses during systemic and local inflammation in rats. Neuroimmunomodulation 9:13–22.

Greenberg AS, Obin MS (2006) Obesity and the role of adipose tissue in inflammation and metabolism. Am J Clin Nutr 83:461S–465S.

Greenwood BN, Strong PV, Dorey AA, Fleshner M (2007a) Therapeutic effects of exercise: wheel running reverses stress-induced interference with shuttle box escape. Behav Neurosci 121:992–1000.

Greenwood BN, Strong PV, Foley TE, Thompson RS, Fleshner M (2007b) Learned helplessness is independent of levels of brain-derived neurotrophic factor in the hippocampus. Neuroscience 144:1193–1208.

Greenwood BN, Foley TE, Day HE, Campisi J, Hammack SH, Campeau S, Maier SF, Fleshner M (2003) Freewheel running prevents learned helplessness/behavioral depression: role of dorsal raphe serotonergic neurons. J Neurosci 23:2889–2898.

Grohmann U, Fallarino F, Puccetti P (2003) Tolerance, DCs and tryptophan: much ado about IDO. Trends Immunol 24:242–248.

Hachinski V (1999) Post-stroke depression, not to be underestimated. Lancet 353:1728.

Harden LM, du Plessis I, Poole S, Laburn HP (2006) Interleukin-6 and leptin mediate lipopolysaccharide-induced fever and sickness behavior. Physiol Behav 89:146–155.

Harden LM, Plessis ID, Poole S, Laburn HP (2008, Aug) Interleukin (IL)-6 and IL-1beta act synergistically within the brain to induce sickness behavior and fever in rats. Brain Behav Immun. 22(6):838–49. Epub 2008 Feb 5.

Hart BL (1988) Biological basis of the behavior of sick animals. Neurosci Biobehav Rev 12:123–137.

Hasday JD, Garrison A (2000) Antipyretic therapy in patients with sepsis. Clin Infect Dis 31 Suppl 5:S234–241.

Hasday JD, Fairchild KD, Shanholtz C (2000) The role of fever in the infected host. Microbes Infect 2:1891–1904.

Hauser P, Soler R, Reed S, Kane R, Gulati M, Khosla J, Kling MA, Valentine AD, Meyers CA (2000) Prophylactic treatment of depression induced by interferon-alpha. Psychosomatics 41:439–441.

Heyen JR, Ye S, Finck BN, Johnson RW (2000) Interleukin (IL)-10 inhibits IL-6 production in microglia by preventing activation of NF-kappaB. Brain Res Mol Brain Res 77:138–147.

Holmes TE, Miller NE (1963) Effects of bacterial endotoxin on water intake, food intake and body temperature in the albino rat. J Exp Med 118: 649–658.

Hopkins SJ (2007) Central nervous system recognition of peripheral inflammation: a neural, hormonal collaboration. Acta Biomed 78 Suppl 1:231–247.
Hunsberger JG, Newton SS, Bennett AH, Duman CH, Russell DS, Salton SR, Duman RS (2007, Dec). Antidepressant actions of the exercise-regulated gene VGF. Nat Med 13(12):1476–82. Epub 2007 Dec 2.
Intiso D, Zarrelli MM, Lagioia G, Di Rienzo F, Checchia De Ambrosio C, Simone P, Tonali P, Cioffi Dagger RP (2004) Tumor necrosis factor alpha serum levels and inflammatory response in acute ischemic stroke patients. Neurol Sci 24:390–396.
Irwin MR, Miller AH (2007) Depressive disorders and immunity: 20 years of progress and discovery. Brain Behav Immun 21:374–383.
Jesberger JA, Richardson JS (1988) Brain output dysregulation induced by olfactory bulbectomy: an approximation in the rat of major depressive disorder in humans? Int J Neurosci 38: 241–265.
Johnson RW, Gheusi G, Segreti S, Dantzer R, Kelley KW (1997) C3H/HeJ mice are refractory to lipopolysaccharide in the brain. Brain Res 752:219–226.
Katsuura G, Arimura A, Koves K, Gottschall PE. (1990, Jan) Involvement of organum vasculosum of lamina terminalis and preoptic area in interleukin 1 beta–induced ACTH release. Am J Physiol 258(1 Pt 1):E163–71.
Katz RJ (1982) Animal model of depression: pharmacological sensitivity of a hedonic deficit. Pharmacol Biochem Behav 16:965–968.
Kelley KW, Bluthe RM, Dantzer R, Zhou JH, Shen WH, Johnson RW, Broussard SR (2003) Cytokine-induced sickness behavior. Brain Behav Immun 17 Suppl 1:S112–118.
Kenis G, Maes M (2002) Effects of antidepressants on the production of cytokines. Int J Neuropsychopharmacol 5:401–412.
Kent S, Kelley KW, Dantzer R (1992a) Effects of lipopolysaccharide on food-motivated behavior in the rat are not blocked by an interleukin-1 receptor antagonist. Neurosci Lett 145:83–86.
Kent S, Bluthe RM, Kelley KW, Dantzer R (1992b) Sickness behavior as a new target for drug development. Trends Pharmacol Sci 13:24–28.
Kent S, Rodriguez F, Kelley KW, Dantzer R (1994) Reduction in food and water intake induced by microinjection of interleukin-1 beta in the ventromedial hypothalamus of the rat. Physiol Behav 56:1031–1036.
Kent S, Bluthe RM, Dantzer R, Hardwick AJ, Kelley KW, Rothwell NJ, Vannice JL (1992c) Different receptor mechanisms mediate the pyrogenic and behavioral effects of interleukin 1. Proc Natl Acad Sci USA 89:9117–9120.
Kim L, Butcher BA, Denkers EY (2004) Toxoplasma gondii interferes with lipopolysaccharide-induced mitogen-activated protein kinase activation by mechanisms distinct from endotoxin tolerance. J Immunol 172:3003–3010.
Kinder LS, Carnethon MR, Palaniappan LP, King AC, Fortmann SP (2004) Depression and the metabolic syndrome in young adults: findings from the Third National Health and Nutrition Examination Survey. Psychosom Med 66:316–322.
King NJ, Thomas SR (2007) Molecules in focus: indoleamine 2,3-dioxygenase. Int J Biochem Cell Biol 39:2167–2172.
Kluger MJ (1979) Temperature regulation, fever, and disease. Int Rev Physiol 20:209–251.
Kluger MJ (1986) Is fever beneficial? Yale J Biol Med 59:89–95.
Kluger MJ (1991) Fever: role of pyrogens and cryogens. Physiol Rev 71:93–127.
Komaki G, Arimura A, Koves K (1992) Effect of intravenous injection of IL-1 beta on PGE2 levels in several brain areas as determined by microdialysis. Am J Physiol 262:E246–251.
Konsman JP, Kelley K, Dantzer R (1999) Temporal and spatial relationships between lipopolysaccharide-induced expression of Fos, interleukin-1beta and inducible nitric oxide synthase in rat brain. Neuroscience 89:535–548.
Konsman JP, Tridon V, Dantzer R (2000) Diffusion and action of intracerebroventricularly injected interleukin-1 in the CNS. Neuroscience 101:957–967.
Larson SJ, Dunn AJ (2001) Behavioral effects of cytokines. Brain Behav Immun 15:371–387.

Laye S, Gheusi G, Cremona S, Combe C, Kelley K, Dantzer R, Parnet P (2000) Endogenous brain IL-1 mediates LPS-induced anorexia and hypothalamic cytokine expression. Am J Physiol Regul Integr Comp Physiol 279:R93–98.
Laye S, Bluthe RM, Kent S, Combe C, Medina C, Parnet P, Kelley K, Dantzer R (1995) Subdiaphragmatic vagotomy blocks induction of IL-1 beta mRNA in mice brain in response to peripheral LPS. Am J Physiol 268:R1327–1331.
Lee YH, Pratley RE (2005) The evolving role of inflammation in obesity and the metabolic syndrome. Curr Diab Rep 5:70–75.
Lenczowski MJ, Bluthe RM, Roth J, Rees GS, Rushforth DA, van Dam AM, Tilders FJ, Dantzer R, Rothwell NJ, Luheshi GN (1999) Central administration of rat IL-6 induces HPA activation and fever but not sickness behavior in rats. Am J Physiol 276:R652–658.
Leon LR, Kozak W, Rudolph K, Kluger MJ (1999) An antipyretic role for interleukin-10 in LPS fever in mice. Am J Physiol 276:R81–89.
Lestage J, Verrier D, Palin K, Dantzer R (2002) The enzyme indoleamine 2,3-dioxygenase is induced in the mouse brain in response to peripheral administration of lipopolysaccharide and superantigen. Brain Behav Immun 16:596–601.
Linthorst AC, Flachskamm C, Muller-Preuss P, Holsboer F, Reul JM (1995) Effect of bacterial endotoxin and interleukin-1 beta on hippocampal serotonergic neurotransmission, behavioral activity, and free corticosterone levels: an in vivo microdialysis study. J Neurosci 15: 2920–2934.
Littlejohn TK, Takikawa O, Truscott RJ, Walker MJ (2003) Asp274 and his346 are essential for heme binding and catalytic function of human indoleamine 2,3-dioxygenase. J Biol Chem 278:29525–29531.
Loftis JM, Huckans M, Ruimy S, Hinrichs DJ, Hauser P (2008) Depressive symptoms in patients with chronic hepatitis C are correlated with elevated plasma levels of interleukin-1beta and tumor necrosis factor-alpha. Neurosci Lett 430:264–268.
Lowe B, Hochlehnert A, Nikendei C (2006) [Metabolic syndrome and depression]. Ther Umsch 63:521–527.
Lyketsos CG, Olin J (2002) Depression in Alzheimer's disease: overview and treatment. Biol Psychiatry 52:243–252.
Lyketsos CG, Lee HB (2004) Diagnosis and treatment of depression in Alzheimer's disease. A practical update for the clinician. Dement Geriatr Cogn Disord 17:55–64.
Maes M, Smith R, Scharpe S (1995) The monocyte-T-lymphocyte hypothesis of major depression. Psychoneuroendocrinology 20:111–116.
Maes M, Bosmans E, Suy E, Vandervorst C, DeJonckheere C, Raus J (1991) Depression-related disturbances in mitogen-induced lymphocyte responses and interleukin-1 beta and soluble interleukin-2 receptor production. Acta Psychiatr Scand 84:379–386.
Maes M, Lambrechts J, Bosmans E, Jacobs J, Suy E, Vandervorst C, de Jonckheere C, Minner B, Raus J (1992) Evidence for a systemic immune activation during depression: results of leukocyte enumeration by flow cytometry in conjunction with monoclonal antibody staining. Psychol Med 22:45–53.
Maier SF. (1984) Learned helplessness and animal models of depression. Prog neuropsychopharmacol Biol Psychiatry 8(3):435–46.
Maness LM, Kastin AJ, Banks WA (1998) Relative contributions of a CVO and the microvascular bed to delivery of blood-borne IL-1alpha to the brain. Am J Physiol 275:E207–212.
Martinsen EW (1994) Physical activity and depression: clinical experience. Acta Psychiatr Scand Suppl 377:23–27.
Massol J, Martin P, Belon JP, Puech AJ, Soubrie P (1989) Helpless behavior (escape deficits) in streptozotocin-diabetic rats: resistance to antidepressant drugs. Psychoneuroendocrinology 14:145–153.
McArthur R, Borsini F (2006) Animal models of depression in drug discovery: a historical perspective. Pharmacol Biochem Behav 84:436–452.
McCaffery JM, Frasure-Smith N, Dube MP, Theroux P, Rouleau GA, Duan Q, Lesperance F (2006) Common genetic vulnerability to depressive symptoms and coronary artery disease: a

review and development of candidate genes related to inflammation and serotonin. Psychosom Med 68:187–200.
McCusker RH, McCrea K, Zunich S, Dantzer R, Broussard SR, Johnson RW, Kelley KW (2006) Insulin-like growth factor-I enhances the biological activity of brain-derived neurotrophic factor on cerebrocortical neurons. J Neuroimmunol 179:186–190.
McElroy SL, Kotwal R, Malhotra S, Nelson EB, Keck PE, Nemeroff CB (2004) Are mood disorders and obesity related? A review for the mental health professional. J Clin Psychiatry 65:634–651.
McIntyr RS, Soczynska JK, Konarski JZ, Woldeyohannes HO, Law C, Miranda A, Fulgosi D, Kennedy SH (2007, Oct-Dec). Should Depressive Syndromes Be Reclassified as "Metabolic Syndrome Type II"? Ann Clin Psychiatry 19(4):257–64.
McMahon EJ, Bailey SL, Castenada CV, Waldner H, Miller SD (2005) Epitope spreading initiates in the CNS in two mouse models of multiple sclerosis. Nat Med 11:335–339.
Mellor AL, Munn DH (1999) Tryptophan catabolism and T-cell tolerance: immunosuppression by starvation? Immunol Today 20:469–473.
Mensah GA, Mokdad AH, Ford E, Narayan KM, Giles WH, Vinicor F, Deedwania PC (2004) Obesity, metabolic syndrome, and type 2 diabetes: emerging epidemics and their cardiovascular implications. Cardiol Clin 22:485–504.
Merali Z, Lacosta S, Anisman H (1997) Effects of interleukin-1beta and mild stress on alterations of norepinephrine, dopamine and serotonin neurotransmission: a regional microdialysis study. Brain Res 761:225–235.
Meyers CA (1999) Mood and cognitive disorders in cancer patients receiving cytokine therapy. Adv Exp Med Biol 461:75–81.
Meyers CA, Valentine AD (1995) Neurologic and psychiatric adverse effects of immunological therapy. CNS Drugs 3: 56–68.
Miller NE (1964), Some psychophysiological studies of motivation and the behavioral effects of illness. Bull Br Psychol Soc 17:1–20.
Miller SD, McMahon EJ, Schreiner B, Bailey SL (2007) Antigen presentation in the CNS by myeloid dendritic cells drives progression of relapsing experimental autoimmune encephalomyelitis. Ann NY Acad Sci 1103:179–191.
Milligan ED, Soderquist RG, Malone SM, Mahoney JH, Hughes TS, Langer SJ, Sloane EM, Maier SF, Leinwand LA, Watkins LR, Mahoney MJ (2006) Intrathecal polymer-based interleukin-10 gene delivery for neuropathic pain. Neuron Glia Biol 2:293–308.
Miyaoka H, Otsubo T, Kamijima K, Ishii M, Onuki M, Mitamura K (1999) Depression from interferon therapy in patients with hepatitis C. Am J Psychiatry 156:1120.
Moreau M, André C, O'Connor JC, Dumich SA, Woods JA, Kelley KW, Dantzer R, Lestage J, Castanon N (2008), Inoculation of Bacillus Calmette-Guerin to mice induces an acute episode of sickness behavior followed by chronic depressive-like behavior, Brain Behav Immun (submitted)
Monleon S, D'Aquila P, Parra A, Simon VM, Brain PF, Willner P (1995) Attenuation of sucrose consumption in mice by chronic mild stress and its restoration by imipramine. Psychopharmacology (Berl) 117:453–457.
Mormède C, Castanon N, Mèdina C, Moze E, Lestage J, Neveu PJ, Dantzer R (2002–2003). Chronic mild stress in mice decreases peripheral cytikine and increases central cytokine expression independently of IL-10 regulation of the cytokine network. Neuroimmunomodulation 10(6):359–66.
Moraska A, Fleshner M (2001) Voluntary physical activity prevents stress-induced behavioral depression and anti-KLH antibody suppression. Am J Physiol Regul Integr Comp Physiol 281:R484–489.
Morikawa O, Sakai N, Obara H, Saito N (1998) Effects of interferon-alpha, interferon-gamma and cAMP on the transcriptional regulation of the serotonin transporter. Eur J Pharmacol 349:317–324.
Mossner R, Daniel S, Albert D, Heils A, Okladnova O, Schmitt A, Lesch KP (2000) Serotonin transporter function is modulated by brain-derived neurotrophic factor (BDNF) but not nerve growth factor (NGF). Neurochem Int 36:197–202.

Muller N, Schwarz MJ (2007) The immune-mediated alteration of serotonin and glutamate: towards an integrated view of depression. Mol Psychiatry 12:988–1000.
Muller N, Schwarz MJ, Dehning S, Douhe A, Cerovecki A, Goldstein-Muller B, Spellmann I, Hetzel G, Maino K, Kleindienst N, Moller HJ, Arolt V, Riedel M (2006) The cyclooxygenase-2 inhibitor celecoxib has therapeutic effects in major depression: results of a double-blind, randomized, placebo controlled, add-on pilot study to reboxetine. Mol Psychiatry 11:680–684.
Muscat R, Papp M, Willner P (1992) Antidepressant-like effects of dopamine agonists in an animal model of depression. Biol Psychiatry 31:937–946.
Musselman DL, Lawson DH, Gumnick JF, Manatunga AK, Penna S, Goodkin RS, Greiner K, Nemeroff CB, Miller AH (2001) Paroxetine for the prevention of depression induced by high-dose interferon alfa. N Engl J Med 344:961–966.
Musso T, Gusella GL, Brooks A, Longo DL, Varesio L (1994) Interleukin-4 inhibits indoleamine 2,3-dioxygenase expression in human monocytes. Blood 83:1408–1411.
Nakamura S (1991) Axonal sprouting of noradrenergic locus coeruleus neurons following repeated stress and antidepressant treatment. Prog Brain Res 88:587–598.
Neveu PJ, Bluthe RM, Liege S, Moya S, Michaud B, Dantzer R (1998) Interleukin-1-induced sickness behavior depends on behavioral lateralization in mice. Physiol Behav 63:587–590.
Nutt DJ (2002) The neuropharmacology of serotonin and noradrenaline in depression. Int Clin Psychopharmacol 17 Suppl 1:S1–12.
Nys GM, van Zandvoort MJ, van der Worp HB, de Haan EH, de Kort PL, Kappelle LJ (2005) Early depressive symptoms after stroke: neuropsychological correlates and lesion characteristics. J Neurol Sci 228:27–33.
O'Connor JC, Lawson MA, André C, Moreau M, Lestage J, Castanon N, Kelley KW, Dantzer R (2008, Jan 15) Lipopolysaccharide-induced depressive-like behavior is mediated by indoleamine 2,3-dioxygenase activation in mice. Mol Psychiatry. [Epub ahead of print]
Owe-Young R, Webster NL, Mukhtar M, Pomerantz RJ, Smythe G, Walker D, Armati PJ, Crowe SM, Brew BJ (2008, May) Kynurenine pathway metabolism in human blood-brain-barrier cells: implications for immune tolerance & neurotoxicity. J Neurochem. 105(4):1346–57. Epub 2008 Jan 21.
Palin K, Bluthe RM, McCusker RH, Moos F, Dantzer R, Kelley KW (2007) TNFalpha-induced sickness behavior in mice with functional 55 kD TNF receptors is blocked by central IGF-I. J Neuroimmunol 187:55–60.
Palin K, Moreau MM, Orcel H, Duvoid-Guillou A, Rabie A, Kelley KW, Moos F (2008) Age impaired fluid homeostasis depends on the balance of IL-6/IGF-1 in the supraoptic nuclei of rats. Neurobiology of Aging. (Epub ahead of print)
Pan W, Kastin AJ (2001) Upregulation of the transport system for TNFalpha at the blood-brain barrier. Arch Physiol Biochem 109:350–353.
Peruzzo B, Pastor FE, Blazquez JL, Schobitz K, Pelaez B, Amat P, Rodriguez EM (2000) A second look at the barriers of the medial basal hypothalamus. Exp Brain Res 132:10–26.
Plotkin SR, Banks WA, Kastin AJ (1996) Comparison of saturable transport and extracellular pathways in the passage of interleukin-1 alpha across the blood-brain barrier. J Neuroimmunol 67:41–47.
Pohjasvaara T, Leppavuori A, Siira I, Vataja R, Kaste M, Erkinjuntti T (1998) Frequency and clinical determinants of poststroke depression. Stroke 29:2311–2317.
Pollak Y, Ovadia H, Goshen I, Gurevich R, Monsa K, Avitsur R, Yirmiya R (2000, Apr 3). Behavioral aspects of experimental autoimmune encephalomyelitis. J Neuroimmunol 104(1):31–6.
Pollak Y, Yirmiya R (2002) Cytokine-induced changes in mood and behaviour: implications for 'depression due to a general medical condition', immunotherapy and antidepressive treatment. Int J Neuropsychopharmacol 5:389–399.
Porsolt RD, Bertin A, Jalfre M (1977a) Behavioral despair in mice: a primary screening test for antidepressants. Arch Int Pharmacodyn Ther 229:327–336.
Porsolt RD, Le Pichon M, Jalfre M (1977b) Depression: a new animal model sensitive to antidepressant treatments. Nature 266:730–732.

Quan N, Banks WA (2007) Brain-immune communication pathways. Brain Behav Immun 21:727–735.

Quan N, Whiteside M, Herkenham M (1998) Time course and localization patterns of interleukin-1beta messenger RNA expression in brain and pituitary after peripheral administration of lipopolysaccharide. Neuroscience 83:281–293.

Quan N, Whiteside M, Kim L, Herkenham M (1997) Induction of inhibitory factor kappaBalpha mRNA in the central nervous system after peripheral lipopolysaccharide administration: an in situ hybridization histochemistry study in the rat. Proc Natl Acad Sci USA 94:10985–10990.

Quan N, Stern EL, Whiteside MB, Herkenham M (1999) Induction of pro-inflammatory cytokine mRNAs in the brain after peripheral injection of subseptic doses of lipopolysaccharide in the rat. J Neuroimmunol 93:72–80.

Quan N, He L, Lai W, Shen T, Herkenham M (2000) Induction of IkappaBalpha mRNA expression in the brain by glucocorticoids: a negative feedback mechanism for immune-to-brain signaling. J Neurosci 20:6473–6477.

Quan N, Herkenham M (2002) Connecting cytokines and brain: a review of current issues. Histol Histopathol 17:273–288.

Raison CL, Capuron L, Miller AH (2006) Cytokines sing the blues: inflammation and the pathogenesis of depression. Trends Immunol 27:24–31.

Renault PF, Hoofnagle JH, Park Y, Mullen KD, Peters M, Jones DB, Rustgi V, Jones EA (1987) Psychiatric complications of long-term interferon alfa therapy. Arch Intern Med 147:1577–1580.

Rhodes JS, Ryabinin AE, Crabbe JC (2005) Patterns of brain activation associated with contextual conditioning to methamphetamine in mice. Behav Neurosci 119:759–771.

Ridker PM, Hennekens CH, Buring JE, Rifai N (2000) C-reactive protein and other markers of inflammation in the prediction of cardiovascular disease in women. N Engl J Med 342:836–843.

Rivest S (2001) How circulating cytokines trigger the neural circuits that control the hypothalamic-pituitary-adrenal axis. Psychoneuroendocrinology 26:761–788.

Rivest S (2003) Molecular insights on the cerebral innate immune system. Brain Behav Immun 17:13–19.

Rodriguez EM, Blazquez JL, Pastor FE, Pelaez B, Pena P, Peruzzo B, Amat P (2005) Hypothalamic tanycytes: a key component of brain-endocrine interaction. Int Rev Cytol 247:89–164.

Romeo HE, Tio DL, Rahman SU, Chiappelli F, Taylor AN (2001) The glossopharyngeal nerve as a novel pathway in immune-to-brain communication: relevance to neuroimmune surveillance of the oral cavity. J Neuroimmunol 115:91–100.

Rosenberg PB (2005) Clinical aspects of inflammation in Alzheimer's disease. Int Rev Psychiatry 17:503–514.

Roth J, De Souza GE (2001) Fever induction pathways: evidence from responses to systemic or local cytokine formation. Braz J Med Biol Res 34:301–314.

Rozanski A, Blumenthal JA, Davidson KW, Saab PG, Kubzansky L (2005, Mar 1). The epidemiology, pathophysiology, and management of psychosocial risk factors in cardiac practice: the emerging field of behavioral cardiology. J Am Coll Cardiol 45(5):637–51.

Rudisch B, Nemeroff CB (2003) Epidemiology of comorbid coronary artery disease and depression. Biol Psychiatry 54:227–240.

Rupniak NM (2003) Animal models of depression: challenges from a drug development perspective. Behav Pharmacol 14:385–390.

Schiepers OJ, Wichers MC, Maes M (2005) Cytokines and major depression. Prog Neuropsychopharmacol. Biol Psychiatry 29:201–217.

Schrocksnadel K, Wirleitner B, Winkler C, Fuchs D (2006) Monitoring tryptophan metabolism in chronic immune activation. Clin Chim Acta 364:82–90.

Schwarcz R, Pellicciari R (2002) Manipulation of brain kynurenines: glial targets, neuronal effects, and clinical opportunities. J Pharmacol Exp Ther 303:1–10.

Selye, H (1936), A syndrome produced by diverse nocuous agents. Nature 138 (4, July)

Shen Y, Connor TJ, Nolan Y, Kelly JP, Leonard BE (1999) Differential effect of chronic antidepressant treatments on lipopolysaccharide-induced depressive-like behavioural symptoms in the rat. Life Sci 65:1773–1786.

Sheng H, Shao J, Dixon DA, Williams CS, Prescott SM, DuBois RN, Beauchamp RD (2000) Transforming growth factor-beta1 enhances Ha-ras-induced expression of cyclooxygenase-2 in intestinal epithelial cells via stabilization of mRNA. J Biol Chem 275:6628–6635.
Sherwin CM (1998) Voluntary wheel running: a review and novel interpretation. Anim Behav 56:11–27.
Shimbo D, Chaplin W, Crossman D, Haas D, Davidson KW (2005) Role of depression and inflammation in incident coronary heart disease events. Am J Cardiol 96:1016–1021.
Siegert RJ, Abernethy DA (2005) Depression in multiple sclerosis: a review. J Neurol Neurosurg Psychiatry 76:469–475.
Skilton MR, Moulin P, Terra JL, Bonnet F (2007) Associations between anxiety, depression, and the metabolic syndrome. Biol Psychiatry 62:1251–1257.
Smith RS (1991) The macrophage theory of depression. Med Hypotheses 35:298–306.
Song C, Leonard BE (2005). The olfactory bulbectomised rat as a model of depression. Neurosci Biobehav Rev 29(4-5):627–47. Epub 2005 Apr 25.
Sola S, Mir MQ, Lerakis S, Tandon N, Khan BV (2006) Atorvastatin improves left ventricular systolic function and serum markers of inflammation in nonischemic heart failure. J Am Coll Cardiol 47:332–337.
Spalletta G, Bossu P, Ciaramella A, Bria P, Caltagirone C, Robinson RG (2006) The etiology of poststroke depression: a review of the literature and a new hypothesis involving inflammatory cytokines. Mol Psychiatry 11:984–991.
Sparkman NL, Buchanan JB, Heyen JR, Chen J, Beverly JL, Johnson RW (2006) Interleukin-6 facilitates lipopolysaccharide-induced disruption in working memory and expression of other proinflammatory cytokines in hippocampal neuronal cell layers. J Neurosci 26:10709–10716.
Starkstein SE, Mizrahi R (2006) Depression in Alzheimer's disease. Expert Rev Neurother 6:887–895.
Steru L, Chermat R, Thierry B, Simon P (1985) The tail suspension test: a new method for screening antidepressants in mice. Psychopharmacology (Berl) 85:367–370.
Stone TW, Forrest CM, Mackay GM, Stoy N, Darlington LG (2007) Tryptophan, adenosine, neurodegeneration and neuroprotection. Metab Brain Dis 22:337–352.
Strine TW, Mokdad AH, Dube SR, Balluz LS, Gonzalez O, Berry JT, Manderscheid R, Kroenke K (2008) The association of depression and anxiety with obesity and unhealthy behaviors among community-dwelling US adults. Gen Hosp Psychiatry 30:127–137.
Strle K, McCusker RH, Johnson RW, Zunich SM, Dantzer R, Kelley KW (2008, Apr) The Prototypical Anti-Inflammatory Cytokine, IL-10, Prevents Loss of IGF-I-Induced Myogenin Protein Expression Caused by IL-1beta. Am J Physiol Endocrinol Metab 294(4):E709–18. Epub 2008 Feb 12.
Strle K, Zhou JH, Shen WH, Broussard SR, Johnson RW, Freund GG, Dantzer R, Kelley KW (2001) Interleukin-10 in the brain. Crit Rev Immunol 21:427–449.
Stunkard AJ, Faith MS, Allison KC (2003) Depression and obesity. Biol Psychiatry 54:330–337.
Sugimoto H, Oda S, Otsuki T, Hino T, Yoshida T, Shiro Y (2006) Crystal structure of human indoleamine 2,3-dioxygenase: catalytic mechanism of O2 incorporation by a heme-containing dioxygenase. Proc Natl Acad Sci USA 103:2611–2616.
Sutherland JP, McKinley B, Eckel RH (2004) The metabolic syndrome and inflammation. Metab Syndr Relat Disord 2:82–104.
Swiergiel AH, Smagin GN, Dunn AJ (1997) Influenza virus infection of mice induces anorexia: comparison with endotoxin and interleukin-1 and the effects of indomethacin. Pharmacol Biochem Behav 57:389–396.
Takikawa O, Yoshida R, Kido R, Hayaishi O (1986) Tryptophan degradation in mice initiated by indoleamine 2,3-dioxygenase. J Biol Chem 261:3648–3653.
Taniguchi T, Sono M, Hirata F, Hayaishi O, Tamura M, Hayashi K, Iizuka T, Ishimura Y (1979) Indoleamine 2,3-dioxygenase. Kinetic studies on the binding of superoxide anion and molecular oxygen to enzyme. J Biol Chem 254:3288–3294.
Tayal V, Kalra BS (2008) Cytokines and anti-cytokines as therapeutics – An update. Eur J Pharmacol 579:1–12.

Taylor MW, Feng GS (1991) Relationship between interferon-gamma, indoleamine 2,3-dioxygenase, and tryptophan catabolism. Faseb J 5:2516–2522.
Tazi A, Dantzer R, Crestani F, Le Moal M (1988) Interleukin-1 induces conditioned taste aversion in rats: a possible explanation for its pituitary-adrenal stimulating activity. Brain Res 473:369–371.
Telner JI, Singhal RL (1981) Effects of nortriptyline treatment on learned helplessness in the rat. Pharmacol Biochem Behav 14:823–826.
Telner JI, Singhal RL, Lapierre YD (1981) Reversal of learned helplessness by nortriptyline. Prog Neuropsychopharmacol 5:587–590.
Teste JF, Pelsy-Johann I, Decelle T, Boulu RG (1993) Anti-immobility activity of different antidepressant drugs using the tail suspension test in normal or reserpinized mice. Fundam Clin Pharmacol 7:219–226.
Thomas WE (1999) Brain macrophages: on the role of pericytes and perivascular cells. Brain Res Brain Res Rev 31:42–57.
Tune LE (1998) Depression and Alzheimer's disease. Depress Anxiety 8 Suppl 1:91–95.
Tuppo EE, Arias HR (2005) The role of inflammation in Alzheimer's disease. Int J Biochem Cell Biol 37:289–305.
van der Bruggen T, Nijenhuis S, van Raaij E, Verhoef J, van Asbeck BS (1999) Lipopolysaccharide-induced tumor necrosis factor alpha production by human monocytes involves the raf-1/MEK1-MEK2/ERK1-ERK2 pathway. Infect Immun 67:3824–3829.
Vargas-Perez H, Borrelli E, Diaz JL (2004) Wheel running use in dopamine D2L receptor knockout mice. Neurosci Lett 366:172–175.
Vila N, Castillo J, Davalos A, Chamorro A (2000) Proinflammatory cytokines and early neurological worsening in ischemic stroke. Stroke 31:2325–2329.
Vitkovic L, Konsman JP, Bockaert J, Dantzer R, Homburger V, Jacque C (2000) Cytokine signals propagate through the brain. Mol Psychiatry 5:604–615.
Watkins LR, Maier SF (2000) The pain of being sick: implications of immune-to-brain communication for understanding pain. Annu Rev Psychol 51:29–57.
Watkins LR, Maier SF, Goehler LE (1995a) Cytokine-to-brain communication: a review & analysis of alternative mechanisms. Life Sci 57:1011–1026.
Watkins LR, Maier SF, Goehler LE (1995b) Immune activation: the role of pro-inflammatory cytokines in inflammation, illness responses and pathological pain states. Pain 63:289–302.
Watkins LR, Hutchinson MR, Ledeboer A, Wieseler-Frank J, Milligan ED, Maier SF (2007) Norman Cousins Lecture. Glia as the "bad guys": implications for improving clinical pain control and the clinical utility of opioids. Brain Behav Immun 21:131–146.
Watkins LR, Goehler LE, Relton JK, Tartaglia N, Silbert L, Martin D, Maier SF (1995c) Blockade of interleukin-1 induced hyperthermia by subdiaphragmatic vagotomy: evidence for vagal mediation of immune-brain communication. Neurosci Lett 183:27–31.
Weinstock-Guttman B, Ransohoff RM, Kinkel RP, Rudick RA (1995) The interferons: biological effects, mechanisms of action, and use in multiple sclerosis. Ann Neurol 37:7–15.
Weisman D, Hakimian E, Ho GJ (2006) Interleukins, inflammation, and mechanisms of Alzheimer's disease. Vitam Horm 74:505–530.
Wichers MC, Koek GH, Robaeys G, Verkerk R, Scharpe S, Maes M (2005) IDO and interferon-alpha-induced depressive symptoms: a shift in hypothesis from tryptophan depletion to neurotoxicity. Mol Psychiatry 10:538–544.
Wichers MC, Maes M (2004, Jan). The role of indoleamine 2,3- dioxygenase (IDO) in the pathophysiology of interferon-alpha-induced depression. J Psychiatry Neurosci. 29(1):11–7.
Widner B, Laich A, Sperner-Unterweger B, Ledochowski M, Fuchs D (2002) Neopterin production, tryptophan degradation, and mental depression – what is the link? Brain Behav Immun 16:590–595.
Willner P (1995) Animal models of depression: validity and applications. Adv Biochem Psychopharmacol 49:19–41.
Willner P (2005) Chronic mild stress (CMS) revisited: consistency and behavioural-neurobiological concordance in the effects of CMS. Neuropsychobiology 52:90–110.

Willner P, Towell A, Sampson D, Sophokleous S, Muscat R (1987). Reduction of sucrose preference by chronic unpredictable mild stress, and its restoration by a tricyclic antidepressant. Psychopharmacology (Berl) 93(3):358–64.

Xia Z, DePierre JW, Nassberger L (1996) Tricyclic antidepressants inhibit IL-6, IL-1 beta and TNF-alpha release in human blood monocytes and IL-2 and interferon-gamma in T cells. Immunopharmacology 34:27–37.

Yirmiya R (1996) Endotoxin produces a depressive-like episode in rats. Brain Res 711:163–174.

Yirmiya R, Barak O, Avitsur R, Gallily R, Weidenfeld J (1997) Intracerebral administration of Mycoplasma fermentans produces sickness behavior: role of prostaglandins. Brain Res 749:71–81.

Yirmiya R, Pollak Y, Barak O, Avitsur R, Ovadia H, Bette M, Weihe E, Weidenfeld J (2001) Effects of antidepressant drugs on the behavioral and physiological responses to lipopolysaccharide (LPS) in rodents. Neuropsychopharmacology 24:531–544.

Yirmiya R, Pollak Y, Morag M, Reichenberg A, Barak O, Avitsur R, Shavit Y, Ovadia H, Weidenfeld J, Morag A, Newman ME, Pollmacher T (2000) Illness, cytokines, and depression. Ann NY Acad Sci 917:478–487.

Young-Xu Y, Chan KA, Liao JK, Ravid S, Blatt CM (2003) Long-term statin use and psychological well-being. J Am Coll Cardiol 42:690–697.

Yuan W, Collado-Hidalgo A, Yufit T, Taylor M, Varga J (1998) Modulation of cellular tryptophan metabolism in human fibroblasts by transforming growth factor-beta: selective inhibition of indoleamine 2,3-dioxygenase and tryptophanyl-tRNA synthetase gene expression. J Cell Physiol 177:174–186.

Yudkin JS (2007) Inflammation, obesity, and the metabolic syndrome. Horm Metab Res 39:707–709.

Zaremba J, Losy J (2001) Early TNF-alpha levels correlate with ischaemic stroke severity. Acta Neurol Scand 104:288–295.

Zaremba J, Skrobanski P, Losy J (2001) Tumour necrosis factor-alpha is increased in the cerebrospinal fluid and serum of ischaemic stroke patients and correlates with the volume of evolving brain infarct. Biomed Pharmacother 55:258–263.

Effect of Systemic Challenge with Bacterial Toxins on Behaviors Relevant to Mood, Anxiety and Cognition

Rachel A. Kohman, Joanne M. Hash-Converse, and Alexander W. Kusnecov

Abstract Activation of the immune system is known to induce functional alterations in the central nervous system that subsequently modify behavior. However, the degree of influence the immune system has on mood and cognitive function remains in question. The present chapter begins with a focused discussion of the behavioral and neuroendocrine effects of superantigen exposure, a model that activates the immune system in a T-cell dependent manner. Administration of a superantigen induces gustatory neophobia, activation of the hypothalamic-pituitary adrenal axis, and activates brain regions associated with emotional behavior. The chapter ends with a review of the current literature on the effects of immune activation on learning and memory processes, in which most work has used the T cell independent endotoxin model and interleukin-1 administration. Whether cytokine production facilitates or disrupts cognitive function appears to be dependent on, among other factors, the dose, timing of immune challenge, and test procedure employed. Throughout the chapter is a discussion of the potential neural mechanisms of the mood and cognitive alterations association with an immunological challenge.

Keywords Anxiety · Learning · Memory · Superantigens · LPS · Lipopolysaccharide · Cytokine · T cells

1 Introduction

The immune system comprises morphologically and functionally distinct cells that process, communicate, and retain information in a manner not unlike the brain. These cells are distributed widely in primary and secondary lymphoid

---

R.A. Kohman (✉)
Department of Psychology, Rutgers University, 152 Frelinghuysen Road, Piscataway, NJ 08855, USA

Supported by Grants MH60706, NIEHS P30 ES05022 and NIEHS Graduate Training grant 5T32 E507148.

A. Siegel, S.S. Zalcman (eds.), *The Neuroimmunological Basis of Behavior and Mental Disorders*, DOI 10.1007/978-0-387-84851-8_10,

organs such as the thymus, lymph nodes, spleen, gut and lungs, and are in dynamic flux throughout these organs via a specialized system of lymphatic vessels interconnected with the cardiovascular circulatory system. Therefore, the immune response, and the inflammation that emerges from its induction, is not confined to any particular anatomical region, which renders most tissues susceptible to potential disruption of function during a prolonged immune response. As a result, regulatory mechanisms have evolved to ensure rapid resolution of the immune response before the development of tissue pathology. Indeed, self-regulation is a hallmark feature of the immune system, and has always been conceived in terms of autocrine and paracrine interactions between cells and their soluble products, most notably the cytokines and chemokines. However, it has become increasingly clear in recent decades that the idea of "self-regulation" by the immune system encompasses more than the intrinsic variables inherent to the immune apparatus in and of itself, but extends also to the activities and functions of the central nervous system (CNS).

This notion emerged from studies demonstrating that multiple facets of CNS function change significantly following immunological activation by various antigens, including sheep red blood cells, benign proteins (e.g., TNP-KLH), and viral and bacterial toxins. Most prominent has been the study of neuroendocrine and neurochemical alterations pursuant to immune challenge, which can serve to regulate ongoing immune responses. For example, it has been documented that increased concentrations of peripheral hormones (e.g., glucocorticoids) and neurotransmitters (e.g., norepinephrine, acetylcholine) following immunological challenge represent a higher-order level of immunological regulation that results in attenuation of immunopathological conditions such as autoimmune disease and toxic shock (Nave et al., 2004; Hernandez et al., 2007). Furthermore, regulation occurs through adrenergic, cholinergic, glucocorticoid, and a number of other neuropeptide and hormone receptors expressed on lymphocytes, monocytes, polymorphonuclear phagocytic cells (macrophages), and dendritic cells. This information has already been thoroughly reviewed (Ader et al., 1995), substantiating the notion that physical interactions between cells of the immune system and molecules produced by CNS activation represent a second tier of functional control over immune responses to antigens.

2 Behavior and the Immune System

Since immune responses produce neurobiological changes, it is not unexpected that behavior, as an emergent property of the brain, should also be modified. In this regard, the pattern of behavior change after immunologic challenge conforms to a "sickness behavior" syndrome, the primary features of which consist of motoric and motivational changes characterized by reduced locomotion, food intake, and social exploration, as well as altered patterns of sleeping behavior (Hart, 1988). In animal studies, these effects have been observed using primarily proinflammatory stimuli, such as the endotoxin lipopolysaccharide (LPS), as well as the inflammatory cytokines interleukin-1 (IL-1) and tumor necrosis factor (TNF), and to a lesser extent other cytokines, such as IL-2, IL-6, and interferon (Anisman and Merali, 1999; Dantzer, 2004; Anisman et al., 2005). Induction of reduced food intake is one of the key features of immunologic stimulation with LPS, although other behavioral changes, such as increased somnolence, anhedonia, and altered nociception, have been observed (Mason, 1993; Borowski et al., 1998; Lacosta et al., 1999). Moreover, as will be discussed in this chapter, LPS and specific cytokines alter behavior in tests of learning and memory. Therefore, activation of the immune system produces a considerable range of behavioral changes that are consistent with the underlying activation of neural substrates mediating cognitive and motivational behaviors. Such effects have been compared to the behavioral profile presented by clinically depressed individuals, leading to suggestions for a potential role of the immune system in depression. However, critical evidence implicating the immune system as a causal factor in the appearance of clinical depression that is unrelated to comorbid disease states (e.g., cancer and multiple sclerosis) has remained elusive. Furthermore, disease states characterized by immune dysregulation (e.g., rheumatoid arthritis, multiple sclerosis) are not invariably accompanied by depression. Therefore, the induction of depressive-like symptomatology by immunologic factors, while compelling, does not necessarily argue for an immune aetiology in depression, since depressive states can be induced by non-immunologic conditions, including metabolic alterations, exposure to psychogenic stressors, as well as degenerative neuropathology. Nonetheless, the immune system may interact with these and other factors contributing to depression, and therefore, given that the immune response engages neural substrates involved in mood and cognition, it is important to incorporate immunologic processes in explanatory models of depression.

In the remainder of this chapter, we will discuss the impact of immunologic challenge on cognitive and motivational behavior. Initially, we will focus on literature documenting the neuroendocrine, neuroanatomical, and behavioral effects of challenge with bacterial superantigens. We do so because there has been a tendency to conceptualize the CNS impact of the immune response in terms of studies that have mainly used cytokine administration or administration of LPS. However, in the same way that behavior and CNS function differs in response to a variety of different psychogenic and physical (i.e., non-immune) stressors, the same can be said of the particular outcomes that follow exposure to the multiple types of pathogens that stimulate the immune system. Finally, after discussing the neural and behavioral effects of superantigens, we will turn to a discussion of the relationship of the immune system to learning and memory. In so doing, we will examine literature that has primarily used endotoxin (i.e., LPS) challenge as the principle model of immunologic stimulation. However, we will not review in detail the wealth of research examining the many different facets of sickness behavior, since this has been thoroughly reviewed in recent years by others (Szelenyi and Szekely, 2004; Dantzer, 2006).

3 Superantigens and the CNS

Superantigens (SAgs) are a distinct class of immunogenic molecules that are produced largely by gram positive bacteria (e.g., *Staphylococcal aureus and Streptococci*), but can also be derived from viruses (e.g., mammary tumor virus; Proft and Fraser, 2003; Wang et al., 2004). The term superantigen was coined to contrast the effects of conventional antigens, which are digested and presented as peptide–MHC complexes (pMHC) to antigen-specific T cell receptors (TCRs), and therefore stimulate only 0.002% of the T cell population (Zamoyska, 2006). In contrast, SAgs can stimulate up to 10–20% of T cells in an MHC-dependent manner and drive them into an extensive lymphoproliferative phase that is associated with high levels of cytokine plasma concentrations (Gonzalo et al., 1993). Such changes can easily be detected without resorting to highly sensitive and detailed methodology, such as limiting dilution analysis or polymerase chain reaction (PCR) analysis of cytokine mRNA; moreover, the magnitude change in the circulating cytokine concentration is likely the basis of the modified CNS activity that takes place after SAg challenge.

T cell stimulation by SAgs is oligoclonal in nature, due to the selectivity of different SAgs for unique motifs on the variable region of the TCR beta chain (Vβ). In mice, rats, and humans a considerable number of different Vβ genes have been identified, and these have been numerically classified (e.g., Vβ1, Vβ2, etc.). The product of the same Vβ gene can be present on TCRs with multiple peptide specificities, such that a given SAg with an affinity for a particular Vβ region can stimulate many T cells, each with different clonal specificity. Therefore, organismic exposure to a SAg selective for only one particular Vβ motif (e.g., Vβ3), will result only in a subset of the pool of total T cells being activated. This activation is MHC-dependent, in that SAgs are known to cross-link the outer portion of the MHC class II molecule with the Vβ region of the TCR. Interestingly, while both CD4 and CD8 T cells are engaged by SAgs, the initial activation is not based on the canonical Lck-dependent signal transduction pathway for TCR signaling, and instead involves an alternative pathway (Zamoyska, 2006). However, the eventual result is the same, in that considerable IL-2 production and cell proliferation follows SAg stimulation (Calandra et al., 1998; Luxembourg and Grey, 2002; Proft and Fraser, 2003).

3.1 Staphylococcal SAgs

Immunological studies of SAg effects have focused largely on the staphylococcal enterotoxins (SEs), which are exotoxins produced by the gram positive bacteria *S. aureus*. These exotoxins are serologically distinct, and have been given an alphabetical nomenclature (viz., SEA [for staphylococcal enterotoxin A], SEB, SEC, and so on), with some exceptions (e.g., toxic shock syndrome toxin-1 [TSST-1]). Analysis of the Vβ specificity of SEs has revealed considerable heterogeneity in their affinity to the full range of known Vβ genes, which has been reviewed by Proft and Fraser (2003), and applies also to the many streptococcal SAgs that are being investigated.

In the mouse, the most commonly studied staphylococcal SAgs are SEA and SEB, which preferentially stimulate Vβ3 and Vβ11 (SEA), and Vβ8 (SEB) T cells, respectively. The relative percentage of T cells bearing any one particular member of the Vβ family varies among inbred mouse strains, resulting in a differential reliance on C57BL/6 mice for studies involving SEA, while SEB is typically administered to BALB/c mice (Liang et al., 1994). However, rats also have been shown to respond to these SAgs, although, as discussed below, SEB has been the main agent used to study neurobiological effects (Gold et al., 1994).

3.2 Neuroendocrine Activation by SEA and SEB

Gonzalo et al. (1993) first showed that challenge of C57BL/6 and BALB/c mice with SEA and SEB, respectively, resulted in elevated plasma concentrations of corticosterone. This suggested activation of the hypothalamic-pituitary-adrenal axis (HPA) axis, and potential reactivity of higher-order neural mechanisms regulating neuroendocrine activity. Indeed, Shurin et al. (1997) confirmed that the elevated corticosterone response to SEB was associated with elevated adrenocorticotropic hormone (ACTH) levels, and furthermore was associated with increased expression of the immediate early gene *c-fos* in the paraventricular nucleus (PVN) of the hypothalamus. Additional support for activation of the PVN came from studies in rats challenged with SEB (Goehler et al., 2001), although this was not confirmed in a recent study by Serrats and Sawchenko (2006). No clear explanation for this discrepancy is currently available, although the latter study did observe a significant increase in plasma ACTH. Moreover, challenge of C57BL/6 mice with a single, acute injection of SEA increased plasma corticosterone and ACTH (Kaneta and Kusnecov, 2005), which at least for measures of corticosterone, persists after three, but not four, SEA challenges (Urbach-Ross et al., 2008). Consequently, there appears to be little dispute that SEA and SEB activate the pituitary-adrenal axis in mice and rats.

The cellular basis for the activation of the HPA axis by SAgs appears to rely on T lymphocytes. This has been demonstrated in various ways. Challenge of nude BALB/c mice with SEB failed to increase corticosterone levels, unless animals were reconstituted with T cells (Williams et al., 1994). Further, the corticosterone response to SEB was abrogated in animals that were treated with the T cell immunosuppressant, cyclosporine A (Shurin et al., 1997). This did not appear to be related to a pharmacologic effect on macrophages, since animals that were not immunosuppressed with cyclosporine, but were subjected to systemic macrophage depletion, continued to show an increased corticosterone response to SEB (Shurin et al., 1997). Interestingly, in this latter study, macrophage depletion attenuated the corticosterone response to LPS, supporting other findings in rats that LPS-induced HPA axis activation is dependent on tissue macrophages (Shurin et al., 1997). With respect to challenge with SEA, it was shown that Rag-1 knockout mice that fail to develop mature, functional T cells do not display an increased corticosterone response following SEA administration (Kawashima and Kusnecov, 2002). These results support a T cell

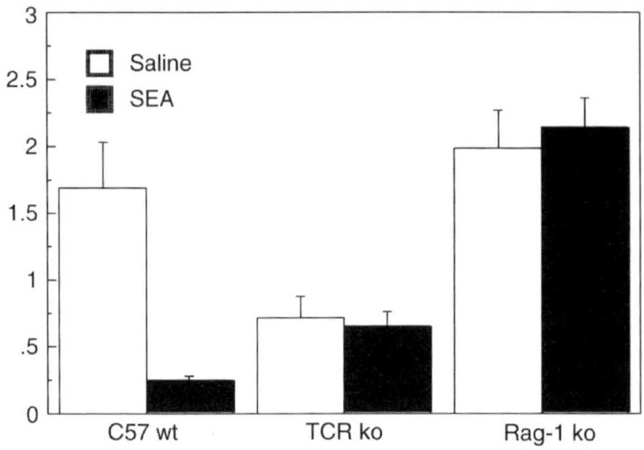

Fig. 1 Effect of 10μg staphylococcal enterotoxin A (SEA) on consumption of a liquid diet in wildtype, T cell receptor ($TCR^{-/-}$) and Rag-$1^{-/-}$ mice (n=4/group)

dependency for superantigenic activation of adrenocortical responses, although changes at the CNS level have yet to be confirmed following manipulation of T cells (however, see Fig. 1 below regarding behavioral effects of SEA).

The ACTH response is driven by a number of secretagogues released by neurosecretory cells from the hypothalamus, in particular those emanating from the PVN. Corticotropin releasing hormone (CRH; also known as corticotropin releasing factor, CRF) is considered the primary stimulus for pituitary ACTH-secreting cells during conditions of acute stress, although arginine vasopressin (AVP), has been shown to contribute to the neuroendocrine control of ACTH production, most likely under chronic stress conditions (for reviews see Aguilera et al., 2007, and Makara et al., 2004). The activating effects of cytokines, such as interleukin-1, were demonstrated some time ago to rely on hypothalamic release of CRH (Veening et al., 1993). Fewer studies have been reported in which pharmacological or immunoneutralization procedures have confirmed a role for central CRH release following systemic immunological challenge (Luheshi et al., 1996).

A number of studies suggest that in mice, SAg effects on the HPA axis also appear to rely on CRH release. In BALB/c mice challenged with SEB, it was shown that mRNA levels for CRH were increased in the PVN and central nucleus of the amygdala (Kusnecov et al., 1999). Furthermore, in the same study, administration of antiserum to CRH significantly attenuated the ACTH response to SEB, suggesting that the increased CRH mRNA reflected increased release of translated peptide. More recently, the increased corticosterone response to SEA in C57BL/6 mice was blocked by pretreatment with astressin, a selective CRH receptor 1 antagonist (Rossi-George et al., 2005). This provides further support that the HPA axis response to acute injections of SAgs involves the release of CRH. Whether this also applies to the use of other SAgs, under conditions of repeated administration (Urbach-Ross et al., 2008), as well as to other animal models, such as rats and primates, remains to be determined.

3.3 Behavioral Effects of Bacterial SAgs

As discussed earlier, the cardinal features of depression, such as anorexia, anhedonia, impaired somnolence, and cognitive deficits have been observed following certain forms of immunologic challenge (Anisman et al., 2005; Dantzer et al., 2008). Such behavioral changes likely reflect a motivational shift that aligns the goals of the organism with the effector state of the immune system. That is, the elimination of pathogens at the cellular level requires a general systemic adjustment that would restrict behaviors that otherwise would compromise the efforts of the immune system. To this end, as already reviewed, the increased activation of the HPA axis has been hypothesized to regulate ongoing immune responses (Besedovsky and del Rey, 2002), and there is some evidence to support this notion (Besedovsky and del Rey, 2000). Further, as judged by immediate early gene mapping studies, at higher-order levels of CNS function there is immunologically induced activation of areas that serve as substrates for cognitive and emotional behavior (e.g., Rossi-George et al., 2005). Therefore, sickness behavior, with its attendant decline in mobility and exploration, does not necessarily imply a state of behavioral inertia.

Interestingly, mice challenged with bacterial SAgs do not show frank signs of illness as might occur in response to LPS. This is not to say that overt signs of illness cannot be induced by SAgs, since administration of SEB has been shown to produce septic shock and increased mortality (Gonzalo et al., 1994). However, these effects are observed using unique manipulations such as D-galactosamine treatment, which sensitizes animals to the SAg-induced cytokine responses (Aoki et al., 1995). Moreover, staphylococcal enterotoxins delivered enterically can produce malaise, although this does not correlate with T cell activation in mice (Harris et al., 1993). Aside from these special conditions of administration, mice given bolus intraperitoneal (i.p.) injections of SEA or SEB alone at doses causing elevations in circulating IL-2 and TNFα, as well as activation of the HPA axis, do not exhibit classical signs of infectious-like illness, such as piloerection, immobility, diarrhea, and body weight loss. However, a notable and reliable effect of SEA or SEB treatment is reduced ingestion of food, whether it be a liquid diet or commercial food pellets (Kusnecov et al., 1999; Rossi-George et al., 2005). This was found to be more pronounced if food was novel, rather than familiar (Kusnecov et al., 1999; Kawashima and Kusnecov, 2002; Rossi-George et al., 2005). Furthermore, when animals have been sufficiently familiarized with a given food in an operant chamber where nose-pokes deliver food pellets, challenge with SEA does not disrupt performance or motivation to ingest food (Kusnecov laboratory, unpublished observations). Consequently, there is no indication that a serious form of malaise is induced by challenge with SEA or SEB.

These data raised the question of whether immunologic stimulation that is sufficient to stimulate the CNS, but not to a degree that arrests behavior and creates somatic illness or malaise, augments behavioral inhibition in the presence of novel stimuli. Behavioral inhibition is observed in young children, and is thought to be a precursor or component of social anxiety (Hirshfeld et al., 1997). Furthermore, it is associated with increased activity in the central nucleus of the amygdala (Kubota et al., 2004), and as the basis for anxiety, may also involve impaired feedback regulation of the amygdala by the prefrontal cortex (Berkowitz et al., 2007). It is notable, therefore, that challenge with SEA and SEB produces increased *c-fos* expression in the central nucleus of the amygdala, as well as other areas of the brain relevant to emotion regulation, such as the bed nucleus of the stria terminalis, septum, and prefrontal cortex (Serrats and Sawchenko, 2006). These data are consistent with increased HPA axis activation after SAg challenge. More importantly, however, they suggest that the reduced ingestion of novel food substances is associated with increased neuronal activity in regions of the brain that are engaged by novelty.

Besides gustatory neophobic responses, challenge with SEA was shown to reduce interactions with a novel object (Kawashima and Kusnecov, 2002). In this test, animals initially explore an empty open field environment for 10 min, after which an unfamiliar cylindrical object is placed in the center of the field. Animals react dramatically to the introduction of the object, confining their movements to the perimeter of the field, gradually developing approach behaviors that bring them closer to the object. In most cases, animals will eventually make contact with the object through nose-touches and hind-limb rearing. This test is particularly useful as an index of anxiety-like behavior in mice (Dulawa et al., 1999; Henry et al., 2006). Therefore, the reduction of novel object contact after animals were given acute injections of SEA (Kawashima and Kusnecov, 2002), likely reflected increased anxiety and/or neophobic behavior. This effect is not due to impaired movement, since both SEA and SEB do not produce a reduction in locomotor behavior (Kawashima and Kusnecov, 2002; Rossi-George et al., 2004).

Interestingly, while these data suggest anxiety-like changes following SAg administration, the display of such changes appears to be limited to specific tests. The elevated plus maze (EPM) is a commonly used test of anxiety in rats and mice, in which animals show preference for either of two closed (i.e., containing walls) arms, and avoid either of the two open, unprotected (i.e., no walls) arms. When C57BL/6 and BALB/c mice were challenged with SEA or SEB, respectively, behavior in the EPM was significantly biased towards increased number of entries into and time spent in the open arms (Rossi-George et al., 2004). An additional test of anxiety, behavior in the light–dark box, assesses avoidance of a brightly lit area with preference given to a dark compartment; increased time spent in the dark compartment is interpreted as signifying an increased anxiety-like state (Ballaz et al., 2007). When BALB/c mice were challenged with SEB, and subsequently tested in the light–dark box, there were no changes in latency to exit from the dark compartment, number of light–dark transitions, nor total time spent in the illuminated area (Rossi-George et al., 2004). When compared to the EPM data, where there was increased percent time spent in the open arms by SEB challenged mice, it is interesting to note that no behavioral change from saline-injected control animals was observed in the light–dark box. It has been suggested that the light–dark box and EPM are assessing different aspects of anxiety (Holmes et al., 2001), and therefore, the differential changes due to SEB may reflect underlying differences in reactivity to the unique stimulus conditions of each test. Whether "anxiety" is discernable in the context of SEB (or SEA) challenge, however, is not readily apparent, given the formulation of what constitutes "anxiety" in these tests. However, it is not unusual for these tests to fail in screening for enhanced anxiety subsequent to anxiogenic treatments (e.g., discussion in Rossi-George et al., 2004). Therefore, anxiety-like states induced by immunologic challenges may only be detected under specific situational demands, reinforcing the need for multiple test batteries. Moreover, careful consideration should be given to the stressful nature of many behavioral tests. For example, given that traditional animal assessments of anxiety (e.g., EPM) induce states of arousal and limbic brain activation by virtue of their contextual novelty (Nguyen et al., 2006), behaviorally and/or neuroanatomically this may not add further to the pre-existing stimulation by immunologic stimuli, such as SEA or SEB. Alternatively, sequential behavioral testing in varying contexts may serve to discriminate between animals on the basis of pretreatment with these SAgs. This idea was suggested by the data reported by Rossi-George et al. (2004), who observed that if SEB-treated BALB/c mice were placed in a novel test cage for assessment of food neophobia, subsequent testing in the light–dark box produced significantly less entries and time in the illuminated area.

Collectively, these data show that depending on what types of behavioral tests are administered, and the circumstances under which they are given, evidence of a behavioral alteration suggestive of a increased anxiety-like state can vary. What is fortunate about the effects of SAg challenge at the doses that have been used in mice (viz., 200–400 μg/Kg for SEA; 2 mg/Kg for SEB), is that there is no significant impact on locomotor behavior that would mitigate against the use of behavioral tests that depend on motor activity (a prerequisite for assessing most rodent behaviors). Consequently, given the existence of increased neuronal activity in limbic brain regions (Serrats and Sawchenko, 2006), and increased HPA axis activity (Kusnecov et al., 1999), it is possible to gain a better sense of precisely what types of behaviors are either disrupted or augmented by SAg challenge.

The central and peripheral mechanisms for the behavioral effects of SAg administration are only beginning to be explored. Centrally, there is good reason to believe that the induction of CRH promotes at least the effects of SEA on reduced food ingestion. The amount of food or water consumed in a novel environment can be modulated by CRH, as well as anxiogenic and anxiolytic drugs (Koob and Heinrichs, 1999). Furthermore, it is now known that there are at least two major types of receptors for CRH, CRH-R1 and CRH-R2 (Liebsch et al., 1999), with the former believed to mediate the anxiogenic effects of CRH (Steckler and Holsboer, 1999). Alternatively, the anorexic effects of CRH likely involve CRH-R2, which is more selectively engaged by the more recently discovered peptide, Urocortin (UCN). With this information in mind, Kaneta and Kusnecov (2005) delivered one of two different CRH receptor antagonists intracerebroventricularly (i.c.v) into C57BL/6 mice. Administration of the non-selective antagonist α-helical CRF blocked the anorexic effect of SEA, whereas administration of the selective CRH-R2 antagonist, astressin-2B, failed to alter the reduction in food intake. This suggested that the anorexic effects of SEA are mediated by CRH-R1 stimulation, and provides support for the hypothesis that an underlying state of anxiety may drive the suppression of food intake. Furthermore, the data are in keeping with the previously reviewed evidence for a CRH-mediated effect on HPA axis activation, and support the notion that SEA treatment increases the release of CRH in the brain.

In pursuing the question of how behavioral effects are mediated following SAg-induced T cell activation, we initially tested whether T cells were required for the behavioral effects of SEA challenge in C57BL/6 mice. Previously, we had reported that Rag-1 knockout mice (which lack T and B lymphocytes) failed to show a corticosterone response to SEA challenge (Kawashima and Kusnecov, 2002). Similarly, in Rag-1 or TCR knockout mice reduction of food intake was not observed when testing for food ingestion was initiated 1.5 h after challenge with 10 μg SEA (Fig. 1). All mice were purchased from Jackson laboratories, while the TCR deficient mice were kindly provided by Dr. Yufang Shi (University of Medicine and Dentistry, New Jersey, – unlike Rag-$1^{-/-}$ mice, $TCR^{-/-}$ mice are deficient only for T lymphocytes; Mombaerts et al., 1992). Note in Fig. 1 that the TCR negative mice displayed lower consumption overall than the wildtype and Rag-1 knockout mice ($p<0.001$). This may be an instance of enhanced food neophobia and/or stress reactivity due to the isolation required for testing (which was performed as previously described by Kawashima and Kusnecov, 2002). Nonetheless, a sufficient amount was consumed (almost 1 gram) to allow for any detectable effects of SEA-induced food avoidance.

Blood was collected from these strains after consumption (2.5 h after SEA challenge). Plasma was measured for TNFα (e.g., as per standard ELISA methods – see Urbach-Ross et al., 2008), which was not detected in saline-injected mice. However, TNFα was increased in wild-type mice injected with SEA (mean ± SE: 106.9 ± 7.4 pg/ml). In contrast, no TNFα was detected in the SEA-challenged Rag-$1^{-/-}$ and $TCR^{-/-}$ mice (< 0.001 pg/ml, equivalent to negative control used in the ELISA). Challenge of Rag-$1^{-/-}$ mice ($N=6$) with 1 μg LPS, followed by sacrifice 2 h later, yielded plasma TNFα (146.1 ± 19.8 pg/ml). Therefore, failure of SEA to stimulate TNFα production in Rag-$1^{-/-}$ mice was not an inherent failure to produce TNFα.

These data showed that T cells are required for the anorexic response to SEA challenge, and furthermore, lack of T cells prevents plasma elevations in TNFα. While this does not show that T cells are in fact producing TNFα after SEA challenge, it does show that TNFα is not produced by non-T cells that may be stimulated by SEA. Given that TNFα has been observed to produce anorexia, HPA axis activation, and central *c-fos* induction in limbic brain regions, it was pertinent to determine whether TNFα was in fact mediating the effects of SEA T cell activation on the CNS (Rossi-George et al., 2005). This decision was made after partially ruling out IL-1β, since its circulating levels are modest if at times undetectable after a single administration of SEA (Urbach-Ross et al., 2008), and IL-1RI knockout mice show normal anorexic responses to SEA (Rossi-George et al., 2005).

Determination of the role of TNFα was initially conducted in TNF knockout mice that were challenged with SEA or Saline. Assessment for the number of immediate early gene (viz., *c-fos*) expressing cells in wild-type mice revealed a significant increase in SEA-induced activation in a number of limbic brain regions, including hypothalamic areas such as the PVN and arcuate nucleus, central nucleus of the amygdala and septum (Rossi-George et al., 2005). However, in TNF-deficient animals given SEA, this increase was significantly attenuated. This was not due to deficient stress reactivity in TNF knockout mice, since open field exposure drove up *c-fos* expression in limbic brain regions (Rossi-George et al., 2005; and unpublished observations). In the same paper, this lack of CNS engagement was extended to an assessment of behavior and corticosterone assessment, which also revealed that TNF-deficient mice failed to show anorexic and corticosterone responses to SEA. Furthermore, immunoneutralization of circulating TNFα in C57BL/6 mice administered SEA produced a similar effect (Rossi-George et al., 2005).

Therefore, it appears that the CNS response to SEA treatment is dependent on systemic production of TNFα. Whether this is a direct effect of TNFα on central TNF receptors, or a peripheral TNF receptor-mediated effect that mobilizes or promotes the actions of other cytokines, remains to be determined. Furthermore, it is not known whether the TNF-dependent effects of SEA are generalizable to other SAgs, such as SEB, which also significantly elevates TNFα levels. However, at the very least, the dependence of SEA-induced CNS activation on TNFα is consistent with the known behavioral and neuroendocrine effects of exogenous recombinant TNFα treatment in rats and mice, in that TNF administration increases plasma corticosterone, promotes anorexia, and activates *c-fos* expression in the limbic brain regions (Hayley et al., 2003).

4 Effects of Immunologic Challenge on Cognitive Behavior

The foregoing discussion focused on T cell activation by bacterial SAgs, with emphasis on basic motivational behaviors that suggest the operation of anxiogenic processes. At present no data exists for the effects of SAgs on learning and memory behaviors. However, considerable work has been conducted assessing cognition in relation to a variety of conditions, including Alzheimer's disease, depression, endotoxin challenge, cytokine therapy treatment, and infection, which are characterized by elevated levels of proinflammatory cytokines (Anisman and Merali, 1999; Kronfol and Remick, 2000; Reichenberg et al., 2001; Wilson et al., 2002; Maier and Watkins, 2003a). The occurrence of cognitive impairment in varied conditions suggests that it may result from a common cause, namely from the immune response itself.

The effects of immune activation on cognitive function are divergent. In some cases, learning decrements are observed after LPS or cytokine treatment, while others have only found evidence for performance effects (Gahtan and Overmier, 2001; Sparkman et al., 2004); still others report facilitation of learning/memory following low-dose cytokine administration (Yirmiya et al., 2002; Brennan et al., 2003, 2004; Brennan and Tieder, 2006). The disparity in findings underscores the potential influence of gender, timing of administration, behavioral paradigm employed, and dose on behavioral outcome. Despite some disagreement in the literature, in general, reports suggest that activation of the immune system, via the activity of cytokines, may impair specific types of memory.

Evaluating the effects of immune activation on cognition can be confounded by a variety of factors including alterations in locomotor activity, motivation, emotionality, attention, and pain sensitivity. Given that immune activation can cause alterations in all of these factors, consideration of each is necessary for proper interpretation of any cytokine-induced cognitive deficits (for review, see Dantzer, 2004; Swiergiel and Dunn, 2007). The biggest obstacle in using behavioral tests to evaluate LPS- or cytokine-induced cognitive deficits in laboratory animals is retardation of motor function and/or alterations in motivational processes that require goal-oriented locomotor effort. For example, reduced activity in LPS-treated animals does not result from a motor deficit, but rather altered motivation, as the expression of sickness behavior is dependent on environmental conditions (Aubert et al., 1997). Therefore, motivational changes render the use of appetitive tasks difficult as LPS-treated animals may not respond for a food reward, due to the anorexic and/or anhedonic effects of immune activation. Indeed, human endotoxin-challenge studies have begun to address this issue by administering doses of LPS (″0.8 ng/kg body weight) high enough to elicit cytokine production, but without producing subjective illness reports (Reichenberg et al., 2001, 2002; Cohen et al., 2003; Krabbe et al., 2005). Under these conditions, healthy young men given intravenous LPS showed cytokine-dependent anorexia (Reichenberg et al., 2002) and depression (Reichenberg et al., 2001), further emphasizing the unique challenge of implementing appetitive paradigms to test for deficits in cognitive processes.

Therefore, LPS- or cytokine-induced cognitive effects are often investigated in behavioral paradigms that motivate animals by employing negative reinforcement, such as escape from cool water (e.g., water maze) or a mild footshock (e.g., conditioned fear paradigms). Under such circumstances, involving sensory provocation, LPS-treated subjects are driven into active behavioral states (e.g., swimming or shuttling between chambers), masking or overcoming any signs of sickness behavior. Still, the potential influence of motivational changes and reduced locomotor activity cannot be completely ruled out and must be taken into consideration when learning is evaluated *following* immune activation. Moreover, it is always possible that in tests of animal learning that involve aversive stimulation, there is synergy with the similarly stressful effects of the immunologic challenge, as exemplified by increased limbic brain region and HPA axis activation observed after LPS challenge (Serrats and Sawchenko, 2006).

Clinical evidence from patients suffering from chronic inflammatory disease or undergoing cytokine therapy provides evidence that immune dysregulation can impact cognitive function (refer to chapters within this text for detailed discussion). Animal models have allowed researchers to better characterize the cognitive deficits and explore the potential mechanisms behind these effects. One of the most widely used rodent tests of spatial learning is the Morris water maze (MWM), in which animals are required to locate a submerged platform in a pool of opaque water. With repeated numbers of trials, successful navigation to the platform is based on the use of spatial cues (i.e., visual markers) located around the pool (Morris, 1984), and is dependent on the hippocampus, a necessary neural substrate for spatial and contextual learning (for review, see Martin and Clark, 2007). Although there are numerous testing procedures, animals typically receive 2–4 trials (each 60–90 sec in duration) per day across multiple days of training. Animals that have acquired the maze will show a reduction in latency (i.e., time) to locate the platform and will swim a shorter path length (i.e., distance) to the platform.

Peripheral administration of LPS was found to disrupt spatial learning in the MWM (Arai et al., 2001; Shaw et al., 2001, 2005; Sparkman et al., 2005a), with treatment prior to the first training day impairing acquisition on later days, as measured by increased latency and swimming distance to the platform (Shaw et al., 2001, 2005; Sparkman et al., 2005a). This effect was consistent after a single exposure to LPS, but varied under conditions of daily LPS administration, meant to model chronic inflammation. For example, Shaw et al. (2001) reported that repeated LPS had no effect on spatial learning, although Sparkman et al. (2005a) found that both repeated and acute LPS treatments were similar in increasing swimming distance to the platform. The disparity in findings may result from a pre-existing difference between treatment groups in the Shaw et al. (2001) study, since animals given daily LPS injections were swimming a shorter distance to the platform on the first day of training than the saline-treated subjects. With respect to sex differences, only one study has tested females in the MWM, noting that female mice had no deficit in learning after either single or repeated exposure to LPS (Sparkman et al., 2004). This contrasted with data in male mice, but might be explained by lower levels of LPS-induced cytokine production and/or higher levels of corticosterone in females (Frederic et al., 1993; Marriott et al., 2006; Ashdown et al., 2007).

The behavioral changes following LPS administration cannot be attributed to LPS-induced sickness behavior. During initial days of testing the LPS-treated subjects take significantly longer to escape than saline-treated subjects, however, this increase in the latency to find the platform is due to the LPS-treated subjects swimming slower rather than swimming a longer distance to the platform (Shaw et al., 2001; Sparkman et al., 2004, 2005a). However, during later days of testing, when the learning decrements became apparent (as indicated by increases in path length) LPS-treated subjects showed no differences in swimming speed or swam faster than saline controls. Relying solely on the latency to locate the platform in the absence of a concurrent measure of distance or swim speed may misrepresent the learning decrements induced by immune activation. Therefore, these results highlight the need for careful interpretation of behavioral data, to ensure that the interpretation of cognitive deficits can be dissociated from a performance deficit (i.e., due to illness).

To circumvent the potential confounding influence of sickness behavior, researchers often administer LPS following training, allowing for investigation of the effects on memory independent of illness. A number of reports suggest that administration of LPS, cytokines (particularly IL-1β), and infection after training can disrupt memory consolidation (Pugh et al., 1998; Banks et al., 2001; Barrientos et al., 2002; Holden et al., 2004; Thomson and Sutherland, 2005; Noble et al., 2007). As an example of this body of work, research by Pugh et al. (1998) demonstrated that LPS administration selectively impairs contextual fear conditioning in rats, whereas there is no effect on cue-specific auditory fear conditioning. Fear of a context is developed when the context becomes associated with an aversive event (e.g., footshock). For cue-specific fear conditioning, rats were placed in a novel conditioning chamber and administered two tone/shock pairings. In the study by Pugh et al. (1998), rats were i.p. administered various doses of LPS (0, 0.125, 0.25, or 0.5 mg/kg) after the conditioning session. Assessment of conditioned fear occurred 48 h after training, through observation of freezing behavior, an index of learned fear. This behavior occurs in response to both the context and tone that has been associated with electric shock. Administration of LPS at doses of 0.125 and 0.25 mg/kg, but not 0.5 mg/kg, decreased freezing in response to the context, but had no effect on auditory-cue fear conditioning. A replication of these findings was recently conducted by Thomson and Sutherland (2005), who found that LPS administration disrupted contextual fear conditioning, but had no effect on conditioning to an auditory stimulus. These data suggest that stimulating the immune system disrupted the consolidation process for contextual information.

The LPS-induced disruption of memory consolidation may have resulted from elevated levels of brain cytokines which impair hippocampal function. Indeed, the hippocampus expresses a high density of receptors for the proinflammatory cytokine IL-1 (Ban, 1994). Moreover, administration of an IL-1 receptor antagonist (IL-1ra) prevented deficits in hippocampal learning, while infusion of IL-1β into the dorsal hippocampus led to similar memory deficits as LPS (Pugh et al., 1998; Barrientos et al., 2002). Furthermore, IL-1 inhibits long-term potentiation (LTP) in the hippocampus, which has been hypothesized to be a neurophysiological correlate of learning (Bellinger et al., 1993; Murray and Lynch, 1998; Kelly et al., 2001). Systemically, however, increased levels of IL-1β appear to be insufficient to disrupt memory consolidation (Thomson and Sutherland, 2005), supporting the notion that LPS-induced impairment of contextual memory may be mediated by central cytokine production.

As mentioned, contextual fear conditioning is a hippocampal-dependent task, whereas auditory-cue fear conditioning is less dependent on the hippocampus (Phillips and LeDoux, 1995; Anagnostaras et al., 2001), suggesting that LPS may interfere with different memory processes. An additional experiment by Pugh et al. (1998), suggests that post-training administration of LPS does not impair the formation of the association between the conditioned stimulus (CS; i.e., context) and the unconditioned stimulus (US, i.e., footshock); rather, LPS selectively disrupts processing and consolidation of contextual information. Pre-exposing the rats to the testing chamber and allowing them to process the context, prevented the LPS-induced deficits in contextual fear conditioning. Therefore, once a memory of the context is formed, LPS administration does not disrupt the formation of an association between the context and the US.

Whether these post-training effects of LPS treatment operate through similar mechanisms, as those observed when LPS is given *prior* to training is presently unknown; although, learning decrements based on pre- versus post-training LPS administration appear to be distinct. The deficits induced by pre-training immune activation may reflect impaired encoding and/or impaired transfer of new information into long-term memory. Several reports have found that administration of LPS prior to testing induces learning deficits, but has little effect if given after acquisition has occurred, by disrupting the formation of new associations (Aubert et al., 1997; Mallon et al., 2003; Sparkman et al., 2004, 2005b, a; Kohman et al., 2007a, b). For example, Aubert et al. (1995) evaluated whether administration of LPS affected the rate of acquisition in an autoshaping task, where subjects must learn to press a lever to obtain a food reward. Following the presentation of the lever, subjects had 15 sec to respond before the lever was retracted and the opportunity for a reward was lost. Successful acquisition of the task was displayed by fast response rates to obtain more food rewards. However, when animals were administered LPS prior to the second of 15 days of training, significantly longer response latencies were observed, suggesting that LPS administration impaired acquisition of this task. As noted previously, LPS administration can decrease food consumption. Aubert et al. (1995) reported that LPS treatment did not affect pellet consumption. Furthermore, a second experiment found that LPS given following acquisition of the task did not affect response latency nor pellet consumption, suggesting that activation of the immune system does not affect memory recall, but rather disrupts acquisition of new information.

Comparable LPS-induced learning deficits were found using the two-way active avoidance conditioning paradigm (Sparkman et al., 2005b; Kohman et al., 2007a, b). This task, unlike the paradigm employed by Aubert et al. (1995), uses negative reinforcement to motivate responding. Subjects were tested in an apparatus that has two equal-sized compartments with a doorway that separates the two sides. Acquisition of the paradigm requires training across multiple days of testing with numerous trials per day. Each trial consisted of the presentation of a light CS followed by the presentation of the US (mild footshock), with animals learning to prevent US onset by crossing over to the adjacent chamber before CS termination. In this paradigm, avoiding shock is distinct from escape behavior, which involves crossing after US onset. The optimal response in this paradigm is an avoidance response, reflecting the formation of an association between the CS and impending shock. Several reports suggest that administration of LPS prior to the initial day of training impairs acquisition of the CS–US association in the active avoidance paradigm (Sparkman et al., 2005b; Kohman et al., 2007a, b). That is, LPS-treated mice performed fewer avoidance responses and show diminished response efficiency. These learning decrements were observed during later days of testing, and persist up to 10 days after LPS administration, well after the physiological effects of LPS (Sparkman et al., 2005b), and are similar to the results of Aubert et al. (1995). Additionally, Sparkman et al. (2005a) reported that administration of LPS prior to the fourth day of training, after subjects have acquired the task, has no effect on performance. In conjunction with the work of Aubert et al. (1995) these data suggest that activation of the immune system disrupts learning of new information, but does not influence previously learned associations.

The specific mechanisms responsible for the LPS-induced effects on learning are undetermined. However, one hypothesis is that peripherally or centrally released cytokines and subsequent interactions with the CNS may be involved. Systemic administration of LPS significantly increases splenic, blood, and CNS production of IL-β (for review see Maier and Watkins, 2003b; Dantzer, 2004). Pretreatment with a CRH receptor-1 antagonist prevented the LPS-induced learning decrements and expression of IL-1β within the hippocampus, without having an effect on corticosterone release, suggesting that the deficits are related to cytokine-induced alterations in hippocampal function (Kohman et al., 2007b). Furthermore, it has been argued that the increased crossing behavior, observed in the LPS-treated subjects (Sparkman et al., 2005a), in the active avoidance paradigm may involve the hippocampus, as lesions of the hippocampus facilitate acquisition of two-way active avoidance task (Good and Honey, 1997; Guillazo-Blanch et al., 2002). However, the seemingly improved performance in hippocampal lesioned subjects may result from behavioral disinhibition, as lesioned subjects simply cross more and artificially inflate the number of avoidance responses (Good and Honey, 1997; Guillazo-Blanch et al., 2002). Furthermore, Ma and Zhu (1997) found that intrahippocampal administration of LPS impaired acquisition and retention in a one-way active avoidance conditioning task, suggesting that alterations in hippocampal function may produce LPS-induced learning deficits. These LPS-induced deficits in learning may, in part, be explained by disruption of memory consolidation, as observed in the contextual fear conditioning paradigm. However, single administration of LPS produces long-term behavioral deficits, when the subjects no longer display overt sickness behavior, yet are unable to acquire the task days after LPS was initially given (Sparkman et al., 2005a).

The use of aversive stimuli, particularly footshock, brings up another potential confounding factor since LPS can alter pain sensitivity (Meller et al., 1994; Thomazzi et al., 1997; Estudante et al., 1998). The effects of immune activation on nociceptive thresholds are rather complex and depend on a number of factors such as dose, route, and timing of administration (Hori et al., 2000). LPS administration has biphasic effects on pain sensitivity. Initially, LPS administration induces hyperalgesic effects that last up to two hours (Mason, 1993; Watkins et al., 1994; Yirmiya et al., 1994; Abe et al., 2001; Kawashima et al., 2002), but with the onset of fever, there is either onset of analgesia or no change in pain sensitivity (Mason, 1993; Yirmiya et al., 1994; Abe et al., 2001). The LPS-induced learning decrements observed in the two-way active avoidance paradigm are unlikely to result from altered pain sensitivity, as testing occurred four hours following LPS administration a time when others report LPS has no hyperalgesic effects or may induce analgesia (Yirmiya et al., 1994; Abe et al., 2001). If the LPS-treated subjects had higher pain thresholds than saline-treated subjects this may actually facilitate learning the CS–US association, as research has shown that acquisition of two-way active avoidance conditioning is faster at lower shock intensities (Archer, 1982). Nonetheless, the potential interference of altered nociception, whether increased or decreased sensitivity, must be considered in studies that involve aversive stimulation.

As noted, the induction of cytokines, particularly IL-1β, within the CNS is believed to disrupt learning and memory. However, there is some evidence to suggest that IL-1 may play a role in normal memory processes (Yirmiya et al., 2002; Avital et al., 2003; Depino et al., 2004). For example, Avital et al. (2003) report that IL-1 receptor knockout mice show impaired spatial learning in the MWM, diminished contextual fear conditioning, and a complete absence of LTP in the CA1 region of the hippocampus relative to wild-type controls, suggesting that IL-1 contributes to normal learning and memory. Whether cytokine production is beneficial or detrimental to cognitive function appears to be dependent on the dose administered. Oitzl et al. (1993) reported that i.c.v administration of 100 ng of IL-1β 60 min prior to training disrupts spatial learning, suggesting that increased levels of IL-1β within the CNS impairs learning. Whereas Yirmiya et al. (2002) later reported that i.c.v administration of 10 ng IL-1β immediately after training had no effect on spatial learning in the MWM, but that inhibiting the activity of IL-1, via administration of an IL-1 receptor antagonist (IL-1ra), impaired performance. Additionally, central and peripheral administration of low doses of IL-1 and TNF-α have been reported to facilitate acquisition of avoidance learning (Gibertini, 1998; Yirmiya et al., 2002; Brennan et al., 2003, 2004; Brennan and Tieder, 2006). Taken together, these studies suggest that high levels of central IL-1β may disrupt memory processes, whereas administration of smaller doses may facilitate learning (Oitzl et al., 1993; Gibertini, 1998; Barrientos et al., 2002; Yirmiya et al., 2002; Brennan et al., 2003; Depino et al., 2004). Similar inconsistent findings have been reported following peripheral administration of IL-1β. Gibertini et al. (1995) found that peripheral administration of IL-1β prior to the first day of testing impaired acquisition of the MWM task, as IL-1β treated mice had longer latencies to locate the platform than saline-treated subjects on day 2. However, a recent report found no evidence that peripheral IL-1β administration impaired acquisition, retention, or consolidation in the MWM (Thomson and Sutherland, 2006). Taken together, these reports suggest that peripheral IL-1β may be insufficient to produce cognitive deficits; however, high levels of IL-1β within the CNS appear to disrupt learning.

It is important to emphasize that activation of the immune system does not appear to disrupt all types of learning and memory, but rather seems to have specific effects on hippocampus- dependent memory processes. As noted previously, activation of the immune system impairs contextual fear conditioning, but has no affect on auditory fear conditioning, a hippocampus-independent form of learning (Pugh et al., 1998; Thomson and Sutherland, 2005). Additional evidence can be extracted from the condition taste aversion (CTA) paradigm, in which subjects are trained to develop an aversion to a novel taste by pairing it with the induction of nausea. Administration of IL-1 or LPS following exposure to a novel taste leads to avoidance of the gustatory CS upon re-exposure (Tazi et al., 1988; Mormede et al., 2003; Cross-Mellor et al., 2005). CTA is a form of non-declarative memory, and is dependent on the insular cortex and other subcortical regions (for review see Wang et al., 2006). These findings show that activation of the immune system does not appear to disrupt formation of non-declarative memories, likely because they are formed in a hippocampal-independent manner. However, it would be of interest to determine whether contextual elements inherent in a CTA paradigm (e.g., location of novel gustatory CS) are processed poorly following administration of an immunologic stimulus.

The majority of research on immune responses and cognition has focused on IL-1β as the primary cause of inflammation related to cognitive deficits. Following activation, the immune system releases multiple proinflammatory cytokines (i.e., IL-1β, IL-6, and TNF-α) that have been shown to have similar effects, but which may vary in degrees of potency (Besedovsky et al., 1991), and also most likely operate in an additive and synergistic fashion (Brebner et al., 2000). A recent report (Sparkman et al., 2006) suggests that IL-6 is involved in LPS-induced working memory deficits, specifically a delayed matching to place version of the MWM. No deficits were observed in LPS-treated IL-6 knockout mice, suggesting that endogenous IL-6 production at the time of testing is necessary for the occurrence of the memory deficit. In contrast, Oitzl et al. (1993) reported that while IL-1β administration impaired spatial learning, no deficits were observed following administration of IL-6. While these results appear to be in disagreement, the difference is largely methodological and conceptual, since LPS induces endogenous IL-6 over a period of hours, and in conjunction with LPS-induced production of other factors (including IL-1 and TNF by microglial cells or astrocytes). This is markedly different from exogenous administration of a bolus infusion of IL-6 acting independently of an endogenous cytokine cascade.

The neural mechanisms behind these inflammation related cognitive deficits remains unknown. Much of the research suggests that the deficits result from alterations in neural function, particularly in the hippocampus, although the precise nature of these alterations is unknown. Indeed, the process of memory formation in and of itself is not fully understood. However, factors that are relevant to memory formation can be altered by immune activation. For example, IL-1β has been found to decrease the presynaptic release of acetylcholine (Rada et al., 1991), a neurotransmitter critical for memory formation (Hasselmo, 2006). Furthermore, Matsumoto et al. (2001) report that activation of the NMDA receptor following intrahippocampal IL-1β administration prevented the IL-1β-induced disruptions in working memory, which is consistent with evidence for IL-1β inhibiting glutamatergic transmission (Coogan and O'Connor, 1997). Therefore, IL-1β may disrupt hippocampal function via reduced engagement of NMDA receptors.

There is also evidence to suggest that cytokines interfere with the production and actions of neurotrophins, such as BDNF (brain derived neurotrophic factor). BDNF is critical for neural development and survival, and is believed to be important for certain types of learning and memory, as it plays a role in synaptic plasticity and LTP (for review see McAllister, 2001; Bekinschtein et al., 2007). Research on the effects of LPS administration on BDNF expression has show conflicting results as some reports show that LPS decreases (Shaw et al., 2005), increases (Miwa et al., 1997), or has no effect (Elkabes et al., 1998; Shaw et al., 2005) on BDNF production. These results may be related to methodological variations, as LPS is reported to have no effect on BDNF levels 10 or 11 days later (Shaw et al., 2001, 2005), but in culture, LPS stimulation decreases BDNF production (Shaw et al., 2005). Additionally, intrahippocampal administration of IL-1β decreased BDNF mRNA expression normally observed after training for contextual fear conditioning (Barrientos et al., 2004). A recent report by Tong et al. (2007) suggests IL-1β may interfere with BDNF-mediated cell survival. Furthermore, IL-1 decreased BDNF-induced activation of MAPK/ERK and CREB (Ca^{2+}/cAMP response element-binding protein).

Both MAPK/ERK and CREB have been implicated in learning and memory, and alter synaptic plasticity (Abel and Kandel, 1998; Silva et al., 1998; Sweatt, 2001; Thomas and Huganir, 2004). Therefore, collectively these findings suggest that cytokine-induced learning decrements may result from diminished production of BDNF and/or disruption of BDNF signaling.

5 Concluding Comments

When attempting to reconcile the sometimes disparate behavioral sequelae of immune activation, it is important to consider findings in the framework of adaptive significance. The malaise, lethargy, and anhedonia characteristic of sickness behavior act to conserve an organism's resources during a period of stress, which has consistently been demonstrated through measures of central changes in monoamine neurotransmitters, immediate early gene activation, and HPA axis activation. This may promote recuperative processes, as well as preparation for and minimization of threat exposure. That immune activation may exacerbate anxiety-like processes likewise has the potential to prevent further risk exposure. The adaptive nature of gustatory neophobia can be understood in the context of the reluctance to consume a novel source of food, which may have initiated the ongoing illness, and serves to protect the organism from further pathogens. Coincident with any anxiety-like states and changes in mood are changes in the capacity for learning and memory formation. The challenge lies in understanding the adaptive significance, if any, of the cognitive deficits associated with immune activation. As was reviewed above, consolidation of new contextual information is impaired by immunologic challenge, while stimulus-specific information predicting threat (e.g., auditory and gustatory CSs in the conditioned fear and taste aversion paradigms) remains unaffected. Therefore, it could be hypothesized that this dissociation reflects changes in attention to selective predictors of threat to biological health, wherein context, which contains multiple attributes, cannot be processed adequately by a cognitive network disrupted by excessive immunologically mediated arousal. Indeed, both the work on SAgs and LPS has demonstrated that inflammatory immune challenge creates increased attention to novel stimuli, and which is supported by evidence for increased immune-mediated activation of limbic circuits typically engaged by arousing stimuli (Lacosta et al., 1999; Kawashima and Kusnecov, 2002; Rossi-George et al., 2005; Serrats and Sawchenko, 2006).

The animal work reviewed above appears to support the notion that cytokine recruitment leads to impaired learning and cognitive processes, although recent human studies may shed some light on this seemingly evolutionary contradiction. At a dose of LPS high enough to induce cytokine production but not illness, endotoxin-induced cytokine activation brought about cortisol-independent memory deficits and increased anxiety in healthy young men (Reichenberg et al., 2001). The same dose of LPS has also been shown to impair declarative memory commensurate with a large increase in IL-6, while a small increase in this cytokine improved declarative memory (Krabbe et al., 2005), indicating that alterations in memory performance are contingent on robustness of the proinflammatory cytokine response. Cohen et al. (2003) delved further into endotoxin-induced memory deficits by examining whether different types of memory may be affected by low-dose administration of LPS. While they found that endotoxin-induced cytokine production caused a reduction in declarative memory, they found an unexpected *enhancement* in working memory (Cohen et al., 2003). As working memory may be more critical for survival when organisms experience threat, it is possible that cognitive resources are redirected toward the mechanisms that best ensure survival. Finally, as with all animal behavioral paradigms, the construct validity of the tests employed must be considered when forming conclusions about the precise impact of immune activation on these complex cognitive and emotional phenomena. In spite of these caveats and problems, there is sufficient evidence to demonstrate that immunological challenge can affect mood and cognitive processes, although the precise mechanisms and significance for these effects remains to be elucidated.

References

Abe M, Oka T, Hori T, Takahashi S (2001) Prostanoids in the preoptic hypothalamus mediate systemic lipopolysaccharide-induced hyperalgesia in rats. Brain Res 916:41–49.

Abel T, Kandel E (1998) Positive and negative regulatory mechanisms that mediate long-term memory storage. Brain Res Brain Res Rev 26:360–378.

Ader R, Cohen N, Felten D (1995) Psychoneuroimmunology – Interactions between the Nervous-System and the Immune-System. Lancet 345:99–103.

Aguilera G, Kiss A, Liu Y, Kamitakahara A (2007) Negative regulation of corticotropin releasing factor expression and limitation of stress response. Stress 10:153–161.

Anagnostaras SG, Gale GD, Fanselow MS (2001) Hippocampus and contextual fear conditioning: recent controversies and advances. Hippocampus 11:8–17.

Anisman H, Merali Z (1999) Anhedonic and anxiogenic effects of cytokine exposure. Adv Exp Med Biol 461:199–233.

Anisman H, Merali Z, Poulter MO, Hayley S (2005) Cytokines as a precipitant of depressive illness: animal and human studies. Curr Pharm Des 11:963–972.

Aoki Y, Hiromatsu K, Kobayashi N, Hotta T, Saito H, Igarashi H, Niho Y, Yoshikai Y (1995) Protective effect of Granulocyte-Colony-Stimulating factor against T-Cell-Meditated Lethal Shock Triggered by Superantigens. Blood 86:1420–1427.

Arai K, Matsuki N, Ikegaya Y, Nishiyama N (2001) Deterioration of spatial learning performances in lipopolysaccharide-treated mice. Jpn J Pharmacol 87:195–201.

Archer T (1982) DSP4 (N-2-chloroethyl-N-ethyl-2-bromobenzylamine), a new noradrenaline neurotoxin, and stimulus conditions affecting acquisition of two-way active avoidance. J Comp Physiol Psychol 96:476–490.

Ashdown H, Poole S, Boksa P, Luheshi GN (2007) Interleukin-1 receptor antagonist as a modulator of gender differences in the febrile response to lipopolysaccharide in rats. Am J Physiol Regul Integr Comp Physiol 292:R1667–1674.

Aubert A, Vega C, Dantzer R, Goodall G (1995) Pyrogens specifically disrupt the acquisition of a task involving cognitive processing in the rat. Brain Behav Immun 9:129–148.

Aubert A, Goodall G, Dantzer R, Gheusi G (1997) Differential effects of lipopolysaccharide on pup retrieving and nest building in lactating mice. Brain Behav Immun 11:107–118.

Avital A, Goshen I, Kamsler A, Segal M, Iverfeldt K, Richter-Levin G, Yirmiya R (2003) Impaired interleukin-1 signaling is associated with deficits in hippocampal memory processes and neural plasticity. Hippocampus 13:826–834.
Ballaz SJ, Akil H, Watson SJ (2007) Previous experience affects subsequent anxiety-like responses in rats bred for novelty seeking. Behavioral Neuroscience 121:1113–1118.
Ban EM (1994) Interleukin-1 receptors in the brain: characterization by quantitative in situ autoradiography. Immunomethods 5:31–40.
Banks WA, Farr SA, La Scola ME, Morley JE (2001) Intravenous human interleukin-1alpha impairs memory processing in mice: dependence on blood-brain barrier transport into posterior division of the septum. J Pharmacol Exp Ther 299:536–541.
Barrientos RM, Higgins EA, Sprunger DB, Watkins LR, Rudy JW, Maier SF (2002) Memory for context is impaired by a post context exposure injection of interleukin-1 beta into dorsal hippocampus. Behav Brain Res 134:291–298.
Barrientos RM, Sprunger DB, Campeau S, Watkins LR, Rudy JW, Maier SF (2004) BDNF mRNA expression in rat hippocampus following contextual learning is blocked by intrahippocampal IL-1beta administration. J Neuroimmunol 155:119–126.
Bekinschtein P, Cammarota M, Izquierdo I, Medina JH (2007) BDNF and Memory Formation and Storage. Neuroscientist 142:147–156.
Bellinger FP, Madamba S, Siggins GR (1993) Interleukin 1 beta inhibits synaptic strength and long-term potentiation in the rat CA1 hippocampus. Brain Res 628:227–234.
Berkowitz RL, Coplan JD, Reddy DP, Gorman JM (2007) The human dimension: How the prefrontal, cortex modulates the subcortical fear response. Rev Neurosci 18:191–207.
Besedovsky HO, del Rey A (2000) The cytokine-HPA axis feed-back circuit. Zeitschrift Fur Rheumatologie 59:26–30.
Besedovsky HO, del Rey A (2002) Introduction: Immune-neuroendocrine network. Neuroendocrine-Immune Interactions 29:1–14.
Besedovsky HO, del Rey A, Klusman I, Furukawa H, Monge Arditi G, Kabiersch A (1991) Cytokines as modulators of the hypothalamus-pituitary-adrenal axis. J Steroid Biochem Mol Biol 40:613–618.
Borowski T, Kokkinidis L, Merali Z, Anisman H (1998) Lipopolysaccharide, central in vivo biogenic amine variations, and anhedonia. Neuroreport 9:3797–3802.
Brebner K, Hayley S, Zacharko R, Merali Z, Anisman H (2000) Synergistic effects of interleukin-1 beta, interleukin-6, and tumor necrosis factor-alpha: Central monoamine, corticosterone, and behavioral variations. Neuropsychopharmacology 22:566–580.
Brennan FX, Tieder JB, 3rd (2006) Centrally administered tumor necrosis factor-alpha facilitates the avoidance performance of Sprague-Dawley rats. Brain Res 1109:142–145.
Brennan FX, Beck KD, Servatius RJ (2003) Low doses of interleukin-1beta improve the lever-press avoidance performance of Sprague-Dawley rats. Neurobiol Learn Mem 80:168–171.
Brennan FX, Beck KD, Servatius RJ (2004) Proinflammatory cytokines differentially affect lever-press avoidance acquisition in rats. Behav Brain Res 153:351–355.
Calandra T, Spiegel LA, Metz CN, Bucala R (1998) Macrophage migration inhibitory factor is a critical mediator of the activation of immune cells by exotoxins of Gram-positive bacteria. Proc Natl Acad Sci USA 95:11383–11388.
Cohen O, Reichenberg A, Perry C, Ginzberg D, Pollmacher T, Soreq H, Yirmiya R (2003) Endotoxin-induced changes in human working and declarative memory associate with cleavage of plasma "readthrough" acetylcholinesterase. J Mol Neurosci 21:199–212.
Coogan A, O'Connor JJ (1997) Inhibition of NMDA receptor-mediated synaptic transmission in the rat dentate gyrus in vitro by IL-1 beta. Neuroreport 8:2107–2110.
Cross-Mellor SK, Kavaliers M, Ossenkopp KP (2005) The effects of lipopolysaccharide and lithium chloride on the ingestion of a bitter-sweet taste: comparing intake and palatability. Brain Behav Immun 19:564–573.
Dantzer R (2004) Cytokine-induced sickness behaviour: a neuroimmune response to activation of innate immunity. Eur J Pharmacol 500:399–411.
Dantzer R (2006) Cytokine, sickness behavior, and depression. Neurol Clin 24:441–460.

Dantzer R, O'Connor JC, Freund GG, Johnson RW, Kelley KW (2008) From inflammation to sickness and depression: when the immune system subjugates the brain. Nat Rev Neurosci 9:46–57.

Depino AM, Alonso M, Ferrari C, del Rey A, Anthony D, Besedovsky H, Medina JH, Pitossi F (2004) Learning modulation by endogenous hippocampal IL-1: blockade of endogenous IL-1 facilitates memory formation. Hippocampus 14:526–535.

Dulawa SC, Grandy DK, Low MJ, Paulus MP, Geyer MA (1999) Dopamine D4 receptor-knock-out mice exhibit reduced exploration of novel stimuli. J Neurosci 19:9550–9556.

Elkabes S, Peng L, Black IB (1998) Lipopolysaccharide differentially regulates microglial trk receptor and neurotrophin expression. J Neurosci Res 54:117–122.

Estudante M, Maia H, Simoes C, Mota-Filipe H, Silva-Lima B (1998) Lipopolysaccharide (LPS) increases nitric oxide (NO) production modulates nociception by thermal and mechanical stimuli in mice. Naunyn-Schmiedebergs Arch Pharmacol 358:R312–R312.

Frederic F, Oliver C, Wollman E, Delhaye-Bouchaud N, Mariani J (1993) IL-1 and LPS induce a sexually dimorphic response of the hypothalamo-pituitary-adrenal axis in several mouse strains. Eur Cytokine Netw 4:321–329.

Gahtan E, Overmier JB (2001) Performance more than working memory disrupted by acute systemic inflammation in rats in appetitive tasks. Physiol Behav 73:201–210.

Gibertini M (1998) Cytokines and cognitive behavior. Neuroimmunomodulation 5:160–165.

Gibertini M, Newton C, Friedman H, Klein TW (1995) Spatial learning impairment in mice infected with Legionella pneumophila or administered exogenous interleukin-1-beta. Brain Behav Immun 9:113–128.

Goehler LE, Gaykema RPA, Hansen MK, Kleiner JL, Maier SF, Watkins LR (2001) Staphylococcal enterotoxin B induces fever, brain c-Fos expression, and serum corticosterone in rats. Am J Physiol Reg Integr Comp Physiol 280:R1434–R1439.

Gold DP, Surh CD, Sellins KS, Schroder K, Sprent J, Wilson DB (1994) Rat T-Cell Responses to Superantigens .2. Allelic Differences in V-Beta-8.2 and V-Beta-8.5 Beta-Chains Determine Responsiveness to Staphylococcal-Enterotoxin-B and Mouse Mammary-Tumor Virus-Encoded Products. J Exp Med 179:63–69.

Gonzalo JA, de Alboran IM, Kroemer G (1993) Dissociation of autoaggression and self-superantigen reactivity. Scand J Immunol 37:1–6.

Gonzalo JA, Baixeras E, Gonzalezgarcia A, Georgechandy A, Vanrooijen N, Martinez C, Kroemer G (1994) Differential in-Vivo Effects of a Superantigen and an Antibody-Targeted to the Same T-Cell Receptor – Activation-Induced Cell-Death Vs Passive Macrophage-Dependent Deletion. J Immunol 152:1597–1608.

Good M, Honey RC (1997) Dissociable effects of selective lesions to hippocampal subsystems on exploratory behavior, contextual learning, and spatial learning. Behav Neurosci 111:487–493.

Guillazo-Blanch G, Nadal R, Vale-Martinez A, Marti-Nicolovius M, Arevalo R, Morgado-Bernal I (2002) Effects of fimbria lesions on trace two-way active avoidance acquisition and retention in rats. Neurobiol Learn Mem 78:406–425.

Harris TO, Grossman D, Kappler JW, Marrack P, Rich RR, Betley MJ (1993) Lack of complete correlation between emetic and T-cell-stimulatory activities of staphylococcal enterotoxins. Infect Immun 61:3175–3183.

Hart BL (1988) Biological basis of the behavior of sick animals. Neurosci Biobehav Rev 12: 123–137.

Hasselmo ME (2006) The role of acetylcholine in learning and memory. Current Opinion in Neurobiology 16:710–715.

Hayley S, Merali Z, Anisman H (2003) Stress and cytokine-elicited neuroendocrine and neurotransmitter sensitization: implications for depressive illness. Stress Int J Biol Stress 6:19–32.

Henry B, Vale W, Markou A (2006) The effect of lateral septum corticotropin-releasing factor receptor 2 activation on anxiety is modulated by stress. J Neurosci 26:9142–9152.

Hernandez ME, Becerril L, Alvarez L, Povon-Romero L (2007) Neuroimmunomodulation pathways. Part one. Salud Mental 30:13–19.

Hirshfeld DR, Biederman J, Brody L, Faraone SV, Rosenbaum JF (1997) Expressed emotion toward children with behavioral inhibition: Associations with maternal anxiety disorder. J Am Acad Child Adol Psy 36:910–917.

Holden JM, Overmier JB, Cowan ET, Matthews L (2004) Effects of lipopolysaccharide on consolidation of partial learning in the Y-maze. Integr Physiol Behav Sci 39:334–340.
Holmes A, Iles JP, Mayell SJ, Rodgers RJ (2001) Prior test experience compromises the anxiolytic efficacy of chlordiazepoxide in the mouse light/dark exploration test. Behav Brain Res 122:159–167.
Hori T, Oka T, Hosoi M, Abe M, Oka K (2000) Hypothalamic mechanisms of pain modulatory actions of cytokines and prostaglandin E-2. Neuroimmunomodulation 917:106–120.
Kaneta T, Kusnecov AW (2005) The role of central corticotropin-releasing hormone in the anorexic and endocrine effects of the bacterial T cell superantigen, Staphylococcal enterotoxin A. Brain Behav Immun 19:138–146.
Kawashima N, Kusnecov AW (2002) Effects of staphylococcal enterotoxin A on pituitary-adrenal activation and neophobic behavior in the C57BL/6 mouse. J Neuroimmunol 123:41–49.
Kawashima N, Fugate J, Kusnecov AW (2002) Immunological challenge modulates brain orphanin FQ/nociceptin and nociceptive behavior. Brain Res 949:71–78.
Kelly A, Lynch A, Vereker E, Nolan Y, Queenan P, Whittaker E, O'Neill LA, Lynch MA (2001) The anti-inflammatory cytokine, interleukin (IL)-10, blocks the inhibitory effect of IL-1 beta on long term potentiation. A role for JNK. J Biol Chem 276:45564–45572.
Kohman RA, Tarr AJ, Byler SL, Boehm GW (2007a) Age increases vulnerability to bacterial endotoxin-induced behavioral decrements. Physiol Behav 91:561–565.
Kohman RA, Tarr AJ, Sparkman NL, Day CE, Paquet A, Akkaraju GR, Boehm GW (2007b) Alleviation of the effects of endotoxin exposure on behavior and hippocampal IL-1beta by a selective non-peptide antagonist of corticotropin-releasing factor receptors. Brain Behav Immun 21:824–835.
Koob GF, Heinrichs SC (1999) A role for corticotropin releasing factor and urocortin in behavioral responses to stressors. Brain Res 848:141–152.
Krabbe KS, Reichenberg A, Yirmiya R, Smed A, Pedersen BK, Bruunsgaard H (2005) Low-dose endotoxemia and human neuropsychological functions. Brain Behav Immun 19:453–460.
Kronfol Z, Remick DG (2000) Cytokines and the brain: implications for clinical psychiatry. Am J Psychiatry 157:683–694.
Kubota O, Hattori K, Hashimoto K, Yagi T, Sato T, Iyo M, Yuasa S (2004) Auditory-conditioned-fear-dependent c-Fos expression is altered in the emotion-related brain structures of Fyn-deficient mice. Mol Brain Res 130:149–160.
Kusnecov AW, Liang R, Shurin G (1999) T-lymphocyte activation increases hypothalamic and amygdaloid expression of CRH mRNA and emotional reactivity to novelty. J Neurosci 19:4533–4543.
Lacosta S, Merali Z, Anisman H (1999) Behavioral and neurochemical consequences of lipopolysaccharide in mice: anxiogenic-like effects. Brain Res 818:291–303.
Liang HE, Chen CC, Chou DL, Lai MZ (1994) Flexibility of the T-Cell Receptor Repertoire. European J Immunol 24:1604–1611.
Liebsch G, Landgraf R, Engelmann M, Lorscher P, Holsboer F (1999) Differential behavioural effects of chronic infusion of CRH1 and CRH2 receptor antisense oligonucleotides into the rat brain. J Psychiat Res 33:153–163.
Luheshi G, Miller AJ, Brouwer S, Dascombe MJ, Rothwell NJ, Hopkins SJ (1996) Interleukin-1 receptor antagonist inhibits endotoxin fever and systemic interleukin-6 induction in the rat. Am J Physiol 270:E91–95.
Luxembourg A, Grey H (2002) Strong induction of tyrosine phosphorylation, intracellular calcium, nuclear transcription factors and interferongamma, but weak induction of IL-2 in naive T cells stimulated by bacterial superantigen. Cell Immunol 219:28–37.
Ma TC, Zhu XZ (1997) Suppression of lipopolysaccharide-induced impairment of active avoidance and interleukin-6-induced increase of prostaglandin E2 release in rats by indometacin. Arzneimittelforschung 47:595–597.
Maier SF, Watkins LR (2003a) Immune-to-central nervous system communication and its role in modulating pain and cognition: Implications for cancer and cancer treatment. Brain Behav Immun 17:S125–S131.

Maier SF, Watkins LR (2003b) Immune-to-central nervous system communication and its role in modulating pain and cognition: Implications for cancer and cancer treatment. Brain Behav Immun 17 Suppl 1:S125–131.

Makara GB, Mergl Z, Zelena D (2004) The role of vasopressin in hypothalamo-pituitary-adrenal axis activation during stress: an assessment of the evidence. Ann N Y Acad Sci 1018:151–161.

Mallon EB, Brockmann A, Schmid-Hempel P (2003) Immune response inhibits associative learning in insects. Proc Biol Sci 270:2471–2473.

Marriott I, Bost KL, Huet-Hudson YM (2006) Sexual dimorphism in expression of receptors for bacterial lipopolysaccharides in murine macrophages: a possible mechanism for gender-based differences in endotoxic shock susceptibility. J Reprod Immunol 71:12–27.

Martin SJ, Clark RE (2007) The rodent hippocampus and spatial memory: from synapses to systems. Cell Mol Life Sci. 64:401–431.

Mason P (1993) Lipopolysaccharide induces fever and decreases tail flick latency in awake rats. Neurosci Lett 154:134–136.

Matsumoto Y, Yoshida M, Watanabe S, Yamamoto T (2001) Involvement of cholinergic and glutamatergic functions in working memory impairment induced by interleukin-1beta in rats. Eur J Pharmacol 430:283–288.

McAllister AK (2001) Neurotrophins and neuronal differentiation in the central nervous system. Cell Mol Life Sci 58:1054–1060.

Meller ST, Dykstra C, Grzybycki D, Murphy S, Gebhart GF (1994) The Possible Role of Glia in Nociceptive Processing and Hyperalgesia in the Spinal-Cord of the Rat. Neuropharmacology 33:1471–1478.

Miwa T, Furukawa S, Nakajima K, Furukawa Y, Kohsaka S (1997) Lipopolysaccharide enhances synthesis of brain-derived neurotrophic factor in cultured rat microglia. J Neurosci Res 50:1023–1029.

Mombaerts P, Iacomini J, Johnson RS, Herrup K, Tonegawa S, Papaioannou VE (1992) RAG-1-deficient mice have no mature B and T lymphocytes. Cell 68:869–877.

Mormede C, Castanon N, Medina C, Dantzer R (2003) Conditioned place aversion with interleukin-1beta in mice is not associated with activation of the cytokine network. Brain Behav Immun 17:110–120.

Morris R (1984) Developments of a water-maze procedure for studying spatial learning in the rat. J Neurosci Method 11:47–60.

Murray CA, Lynch MA (1998) Evidence that increased hippocampal expression of the cytokine interleukin-1 beta is a common trigger for age- and stress-induced impairments in long-term potentiation. J Neurosci 18:2974–2981.

Nave H, Bedoui S, Moenter F, Steffens J, Felies M, Gebhardt T, Straub RH, Pabst R, Dimitrijevic M, Stanojevic S, von Horsten S (2004) Reduced tissue immigration of monocytes by neuropeptide Y during endotoxemia is associated with Y-2 receptor activation. J Neuroimmunol 155:1–12.

Nguyen NK, Keck ME, Hetzenauer A, Thoeringer CK, Wurst W, Deussing JM, Holsboer F, Muller MB, Singewald N (2006) Conditional CRF receptor 1 knockout mice show altered neuronal activation pattern to mild anxiogenic challenge. Psychopharmacology 188:374–385.

Noble F, Rubira E, Boulanouar M, Palmier B, Plotkine M, Warnet JM, Marchand-Leroux C, Massicot F (2007) Acute systemic inflammation induces central mitochondrial damage and mnesic deficit in adult Swiss mice. Neurosci Lett 424:106–110.

Oitzl MS, van Oers H, Schobitz B, de Kloet ER (1993) Interleukin-1 beta, but not interleukin-6, impairs spatial navigation learning. Brain Res 613:160–163.

Phillips RG, LeDoux JE (1995) Lesions of the fornix but not the entorhinal or perirhinal cortex interfere with contextual fear conditioning. J Neurosci 15:5308–5315.

Proft T, Fraser JD (2003) Bacterial superantigens. Clin Exp Immunol 133:299–306.

Pugh CR, Kumagawa K, Fleshner M, Watkins LR, Maier SF, Rudy JW (1998) Selective effects of peripheral lipopolysaccharide administration on contextual and auditory-cue fear conditioning. Brain Behav Immun 12:212–229.

Rada P, Mark GP, Vitek MP, Mangano RM, Blume AJ, Beer B, Hoebel BG (1991) Interleukin-1 beta decreases acetylcholine measured by microdialysis in the hippocampus of freely moving rats. Brain Res 550:287–290.

Reichenberg A, Kraus T, Haack M, Schuld A, Pollmacher T, Yirmiya R (2002) Endotoxin-induced changes in food consumption in healthy volunteers are associated with TNF-alpha and IL-6 secretion. Psychoneuroendocrinology 27:945–956.

Reichenberg A, Yirmiya R, Schuld A, Kraus T, Haack M, Morag A, Pollmacher T (2001) Cytokine-associated emotional and cognitive disturbances in humans. Arch Gen Psychiatry 58:445–452.

Rossi-George A, LeBlanc F, Kaneta T, Urbach D, Kusnecov AW (2004) Effects of bacterial superantigens on behavior of mice in the elevated plus maze and light-dark box. Brain Behav Immun 18:46–54.

Rossi-George A, Urbach D, Colas D, Goldfarb Y, Kusnecov AW (2005) Neuronal, endocrine, and anorexic responses to the T-cell superantigen staphylococcal enterotoxin A: dependence on tumor necrosis factor-alpha. J Neurosci 25:5314–5322.

Serrats J, Sawchenko PE (2006) CNS activational responses to staphylococcal enterotoxin B: T-lymphocyte-dependent immune challenge effects on stress-related circuitry. J Comp Neurol 495:236–254.

Shaw KN, Commins S, O'Mara SM (2001) Lipopolysaccharide causes deficits in spatial learning in the watermaze but not in BDNF expression in the rat dentate gyrus. Behav Brain Res 124:47–54.

Shaw KN, Commins S, O'Mara SM (2005) Cyclooxygenase inhibition attenuates endotoxin-induced spatial learning deficits, but not an endotoxin-induced blockade of long-term potentiation. Brain Res 1038:231–237.

Shurin G, Shanks N, Nelson L, Hoffman G, Huang L, Kusnecov AW (1997) Hypothalamic-pituitary-adrenal activation by the bacterial superantigen staphylococcal enterotoxin B: Role of macrophages and T cells. Neuroendocrinology 65:18–28.

Silva AJ, Kogan JH, Frankland PW, Kida S (1998) CREB and memory. Annu Rev Neurosci 21:127–148.

Sparkman NL, Martin LA, Calvert WS, Boehm GW (2004) Effects of intraperitoneal lipopolysaccharide on Morris maze performance in year-old and 2-month-old female C57BL/6J mice. Behav Brain Res 159:145–151.

Sparkman NL, Kohman RA, Scott VJ, Boehm GW (2005a) Bacterial endotoxin-induced behavioral alterations in two variations of the Morris water maze. Physiol Behav 86:244–251.

Sparkman NL, Kohman RA, Garcia AK, Boehm GW (2005b) Peripheral lipopolysaccharide administration impairs two-way active avoidance conditioning in C57BL/6J mice. Physiol Behav 85:278–288.

Sparkman NL, Buchanan JB, Heyen JR, Chen J, Beverly JL, Johnson RW (2006) Interleukin-6 facilitates lipopolysaccharide-induced disruption in working memory and expression of other proinflammatory cytokines in hippocampal neuronal cell layers. J Neurosci 26:10709–10716.

Steckler T, Holsboer M (1999) Corticotropin-releasing hormone receptor subtypes and emotion. Biol Psychiat 46:1480 1508.

Sweatt JD (2001) The neuronal MAP kinase cascade: a biochemical signal integration system subserving synaptic plasticity and memory. J Neurochem 76:1–10.

Swiergiel AH, Dunn AJ (2007) Effects of interleukin-1beta and lipopolysaccharide on behavior of mice in the elevated plus-maze and open field tests. Pharmacol Biochem Behav 86:651–659.

Szelenyi Z, Szekely M (2004) Sickness behavior in fever and hypothermia. Front Biosci 9: 2447–2456.

Tazi A, Dantzer R, Crestani F, Le Moal M (1988) Interleukin-1 induces conditioned taste aversion in rats: a possible explanation for its pituitary-adrenal stimulating activity. Brain Res 473:369–371.

Thomas GM, Huganir RL (2004) MAPK cascade signalling and synaptic plasticity. Nat Rev Neurosci 5:173–183.

Thomazzi SM, Ribeiro RA, Campos DI, Cunha FQ, Ferreira SH (1997) Tumor necrosis factor, interleukin-1 and interleukin-8 mediate the nociceptive activity of the supernatant of LPS-stimulated macrophages. Mediat Inflamm 6:195–200.

Thomson LM, Sutherland RJ (2005) Systemic administration of lipopolysaccharide and interleukin-1beta have different effects on memory consolidation. Brain Res Bull 67:24–29.

Thomson LM, Sutherland RJ (2006) Interleukin-1beta induces anorexia but not spatial learning and memory deficits in the rat. Behav Brain Res 170:302–307.

Tong L, Balazs R, Soiampornkul R, Thangnipon W, Cotman CW (2007) Interleukin-1beta impairs brain derived neurotrophic factor-induced signal transduction. Neurobiol aging 29:1380–1393.

Urbach-Ross D, Crowell B, Kusnecov AW (2008) Relationship of varying patterns of cytokine production to the anorexic and neuroendocrine effects of repeated Staphylococcal enterotoxin A exposure. J Neuroimmunol 196:49–59.

Veening JG, Vandermeer MJM, Joosten H, Hermus ARMM, Rijnkels CEM, Geeraedts LM, Sweep CGJF (1993) Intravenous Administration of Interleukin-1-Beta Induces Fos-Like Immunoreactivity in Corticotropin-Releasing Hormone Neurons in the Paraventricular Hypothalamic Nucleus of the Rat. J Chem Neuroanat 6:391–397.

Wang H, Hu Y, Tsien JZ (2006) Molecular and systems mechanisms of memory consolidation and storage. Prog Neurobiol 79:123–135.

Wang Y, Jiang JD, Xu D, Li Y, Qu C, Holland JF, Pogo BG (2004) A mouse mammary tumor virus-like long terminal repeat superantigen in human breast cancer. Cancer Res 64:4105–4111.

Watkins LR, Wiertelak EP, Goehler LE, Smith KP, Martin D, Maier SF (1994) Characterization of cytokine-induced hyperalgesia. Brain Res 654:15–26.

Williams O, Aroeira LS, Martinez C (1994) Absence of peripheral clonal deletion and anergy in immune responses of T cell-reconstituted athymic mice. Eur J Immunol 24:579–584.

Wilson CJ, Finch CE, Cohen HJ (2002) Cytokines and cognition – the case for a head-to-toe inflammatory paradigm. J Am Geriatr Soc 50:2041–2056.

Yirmiya R, Winocur G, Goshen I (2002) Brain interleukin-1 is involved in spatial memory and passive avoidance conditioning. Neurobiol Learn Mem 78:379–389.

Yirmiya R, Rosen H, Donchin O, Ovadia H (1994) Behavioral effects of lipopolysaccharide in rats: involvement of endogenous opioids. Brain Res 648:80–86.

Zamoyska R (2006) Superantigens: supersignalers? Sci STKE 2006:pe45.

Cytokines, Immunity and Sleep

Francesca Baracchi and Mark R. Opp

Abstract Sleep is a fundamental process necessary for physical and mental health. Although the functions of sleep remain to be fully elucidated, data obtaining during the last 30 years demonstrate important bi-directional links between sleep and the immune system. Sleep–wake behavior is altered by the activation of the immune system during infections and, conversely, sleep loss is often concomitant with pathologies associated with an increase of inflammatory mediators. Increasing our knowledge of the molecular and cellular pathways by which sleep and immune system interact should provide a better understanding of the benefits of sleep and new insights into factors that result in a healthy immune system.

Keywords Sleep · Immunity · Brain · IL-1β · TNFα

1 Introduction

Sleep is a complex behavior observed throughout the animal kingdom. For many years sleep was considered a passive process, characterized by a lack of wakefulness due to a progressive reduction in activity of sensory systems. However, during the Twentieth century several neuroanatomic substrates were determined to be involved in either the control of wakefulness or the active promotion of sleep. Today, sleep is considered an active physiological process involving components of both the central nervous and the autonomic nervous systems. However, although we now know much about the neurochemical and neuroanatomic substrates involved in the regulation of arousal state, the reason why we spend one third of our lives sleeping is still not understood. Multiple functional theories for sleep, such as memory formation, energy conservation, neuronal repair and reorganization, have been advanced, but none have proved definitive.

M.R. Opp (✉)
Department of Anesthesiology, Department of Molecular and Integrative Physiology, Neuroscience Graduate Program, University of Michigan, 7422 Medical Sciences Building I, 1150 W. Medical enter Drive, Ann Arbor, MI 48109–5615, USA
e-mail: mopp@med.umich.edu

A. Siegel, S.S. Zalcman (eds.), *The Neuroimmunological Basis of Behavior and Mental Disorders*, DOI 10.1007/978-0-387-84851-8_11,

Sleep loss and sleep disruption are common features of modern society, whether due to shift work, medical conditions, or the distractions of electronic media and entertainment. The impact of sleep loss and sleep disruption on public health has recently been highlighted in a report issued by the Institute of Medicine entitled *Sleep Disorders and Sleep Deprivation: An Unmet Public Health Need* (Committee on Sleep Medicine and Research 2006). Cumulative long-term effects of sleep loss on public health include increased risk of hypertension, diabetes, obesity, depression, heart attack, and stroke. As such, it is clear that besides mental health, sleep is critical to physical health and well-being.

During the last 30 years, incontrovertible evidence has been obtained indicating that the central nervous system (CNS) and the immune system are intimately linked and exert important influences on each other (Blalock 1989; Dantzer 2001; Felten & Felten 1987; Krueger & Toth 1994). During this same period, systematic investigations demonstrate bidirectional interactions between sleep and the immune system: sleep is altered during immune activation and sleep loss alters immune function (Bryant et al. 2004; Opp 2006; Opp et al. 2007).

It is common to experience excess sleepiness and fatigue during the onset of an infection. Changes in sleep are, in fact, prominent among alterations in behavior collectively referred to as "sickness behavior." Sickness behavior is an adaptive response to infection that is triggered by microbial pathogen-induced activation of the peripheral immune system (Dantzer et al. 2008). Conversely, there is also accumulating evidence that sleep loss can impair immune function. Studies have combined sleep deprivation with measurement of one or more parameters of the immune response and demonstrate that immune parameters change after sleep loss (Irwin et al. 1994, 1999; Matsumoto et al. 2001; Opp 2006) and that prolonged sleep deprivation of rats can lead to the development of sepsis and eventually to death (Everson et al. 1989; Everson & Toth 2000; Rechtschaffen et al. 1989).

The molecular mechanism responsible for the changes in sleep during infection and changes in immune function induced by sleep loss are beginning to be understood. Studies focusing on these interactions implicate pro-inflammatory cytokines as important mediators (Krueger et al. 2001; Opp 2005, 2006). Among cytokines, at least two, interleukin-1β (IL-1β) and tumor necrosis factor α (TNF-α), are involved in the physiological regulation of sleep in the absence of an immune challenge (Krueger et al. 2001; Opp 2005, 2006).

In this chapter we will address the bidirectional communication between the CNS and the immune system emphasizing how modifications and alterations in sleep can affect immune function and, vice versa i.e., how the activation of the immune system alters sleep–wake behavior.

2 Basics of Sleep

Sleep is a complex behavior characterized by neurophysiological, autonomic, endocrine, and vegetative changes in the organism. Sleep can be defined in behavioral terms as a reversible, rhythmic, and homeostatically regulated state of perceptual disengagement from the environment characterized by a cyclic succession of different psychophysiological phenomena.

Sleep is not a unitary process, and, in fact, comprises two different phases: one phase, non rapid eye movement (NREM) sleep is characterized by electroencephalogram (EEG) activity of high amplitude and low frequency waveforms. These high amplitude, low frequency waves resulting from the synchronous firing of cortical neurons, are the characteristic feature of this sleep phase that lends the name slow wave sleep. The second phase of sleep, called rapid eye movement (REM) sleep, is characterized by low voltage and high frequency EEG activity similar to that observed during active wakefulness. For this reason, this phase of sleep is also referred to as desynchronized sleep or paradoxical sleep.

During NREM sleep, parasympathetic activity predominates and heart rate, blood pressure, muscle tone, and temperature decrease, resulting in reduced metabolic activity. Spectral analysis of the sleep EEG indicates that the most predominant frequency component of the EEG during NREM sleep is a frequency band ranging from 0.5 to 4 Hz, referred to as the delta frequency band. Spectral power in the delta frequency band during NREM sleep, often referred to as delta power or slow wave activity, is widely regarded as a measure of intensity or depth of sleep in both humans and animals (Borbély 1982). The threshold for arousal from NREM sleep increases when delta power increases.

The EEG during REM sleep is characterized by cortical desynchrony resulting in a waveform of higher frequency and lower amplitude than that observed during NREM sleep. Although the cortical EEG is desynchronized, the hippocampal EEG is highly synchronized within the 4–9 Hz frequency band, referred to as the theta frequency band. During REM sleep, parasympathetic activity is tonically preponderant (Baust et al. 1969; Berlucchi et al. 1964), even though a highly variable sympathetic activity can induce phasic events in blood pressure, cardiovascular, and respiratory activity.

The two phases of sleep alternate through the course of a sleep period. For example, undisturbed sleep of humans typically consists of four to five NREM sleep–REM sleep cycles, each of about 90–100 min duration. NREM sleep is more abundant during the first half of the night, whereas most REM sleep occurs in the latter part of the night. In healthy young adults, NREM sleep occupies about 75–80% of sleep time and REM sleep occupies approximately 20–25% of the sleep time.

Sleep amount and architecture differ across species, in part because of ecological adaptation (Siegel 2005; Tobler 1988, 1995). Laboratory rats and mice (the most commonly used species in sleep research) when entrained to a 12:2 h light–dark cycle, spend about 70% of the light period asleep compared to about 30% of the dark period. As such, rats and mice are considered nocturnal. Relative to humans, the NREM sleep–REM sleep cycles of rats and mice are much shorter, lasting about 8–12 min, and more numerous. Under normal conditions the sequence of these cycles is the same in humans and rodents; there are transitions from wakefulness to NREM sleep to REM sleep. Irrespective of sleep pattern and amount, humans and nonhuman animals exhibit a strict periodicity of sleep throughout the life cycle. The sleep–wake cycle and other physiological variables, such as body temperature and endocrine activity, are exquisitely synchronized to the day–night cycle.

Sleep is a homeostatically regulated process: sleep loss is compensated by increases in the duration and/or quality of subsequent sleep (Cerri et al. 2005; Franken 2002; Rechtschaffen et al. 1999). Numerous studies of responses to sleep loss have been published. Data from rats and mice demonstrate that even short or partial sleep deprivation alters the EEG, neurochemical balance, and gene expression in brain. In addition, short or partial sleep deprivation induces changes in performance and some aspects of learning and memory. Prolonged, long-term sleep deprivation of rats induces a unique clinical profile, with abnormalities in metabolic, immune, and endocrine function that ultimately result in death (Everson 2005; Everson & Toth 2000; Rechtschaffen et al. 1989). Additional details are given in Section 4.

This brief overview serves to remind the reader that sleep is sensitive to endogenous and exogenous stimuli, and is essential for physical and mental well-being. For these reasons, sleep is highly regulated by multiple, redundant neurochemical and neuroanatomic systems. A review of mechanisms responsible for the regulation of sleep–wake behavior is beyond the scope of this chapter, and interested readers are referred to recent reviews (Jones 2005; McCarley 2007).

3 Cytokines and Sleep

The CNS utilizes numerous transmitter substances for communication among neurons. When released, these substances may act locally at the synaptic level (neurotransmitters) or on more distant targets (peptides and hormones). As briefly stated in the previous section, there are numerous, redundant and overlapping neurochemical and neuroanatomic systems involved in the regulation of sleep–wake behavior. Some transmitter systems are involved in the regulation of NREM sleep, whereas others are involved in the regulation of REM sleep, and yet others in the regulation of wakefulness (see Jones 2005 for a review). Interactions among these various systems are essential for normal transitions from one arousal state to another, and for the overall pattern and timing of sleep cycles. As such, there are multiple characteristics by which one may identify a transmitter substance as being involved in the regulation of either NREM or REM sleep phases. Several lists of criteria that must be met for a transmitter substance to be considered involved in the regulation of sleep–wake behavior have been published (e.g., Borbély 1990; Krueger et al. 1999). An abbreviated list of these criteria includes the following: (1) the substance and its receptors must be found in the brain, (2) the substance, when administered exogenously or when produced endogenously, should increase time spent in the sleep phase for which that transmitter has been implicated, (3) inhibition or inactivation of the transmitter, or blockade of its receptors, should reduce the amount of time spent in the sleep phase for which that transmitter has been implicated, (4) concentrations or turnover rates of the transmitter should vary with the sleep–wake cycle, (5) the transmitter should increase with prior wakefulness, and (6) increases in sleep phases that occur after prolonged wakefulness, during infection, or after mild increases in ambient temperature should be reduced and/or blocked if the transmitter or its receptor is antagonized.

The list of cytokines and chemokines studied in laboratory animals or human subjects and demonstrated to affect sleep is extensive and includes: IL-1α, IL-1β, IL-2, IL-4, IL-6, IL-8, IL-10, IL-13, IL-15, IL-18, TNF-α, TNF-β, interferon α (IFN-α), IFN-β, INF-γ, and MIP-1β/CCL4. Among these substances only two, IL-1β and TNF-α, fulfill all the criteria previously enumerated and may be considered sleep regulatory substances (Krueger et al. 2001; Opp 2005). Evidence for a role for IL-1β and TNF-α in the regulation of physiological (spontaneous) sleep has been derived from numerous electrophysiological, biochemical, and molecular genetic studies. In addition, effects on sleep of alterations in these two cytokine systems are predictable on the basis of known signaling pathways, which may also be targeted for intervention.

Although most cytokines have been discovered in the peripheral immune system, the presence of several cytokines and their receptors within the CNS has been amply demonstrated (Benveniste 1992; Breder et al. 1988; Eriksson et al. 2000; Schöbitz et al. 1994). The CNS detects peripheral immune activation by mechanisms that include, among others, cytokine-induced stimulation of the vagus nerve, by actions of circulating cytokines at the circumventricular organs, or by active transport of cytokines from the periphery into the CNS (reviewed (Dantzer 1994), and see other chapters in this volume). However, cytokines are also synthesized de novo and released within the CNS by both neurons (Breder et al. 1988; Ignatowski et al. 1997; Marz et al. 1998) and glia (Frei et al. 1989; Giulian et al. 1986; Sawada et al. 1989). Neurons immunoreactive for IL-1β and TNF-α are located in brain regions implicated in the regulation of sleep–wake behavior, notably the hypothalamus, hippocampus, and brain stem (Breder et al. 1993, 1988). Signaling receptors for both IL-1β and TNF-α are also present in several areas of the normal (nonpathologic) brain, such as the choroid plexus, hippocampus, hypothalamus, brain stem, and cortex and are expressed by both neurons and astrocytes (Ban 1994; Bette et al. 2003).

IL-1β and TNF-α increase NREM sleep in several species (rat, mouse, monkey, cat, rabbit, sheep) irrespective of the route of administration (e.g., Deboer et al. 2002; Dickstein et al. 1999; Fang et al. 1997; Krueger et al. 1984; Kubota et al. 2002; Olivadoti & Opp 2008; Opp et al. 1991; Reite et al. 1987; Shoham et al. 1987; Susic & Totic 1989; Tobler et al. 1984). The increase in NREM sleep that follows the administration of IL-1β or TNF-α has some characteristics of physiological sleep in the sense that sleep remains episodic and is easily reversible when animals are stimulated. However, effective doses of IL-1β and TNF-α generally fragment NREM sleep and suppress REM sleep. The effects of IL-1β on NREM sleep are also dose (Olivadoti & Opp 2008; Opp et al. 1991) and time (Lancel et al. 1996; Opp et al. 1991) dependent.

Delta power during NREM sleep is also affected by IL-1β and TNF-α, but the effects of these cytokines on this sleep parameter are complex and not well understood. IL-1β increases the amplitude of EEG slow waves during NREM sleep in rabbit and rat, but these changes are dependent on the dose and time of administration (Lancel et al. 1996; Opp et al. 1991; Shoham et al. 1987; Tobler et al. 1984). Although central administration of TNF-α enhances slow wave oscillations during NREM sleep (Takahashi et al. 1996a, b, 1997), intraperitoneal administration increases NREM sleep duration without a concomitant increase of EEG delta power (Fang et al. 1997, 1998; Kubota et al. 2002). Moreover, both IL-1β and TNF-α enhance EEG delta power in a dose-dependent manner when applied locally to the surface of the cerebral cortex, but do not alter NREM sleep duration under these conditions (Yasuda et al. 2005; Yoshida et al. 2004). Taken together, these observations suggest that the effects of these two cytokines on delta power are complex and may be dissociated from the duration of time spent in NREM sleep.

Little is known about the mechanisms underlying the effects of cytokines on REM sleep. Administration of effective doses of cytokines such as IL-1, IL-2, IL-15, IL-18, and TNF-α inhibit REM sleep (Imeri et al. 1999; Kapás et al. 1992; Kubota et al. 2001a, b, c; Opp & Imeri 2001; Opp et al. 1991). Antagonizing endogenous cytokines in healthy animals with receptor antagonists, soluble receptors or antibodies either has no effect on REM sleep (Opp & Krueger 1994a; Takahashi et al. 1995a, 1997) or only slightly reduces REM sleep (Takahashi et al. 1995b). It has been hypothesized that reduced REM sleep of mice lacking the TNF receptor 1 (i.e., TNFR1 KO; Fang et al. 1997) is an indirect result of the concomitant reduction in NREM sleep, which limits the potential to enter REM sleep under normal circumstances. Recently it has been shown that alterations in REM sleep of mice lacking both IL-1β receptor 1 and TNF-α receptor 1 can occur independently from changes in NREM sleep, suggesting that IL-1β and TNF-α influence REM sleep regulation by mechanisms that are independent of those involved in NREM sleep regulation (Baracchi & Opp 2008).

In contrast to the increase in NREM sleep after administration of IL-1β or TNF-α, antagonizing either of these cytokine systems reduces spontaneous NREM sleep. For example, inactivating or interfering with the normal action of IL-1β or TNF-α by means of antibodies, antagonists or soluble receptors (Krueger et al. 2001; Opp et al. 1992; Opp & Krueger 1994a; Takahashi et al. 1996a, b) reduces spontaneous NREM sleep expression and the intensity of the sleep response to manipulations that result in increased NREM sleep, such as sleep deprivation, excessive food intake, and acute elevation of ambient temperature, which are associated with enhanced production of either IL-1β or TNF-α (Mackiewicz et al. 1996; Taishi et al. 1998). Moreover knockout mice that lack the type 1 IL-1 receptor (Fang et al. 1997), or the type 1 TNF receptor (Fang et al. 1998) or both (Baracchi & Opp 2008) spend less time in NREM sleep than do control strains of mice.

Finally, the fact that diurnal rhythms of IL-1β and TNF-α vary with the wake–sleep cycle provides further evidence for the involvement of IL-1β and TNF-α in physiological sleep regulation. In rats, IL-1β and TNF-α mRNA in brain exhibit a diurnal rhythm with peaks that occur at light onset; the light period in these rodents is the time when NREM sleep propensity is maximum (Bredow et al. 1997; Cearley et al. 2003; Taishi et al. 1997, 1998). In humans, IL-1β plasma levels are highest at the onset of sleep (Moldofsky et al. 1986), and in cat cerebrospinal fluid IL-1 levels vary with the sleep–wake cycle (Lue et al. 1988). Furthermore, TNF-α

protein levels in rat also have a sleep-associated diurnal rhythm in several brain areas (Floyd & Krueger 1997).

There is accumulating evidence that IL-6 may play a role in the excessive daytime sleepiness associated with sleep disorders such as sleep apnea, narcolepsy, and primary insomnia (Burgos et al. 2006; Okun et al. 2004; Vgontzas et al. 1997, 2002). IL-6 is normally present in brain and its expression in brain (Guan et al. 2003) and the periphery (Guan et al. 2003; Shearer et al. 2001) is affected by sleep deprivation. However, although administration of IL-6 increases NREM sleep of rats, antagonizing IL-6 in the rat brain has no effect on NREM sleep (Hogan et al. 2003). Furthermore, NREM sleep of mice lacking IL-6 is normal (Morrow & Opp 2005). Collectively, these data suggest that although IL-6 may not be involved in the regulation of physiological NREM sleep, elevated levels during pathology may contribute to alterations of some facets of NREM sleep. At least one other cytokine exhibits similar features with respect to sleep. IL-18 enhances NREM sleep of rats, whereas anti-rat IL-18 antibodies do not affect spontaneous sleep (Kubota et al. 2001b). With respect to the other cytokines, chemokines, and growth factors studied thus far, there are insufficient data to state that any are involved in the regulation of spontaneous NREM sleep, although this possibility cannot be ruled out. It is clear, however, that numerous immunomodulators modulate sleep and may contribute to the alterations in sleep that occur during immune challenge.

4 Effects of Sleep Loss on Immunity

The importance of sleep for host functions is usually studied by examining the effects of either acute or prolonged periods of sleep loss. The quantification of the impact of sleep loss on the immune system is difficult because of the excessively large number of parameters that may be measured. Determination of the impact of sleep loss on immune function is further compounded because sleep deprivation is usually associated with stress, changes in locomotor activity, feeding, hormonal secretion, and body temperature, each of which may directly or indirectly affect the immune system. As such, isolating from other confounding responses the effects on immune function of sleep loss per se has proven difficult. Many studies of the impact of sleep loss on immune function have been conducted, using human volunteers or laboratory animals, with varied and often contradictory results. Another important factor contributing to the variability of reported responses to sleep deprivation is the lack of consistency between studies in methods used: i.e., differences in sleep deprivation methods and duration, timing of blood sampling, choice of immune parameters measured and the type of assay used to measure these parameters.

The most common assessment of the impact of sleep loss on the immune system has been determination of changes in cytokine protein profiles and/or cytokine mRNA. Sleep deprivation alters circulating cytokines and the production of cytokines by stimulated peripheral blood lymphocytes obtained from sleep-deprived human volunteers. In 1989, Moldofsky et al. (1989) reported that 40 h of sleep deprivation enhanced plasma IL-1-like and IL-2-like activity. Plasma IL-1β and TNF-α concentrations in healthy women increased after 42h of sleep deprivation (Altemus et al. 2001). One night of sleep deprivation increases IL-1β and TNF-α, and decreases IL-2 in whole blood samples obtained during nighttime sleep and stimulated with mitogens (Born et al. 1997). IL-6 is elevated in sleep-deprived or sleep-restricted volunteers (Frey et al. 2007; Shearer et al. 2001; Vgontzas et al. 2004).

Relative to morning levels following uninterrupted nighttime sleep, one night of sleep deprivation of human volunteers increases production of IL-6 and TNF-α by peripheral blood monocyte populations that have been stimulated by lipopolysaccharide (Irwin et al. 2006). In addition, in this study sleep loss for one night induced a more than 3-fold increase in transcription of IL-6 mRNA and a 2-fold increase in TNF-α mRNA (Irwin et al. 2006). Prolonged sleep deprivation does not induce consistent changes in T-cell derived cytokines, such as IL-2, IL-4, and IFNγ(Boyum et al. 1996; Dinges et al. 1995; Everson 2005).

Sleep deprivation-induced alterations in cytokine message and protein have also been demonstrated in animal models. In rats, IL-1β and TNF-α mRNA increase in the hypothalamus, hippocampus, cerebral cortex, and brain stem after sleep deprivation (Bredow et al. 1997; Mackiewicz et al. 1996). In rats and rabbits, central administration of anti-IL-1β antibodies attenuates the increase in NREM sleep that follows sleep deprivation (Opp & Krueger 1994a, b), suggesting that increased NREM sleep after sleep deprivation is mediated at least in part by elevated IL-1β. Moreover, 36h of sleep deprivation of mice increases serum levels of IL-1β, IL-6, and TNF-α (Hu et al. 2003). Additional evidence that IL-1β and TNF-α play a role in the increase in NREM sleep after sleep deprivation is the recent observation that mice in which both the IL-1β receptor type 1 and the TNF-α receptor type 1 have been genetically ablated (knocked out) exhibit less NREM sleep during the recovery period after sleep deprivation than do genetically intact control animals (Baracchi & Opp 2008).

Several studies have investigated the effects of sleep deprivation on immune-cell number and/or function (reviewed, Opp et al. 2007). Leukocyte numbers and cell types change in response to sleep deprivation (Dinges et al. 1995). For example, after one night of sleep deprivation of humans, the number of granulocytes significantly increases but lymphocytes and B cells are unaffected (Heiser et al. 2000). In contrast to unaltered lymphocyte subpopulations observed after only one night of sleep deprivation, 64h of sleep loss increases the number of granulocytes concomitantly with reductions in numbers of CD4$^+$, CD16$^+$, CD56$^+$, and CD57$^+$ lymphocytes (Dinges et al. 1994). In a recent study, two different methods of sleep deprivation revealed distinct effects on the immune system. Partial sleep deprivation of rats for 24h caused only a decrease in blood lymphocytes, whereas sleep restriction to 6h a day for 21 days led to a decrease in total leukocytes and lymphocytes and to an increase in immunoglobulin M production (Zager et al. 2007). Effects of sleep deprivation on natural killer (NK) cell numbers also have been determined. Some studies demonstrate decreased NK cell numbers after sleep deprivation (Heiser et al. 2000; Irwin et al. 1996), whereas in others sleep deprivation led to increased NK cells numbers and activity (Born et al. 1997; Matsumoto et al. 2001).

Preclinical animal studies demonstrate that long-term sleep deprivation of rats results in death (Everson 1995; Rechtschaffen et al. 1989). In these studies, rats were sleep deprived using the disk-over-water method, which consists in an apparatus comprising a flat disk divided into two halves positioned over a pan containing water. One rat is positioned on each half of the disk and one of the two animals is designated as the experimental subject while the other one is considered the control. Every time the experimental animal falls asleep, the disk is slowly rotated until the animal wakes up. If the animal keeps sleeping, the rotation of the disk forces it against the wall and then into the water. The control animal may partially be sleep deprived but it is able to sleep whenever the experimental animal is awake (Rechtschaffen et al. 1983). Because the physical stimulation is equally administered to experimental and control animal, the disk-over-water is considered to be a method that controls for nonspecific stress responses. With this procedure, control rats show either minimal or no stress indicators. On the other hand, sleep-deprived rats manifest a unique syndrome presenting with weight loss despite progressively higher rates of food intake, distinctive skin lesions, thermoregulatory changes, and eventual death.

Rats sleep deprived by the disk-over-water method develop a bloodstream infection by bacteria generally ascribed to be of gut origin (Everson 1993). These gut bacteria, once translocated, are postulated to trigger a hypermetabolic and systematic inflammatory state resulting in sepsis without a septic focus (Nieuwenhuijzen et al. 1996). Moreover, aerobic and anaerobic bacteria are detectable in mesenteric lymph nodes after only 5 days of sleep deprivation. The migration of these bacteria from the lymph nodes is evidenced by transient and polymicrobial infections in major organs (Everson & Toth 2000). The penetration of bacteria into normally sterile tissues during sleep deprivation implies the development of immune insufficiency and abnormal host defense (Everson 2005). In recent studies (Everson 2005; Zager et al. 2007), the measurement of several clinical immune parameters, such as the composition of the circulating leukocyte pool, and detection of chemokines, cytokines and immunoglobulins, suggests that sleep deprivation activates mechanisms associated with innate immunity and responses by B lymphocytes that are consistent with polyclonal activation. However, these immune responses seem to be ineffective since there is a failure in eradicating invading bacteria (Everson 2005). These data indicate that sleep deprivation might induce a chronic infectious and antigenic state that precedes the outward signs of poor health. As such, the clinical outcome of chronically sleep-deprived animals may be due to an inability of the immune system to combat an infection (sepsis) resulting from increased translocation of bacteria across the gut.

Another approach that has been used to investigate the impact of sleep loss on the immune system is the infection of sleep-deprived animals with replicating pathogens. Results of these studies in some instances are also contradictory. For example, a preclinical study from Brown et al. (1989) showed that mice that were sleep deprived before inoculation with influenza virus did not clear the virus as efficiently as control animals. However subsequent studies either failed to detect any difference in viral clearance between rested and sleep-deprived mice

(Toth & Rehg 1998), or suggested that sleep deprivation enhanced immune responses to viral infection (Renegar et al. 2000). However, functional consequences of acute sleep loss on responses to vaccines have been demonstrated in humans. The antibody response to influenza vaccination is reduced by about 50% in volunteers with sleep restricted to 4 h a night for 4 nights as compared to control subjects who had been allowed 8 h sleep (Spiegel et al. 2002). Similarly, the antibody response of volunteers to hepatitis A vaccination after one night of total sleep deprivation is about half that of control subjects allowed a full night's sleep (Lange et al. 2003). Finally, studies of rabbits sleep deprived for 4 h prior to inoculation with *Escherichia coli* demonstrate that clinical outcomes are not dramatically altered, although infection-induced alterations in sleep are exacerbated by sleep deprivation (Toth 1995).

5 Effects of Immune Challenge on Sleep

Sleepiness, fatigue, tiredness, and malaise are frequent early responses to infection sickness. The first systematic studies of the impact of infections on sleep were conducted during the pandemic of encephalitis lethargica in the 1920s and 1930s by neurologist Costantin Von Economo. Von Economo was the first to suggest that the hypersomnolence or insomnia resulting from this infection were related to lesions of specific regions of the hypothalamus (von Economo 1930). During the past 20 years, several animal models have been used to determine the impact of infections on sleep, including infections with viral, bacterial, and fungal pathogens, parasites and prion-related diseases. The investigation of this topic in humans is, for obvious reasons, more difficult to approach and only few studies have systematically examined the influence of infections on sleep.

5.1 Viral Infections

The extent to which viral infections alter sleep has been examined in animal models using influenza virus. Multiple studies consistently demonstrate that rabbits (Kimura-Takeuchi et al. 1992a, b) or mice (Fang et al. 1995; Toth 1996; Toth et al. 1995; Toth & Verhulst 2003) infected with influenza spend more time in NREM sleep. Infection-induced increases in NREM sleep of mice are apparent primarily during the dark period, during which these laboratory species are normally more active. The increase in NREM sleep of rabbits infected with influenza, although robust, is of brief duration. In contrast, increases in NREM sleep of mice infected with influenza are prolonged, lasting at least 96 h (Fang et al. 1995; Toth et al. 1995). Differences between rabbits and mice in the duration of sleep alterations during influenza infection may be the result of the degree to which the virus replicates; mouse-adapted strains of influenza undergo complete replication whereas in rabbits the virus undergoes only partial replication. The changes in NREM sleep of mice during influenza infection are likely mediated by an IFN response since the increase in NREM sleep is observed in C57BL/6J mice, which produce relatively high levels of IFNα/β in response to the virus, but not in BALB/c mice, which produce low levels of IFNα/β (Toth 1996). Moreover it has recently been shown that IFN type I receptor-deficient mice have altered clinical symptoms in response to influenza virus. With respect to control mice, time spent in NREM sleep increases more in the knockout mice during the early stages of infection, whereas NREM sleep in these knockouts is reduced during late stages of infections (Traynor et al. 2007).

Infection of humans with influenza or rhinovirus is reported to variably increase or decrease slow wave sleep time. Smith (Smith 1992) demonstrated reductions in total sleep time during the acute incubation (asymptomatic) period of influenza A and B infections; total sleep time increased when subjects were symptomatic. Smith and colleagues (Smith 1992) also reported similar changes in sleep patterns during rhinovirus infection (a cold), total sleep time increased when volunteers were symptomatic. In contrast, Drake and colleagues (Drake et al. 2000) reported that rhinovirus type 23 infection of volunteers does not affect sleep during the acute incubation period but reduces total sleep time during the symptomatic period of the infection.

Characteristic changes in sleep structure also occur during infection with immunodeficiency viruses. Preclinical studies demonstrate that feline immunodeficiency virus (FIV) alters sleep of cats, inducing a gradual change in the normal timing of NREM sleep, increasing arousals, and reducing REM sleep (Prospéro-García et al. 1994). NREM sleep of rats increases after intracerebroventricular administration of HIV glycoproteins 120, 160, and 41 implicating these viral proteins as potential mediators of HIV-induced alterations in sleep of human patients.

In humans, sleep disruptions are reported in individuals who are seropositive for HIV, yet are asymptomatic (Kubicki et al. 1988; Norman et al. 1988, 1992). These disruptions are characterized by increased daytime fatigue and alterations in nighttime sleep, such as increases in NREM sleep stages 3 and 4 (Norman et al. 1987; White et al. 1995). With disease progression, the disruptions to sleep become more severe. However, other studies report decreased NREM sleep in HIV seropositive patients (Reid & Dwyer 2005). Yet other studies failed to detect increases in total amounts of NREM sleep stages 3 and 4, but demonstrated that substantial amounts of NREM sleep stages 3 and 4 occurred late in the night, instead of early in the night as is normal (Norman et al. 1992; Wiegand et al. 1991). Such differences among studies of HIV infections and sleep may well be the result of the variable lengths of time subjects were infected before the sleep study and differing rates of disease progression. However, all studies to date demonstrate alterations in sleep of patients infected with HIV.

Another virus that has been investigated with respect to sleep is the Epstein Barr virus (EBV) a ubiquitous human gammaherpesvirus. Patients infected with EBV typically report a strong feeling of tiredness that may be associated with an increase in sleep time and a shorter sleep latency during daytime (Pollmacher et al. 1995). Animal models for chronic fatigue are generally lacking. However, murine gammaherpesvirus 68 (MuGHV, also known as MHV68 and gammaHV68) has been used in mice as a model for EBV infection due to similarities in immune response, viral genetics, and subsequent development of life-long latency (Flano et al. 2002; Olivadoti et al. 2007).

During the 30 days after inoculation with MuGHV, mice demonstrate increased fatigue and alterations in sleep (data published in abstract form: Olivadoti & Opp 2006).

5.2 Bacterial Infections

The majority of studies of the impact of bacterial infections on sleep have been conducted in rabbits (reviewed in Toth 1999; Toth & Opp 2002). Infection of rabbits with *Staphylococcus aureus* or *Escherichia coli* results in a biphasic change in NREM sleep, with an initial enhancement that is followed by a reduction in NREM sleep amount (Toth & Krueger 1988, 1989). REM sleep is suppressed for the duration of these infections. The development of fever during infection generally occurs with increases in NREM sleep. However, the febrile response of rabbits to *S. aureus* is protracted, and still apparent during the period of infection when NREM sleep is suppressed. These observations, and others (Krueger & Takahashi 1997), indicate that fever and increased NREM sleep may be dissociated, and that alterations is sleep are not a consequence of changes in body temperature.

Bacterial replication in the host is not necessary to elicit alterations in sleep as increases in NREM sleep also are observed in rabbits inoculated with killed bacteria or isolated bacterial components (reviewed by Toth 1999; Toth & Opp 2002). Structure and function studies indicate that particular properties of the bacterial pathogen may be responsible for various aspects of alterations in sleep. For example, NREM sleep of rabbits increases more rapidly, but for a shorter duration, in response to gram-negative bacteria than in response to gram-positive bacteria (reviewed (Toth 1999; Toth & Opp 2002)). Lipid A from gram-negative endotoxin elicits increases in NREM sleep of rabbits within an hour of administration, whereas muramyl dipeptide, a synthetic analog of the monomeric muramyl peptide component of bacterial cell wall peptidoglycan, increases NREM sleep after a longer latency (Krueger et al. 1986; Shoham & Krueger 1988).

5.3 Other Pathogens

Sleep is altered in response to pathogens other than virus and bacteria. For example, the sleep of rabbits is altered by infection with the fungus *Candida albicans* in a manner generally similar to that of gram-positive bacteria (Toth & Krueger 1989).

Prions also disrupt sleep. Rats and cats develop alterations in EEG parameters and sleep–wake behavior after inoculation with scrapie-infected brain homogenates from animals or from human postmortem tissue (Bassant et al. 1984; Gourmelon et al. 1987). Mice in which the prion protein gene has been ablated show alterations in circadian rhythms and sleep (Tobler et al. 1996, 1997). Human fatal familial insomnia is characterized by profound alterations in sleep secondary to prion-induced degeneration of the thalamus (Montagna 2005), whereas Creutzfeldt-Jakob patients have severe sleep EEG abnormalities. Brain autopsy in these patients reveal prominent changes in cortical areas, but only mild changes in the thalamus (Landolt et al. 2006).

Sleep disruptions are severe in patients infected with the protozoan *Trypanosoma brucei*, a parasite transmitted by the tsetse fly that induces a clinical syndrome referred to as "sleeping sickness." Infection with this parasite disrupts nighttime sleep, there are many short bouts of sleep during daytime, and the normal distribution of sleep and wakefulness across the 24 h day disappears (Buguet et al. 2001; Lundkvist et al. 2004). The impact of trypanosomiasis on sleep–wake behavior has been modeled in rabbits and rats (Berge et al. 2005; Darsaud et al. 2004; Lundkvist et al. 2002; Toth et al. 1994). Trypanosome infection of rabbits increases NREM sleep after several days. These increases in NREM sleep are concomitant with fever and other signs of clinical illness. This initial period of increased NREM sleep subsides, but there are repeated episodes of increased NREM sleep that occur in association with recrudescence of the parasite (Toth et al. 1994). As in humans, rats and rabbits lose the circadian rhythms of sleep and body temperature.

6 Mechanisms by Which Cytokines Regulate Sleep

As stated in the introduction, sleep is an active process regulated by several brain regions and neurochemical circuits, the reciprocal activities and interactions of which determine the alternation between wakefulness and sleep. Several arousal systems are present in the brain. For example noradrenergic, dopaminergic and serotonergic neurons, located in different nuclei of the brain stem, fire during wakefulness, decrease firing during NREM sleep and cease firing during REM sleep (reviewed (Jones 2005)). In contrast, acetylcholine (ACh)-containing neurons discharge during waking, decrease firing during NREM sleep and fire at high rates during REM sleep in association with fast cortical activity. Neurons that do not contain ACh, including GABAergic neurons in the basal forebrain and preoptic area (POA) of the hypothalamus, are active in a reciprocal manner to the neurons of the arousal systems, with higher discharge rates during sleep.

Interactions among monoaminergic and cholinergic systems and the immune system have been widely demonstrated (reviewed in Wrona 2006), but specific studies investigating the neuronal circuits involved in sickness-induced alterations in sleep have only recently begun. Immunohistochemical and electrophysiological studies indicate that IL-1β may enhance NREM sleep, in part, by inhibiting the action of wake-promoting nuclei. For instance, in rats the intracerebroventricular administration of IL-1 increases NREM sleep 4–5 h after injection (Opp & Imeri 2001; Opp et al. 1991), and this increase in NREM sleep is associated with increased number of Fos-immunoreactive neurons in the median preoptic nucleus of the hypothalamus (Baker et al. 2005). The POA of the hypothalamus contains wake-related neurons and sleep-related neurons that are activated at the onset of NREM sleep (Sherin et al. 1996; Szymusiak et al. 1998). When perfused via microdialysis to this specific hypothalamic region, IL-1β reduces firing rates of wake-related neurons and increases the firing rate of a subpopulation of sleep-related neurons (Alam et al. 2004). These data suggest that the POA is one site of action for IL-1β in the regulation of NREM sleep.

Another brain area that has been investigated as a potential site upon which IL-1β acts to regulate sleep is the dorsal raphe nucleus (DRN) of the brain stem. This nucleus contains the cell bodies of serotonergic neurons that innervate the entire nervous system (Jacobs & Azmitia 1992). Beisdes being the origin of the major ascending serotonergic system, the DRN contains IL-1 receptors. Microinjection of IL-1β into the DRN of rats increases NREM sleep (Brambilla et al. 2007). The IL-1 and the serotonergic systems influence each other and exhibit a wide range of overlapping activities (Imeri & De Simoni 1999). For example, serotonergic activation by the serotonin precursor L-5-hydroxytryptophan (5-HT)alters IL-1 mRNA expression in discrete brain regions (Gemma et al. 2003), whereas IL-1 stimulates the release of 5-HT in the hypothalamus and other regions of the brain. The full manifestation of IL-1 effects on sleep requires an intact serotonergic system as depletion of brain 5-HT or blockade of 5-HT_2 receptors interferes with IL-1-induced increase in NREM sleep (Imeri et al. 1999, 1997).

Mechanisms by which IL-1 acts on the serotonergic neurons of the DRN have been investigated in vitro. Intracellular recordings from identified serotonergic neurons in the DRN of guinea pig slice preparations indicate that IL-1β reduces the firing rate of these neurons by 50% on average (Manfridi et al. 2003). This effect is reversible, as firing rates return to baseline after washout. A recent study further elucidates potential mechanisms by which IL-1β inhibits the activity of these serotonergic DRN neurons by showing that IL-1β enhances GABAergic inhibitory post-synaptic potentials (Brambilla et al. 2007). Because serotonin promotes wakefulness, collectively these data suggest that one of the mechanisms by which IL-1β enhances NREM sleep may be inhibitory actions on DRN serotonergic neurons.

The observation that IL-1β enhances GABAergic inhibitory postsynaptic potentials of serotonergic neurons in the DRN is in agreement with observations that IL-1 enhances GABA inhibitory effects at both at pre- and postsynaptic levels. IL-1 increases GABA release (Feleder et al. 2000; Tabarean et al. 2006) and GABAergic inhibitory postsynaptic potentials in hippocampal neurons (Luk et al. 1999). Furthermore, IL-1 recruits $GABA_A$ receptors to the cell surface of hippocampal neurons (Serantes et al. 2006), and increases cytosolic Ca^+ in a subpopulation of cultured hypothalamic neurons, some of which are GABAergic (De et al. 2002).

Biochemical and genetic approaches have been widely used to investigate mechanisms by which cytokines regulate sleep–wake behavior. Two different types of membrane receptors have been characterized for both IL-1β and TNF-α. For IL-1β, the type 1 receptor (IL-1R1) is the signaling receptor, whereas the type 2 receptor lacks an intracellular domain and acts as a decoy receptor, likely involved in the regulation of IL-1β activity (Colotta et al. 1993; Sims et al. 1993). The two TNF-α receptors share significant similarities in their extracellular regions, whereas their intracellular domains exhibit striking structural differences reflecting different signaling pathways and functions (Hohmann et al. 1990; Holtmann & Neurath 2004). The IL-1R1 is a member of the IL-1 receptor/Toll-like receptor (TLR) superfamily (O'Neill & Greene 1998). Receptors in the IL-1 receptor/TLR superfamily are involved in host defense and inflammation, and include TLR-4, the receptor for bacterial products such as lipopolysaccharide (Poltorak et al. 1998). TLR-4 shares significant sequence similarity with the IL-1R1 cytosolic region and uses a very similar signaling pathway (Dunne & O'Neill 2003).

Both IL-1β and TNF-α act within a cascade of parallel interacting pathways to regulate NREM sleep (Krueger & Fang 2000). IL-1β and TNF-α induce each other's production and both of them stimulate nuclear factor kappa B (NFκB), a DNA binding protein involved in transcription (Baeuerle & Henkel 1994; Hohmann et al. 1990; Holtmann & Neurath 2004). Production of both IL-1β and TNF-α is enhanced by NFκB activation forming a relatively short-looped positive feedback system. The activation of NFκB also leads to the production of other substances that modulate sleep, such as adenosine, cyclooxygenase-2 (Cox-2 (Tsai et al. 2002), which is involved in the production of prostaglandin D_2 (PGD_2)), and nitric oxide synthase-2 ((Xie et al. 1994), which induces the production of nitric oxide (NO)). Adenosine, PGD_2, and NO are all modulators of NREM sleep (Obal & Krueger 2003).

NFκB seems to be a key element in cytokine-induced alterations in sleep–wake behavior. Inhibition of this transcription factor reduces spontaneous NREM sleep and attenuates IL-1β-induced increases in NREM sleep (Kubota et al. 2000). After short-term sleep deprivation, NFκB is activated in the cortex (Chen et al. 1999), in the basal forebrain (Basheer et al. 2001; Ramesh et al. 2007), and in the lateral hypothalamus (Brandt et al. 2004). In addition, NFκB mediates increases of the adenosine A_1 receptor in the cholinergic basal forebrain that is observed after 24 h of sleep deprivation (Basheer et al. 2007).

The positive short loop between NFκB and IL-1β and TNF-α is controlled and damped by other inhibiting substances. For example, IL-1β induces production of corticotropin releasing hormone (CRH), which in turn inhibits IL-1β production via actions of glucocorticoids on NFκB. CRH increases wakefulness and this effect is dependent on inhibition of IL-1β (Opp et al. 1989). Cox-2 is involved not only in the synthesis of both sleep-promoting PGD_2 but also in the production of sleep-inhibiting PGE_2, which also inhibits IL-1 (Obal et al. 1990).

7 Conclusions

Bidirectional links between the immune system and sleep regulatory systems have been widely investigated during the last 30 years. It is now known that sleep loss affects immune function and that immune activation from infection alters sleep. Adequate sleep is essential for physical and mental health. It is becoming increasingly evident that besides effects on cognition and performance, sleep loss may be a contributing factor to multiple pathologies. Increasing our knowledge of the molecular and cellular pathways by which sleep and the immune system interact should lead to a better understanding of the benefits of sleep and new insights into factors that result in a healthy immune system.

Acknowledgements The authors were supported during the writing of this chapter by the National Institutes of Health grants MH64843, HL80972, and the Department of Anesthesiology, University of Michigan Medical School.

References

Alam, M. N., McGinty, D., Bashir, T., kumar, S., Imeri, L., Opp, M. R., & Szymusiak, R. 2004, "Interleukin-1β modulates state-dependent discharge activity of preoptic area and basal forebrain neurons: role in sleep regulation", *European Journal of Neuroscience*, vol. 20, no. 1, pp. 207–216.

Altemus, M., Rao, B., Dhabhar, F. S., Ding, W., & Granstein, R. D. 2001, "Stress-induced changes in skin barrier function in healthy women", *Journal of Investigative Dermatology*, vol. 117, no. 2, pp. 309–317.

Baeuerle, P. A. & Henkel, T. 1994, "Function and activation of NF-kappa B in the immune system", *Annual Review of Immunology*, vol. 12, pp. 141–179.

Baker, F. C., Shah, S., Stewart, D., Angara, C., Gong, H., Szymusiak, R., Opp, M. R., & McGinty, D. 2005, "Interleukin 1beta enhances non-rapid eye movement sleep and increases c-Fos protein expression in the median preoptic nucleus of the hypothalamus", *American Journal of Physiology – Regulatory, Integrative and Comparative Physiology*, vol. 288, no. 4, pp. R998–R1005.

Ban, E. M. 1994, "Interleukin-1 receptors in the brain: characterization by quantitative in situ autoradiography", *ImmunoMethods*, vol. 5, no. 1, pp. 31–40.

Baracchi, F. & Opp, M. 2008, "Sleep-wake behavior and responses to sleep deprivation of mice lacking both Interleukin-1β receptor 1 and Tumor Necrosis Factor-α receptor 1", *Brain, Behavior and Immunity*, 22: 982–993, 2008.

Basheer, R., Bauer, A., Elmenhorst, D., Ramesh, V., & McCarley, R. W. 2007, "Sleep deprivation upregulates A1 adenosine receptors in the rat basal forebrain", *NeuroReport*, vol. 18, no. 18, pp. 1895–1899.

Basheer, R., Rainnie, D. G., Porkka-Heiskanen, T., Ramesh, V., & McCarley, R. W. 2001, "Adenosine, prolonged wakefulness, and A1-activated NF-κB DNA binding in the basal forebrain of the rat", *Neuroscience*, vol. 104, no. 3, pp. 731–739.

Bassant, M. H., Cathala, F., Court, L., Gourmelon, P., & Hauw, J. J. 1984, "Experimental scrapie in rats: first electrophysiological observations", *Electroencephalography Clinical Neurophysiology*, vol. 57, no. 6, pp. 541–547.

Baust, W., Bohnert, B., & Riemann, O. 1969, "The regulation of the heart rate during sleep", *Electroencephalography Clinical Neurophysiology*, vol. 27, no. 6, p. 626.

Benveniste, E. N. 1992, "Inflammatory cytokines within the central nervous system: sources, function, and mechanisms of action", *American Journal of Physiology*, vol. 263, pp. C1–C16.

Berge, B., Chevrier, C., Blanc, A., Rehailia, M., Buguet, A., & Bourdon, L. 2005, "Disruptions of ultradian and circadian organization of core temperature in a rat model of African trypanosomiasis using periodogram techniques on detrended data", *Chronobiology International*, vol. 22, no. 2, pp. 237–251.

Berlucchi, G., Moruzzi, G., Salvi, G., & Strata, P. 1964, "Pupil behavior and ocular movments during synchronized and desynchronized sleep", *Archives Italiennes De Biologie*, vol. 102, pp. 230–244.

Bette, M., Kaut, O., Schafer, M. K., & Weihe, E. 2003, "Constitutive expression of p55TNFR mRNA and mitogen-specific up-regulation of TNF alpha and p75TNFR mRNA in mouse brain", *Journal of Comparative Neurology*, vol. 465, no. 3, pp. 417–430.

Blalock, J. E. 1989, "A molecular basis for bidirectional communication between the immune and neuroendocrine systems", *Physiological Review*, vol. 69, no. 1, pp. 1–32.

Borbély, A. A. 1982, "A two process model of sleep regulation", *Human Neurobiology*, vol. 1, pp. 195–204.

Borbély, A. A. 1990, "'Sleep substances': criteria and physiological basis," in *Endogenous Sleep Factors*, S. Inoué & J. M. Krueger, eds., SPB Academic Publishing, The Hague, pp. 31–38.

Born, J., Lange, T., Hansen, K., Molle, M., & Fehm, H. L. 1997, "Effects of sleep and circadian rhythm on human circulating immune cells", *Journal of Immunology*, vol. 158, no. 9, pp. 4454–4464.

Boyum, A., Wiik, P., Gustavsson, E., Veiby, O. P., Reseland, J., Haugen, A. H., & Opstad, P. K. 1996, "The effect of strenuous exercise, calorie deficiency and sleep deprivation on white blood cells, plasma immunoglobulins and cytokines", *Scandinavian Journal of Immunology*, vol. 43, no. 2, pp. 228–235.

Brambilla, D., Franciosi, S., Opp, M. R., & Imeri, L. 2007, "Interleukin-1 inhibits firing of serotonergic neurons in the dorsal raphe nucleus and enhances GABAergic inhibitory post-synaptic potentials", *European Journal of Neuroscience*, vol. 26, no. 7, pp. 1862–1869.

Brandt, J. A., Churchill, L., Rehman, A., Ellis, G., Memet, S., Israel, A., & Krueger, J. M. 2004, "Sleep deprivation increases the activation of nuclear factor kappa B in lateral hypothalamic cells", *Brain Research*, vol. 1004, no. 1–2, pp. 91–97.

Breder, C. D., Dinarello, C. A., & Saper, C. B. 1988, "Interleukin-1 immunoreactive innervation of the human hypothalamus", *Science*, vol. 240, pp. 321–324.

Breder, C. D., Tsujimoto, M., Terano, Y., Scott, D. W., & Saper, C. B. 1993, "Distribution and characterization of tumor necrosis factor-alpha-like immunoreactivity in the murine central nervous system", *Journal of Comparative Neurology*, vol. 337, no. 4, pp. 543–567.

Bredow, S., Taishi, P., Guha-Thakurta, N., & Krueger, J. M. 1997, "Diurnal variations of tumor necrosis factor-α mRNA and α-tubulin mRNA in rat brain", *Neuroimmunomodulation*, vol. 4, pp. 84–90.

Brown, R., Pang, G., Husband, A. J., & King, M. G. 1989, "Suppression of immunity to influenza virus infection in the respiratory tract following sleep disturbance", *Regional Immunology*, vol. 2, pp. 321–325.

Bryant, P. A., Trinder, J., & Curtis, N. 2004, "Sick and tired: does sleep have a vital role in the immune system?", *Natural Reviews Immunology*, vol. 4, no. 6, pp. 457–467.

Buguet, A., Bourdon, L., Bouteille, B., Cespuglio, R., Vincendeau, P., Radomski, M. W., & Dumas, M. 2001, "The duality of sleeping sickness: focusing on sleep", *Sleep Medicine Reviews*, vol. 5, no. 2, pp. 139–153.

Burgos, I., Richter, L., Klein, T., Fiebich, B., Feige, B., Lieb, K., Voderholzer, U., & Riemann, D. 2006, "Increased nocturnal interleukin-6 excretion in patients with primary insomnia: a pilot study", *Brain, Behavior, and Immunity*, vol. 20, pp. 246–253.

Cearley, C., Churchill, L., & Krueger, J. M. 2003, "Time of day differences in IL1beta and TNFalpha mRNA levels in specific regions of the rat brain", *Neuroscience Letters*, vol. 352, no. 1, pp. 61–63.

Cerri, M., Ocampo-Garces, A., Amici, R., Baracchi, F., Capitani, P., Jones, C. A., Luppi, M., Perez, E., Parmeggiani, P. L., & Zamboni, G. 2005, "Cold exposure and sleep in the rat: effects on sleep architecture and the electroencephalogram", *Sleep*, vol. 28, no. 6, pp. 694–705.

Chen, Z., Gardi, J., Kushikata, T., & Krueger, J. 1999, "Nuclear factor-k B-like activity increases in murine cerebral cortex after sleep deprivation", *American Journal of Physiology*, vol. 276, pp. R1812–R1818.

Colotta, F., Re, F., Muzio, M., Bertini, R., Polentarutti, N., Sironi, M., Giri, J. G., Dower, S. K., Sims, J. E., & Mantovani, A. 1993, "Interleukin-1 type II receptor: a decoy target for IL-1 that is regulated by IL-4", *Science*, vol. 261, pp. 472–475.

Committee on Sleep Medicine and Research, B. o. H. S. P. 2006, *Sleep Disorders and Sleep Deprivation: an Unmet Public Health Problem,* The National Academies Press, Washington, DC.

Dantzer, R. 1994, "How do cytokines say hello to the brain? neural versus humoral mediation", *European Cytokine Network*, vol. 5, no. 3, pp. 271–273.

Dantzer, R. 2001, "Cytokine-induced sickness behavior: mechanisms and implications", *Annals of the New York Acadamy of Sciences*, vol. 933, pp. 222–234.

Dantzer, R., O'Connor, J. C., Freund, G. G., Johnson, R. W., & Kelley, K. W. 2008, "From inflammation to sickness and depression: when the immune system subjugates the brain", *Nature Reviews Neuroscience*, vol. 9, no. 1, pp. 46–56.

Darsaud, A., Bourdon, L., Mercier, S., Chapotot, F., Bouteille, B., Cespuglio, R., & Buguet, A. 2004, "Twenty-four-hour disruption of the sleep-wake cycle and sleep-onset REM-like episodes in a rat model of African trypanosomiasis", *Sleep*, vol. 27, no. 1, pp. 42–46.

De, A., Churchill, L., Obal, F., Jr., Simasko, S. M., & Krueger, J. M. 2002, "GHRH and IL1beta increase cytoplasmic Ca(2+) levels in cultured hypothalamic GABAergic neurons", *Brain Research*, vol. 949, no. 1–2, pp. 209–212.

Deboer, T., Fontana, A., & Tobler, I. 2002, "Tumor necrosis factor (TNF) ligand and TNF receptor deficiency affects sleep and the sleep EEG", *Journal of Neurophysiology*, vol. 88, no. 2, pp. 839–846.

Dickstein, J. B., Moldofsky, H., Lue, F. A., & Hay, J. B. 1999, "Intracerebroventricular injection of TNF-alpha promotes sleep and is recovered in cervical lymph", *American Journal of Physiology*, vol. 276, no. 4 Pt 2, pp. R1018–R1022.

Dinges, D. F., Douglas, S. D., Hamarman, S., Zaugg, L., & Kapoor, S. 1995, "Sleep deprivation and human immune function", *Advances in Neuroimmunology*, vol. 5, no. 2, pp. 97–110.

Dinges, D. F., Douglas, S. D., Zaugg, L., Campbell, D. E., McMann, J. M., Whitehouse, W. G., Orne, E. C., Kapoor, S. C., Icaza, E., & Orne, M. T. 1994, "Leukocytosis and natural killer cell function parallel neurobehavioral fatigue induced by 64 hours of sleep deprivation", *Journal of Clinical Investigation*, vol. 93, pp. 1930–1939.

Drake, C. L., Roehrs, T. A., Royer, H., Koshorek, G., Turner, R. B., & Roth, T. 2000, "Effects of an experimentally induced rhinovirus cold on sleep, performance, and daytime alertness", *Physiology & Behavior*, vol. 71, pp. 75–81.

Dunne, A. & O'Neill, L. A. 2003, "The interleukin-1 receptor/Toll-like receptor superfamily: signal transduction during inflammation and host defense", *Science's STKE: Signal Transduction Knowledge Environment*, vol. 2003, no. 171, p. re3.

Eriksson, C., Nobel, S., Winblad, B., & Schultzberg, M. 2000, "Expression of interleukin 1α and β, and interleukin 1 receptor antagonist mRNA in the rat central nervous system after peripheral administration of lipopolysaccharides", *Cytokine*, vol. 12, no. 5, pp. 423–431.

Everson, C. A. 1993, "Sustained sleep deprivation impairs host defense", *American Journal of Physiology*, vol. 265, pp. R1148–R1154.

Everson, C. A. 1995, "Functional consequences of sustained sleep deprivation in the rat", *Behavioural Brain Research*, vol. 69, no. 1–2, pp. 43–54.

Everson, C. A. 2005, "Clinical assessment of blood leukocytes, serum cytokines, and serum immunoglobulins as responses to sleep deprivation in laboratory rats", *American Journal of Physiology – Regulatory, Integrative and Comparative Physiology*, vol. 289, no. 4, pp. R1054–R1063.

Everson, C. A., Bergmann, B. M., & Rechtschaffen, A. 1989, "Sleep deprivation in the rat: III. Total sleep deprivation", *Sleep*, vol. 12, no. 1, pp. 13–21.

Everson, C. A. & Toth, L. A. 2000, "Systemic bacterial invasion induced by sleep deprivation", *American Journal of Physiology*, vol. 278, no. R905, p. R916.

Fang, J., Sanborn, C. K., Renegar, K. B., Majde, J. A., & Krueger, J. M. 1995, "Influenza viral infections enhance sleep in mice", *Proceedings of the Society for Experimental Biology and Medicine*, vol. 210, pp. 242–252.

Fang, J., Wang, Y., & Krueger, J. M. 1997, "Mice lacking the TNF 55 kDa receptor fail to sleep more after TNFα treatment", *Journal of Neuroscience*, vol. 17, pp. 5949–5955.

Fang, J., Wang, Y., & Krueger, J. M. 1998, "Effects of interleukin-1 beta on sleep are mediated by the type I receptor", *American Journal of Physiology*, vol. 274, no. 3 Pt 2, pp. R655–R660.

Feleder, C., Arias, P., Refojo, D., Nacht, S., & Moguilevsky, J. 2000, "Interleukin-1 inhibits NMDA-stimulated GnRH secretion: associated effects on the release of hypothalamic inhibitory amino acid neurotransmitters", *Neuroimmunomodulation*, vol. 7, no. 1, pp. 46–50.

Felten, D. L. & Felten, S. Y. 1987, "Immune interactions with specific neural structures", *Brain, Behavior and Immunity*, vol. 1, no. 4, pp. 279–283.

Flano, E., Woodland, D. L., & Blackman, M. A. 2002, "A mouse model for infectious mononucleosis", *Immunologic Research*, vol. 25, no. 3, pp. 201–217.

Floyd, R. A. & Krueger, J. M. 1997, "Diurnal variation of TNFα in the rat brain", *NeuroReport*, vol. 8, pp. 915–918.

Franken, P. 2002, "Long-term vs. short-term processes regulating REM sleep", *Journal of Sleep Research*, vol. 11, no. 1, pp. 17–28.

Frei, K., Malipiero, U. V., Leist, T. P., Zinkernagel, R. M., Schwab, M. E., & Fontana, A. 1989, "On the cellular source and function of interleukin 6 produced in the central nervous system in viral diseases", *European Journal of Immunology*, vol. 19, no. 4, pp. 689–694.

Frey, D. J., Fleshner, M., & Wright, K. P., Jr. 2007, "The effects of 40 hours of total sleep deprivation on inflammatory markers in healthy young adults", *Brain, Behavior and Immunity*, vol. 21, no. 8, pp. 1050–1057.

Gemma, C., Imeri, L., & Opp, M. R. 2003, "Serotonergic activation stimulates the pituitary-adrenal axis and alters interleukin-1 mRNA expression in rat brain", *Psychoneuroendocrinology*, vol. 28, no. 7, pp. 875–884.

Giulian, D., Baker, T. J., Shih, L.-C. N., & Lachman, L. B. 1986, "Interleukin 1 of the central nervous system is produced by ameboid microglia", *Journal of Experimental Medicine*, vol. 164, pp. 594–604.

Gourmelon, P., Amyx, H. L., Baron, H., Lemercier, G., Court, L., & Gibbs, C. J., Jr. 1987, "Sleep abnormalities with REM disorder in experimental Creutzfeldt-Jakob disease in cats: a new pathological feature", *Brain Research*, vol. 411, no. 2, pp. 391–396.

Guan, Z., Omori, T., Vgontzas, A. N., Bixler, E. O., & Fang, J. 2003, *Sleep-dependent alterations of interleukin-6 in the rat brain and peripheral tissues.* 2003 Abstract Viewer/Itinerary Planner, Society for Neuroscience, Washington, DC.Online. Program No. 618.12.

Heiser, P., Dickhaus, B., Schreiber, W., Clement, H.-W., Hasse, C., Hennig, J., Remschmidt, R., Krieg, J. C., Wesemann, W., & Opper, C. 2000, "White blood cells and cortisol after sleep deprivation and recovery sleep in humans", *European Archives of Psychiatry and Clinical Neuroscience,* vol. 250, pp. 16–23.

Hogan, D., Morrow, J. D., Smith, E. M., & Opp, M. R. 2003, "Interleukin-6 alters sleep of rats", *Journal of Neuroimmunology*, vol. 137, no. 1–2, pp. 59–66.

Hohmann, H.-P., Brockhaus, M., Bauerle, P. A., Remy, R., Kolbeck, R., & van Loon, A. P. G. M. 1990, "Expression of the type A and B tumor necrosis factor (TNF) receptor is independently regulated, and both receptors mediate activation of the transcription factor NF-κB", *Journal of Biological Chemistry*, vol. 265, pp. 22409–22417.

Holtmann, M. H. & Neurath, M. F. 2004, "Differential TNF-signaling in chronic inflammatory disorders", *Current Molecular Medicine*, vol. 4, no. 4, pp. 439–444.

Hu, J., Chen, Z., Gorczynski, C. P., Gorczynski, L. Y., Kai, Y., Lee, L., Manuel, J., & Gorczynski, R. M. 2003, "Sleep-deprived mice show altered cytokine production manifest by perturbations in serum IL-1ra, TNFa, and IL-6 levels", *Brain, Behavior, and Immunity*, vol. 17, no. 6, pp. 498–504.

Ignatowski, T. A., Noble, B. K., Wright, J. R., Gorfien, J. L., Heffner, R. R., & Spengler, R. N. 1997, "Neuronal-associated tumor necrosis factor (TNF alpha): its role in noradrenergic functioning and modification of its expression following antidepressant drug administration", *Journal of Neuroimmunology*, vol. 79, no. 1, pp. 84–90.

Imeri, L., Bianchi, S., & Mancia, M. 1997, "Muramyl dipeptide and IL-1 effects on sleep and brain temperature after inhibition of serotonin synthesis", *American Journal of Physiology – Regulatory, Integrative and Comparative Physiology*, vol. 273, no. 5, pp. R1663–R1668.

Imeri, L. & De Simoni, M. G. 1999, "Immune alterations in neurotransmission," in *Handbook of Behavioral State Control: Cellular and Molecular Mechanisms*, R. Lydic & H. A. Baghdoyan, eds., CRC Press, Boca Raton, pp. 659–674.

Imeri, L., Mancia, M., & Opp, M. R. 1999, "Blockade of 5-HT_2 receptors alters interleukin-1-induced changes in rat sleep", *Neuroscience*, vol. 92, pp. 745–749.

Irwin, M., Mascovich, A., Gillian, J. C., Willoughby, R., Pike, J., & Smith, T. L. 1994, "Partial sleep deprivation reduces natural killer cell activity in humans", *Psychosomatic Medicine*, vol. 56, pp. 493–498.

Irwin, M., McClintick, J., Costlow, C., Fortner, M., White, J., & Gillin, J. C. 1996, "Partial night sleep deprivation reduces natural killer and cellular immune responses in humans", *FASEB Journal*, vol. 10, no. 5, pp. 643–653.

Irwin, M., Thompson, J., Miller, C., Gillin, J. C., & Ziegler, M. 1999, "Effects of sleep and sleep deprivation on catecholamine and interleukin-2 levels in humans: clinical implications", *Journal of Clinical Endocrinology and Metabolism*, vol. 84, no. 6, pp. 1979–1985.

Irwin, M. R., Wang, M., Campomayor, C. O., Collado-Hidalgo, A., & Cole, S. 2006, "Sleep deprivation and activation of morning levels of cellular and genomic markers of inflammation", *Archives of Internal Medicine*, vol. 166, no. 16, pp. 1756–1762.

Jacobs, B. L. & Azmitia, E. C. 1992, "Structure and function of the brain serotonin system", *Physiological Reviews*, vol. 72, pp. 165–229.
Jones, B. E. 2005, "From waking to sleeping: neuronal and chemical substrates", *Trends in Pharmacological Sciences*, vol. 26, no. 11, pp. 578–586.
Kapás, L., Hong, L., Cady, A. B., Opp, M. R., Postlethwaite, A. E., Seyer, J. M., & Krueger, J. M. 1992, "Somnogenic, pyrogenic, and anorectic activities of tumor necrosis factor-α and TNF-α fragments", *American Journal of Physiology*, vol. 263, pp. R708–R715.
Kimura-Takeuchi, M., Majde, J. A., Toth, L. A., & Krueger, J. M. 1992a, "Influenza virus-induced changes in rabbit sleep and acute phase response", *American Journal of Physiology*, vol. 263, pp. R1115–R1121.
Kimura-Takeuchi, M., Majde, J. A., Toth, L. A., & Krueger, J. M. 1992b, "The role of double-stranded RNA in induction of the acute phase response in an abortive influenza virus infection model", *Journal of Infectious Diseases*, vol. 166, pp. 1266–1275.
Krueger, J. M. & Fang, J. 2000, "Host defense," in *Principles and Practice of Sleep Medicine*, 3rd edn, M. H. Kryger, T. Roth, & W. C. Dement, eds., W.B Saunders Company, Philadelphia, PA, pp. 255–265.
Krueger, J. M., Kubillus, S., Shoham, S., & Davenne, D. 1986, "Enhancement of slow-wave sleep by endotoxin and lipid A", *American Journal of Physiology*, vol. 251, pp. R591–R597.
Krueger, J. M., Obal, F., Jr., & Fang, J. 1999, "Humoral regulation of physiological sleep: cytokines and GHRH", *Journal of Sleep Research*, vol. 8 Suppl 1, pp. 53–59.
Krueger, J. M., Obal, F. J., Fang, J., Kubota, T., & Taishi, P. 2001, "The role of cytokines in physiological sleep regulation", *Annals of the New York Acadamy of Sciences*, vol. 933, pp. 211–221.
Krueger, J. M. & Takahashi, S. 1997, "Thermoregulation and sleep: closely linked but separable," in *Annals of the New York Academy of Sciences vol. 813; Thermoregulation: Proceedings of the 10th International Symposium on the Pharmacology of Thermoregulation*, C. M. Blatteis, ed., New York Academy of Sciences, New York, pp. 281–286.
Krueger, J. M. & Toth, L. A. 1994, "Cytokines as regulators of sleep", *Annals of the New York Acadamy of Sciences*, vol. 739, pp. 299–310.
Krueger, J. M., Walter, J., Dinarello, C. A., Wolff, S. M., & Chedid, L. 1984, "Sleep-promoting effects of endogenous pyrogen (interleukin-1)", *American Journal of Physiology*, vol. 246, pp. R994–R999.
Kubicki, S., Henkes, H., Terstegge, K., & Ruf, B. 1988, "AIDS related sleep disturbances – a preliminary report," in *HIV and the Nervous System*, S. Kubicki et al., eds., Gustav Fischer, New York, pp. 97–105.
Kubota, T., Brown, R. A., Fang, J., & Krueger, J. M. 2001a, "Interleukin-15 and interleukin-2 enhance non-REM sleep in rabbits", *American Journal of Physiology – Regulatory, Integrative and Comparative Physiology*, vol. 281, pp. R1004–R1012.
Kubota, T., Fang, J., Brown, R. A., & Krueger, J. M. 2001b, "Interleukin-18 promotes sleep in rabbits and rats", *American Journal of Physiology – Regulatory, Integrative and Comparative Physiology*, vol. 281, no. 3, pp. R828–R838.
Kubota, T., Fang, J., Guan, Z., Brown, R. A., & Krueger, J. M. 2001c, "Vagotomy attenuates tumor necrosis factor-α-induced sleep and EEG δ-activity in rats", *American Journal of Physiology – Regulatory, Integrative, Comparative Physiology*, vol. 280, pp. R1213–R1220.
Kubota, T., Kushikata, T., Fang, J., & Krueger, J. M. 2000, "Nuclear factor-kappaB inhibitor peptide inhibits spontaneous and interleukin-1beta-induced sleep", *American Journal of Physiology – Regulatory, Integrative, Comparative Physiology*, vol. 279, no. 2, pp. R404–R413.
Kubota, T., Li, N., Guan, Z., Brown, R. A., & Krueger, J. M. 2002, "Intrapreoptic microinjection of TNF-alpha enhances non-REM sleep in rats", *Brain Research*, vol. 932, no. 1–2, pp. 37–44.
Lancel, M., Mathias, S., Faulhaber, J., & Schiffelholz, T. 1996, "Effect of interleukin-1β on EEG power density during sleep depends on circadian phase", *American Journal of Physiology*, vol. 270, pp. R830–R837.
Landolt, H. P., Glatzel, M., Blattler, T., Achermann, P., Roth, C., Mathis, J., Weis, J., Tobler, I., Aguzzi, A., & Bassetti, C. L. 2006, "Sleep-wake disturbances in sporadic Creutzfeldt-Jakob disease", *Neurology*, vol. 66, no. 9, pp. 1418–1424.

Lange, T., Perras, B., Fehm, H. L., & Born, J. 2003, "Sleep enhances the human antibody response to hepatitis A vaccination", *Psychosomatic Medicine*, vol. 65, no. 5, pp. 831–835.

Lue, F. A., Bail, M., Jephthah-Ochola, J., Carayanniotis, K., Gorczynski, R., & Moldofsky, H. 1988, "Sleep and cerebrospinal fluid interleukin-1-like activity in the cat", *International Journal of Neuroscience*, vol. 42, pp. 179–183.

Luk, W. P., Zhang, Y., White, T. D., Lue, F. A., Wu, C., Jiang, C. G., Zhang, L., & Moldofsky, H. 1999, "Adenosine: a Mediator of Interleukin-1beta-Induced Hippocampal Synaptic Inhibition", *Journal of Neuroscience*, vol. 19, no. 11, pp. 4238–4244.

Lundkvist, G. B., Hill, R. H., & Kristensson, K. 2002, "Disruption of circadian rhythms in synaptic activity of the suprachiasmatic nuclei by African trypanosomes and cytokines", *Neurobiology of Disease*, vol. 11, no. 1, pp. 20–27.

Lundkvist, G. B., Kristensson, K., & Bentivoglio, M. 2004, "Why trypanosomes cause sleeping sickness", *Physiology (Bethesda.)*, vol. 19, pp. 198–206.

Mackiewicz, M., Sollars, P. J., Ogilvie, M. D., & Pack, A. I. 1996, "Modulation of IL-1β gene expression in the rat CNS during sleep deprivation", *NeuroReport*, vol. 7, pp. 529–533.

Manfridi, A., Brambilla, D., Bianchi, S., Mariotti, M., Opp, M. R., & Imeri, L. 2003, "Interleukin-1β enhances non-rapid eye movement sleep when microinjected into the dorsal raphe nucleus and inhibits serotonergic neurons in vitro", *European Journal of Neuroscience*, vol. 18, no. 5, pp. 1041–1049.

Marz, P., Cheng, J. G., Gadient, R. A., Patterson, P. H., Stoyan, T., Otten, U., & Rose-John, S. 1998, "Sympathetic neurons can produce and respond to interleukin 6", *Proceedings of the National Academy of Sciences USA*, vol. 95, no. 6, pp. 3251–3256.

Matsumoto, Y., Mishima, K., Satoh, K., Tozawa, T., Mishima, Y., Shimizu, T., & Hishikawa, Y. 2001, "Total sleep deprivation induces an acute and transient increase in NK cell activity in healthy young volunteers", *Sleep*, vol. 24, no. 7, pp. 804–809.

McCarley, R. W. 2007, "Neurobiology of REM and NREM sleep", *Sleep Medicine*, vol. 8, no. 4, pp. 302–330.

Moldofsky, H., Lue, F. A., Davidson, J. R., & Gorczynski, R. 1989, "Effects of sleep deprivation on human immune functions", *FASEB Journal*, vol. 3, pp. 1972–1977.

Moldofsky, H., Lue, F. A., Eisen, J., Keystone, E., & Gorczynski, R. M. 1986, "The relationship of interleukin-1 and immune functions to sleep in humans", *Psychosomatic Medicine*, vol. 48, no. 5, pp. 309–318.

Montagna, P. 2005, "Fatal familial insomnia: a model disease in sleep physiopathology", *Sleep Medicine Reviews*, vol. 9, no. 5, pp. 339–353.

Morrow, J. D. & Opp, M. R. 2005, "Sleep-wake behavior and responses of interleukin-6-deficient mice to sleep deprivation", *Brain, Behavior and Immunity*, vol. 19, no. 1, pp. 28–39.

Nieuwenhuijzen, G. A., Deitch, E. A., & Goris, R. J. 1996, "The relationship between gut-derived bacteria and the development of the multiple organ dysfunction syndrome", *Journal of Anatomy*, vol. 189, Pt 3, pp. 537–548.

Norman, S. E., Chediak, A. D., Freeman, C., Kiel, M., Mendez, A., Duncan, R., Simoneau, J., & Nolan, B. 1992, "Sleep disturbances in men with asymptomatic human immunodeficiency (HIV) infection", *Sleep*, vol. 15, no. 2, pp. 150–155.

Norman, S. E., Resnik, L., Cohn, M. A., Duara, R., Herbst, J., & Berger, J. R. 1988, "Sleep disturbances in HIV-seropositive patients", *Journal of the American Medical Association*, vol. 260, no. 7, p. 922.

Norman, S., Shaukat, M., Nay, K. N., Cohn, M., & Resnik, L. 1987, "Alterations in sleep architecture in asymptomatic HIV seropositive patients", *Sleep Research* vol. 16, p. 494.

Obal, F., Jr. & Krueger, J. M. 2003, "Biochemical regulation of non-rapid-eye-movement sleep", *Front Bioscience*, vol. 8, pp. d520–d550.

Obal, F., Jr., Opp, M., Cady, A. B., Johannsen, L., Postlethwaite, A. E., Poppleton, H. M., Seyer, J. M., & Krueger, J. M. 1990, "Interleukin 1 alpha and an interleukin 1 beta fragment are somnogenic", *American Journal of Physiology*, vol. 259, no. 3 Pt 2, pp. R439–R446.

O'Neill, L. A. & Greene, C. 1998, "Signal transduction pathways activated by the IL-1 receptor family: ancient signaling machinery in mammals, insects, and plants", *Journal of Leukocyte Biology*, vol. 63, no. 6, pp. 650–657.

Okun, M. L., Giese, S., Lin, L., Einen, M., Mignot, E., & Coussons-Read, M. E. 2004, "Exploring the cytokine and endocrine involvement in narcolepsy", *Brain, Behavior and Immunity*, vol. 18, no. 4, pp. 326–332.

Olivadoti, M. D. & Opp, M. R. 2008, "Effects of i.c.v. administration of interleukin-1 on sleep and body temperature of interleukin-6-deficient mice", *Neuroscience*, vol. 153, no. 1, pp. 338–348.

Olivadoti, M., Toth, L. A., Weinberg, J., & Opp, M. R. 2007, "Murine gammaherpesvirus 68: a model for the study of Epstein-Barr virus infections and related diseases", *Comparative Medicine*, vol. 57, no. 1, pp. 44–50.

Opp, M. R. 2005, "Cytokines and sleep", *Sleep Medicine Reviews*, vol. 9, no. 5, pp. 355–364.

Opp, M. R. 2006, "Sleep and psychoneuroimmunology", *Neurologic Clinics*, vol. 24, no. 3, pp. 493–506.

Opp, M., Born, J., & Irwin, M. 2007, "Sleep and the immune system," in *Psychoneuroimmunology*, 4th edn, R. Ader, ed., Elseveir Academic Press, Burlington, MA, pp. 579–618.

Opp, M. R. & Imeri, L. 2001, "Rat strains that differ in corticotropin-releasing hormone production exhibit different sleep-wake responses to interleukin 1", *Neuroendocrinology*, vol. 73, pp. 272–284.

Opp, M. R. & Krueger, J. M. 1994a, "Anti-interleukin-1β reduces sleep and sleep rebound after sleep deprivation in rats", *American Journal of Physiology*, vol. 266, pp. R688–R695.

Opp, M. R. & Krueger, J. M. 1994b, "Interleukin-1 is involved in responses to sleep deprivation in the rabbit", *Brain Research*, vol. 639, pp. 57–65.

Opp, M. R., Obál, F., & Krueger, J. M. 1991, "Interleukin-1 alters rat sleep: temporal and dose-related effects", *American Journal of Physiology*, vol. 260, pp. R52–R58.

Opp, M. R., Obál, F. Jr., & Krueger, J. M. 1989, "Corticotropin-releasing factor attenuates interleukin-1 induced sleep and fever in rabbits", *American Journal of Physiology*, vol. 257, pp. R528–R535.

Opp, M. R., Postlethwaite, A. E., Seyer, J. M., & Krueger, J. M. 1992, "Interleukin 1 receptor antagonist blocks somnogenic and pyrogenic responses to an interleukin 1 fragment", *Proceedings of the National Academy of Sciences USA*, vol. 89, pp. 3726–3730.

Pollmacher, T., Hinze-Selch, D., Mullington, J., & Holsboer, F. 1995, "Clozapine-induced increase in plasma levels of soluble interleukin-2 receptors", *Archives of General Psychiatry*, vol. 52, no. 10, pp. 877–878.

Poltorak, A., He, X., Smirnova, I., Liu, M. Y., Van Huffel, C., Du, X., Birdwell, D., Alejos, E., Silva, M., Galanos, C., Freudenberg, M., Ricciardi-Castagnoli, P., Layton, B., & Beutler, B. 1998, "Defective LPS signaling in C3H/HeJ and C57BL/10ScCr mice: mutations in Tlr4 gene", *Science*, vol. 282, no. 5396, pp. 2085–2088.

Prospéro-García, O., Herold, N., Waters, A. K., Phillips, T. R., Elder, J. H., & Henriksen, S. J. 1994, "Intraventricular administration of a FIV-envelope protein induces sleep architecture changes in rats", *Brain Research*, vol. 659, pp. 254–258.

Ramesh, V., Thatte, H. S., McCarley, R. W., & Basheer, R. 2007, "Adenosine and sleep deprivation promote NF-kappaB nuclear translocation in cholinergic basal forebrain", *Journal of Neurochemistry*, vol. 100, no. 5, pp. 1351–1363.

Rechtschaffen, A., Bergmann, B. M., Everson, C. A., Kushida, C. A., & Gilliland, M. A. 1989, "Sleep deprivation in the rat: X. Integration and discussion of the findings", *Sleep*, vol. 12, no. 1, pp. 68–87.

Rechtschaffen, A., Bergmann, B. M., Gilliland, M. A., & Bauer, K. 1999, "Effects of method, duration, and sleep stage on rebounds from sleep deprivation in the rat", *Sleep*, vol. 22, no. 1, pp. 11–31.

Rechtschaffen, A., Gilliland, M. A., Bergmann, B. M., & Winter, J. B. 1983, "Physiological correlates of prolonged sleep deprivation in rats", *Science*, vol. 221, no. 4606, pp. 182–184.

Reid, S. & Dwyer, J. 2005, "Insomnia in HIV infection: a systematic review of prevalence, correlates, and management", *Psychosomatic Medicine*, vol. 67, no. 2, pp. 260–269.

Reite, M., Laudenslager, M., Jones, J., Crnic, L., & Kaemingk, K. 1987, "Interferon decreases REM latency", *Biological Psychiatry*, vol. 22, pp. 104–107.

Renegar, K. B., Crouse, D., Floyd, R. A., & Krueger, J. 2000, "Progression of influenza viral infection through the murine respiratory tract: the protective role of sleep deprivation", *Sleep*, vol. 23, no. 7, pp. 859–863.
Sawada, M., Kondo, N., Suzumura, A., & Marunouchi, T. 1989, "Production of tumor necrosis factor-alpha by microglia and astrocyes in culture", *Brain Research*, vol. 491, pp. 394–397.
Schöbitz, B., De Kloet, E. R., & Holsboer, F. 1994, "Gene expression and function of interleukin 1, interleukin 6 and tumor necrosis factor in the brain", *Progress in Neurobiology*, vol. 44, pp. 397–432.
Serantes, R., Arnalich, F., Figueroa, M., Salinas, M., Andres-Mateos, E., Codoceo, R., Renart, J., Matute, C., Cavada, C., Cuadrado, A., & Montiel, C. 2006, "Interleukin-1beta enhances GABAA receptor cell-surface expression by a phosphatidylinositol 3-kinase/Akt pathway: relevance to sepsis-associated encephalopathy", *Journal of Biological Chemistry*, vol. 281, no. 21, pp. 14632–14643.
Shearer, W. T., Reuben, J. M., Mullington, J. M., Price, N. J., Lee, B. N., Smith, E. O., Szuba, M. P., Van Dongen, H. P. A., & Dinges, D. F. 2001, "Soluble TNF-α receptor 1 and IL-6 plasma levels in humans subjected to the sleep deprivation model of spaceflight", *Journal of Allergy and Clinical Immunology*, vol. 107, pp. 165–170.
Sherin, J. E., Shiromani, P. J., McCarley, R. W., & Saper, C. B. 1996, "Activation of Ventrolateral Preoptic Neurons During Sleep", *Science*, vol. 271, no. 5246, pp. 216–219.
Shoham, S., Davenne, D., Cady, A. B., Dinarello, C. A., & Krueger, J. M. 1987, "Recombinant tumor necrosis factor and interleukin 1 enhance slow-wave sleep", *American Journal of Physiology*, vol. 253, pp. R142–R149.
Shoham, S. & Krueger, J. M. 1988, "Muramyl dipeptide-induced sleep and fever: effects of ambient temperature and time of injection", *American Journal of Physiology*, vol. 255, pp. R157–R165.
Siegel, J. M. 2005, "Clues to the functions of mammalian sleep", *Nature*, vol. 437, no. 7063, pp. 1264–1271.
Sims, J. E., Gayle, M. A., Slack, J. L., Alderson, M. R., Bird, T. A., Giri, J. G., Colotta, F., Re, F., Mantovani, A., Shanebeck, K., Grabstein, K. H., & Dower, S. K. 1993, "Interleukin 1 signaling occurs exclusively via the type I receptor", *Proceedings of the National Academy of Sciences USA*, vol. 90, pp. 6155–6159.
Smith, A. 1992, "Sleep, colds and performance," in *Sleep, Arousal, and Performance: A Tribute to Bob Wilkinson*, R. J. Broughton & R. Ogilvie, eds., Birkhauser, Boston, MA, pp. 231–242.
Spiegel, K., Sheridan, J. F., & Van Cauter, E. 2002, "Effect of sleep deprivation on response to immunization", *Journal of the American Medical Association*, vol. 288, no. 12, pp. 1471–1472.
Susic, V. & Totic, S. 1989, "'Recovery' function of sleep: effects of purified human interleukin-1 on sleep and febrile response of cats", *Metabolic Brain Disease*, vol. 4, no. 1, pp. 73–80.
Szymusiak, R., Alam, N., Steininger, T. L., & McGinty, D. 1998, "Sleep-waking discharge patterns of ventrolateral preoptic/anterior hypothalamic neurons in rats", *Brain Research*, vol. 803, pp. 178–188.
Tabarean, I. V., Korn, H., & Bartfai, T. 2006, "Interleukin-1[beta] induces hyperpolarization and modulates synaptic inhibition in preoptic and anterior hypothalamic neurons", *Neuroscience*, vol. 141, no. 4, pp. 1685–1695.
Taishi, P., Bredow, S., Guha-Thakurta, N., & Krueger, J. M. 1997, "Diurnal variations of interleukin-1β mRNA and β-actin mRNA in rat brain", *Journal of Neuroimmunology*, vol. 75, pp. 69–74.
Taishi, P., Chen, Z., Hansen, M. K., Zhang, J., Fang, J., & Krueger, J. M. 1998, "Sleep-associated changes in interleukin-1β mRNA in the brain", *Journal of Interferon and Cytokine Research*, vol. 18, pp. 793–798.
Takahashi, S., Fang, J., Kapás, L., Wang, Y., & Krueger, J. M. 1997, "Inhibition of brain interleukin-1 attenuates sleep rebound after sleep deprivation in rabbits", *American Journal of Physiology*, vol. 273, pp. R677–R682.
Takahashi, S., Kapás, L., Fang, J., & Krueger, J. M. 1995a, "An anti-tumor necrosis factor antibody suppresses sleep in rats and rabbits", *Brain Research*, vol. 690, pp. 241–244.

Takahashi, S., Kapas, L., & Krueger, J. M. 1996a, "A tumor necrosis factor (TNF) receptor fragment attenuates TNF-alpha- and muramyl dipeptide-induced sleep and fever in rabbits", *Journal of Sleep Research*, vol. 5, no. 2, pp. 106–114.

Takahashi, S., Kapás, L., Seyer, J. M., Wang, Y., & Krueger, J. M. 1996b, "Inhibition of tumor necrosis factor attenuates physiological sleep in rabbits", *NeuroReport*, vol. 7, pp. 642–646.

Takahashi, S., Tooley, D. D., Kapás, L., Fang, J., Seyer, J. M., & Krueger, J. M. 1995b, "Inhibition of tumor necrosis factor in the brain suppresses rabbit sleep", *Pflügers Archives of European Journal of Physiology*, vol. 431, no. 2, pp. 155–160.

Tobler, I. 1988, "Evolution and comparative physiology of sleep in animals," in *Clinical Physiology of Sleep*, R. Lydic & J. F. Biebuyck, eds., American Physiological Society, Bethesda, MD, pp. 21–30.

Tobler, I. 1995, "Is sleep fundamentally different between mammalian species?", *Behavioural Brain Research*, vol. 69, pp. 35–41.

Tobler, I., Borbély, A. A., Schwyzer, M., & Fontana, A. 1984, "Interleukin-1 derived from astrocytes enhances slow wave activity in sleep EEG of the rat", *European Journal of Pharmacology*, vol. 104, pp. 191–192.

Tobler, I., Deboer, T., & Fischer, M. 1997, "Sleep and sleep regulation in normal and prion protein-deficient mice", *Journal of Neuroscience*, vol. 17, pp. 1869–1879.

Tobler, I., Gaus, S. E., Deboer, T., Achermann, P., Fischer, M., Rülicke, T., Moser, M., Oesch, B., McBrides, P. A., & Manson, J. C. 1996, "Altered circadian activity rhythms and sleep in mice devoid of prion protein", *Nature*, vol. 380, pp. 639–642.

Toth, L. A. 1995, "Sleep, sleep deprivation and infectious disease: studies in animals", *Advances in Neuroimmunology*, vol. 5, no. 1, pp. 79–92.

Toth, L. A. 1996, "Strain differences in the somnogenic effects of interferon inducers in mice", *Journal of Interferon and Cytokine Research*, vol. 16, pp. 1065–1072.

Toth, L. A. 1999, "Microbial modulation of sleep," in *Handbook of Behavioral State Control: Cellular and Molecular Mechanisms*, R. Lydic & H. A. Baghdoyan, eds., CRC Press, Boca Raton, FL, pp. 641–657.

Toth, L. A. & Krueger, J. M. 1988, "Alterations of sleep in rabbits by *Staphylococcus aureus* infection", *Infection and Immunity*, vol. 56, pp. 1785–1791.

Toth, L. A. & Krueger, J. M. 1989, "Effects of microbial challenge on sleep in rabbits", *FASEB Journal*, vol. 3, pp. 2062–2066.

Toth, L. A. & Opp, M. R. 2002, "Infection and sleep," in *Sleep Medicine*, T. Lee-Chiong, M. A. Carskadon, & M. Sateia, eds., Hanley & Belfus, Inc., Philadelphia, PA, pp. 77–84.

Toth, L. A. & Rehg, J. E. 1998, "Effects of sleep deprivation and other stressors on the immune and inflammatory responses of influenza-infected mice", *Life Sciences*, vol. 63, pp. 701–709.

Toth, L. A., Rehg, J. E., & Webster, R. G. 1995, "Strain differences in sleep and other pathophysiological sequelae of influenza virus infection in naive and immunized mice", *Journal of Neuroimmunology*, vol. 58, pp. 89–99.

Toth, L. A., Tolley, E. A., Broady, R., Blakely, B., & Krueger, J. M. 1994, "Sleep during experimental trypanosomiasis in rabbits", *Proceedings of the Society for Experimental Biology and Medicine*, vol. 205, pp. 174–181.

Toth, L. A. & Verhulst, S. J. 2003, "Strain differences in sleep patterns of healthy and influenza-infected inbred mice", *Behavior Genetics*, vol. 33, no. 3, pp. 325–336.

Traynor, T. R., Majde, J. A., Bohnet, S. G., & Krueger, J. M. 2007, "Interferon type I receptor-deficient mice have altered disease symptoms in response to influenza virus", *Brain, Behavior and Immunity*, vol. 21, no. 3, pp. 311–322.

Tsai, S. H., Liang, Y. C., Chen, L., Ho, F. M., Hsieh, M. S., & Lin, J. K. 2002, "Arsenite stimulates cyclooxygenase-2 expression through activating IkappaB kinase and nuclear factor kappaB in primary and ECV304 endothelial cells", *Journal of Cellular Biochemistry*, vol. 84, no. 4, pp. 750–758.

Vgontzas, A. N., Papanicolaou, D. A., Bixler, E. O., & Chrousos, G. P. 1997, "Elevation of plasma cytokines in disorders of excessive daytime sleepiness: role of sleep disturbance and obesity", *Journal of Clinical Endocrinology and Metabolism*, vol. 82, pp. 1313–1316.

Vgontzas, A. N., Zoumakis, E., Bixler, E. O., Lin, H. M., Follett, H., Kales, A., & Chrousos, G. P. 2004, "Adverse effects of modest sleep restriction on sleepiness, performance, and inflammatory cytokines", *Journal of Clinical Endocrinology Metabolism*, vol. 89, no. 5, pp. 2119–2126.

Vgontzas, A. N., Zoumakis, M., Papanicolaou, D. A., Bixler, E. O., Prolo, P., Lin, H. M., Vela-Bueno, A., Kales, A., & Chrousos, G. P. 2002, "Chronic insomnia is associated with a shift of interleukin-6 and tumor necrosis factor secretion from nighttime to daytime", *Metabolism*, vol. 51, no. 7, pp. 887–892.

von Economo, C. 1930, "Sleep as a problem of localization", *Journal of Nervous and Mental Disease*, vol. 71, no. 3, pp. 249–259.

White, J. L., Darko, D. F., Brown, S. J., Miller, J. C., Hayduk, R., Kelly, T., & Mitler, M. M. 1995, "Early central nervous system response to HIV infection: sleep distortion and cognitive-motor decrements", *AIDS*, vol. 9, pp. 1043–1050.

Wiegand, M., Moller, A. A., Schreiber, W., Krieg, J. C., & Holsboer, F. 1991, "Alterations of nocturnal sleep in patients with HIV infection", *Acta Neurologica Scandinavica*, vol. 83, pp. 141–142.

Wrona, D. 2006, "Neural-immune interactions: an integrative view of the bidirectional relationship between the brain and immune systems", *Journal of Neuroimmunology*, vol. 172, no. 1–2, pp. 38–58.

Xie, Q. W., Kashiwabara, Y., & Nathan, C. 1994, "Role of transcription factor NF-kappa B/Rel in induction of nitric oxide synthase", *Journal of Biological Chemistry*, vol. 269, no. 7, pp. 4705–4708.

Yasuda, T., Yoshida, H., Garcia-Garcia, F., Kay, D., & Krueger, J. M. 2005, "Interleukin-1beta has a role in cerebral cortical state-dependent electroencephalographic slow-wave activity", *Sleep*, vol. 28, no. 2, pp. 177–184.

Yoshida, H., Peterfi, Z., Garcia-Garcia, F., Kirkpatrick, R., Yasuda, T., & Krueger, J. M. 2004, "State-specific asymmetries in EEG slow wave activity induced by local application of TNF[alpha]", *Brain Research*, vol. 1009, no. 1–2, pp. 129–136.

Zager, A., Andersen, M. L., Ruiz, F. S., Antunes, I. B., & Tufik, S. 2007, "Effects of acute and chronic sleep loss on immune modulation of rats", *American Journal of Physiology – Regulatory, Integrative and Comparative Physiology*, vol. 293, no. 1, pp. R504–R509.

Cytokines and Aggressive Behavior

Allan Siegel, Suresh Bhatt, Rekha Bhatt, and Steven S. Zalcman

Abstract Studies conducted in rodents, primates and humans have provided evidence that proinflammatory cytokines may play an important in the regulation of aggression and rage behavior. More recent studies conducted in the cat have generated more direct evidence of cytokine involvement in modulating rage behavior. Activation of IL-I receptors in the medial hypothalamus and periaqueductal gray (PAG) potentiates defensive rage behavior in the cat. Facilitation of defensive rage is mediated through 5-HT2 receptors in the medial hypothalamus and PAG. Activation of IL-2 receptors in the medial hypothalamus and PAG differentially affect defensive rage behavior. In the medial hypothalamus, IL-2 receptors suppress defensive rage and this effect is mediated through GABAA receptors; in the PAG, IL-2 receptors facilitate the occurrence of defensive rage behavior and such effects are mediated through substance P NK1 receptors. With respect to peripheral mechanisms, LPS administration induces the release of a cascade of proinflammatory cytokines. Among the cytokines released, TNF-α appears to play a significant role in the induction of the suppressive effects of LPS upon defensive rage and in sickness behavior in the cat. Concerning the central mechanisms regulating LPS induced suppression of defensive rage and sickness behavior, serotonin 5-HT1A and PGE2 receptors in the medial hypothalamus appears to play key roles in controlling these processes.

Keywords Aggression · Cat · Cytokines · Defensive rage · IL-1 · IL-2 · LPS · Medial hypothalamus · Neurotransmitters · Periaqueductal gray · Serotonin · Substance P

1 Introduction

Interest in the possible role of cytokines in emotional behavior in general and in aggressive behavior in particular is derived from evidence in the literature suggesting a reciprocal relationship between immune function and aggressive behavior.

A. Siegel (✉)
Departments of Neurology & Neuroscience and Psychiatry, UMDNJ – New Jersey Medical School, MSB Room H-592, 185 South Orange Avenue, Newark, NJ 07103, USA
e-mail: Siegel@UMDNJ.edu

A. Siegel, S.S. Zalcman (eds.), *The Neuroimmunological Basis of Behavior and Mental Disorders*, DOI 10.1007/978-0-387-84851-8_12,

A variety of approaches have been utilized to generate this relationship. These include studies describing: (1) the role of social status upon immune function, (2) the effects of aggression upon immune function of the victim or the perpetrator, and (3) the modulating effects of immune functions upon aggressive behavior. Accordingly, the first part of this chapter provides an overview of studies conducted along these lines, while the second part focuses upon the specific role of cytokines in aggression and rage behavior.

2 Social Status and Immunity

Studies involving the social status of animals serve as an introduction to the topic since they relate to intraspecific aggression (i.e., defensive aggression and rage) and their effects upon immunity. A number of studies provide support for this linkage.

In one study, Stefanski (1998) employed a despotic hierarchy in which one subject dominates the other subdominant male rats. Here, the dominant animals are approximately equal and are not aggressive to each other. The authors reported that subdominant animals had deficient immune functions as evidenced by decreased numbers of CD4 and CD8 T cells, as well as by a decreased functional capacity of the T cells as determined by their proliferative response to mitogen ConA. Subdominant animals also displayed enhancement of granulocyte numbers. These changes may have been induced as a result of elevated epinephrine and norepinephrine in subdominant rats, perhaps as a function of an induced stress response. The potential clinical significance of this finding may be extrapolated from a report by Hessing et al. (1994) who demonstrated that the social status of pigs is critical in predicting morbidity and mortality following challenge by a pseudorabies virus in pigs. In this study, it was observed that, in a stable social structure, morbidity and mortality were highest in subordinate animals.

Other approaches have been employed for the study of social status upon immune function. One such approach involves a naturalistic model of observing the behavior of primates in the wild. In a study by Sapolsky and Spencer (1997), a troop of wild baboons (in a national reserve in Kenya) were initially habituated to observation by the investigators for 4 years prior to the initiation of the study. The study examined the importance of dominance hierarchies on anabolic hormone insulin-like growth factor-I (IGF-I; which has been implicated as a possible mediator in a variety of physiological and immunological functions) and observed that IGF-I was suppressed in subordinate male baboons.

Based on the assumption that access to resources is a function in part of dominance hierarchies, a separate approach utilized a model involving competition for limited resources (Boccia et al., 1992). In this study involving a competitive water test, one group of pigtail or bonnet macaques was given access to a single water spout following 24 h of deprivation. After the water was turned on, the order in which the animals drank, their length of time drinking, and displacement of other animals were recorded in these two types of macaques whose social behavior differed. Since pigtail macaques displayed more agonistic responses than bonnet macaques, it was not surprising that pigtail macaques displayed higher natural killer cell (NK) function than bonnet macaques. Moreover, during competition, the two groups of macaques differed in other ways. Specifically, the bonnet macaques formed a line by dominance rank in waiting their turn to drink as opposed to the pigtails that competed with one another for access to the water. During competition for water, the pigtail macaques displayed an increase in NK activity. It is of interest to note the similarities in the findings concerning the pigtail macaques with those described below concerning immune function and aggression in humans.

Another variation in approach has been to examine the immunological consequences of active hierarchy formation with respect to aggressive behavior. In a study aimed to examine this relationship, Gust et al. (1991) observed that, in response to the establishment of a dominance hierarchy, there was a subsequence short-term rise in cortisol and a long-term decrease in immune function (i.e., CD4 and CD8 T cells) with the lowest ranking animals displaying significantly reduced numbers of T cells. In a related study, which involved a more chronic model of social hierarchy disruption in cynomolgus monkeys, Cohen et al. (1997), animals were placed into an unstable social situation for 15 months where they reorganized into new groups every month. Animals were exposed to influenza virus after 8 months and then subsequently infected with adenovirus after 15 months of reorganization. The animals which were exposed to instability manipulations were associated with aggressive behavior. Animals which maintained a low status independent of social reorganization were associated with enhanced probabilities of infection and reduced aggressive behavior, suggesting that reorganization was of lesser importance than social rank with respect to immune competence. Supporting data was obtained by Capitanio et al. (1998) who examined survival rates of male rhesus monkeys in unstable social environments and then exposed to simian immunodeficiency virus (SIV). Here, animals experiencing a disrupted social hierarchy had significantly shorter survival periods were more actively engaged in aggressive behavior, but were less engaged in other types of behaviors such as grooming.

3 Effects of Aggression on Recipient's and Perpetrator's Immune Function

In assessing the relationship of immune function with aggressive behavior, it is of interest to compare the ways in which aggressive behavior impacts upon the immune system of the recipient relative to the aggressor. It should be noted that a number of investigators examining this relationship do not view aggressive encounters as aggression "per se," but as a form of natural stressor. However, concerning the objectives of the present chapter, the relationship of immune function to aggressive behavior will be underscored. While a wide body of studies has been conducted on this general problem area, the present discussion will be limited to the approach which utilized two models: a "social disruption" model, in which an aggressive intruder is placed into an established cage holding other animals, and a "social confrontational" model, in which the animal serves as an "intruder" in a "resident's" home cage.

3.1 Recipient's Immune Function

As noted above with respect to the social disruption model, a dominant animal, which is repeatedly transferred into cages with an established social hierarchy, typically attacks animals that display submissive responses. In cases where a "resident" attacks an "intruder," the intruder is replaced by another until the resident expresses submissive behavior. Utilizing the social disruption model, studies conducted by Avitsur et al. (2001) and Stark et al. (2001) have provided evidence that resident mice display submissive behavior and display a profile which includes glucocorticoid-resistant splenocytes as well as glucocorticoid-resistant macrophages, and increased levels of IL-6, respectively.

The social confrontation (i.e., resident–intruder) model has been employed by a variety of investigators. Because of space limitations, discussion will be limited to several papers that illustrate the efficacy and complexity of the effects of this model. In a study by Dreau et al. (1999), mice exposed to an aggressive resident five times per day for two days resulted in decreased numbers of thymocytes, a decreased in vitro response to lipopolysaccharide (LPS), and decreases in CD4 and CD8 T cells. Stefanski and Engler (1998) attempted to distinguish between acute and chronic exposure to social confrontation with respect to such effects upon immune function. Mice (intruders) were exposed to an aggressive resident for 2 or 48 h in which agonistic encounters ensued. The majority of the immunological variables measured were affected similarly in the two conditions, including increased granulocytes, decreased lymphocytes, T, and B cells. However, the magnitude of the changes appeared to be smaller after 48 than 2 h. Of particular interest was the observation that significant correlations were observed between the degree of submissive behavior displayed by the losers and several of the immunological measures observed after 2 h. The authors suggest that stressful conditions tested over different periods of time may not yield identical effects upon immune functioning.

Further attempts to characterize the status of immune functioning following confrontation were described by Stefanski and Ben-Eliyahu (1996) by examining the development of tumor metastasis in rats. Animals which experienced 7 h of confrontation received mammary tumor cells 1 h following initiation of the confrontation. The results indicated that intruders had increased tumor retention that was related to the degree of submission displayed. This effect could be reduced following administration of butoxamine, a β-adrenergic antagonist, and the effect could be completely eliminated by adrenal demedullation.

While there may have been somewhat a lack of precision in manipulating the social variables, including social interactions, the overall effects suggest that immunosuppression appears to be present in animals that display submissive behavior.

3.2 Perpetrator's Immune Function

The literature concerning the effects of aggression on the perpetrator's immune functions is more extensive and includes models and factors such as: social-isolation, resident–intruder, effects of strain differences, and human relationships involving marital conflict and hostility. The following discussion provides illustrations from the literature with respect to each of these approaches to the study of the effects of aggression upon the perpetrator's immune functions.

In the social-isolation model, the animal is kept in isolation prior to its introduction into a neutral group setting. Utilizing this model, Gryazeva et al. (2001) allowed mice to be exposed to multiple aggressive encounters following isolation. These authors observed that continued aggressive encounters resulted in elevations in the proportion of segmented neutrophils, CD4 and CD8 T cells in spleen. Earlier related findings by Hardy et al. (1990) demonstrated that the direction of the changes in T-cells were dependent upon the dominance status of the mouse (i.e., submissive mice had lower T cell proliferation compared to dominant and non-fighter control animals). When the experiment was repeated in the absence of wounding, dominant mice expressed elevated T-cell proliferation relative to submissive and non-fighter controls, suggesting that the physical consequences of fighting were independent of the psychological components of the aggressive encounters.

The resident–intruder model was employed in social-isolation models and is perhaps, the most popular of the designs used for the study of aggression in rodents. Employing this model, Kavelaars et al. (1999) reported that animals expressing short attack latencies were less resistant to experimental autoimmune encephalomyelitis, indicating that aggressive rats were more susceptible to this form of encephalomyelitis than non-aggressive rats. It is of interest to point out that this observation appears to differ from the general literature which suggests that aggressive encounters generates opposite effects upon immune function. In fact, when both resident and intruders were reared in isolation and then received immunological challenges with sheep red blood cells following exposure to a resident–intruder fighting, a short social challenge resulted in depression in the primary immune response in submissive mice and increase in the primary immune response in dominant mice when exposed to a longer chronic social challenge (Gasparotto et al., 2002).

Concerning studies involving mice bred for different levels of aggression, several findings are of interest. Petitto et al. (1993, 1994) examined immune factors in mice that were selectively bred for different levels of aggressive behavior related to social isolation. The most significant variations were noted when low aggressive line of aggression was compared with an unselected line. The low aggressive line displayed enhanced freezing and immobility, which may be viewed as social isolation, but they did not appear to differ with respect to other functions. These socially inhibited animals exhibited reduced NK activity, lower T-cell proliferation, IL-2 and gamma interferon production, and increased susceptibility to tumor development. In another study, Devoino et al. (1993) provided evidence that responses to immune challenge are dependent upon the level of aggressiveness associated with different strains of mice. Specifically, these authors showed that mice bred for different levels of social isolation related aggression also displayed different immune responses following immunization with sheep red blood cells. Plaque forming cells decreased in submissive mice relative to control animals, while the aggressive strain presented with greater numbers of plaque forming cells.

4 Relationships Between Aggressive Behavior and Immunity in Humans

The discussion provided above considered the interactions between aggressive behavior (or the effects of aggressive encounters) and immune functions. An increasing body of data has emerged detailing the relationships between aggressive behavior in normal and patient populations, the latter of which includes both psychiatric and non-psychiatric medical disorders. The present section provides a brief review of these findings.

With respect to studies in which non-patient populations were employed, Suarez et al. (2004) showed that Cook-Medley Hostility scores were increased following LPS-stimulated production of proinflammatory cytokines by blood monocytes. Findings that were consistent with those of Suarez et al. were reported by Kiecolt-Glaser et al. (1997). These authors reported that higher levels of hostile marital interactions are associated with increased production of plasma proinflammatory cytokines.

Two additional studies provided added support for a relationship between aggression and immune function. One study was conducted by Granger et al. (2000) on men who served in the U.S. Army during the period of 1965–1971 in Vietnam. The measure of aggression used in this study was derived from the term "antisocial personality" as defined in the DSM-III. The results revealed a curvilinear relationship between aggression and immune function in that moderate levels of aggressive behavior were correlated with elevated numbers of CD4+ T cells and B cells. An earlier study by Christensen et al. (1996) examined how cynical hostility could modify the effects of self disclosure of personal information about stressful experiences. Subjects with greater hostility had higher NK activity.

The studies described immediately above considered the relationship between immune function and aggressive behavior in normal human populations. The studies described below involved individuals with psychiatric abnormalities associated with various medical disorders such as cancer, AIDS, and hepatitis C (Capuron et al., 2004). For example, cytokine immunotherapy has been shown to facilitate aggressive behavior (as determined by measures of anger, hostility and irritability (Kraus et al., 2003; McHutchison et al., 1998)). However, interpretation of these data may be somewhat complicated by the fact that such patients also display other psychiatric symptoms such as depression, anxiety, and cognitive problems. Since these conditions may lead to increases in aggressive behavior independent of cytokine treatment, conclusions concerning the effects of cytokines on aggressive behavior in such populations should be interpreted with considerable caution. Further studies are needed to clarify this possible confounding set of variables.

5 Role of Cytokines in Defensive Rage Behavior

5.1 Brief Overview of the Neural Substrates of Aggression and Rage

In attempting to understand how cytokines might affect aggressive processes, it is useful to briefly review the nature of the neural substrates of defensive rage behavior.

Defensive rage behavior in the cat is characterized by pronounced hissing, marked papillary dilatation, piloerection, retraction of the ears, unsheathing of the claws, significant increases in heart rate and blood pressure, and a paw strike at a moving object present in its visual field. It is important to note that this behavior takes place under natural conditions in response to a perceived threat (Leyhausen, 1979). This form of aggression can also be elicited by electrical stimulation of parts of the hypothalamus and midbrain. The specific circuitry relating the medial hypothalamus and periaqueductal gray (PAG) with respect to defensive rage behavior can be summarized in the following manner.

The medial hypothalamus, especially the anterior third of this region, receives significant inputs from ventromedial nucleus (of hypothalamus) and limbic structures such as the medial amygdala and septal area. It thus constitutes a central component of the mechanism for integration of defensive rage behavior, whose underlying neurons are significantly modulated by these inputs (Siegel et al., 1999; Siegel, 2005). The principal descending output of the anterior medial hypothalamus mediating defensive rage behavior is directed to the rostral half of the dorsolateral quadrant of the PAG (Fuchs et al., 1985a, b). The major outputs of the dorsolateral PAG target regions of the lower brain stem linked to autonomic and somatomotor components of the defensive rage response such as reticular formation of the medulla and pons, including the solitary nucleus and associated structures (mediating autonomic responses) and to the motor nuclei of the trigeminal and facial nerves as well as other nuclei of the medulla which project to lower motor neurons of spinal cord (mediating somatomotor responses; Shaikh et al., 1987). Collectively, activation of the PAG (or medial hypothalamus) can synchronously drive these brain stem neurons, which constitutes the underlying neural substrate for the expression of the defensive rage response. The essential relationship between the medial hypothalamus and PAG is underscored by the fact that disruption receptors (Schubert et al., 1996) will eliminate defensive rage behavior elicited from the hypothalamus (Fuchs et al., 1985b).

5.2 Central Nervous System

5.2.1 Interleukin-1

Medial hypothalamus: The discussion above has indicated that the central regions governing the expression of defensive rage behavior in the cat includes the medial hypothalamus and midbrain PAG. Interleukin-1 (IL-1) is an immune and brain-derived cytokine that has been shown to affect functions of the central nervous system (CNS).

For example, IL-1 has been shown to alter appetitive behaviors and that such changes can be blocked by a L-5-hydroxytryptophan (5-HT) synthesis inhibitor (Zubareva et al., 2001). In addition, IL-1 can also modulate sleep functions which are regulated in part by 5-HT_2 receptors (Gemma et al., 1997; Imeri et al., 1999). Further evidence of IL-1 involvement in the hypothalamus is gained from a study by Gemma et al. (2003) who showed that administration of the serotonin precursor, L-5-HT, induces an increase in IL-1β mRNA expression in the hypothalamus.

On the basis of the facts that the medial hypothalamus and PAG are key areas for elicitation of defensive rage behavior and that IL-1 is implicated in functions of the hypothalamus, the following series of studies were conducted. These experiments involved the following strategies, which involved several different stages in experimental design.

A basic approach required that defensive rage behavior be elicited from the PAG and cytokine and neurotransmitter compounds administered into the medial hypothalamus. Thus, in the first stage of the study conducted by Hassanain et al., (2003a, 2005), a stimulating electrode was implanted into a site in the PAG from which defensive rage could be elicited and a cannula electrode was implanted into a site within the medial hypothalamus from which stimulation could elicit the same form of aggression. In order to demonstrate the functional relationship between these two sites, namely, that they reflect two different regions within the same circuit for the expression of defensive rage, a dual stimulation experiment was conducted. In this experiment, the response latency (defined as the time period between onset of stimulation and onset of hissing) following "single" stimulation of the PAG alone was recorded on the first trial, and, on the second trial, response latencies following "dual" stimulation of the PAG and medial hypothalamic site were recorded. This paradigm was repeated for a total of 10 pairs of trials comparing response latencies after "single" vs. "dual" stimulation. In conducting this experiment, the level of current applied to the medial hypothalamus was set at approximately 80% of the minimal current required to elicit defensive rage behavior when applied to the hypothalamic site. In this experiment, whose findings were typical of the ones that follow, dual stimulation facilitated the occurrence of defensive rage as reflected by a mean reduction of 72% in response latencies following onset of stimulation (Fig. 1). A *t*-test for paired observations was significant ($p<.05$),

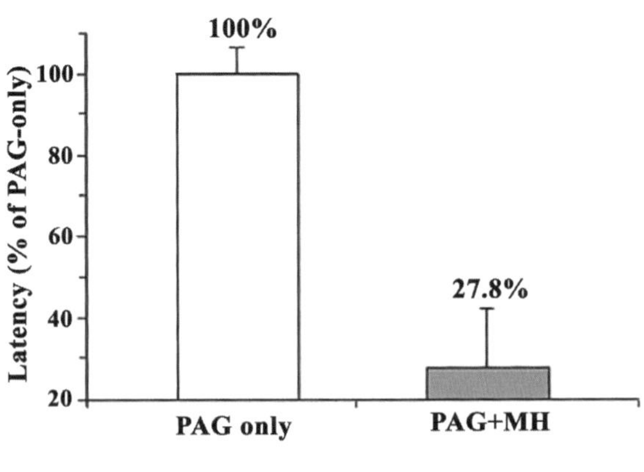

Fig. 1 Comparison of dual and single stimulation. Dual stimulation of the periaqueductal gray (PAG) and medial hypothalamus significantly facilitated (72.2%, $p<.05$) PAG elicited defensive rage relative to single stimulation. (From Hassanain et al., 2005, with permission)

indicating a functional relationship between the medial hypothalamus and dorsal PAG with respect to defensive rage behavior.

In a related experiment which basically tested the same principle, namely, the functional relationship between the PAG and medial hypothalamus, electrical stimulation was applied to PAG sites from which defensive rage was elicited for 6–10 s and repeated at 2-min intervals for 1 h. The animal was then sacrificed and brain tissue was processed for *c-fos* immunocytochemistry. The results indicated that *c-fos* labeling was heaviest in the dorsomedial nucleus and ventromedial hypothalamus on the side of the brain ipsilateral to the site of stimulation in the PAG (Fig. 2). There was also some presence of *fos*-like immunoreactivity in the contralateral hypothalamus indicating that there is some crossed input into this region of hypothalamus following stimulation of the PAG. Thus, this finding supports the data obtained from the dual stimulation experiment linking the PAG and medial hypothalamus as part of the same functional circuit for the expression of defensive rage behavior.

In the next phase of the study, we sought to test the hypothesis that activation of IL-1β receptors in the medial hypothalamus will modulate defensive rage behavior. The paradigm for this experiment involved a pre-drug administration of 5 trials of stimulation of the PAG, using an average intertribal interval of approximately 2 min, in order to establish a baseline latency. Then, IL-1β was administered in doses of 5–50 ng in 0.25 μl into the medial hypothalamic defensive rage site. Five trials of stimulation were then applied in each of the following blocks of time in the post-injection period: 20–35, 60–75, 120–135, 180–195, 240–255, and 300–315 min. The results indicated that microinjections of IL-1β into the medial hypothalamus markedly facilitated defensive rage behavior elicited from the PAG in a dose-dependent manner (Fig. 3). A striking feature of this effect was that following administration of IL-1β, two peaks of facilitation were noted. The first occurred at 60 min and the second at 180 min, post-injection. Presumably, the initial facilitation observed was due to the exogenous cytokine, and the second may have been due to the release of endogenous IL-1 or perhaps another cytokine possibly induced by administration of IL-1β.

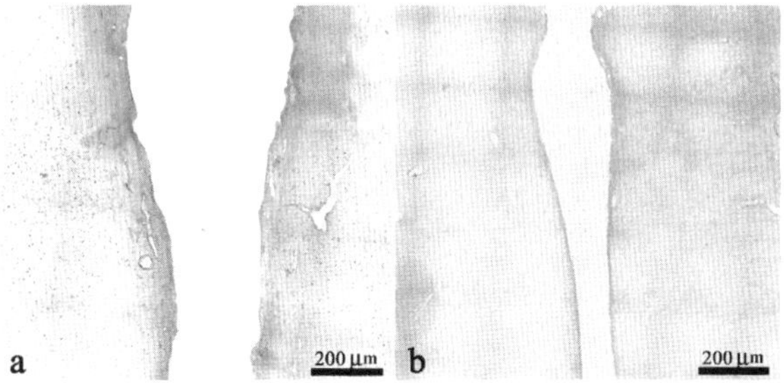

Fig. 2 (a) Fos-like immuno-reactivity was increased in the dorsomedial and ventromedial nuclei both ipsilateral and contralateral to the periaqueductal gray (PAG) stimulation site. (b) Fos-like immunoreactivity was virtually absent in the ventromedial hypothalamus of an unstimulated cat. (From Hassanian et al., 2005, with permission)

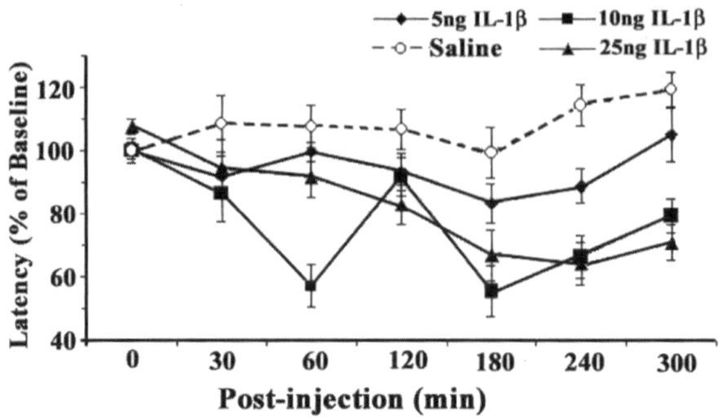

Fig. 3 Effects of low dose IL-1*β* injected directly into the medial hypothalamus facilitates defenseive rage in a dose-dependant manner at 5 and 10 ng. At 10 ng IL-1*β*-mediated facilitation of defensive rage is maximal at 60 and 180 min post-injection. (Hassanain et al., 2003b, with permission)

The results of the findings described above established that IL-1*β* in the medial hypothalamus potentiates defensive rage behavior. What these findings did not reveal is the mechanism underlying how IL-1*β* facilitates defensive rage. The next phase of the study was designed to identify this mechanism. Since it has been shown that: (1) IL-1 modulates brain serotonin activity (Dunn, 1992; Laviano et al., 1999; Merali et al., 1997; Song et al., 1999; Zalcman et al., 1994); (2) immunoreactive 5-HT axons and nerve terminals are present in the dorsomedial and ventromedial hypothalamus (Leger et al., 2001); and (3) activation of 5-HT_2 receptors in the PAG or medial hypothalamus facilitate defensive rage behavior (Shaikh et al., 1997; Hassanain et al., 2003b), the hypothesis was proposed that IL-1 modulation of defensive rage behavior is mediated through 5-HT_2 receptors in the medial hypothalamus. To test this hypothesis, sites in the medial hypothalamus were pretreated with the 5-HT_2 receptor antagonist, 6-methyl-1 -methylpropyl-1-(1-methylethyl)-ergoline-8-carboxylic acid (8*β*)-2-hydroxy-1-methylpropyl ester (Z)-2-butenedioate (LY-53857, Sigma-Aldrich), prior to administration of IL-1*β*. The results, shown in Fig. 4, indicate that pretreatment with the 5-HT_2 receptor antagonist blocked the potentiating effects of IL-1*β*. Administration of the 5-HT_2 receptor antagonist alone had no effect upon defensive rage behavior. *Collectively, the data indicate that IL-β receptors in the medial hypothalamus potentiate defensive rage behavior and that such effects are mediated through 5-HT_2 receptors.*

PAG. The study described above established that IL-1*β* receptors in the medial hypothalamus potentiate defensive rage behavior and that these effects are mediated through 5-HT_2 receptors. Since both the PAG and medial hypothalamus are anatomically and functionally linked for the expression of defensive rage behavior, it was of interest to determine whether IL-1*β* receptors in the PAG modulate defensive rage behavior and whether such modulation is mediated through 5-HT receptors.

In conducting this experiment, the same paradigms that were used for the study of IL-1 in the hypothalamus were applied for the study of IL-1 in the PAG. The only

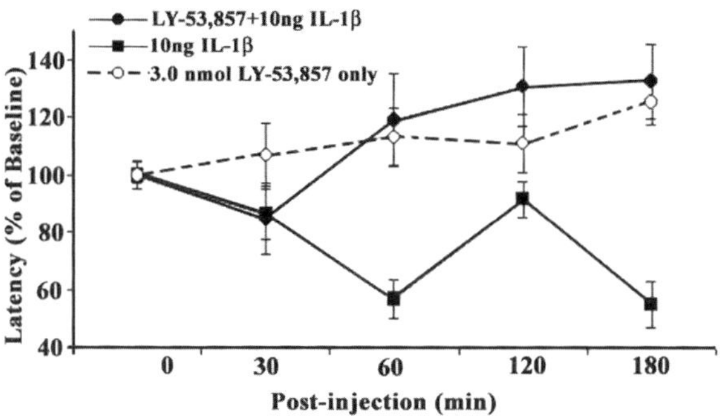

Fig. 4 Blockade of the facilitative effects of IL-1β by a 5-hydroxytryptophan (5-HT_2) receptor antagonist. Pretreatment with the 5-HT_2 receptor antagonist, LY-53857 (3 nmol), completely blocked the facilitative effects of interleukin (IL)-1β on defensive rage ($p<.01$). Administration of the same dose of LY-53857 alone failed to alter response latencies (Hassanain et al., 2003b with permission)

difference was that a stimulating electrode was placed in the hypothalamus at a site from which defensive rage could be elicited and a cannula electrode at a defensive rage site in the PAG where drugs could be infused.

In the first phase of this experiment, which was similar to the previous experiment (Bhatt et al., 2008), a dual stimulation paradigm involving the PAG and medial hypothalamus demonstrated that the PAG facilitated defensive rage elicited from the medial hypothalamus, thus establishing the functional linkage between the two sites with respect to this form of aggressive behavior. In the next phase of the study, microinjections of IL-1β facilitated defensive rage behavior in a dose-dependent manner (Fig. 5) and these effects are completely blocked following pre-

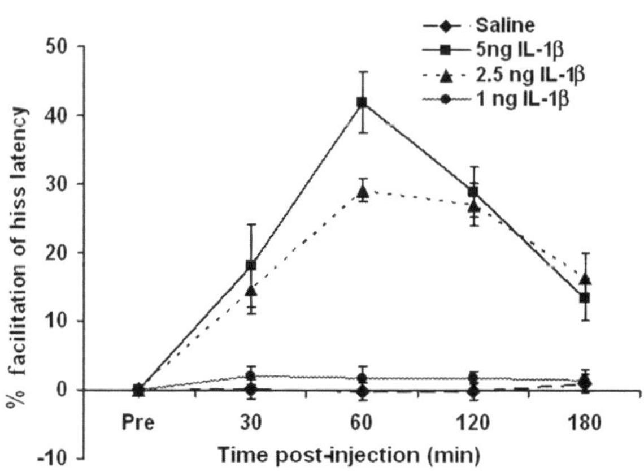

Fig. 5 Microinjections of interleukin (IL)-1β into the periaqueductal gray (PAG) facilitated defensive rage elicited from the medial hypothalamus in a dose- and time-dependant manner. Maximal facilitation of defensive rage was induced by 5 ng of IL-1β at 60 min, post-injection ($p<.0001$, $N=5$). (From Bhatt et al., 2008, with permission)

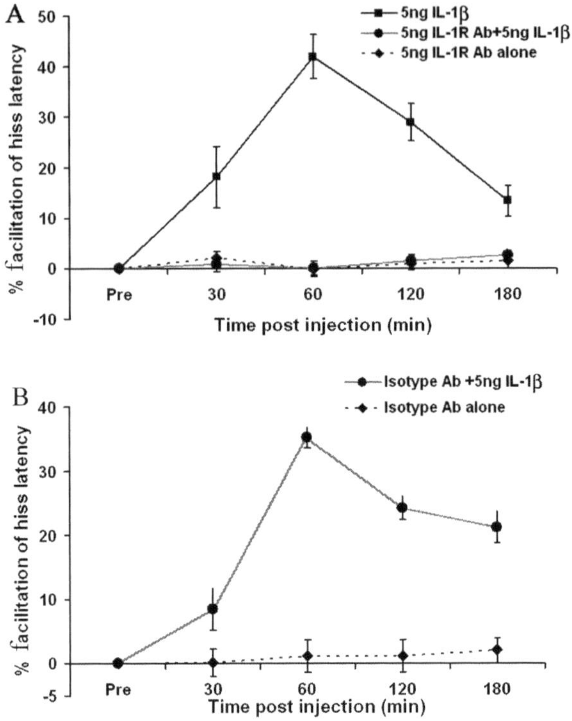

Fig. 6 (A) Pretreatment of the periaqueductal gray (PAG) with an anti-interleukin(IL)-1 receptor antibody completely blocked the facilitating effects of IL-1 upon medial hypothalamically elicited defensive rage ($p<.0001$, $N=5$). Note that administration of the anti-IL-1 receptor antibody alone did not alter response latencies for hissing ($p=.91$, NS, $N=5$). (B) Pretreatment of the defensive rage site in the PAG with an isotype antibody (IgG_{2a}) did not block the potentiating effects of IL-1β upon defensive rage ($p<.0001$, $N=2$) and that administration of the isotype antibody alone did not alter latencies for defensive rage ($p=.95$, NS, $N=2$). (From Bhatt et al., 2008, with permission)

treatment of the PAG with an anti-IL-1 receptor antibody (Fig. 6). At this point, we wished to determine the possible mechanism by which IL-1β facilitated defensive rage. Several neurotransmitter receptors which are known to facilitate defensive rage behavior within the PAG were considered to be logical candidates here. These included 5-HT_2, NK_1, and Cholecystokinin $(CCK)_B$ receptors (Bhatt et al., 2003; Gregg and Siegel, 2003; Luo et al., 1998; Shaikh et al., 1997). The results, shown in Fig. 7, indicate that only the 5-HT_2 receptor antagonist was effective in blocking the potentiating effects of IL-1β. Collectively, these data indicate that IL-1 has a specific potentiating effect upon defensive rage behavior and that this effect is mediated selectively through 5-HT_2 receptors in the PAG. Thus, the potentiating effects of IL-1β upon defensive rage behavior are virtually identical in both the medial hypothalamus and PAG.

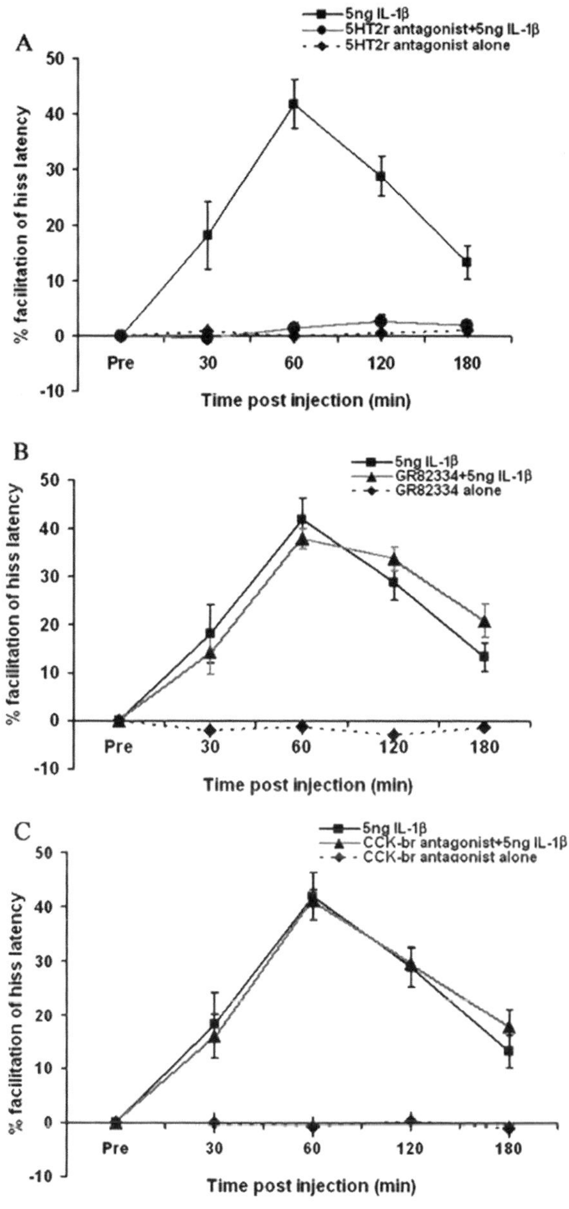

Fig. 7 (A) Pre-treatment with 5-hydroxytryptamine (5-HT_2) recptor antagonist, Ly53857, completely blocked the facilitating effects of IL-1β upon defensive rage behavior elicited from the medial hypothalamus ($p < .0001$, $N = 5$). Administration of the 5-HT_2 receptor antagonist alone had no effect upon defensive rage behavior elicited from the medial hypothalamus ($p = .112$, NS). (B) Pre-treatment with the NK-1 recptor antagonist, GR82334, had no effect upon defensive rage behavior elicited from the medial hypothalamus ($p = .112$, NS). (C) Pre-treatment with the cholecyctokinin (CCK_b) receptor antagonist, CR2945, had no effect upon defensive rage behavior elicited from the medial hypothalamus ($p = .112$, NS). (From Bhatt et al., 2008, with permission)

5.2.2 Interleukin-2

A second proinflammatory cytokine, IL-2, has been studied in our laboratory as well in a manner similar to the analysis applied for the study of IL-1. The findings from experiments conducted with IL-2 are described below.

Hypothalamus. The rationale for the study of IL-2 is based upon observations reported in the literature indicating that IL-2 RNA, protein, and receptor genes have been identified in both glia and neurons in a variety of regions in the CNS, which include the hypothalamus (Araujo et al., 1989; Eizenberg et al., 1995; Hanisch and Quirion, 1996; Petitto and Huang, 1994). Similar to our previous studies, the first phase of the study (Bhatt et al., 2005) established from a dual stimulation experiment that the sites in the medial hypothalamus and PAG associated with defensive rage behavior were part of the same circuitry for the expression of this form of aggression. In the second phase of the study, IL-2 microinjected into the medial hypothalamus suppressed defensive rage in a dose-dependent manner without causing any subsequent change in body temperature (Fig. 8) and this effect could be blocked with pretreatment with an anti-IL-2 antibody, anti-IL2R antibody, or $GABA_A$ receptor antagonist (Fig. 9).

Because GABA neurons and receptors have been shown to be present in the medial hypothalamus (Cheu and Siegel, 1998), we tested the hypothesis that IL-2 suppression of defensive rage behavior is mediated through $GABA_A$ receptors. In this experiment (Bhatt et al., 2005), *pretreatment with the* $GABA_A$ *receptor antago-*

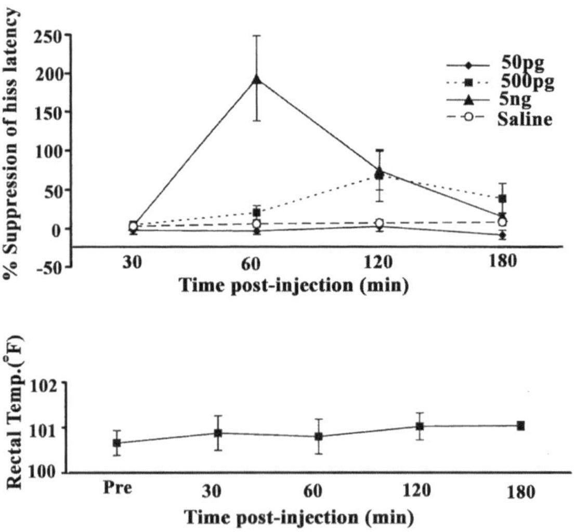

Fig. 8 Effects of microinjections of interleukin (IL)-2 into the medial hypothalamus upon defensive rage. IL-2 dose-dependently suppressed defensive rage elicited from the periaqueductal gray (PAG) in a dose- and time-dependent manner ($p<.001$). Maximal suppression of defensive rage was induced by 5 ng of IL-2 (upper panel). Micorinjections of 5 ng of IL-2 had no effect on rectal temperature (lower panel). (Bhatt et al., 2005, with permission)

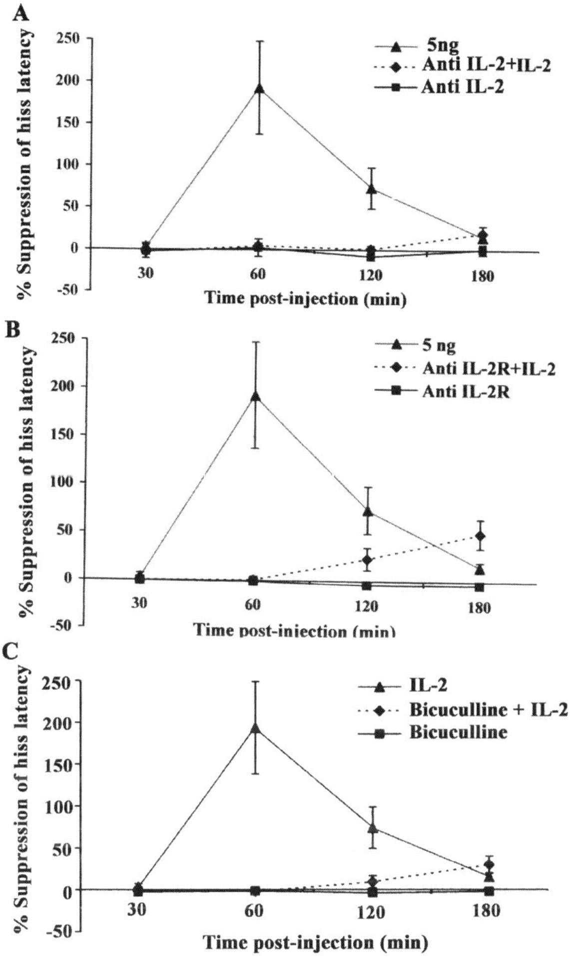

Fig. 9 (A) Effects of an anti interleukin (IL)-2 monoclonal antibody upon defensive rage. Pretreatment with an anti IL-2 antibody blocked the suppressive effects of IL-2 upon periaqueductal gray (PAG)-elicited defensive rage. (B) Effects of an anti IL-2 receptor antibody (anti-IL-2R_{α}) upon defensive rage. Pretreatment with anti IL-2 receptor antibody blocked the suppressive effects of IL-2 upon PAG-elicited defensive rage ($p<.01$). (C) Effects of a $GABA_A$ receptor antagonist upon defensive rage. Pretreatment with the $GABA_A$ receptor antagonist, bicuculline, blocked IL-2s suppressive effects upon PAG-elicited defensive rage ($p<.01$). (From Bhatt et al., 2005, with permission)

nist, bicuculline, completely blocked the suppressive effects of IL-2, thus indicating that these suppressive effects are mediated through $GABA_A$ receptors in the medial hypothalamus. Figure 10 demonstrates the presence of IL-2 receptors proximal to a medial hypothalamic site from which defensive rage behavior had been elicited and IL-2 administered.

Other control experiments established the specificity of the suppressive effects of IL-2. These include the following: (1) microinjections of IL-2 into the lateral

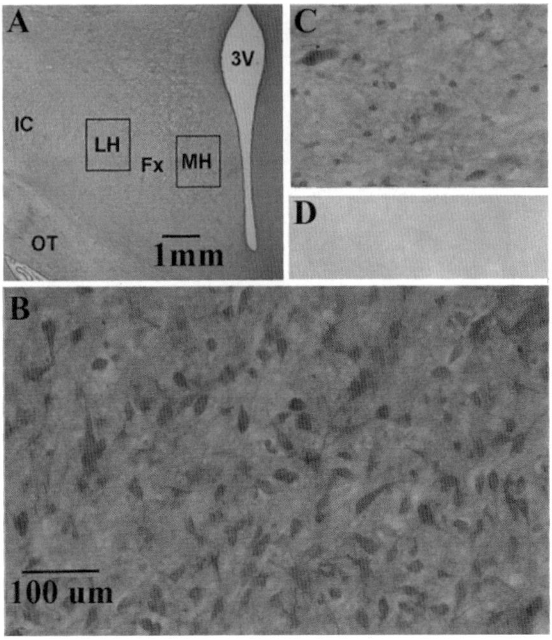

Fig. 10 Immunocytochemistry. Distribution of IL-2 receptors in the medial hypothalamus. (A) Low power view of the anterior hypothalamus indicating the regions from which photomicrographs were taken and shown in panels (B) of the medial hypothalamus, depicting relatively intense and extensive labeling of IL-2 receptors on neurons, and panel (C) of the lateral hypothalamus, depicting relatively sparse labeling on neurons; (D) photomicrograph of a section taken through the medial hypothalamus in which the primary antibody was omitted. Fx, fornix, IC, internal capsule ; LH, lateral hypothalamus; MH, medial hypothalamus; OT, optic tract. Scale bars = 1 μm, shown in panels B, C, and D. (From Bhatt et al., 2005, with permission)

hypothalamus had no effect upon defensive rage elicited from the PAG; (2) pretreatment of the medial hypothalamic defensive rage site with a 5-HT_{1A} receptor antagonist, p-MPPI, also had no effect upon the suppressive effects of IL-2; and (3) microinjections of IL-2 into the medial hypothalamus had no effect upon predatory attack behavior elicited from the lateral hypothalamus.

PAG. In a parallel manner, we further sought to determine whether IL-2 receptor activation in the PAG would affect defensive rage behavior. After the functional relationship between PAG and medial hypothalamic sites were established by dual stimulation procedures, microinjections of IL-2 were placed into defensive rage sites in the dorsolateral PAG. The results, shown in Fig. 11 revealed that in contrast to its suppressive effects when administered into the medial hypothalamus, IL-2 microinjected into the PAG resulted in a dose-dependent facilitation of defensive rage behavior elicited from the PAG (Bhatt and Siegel, 2006). In this study, the facilitating effects of IL-2 upon defensive rage were also blocked following pretreatment with an anti-IL-2 monoclonal antibody. Since it was previously shown that substance P-NK_1 receptors in the PAG facilitate defensive rage behavior (Gregg and Siegel, 2003), we

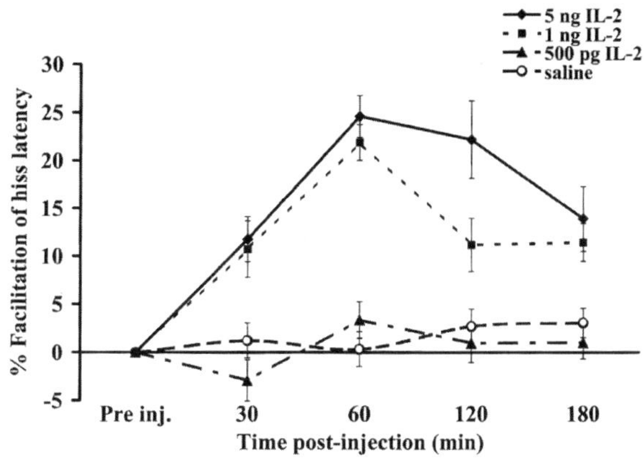

Fig. 11 Effects of microinjections of interleukin (IL)-2 into the periaqueductal gray (PAG) upon defensive rage. IL-2 facilitated defensive rage elicited from the medial hypothalamus in a dose- and time-dependant manner ($p<.0001$, $N=5$). Maximal facilitation of defensive rage was induced by 5 ng of IL-2. (From Bhatt et al., 2006, with permission)

tested the hypothesis that IL-2 potentiation of defensive rage behavior was mediated through NK_1 receptors in the PAG. The hypothesis was tested by pretreating the PAG defensive rage site with the selective NK_1 antagonist, GR82334, prior to administration of IL-2. The results indicated that pretreatment with GR82334 blocked the potentiating effects of IL-2 (Fig. 12) *thus establishing that IL-2 potentiation of defensive rage behavior in the PAG is mediated through NK_1 receptors. Moreover, a critical point here is that the findings demonstrate that the same proinflammatory cytokines differentially affects defensive rage behavior and the directionality of the changes specifically relate to the anatomical locus of the cytokine receptor.*

One final aspect of the modulatory properties of IL-2 upon defensive rage behavior needed to be addressed. Namely, how can one account for the fact that IL-2 suppressed

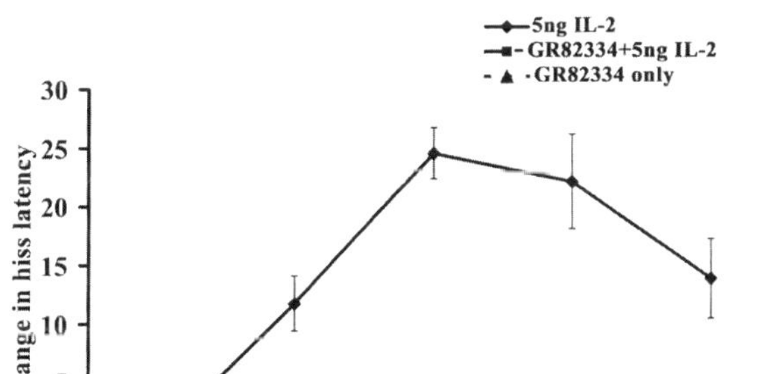

Fig. 12 Effects of natural killer $(NK)_1$ receptor antagonist upon defensive rage. Pre-treatment with the NK_1 receptor antagonist GR82334, completely blocked the facilitating effects of IL-2 upon defensive rage behavior elicited from the medial hypothalamus ($p<.0001$, $N=5$). Administration of the NK_1 receptor antagonist alone had no effect upon defensive rage behavior elicited from the medial hypothalamus ($p=.96$, NS). (From Bhatt et al., 2006, with permission)

defensive rage behavior in the medial hypothalamus while potentiating this form of aggression in the PAG, especially since both the medial hypothalamus and PAG contain the relatively similar distributions of populations of neurotransmitter-receptors? In attempting to answer this question, western blotting was applied to medial hypothalamic and PAG brain tissue in order to discern whether there were any differences in the presence of $GABA_A$, NK_1, and IL-2 receptor subunits in these regions, which could account for these differential effects upon defensive rage. The results of this experiment, shown in Fig. 13 indicated that while receptor subunits for IL-2 and NK_1 did not differ between the hypothalamus and PAG (Fig. 13a, b), the receptor subunits for $GABA_A$ revealed that the ε subunit chain in the medial hypothalamus was absent (Fig. 13c). This finding is consistent with a recent finding by Irnaten et al. (2002) who demonstrated that the suppressive effects of pentobarbital in a parasympathetic neuronal preparation require the absence of the ε subunit. The immediate conclusion to be drawn from this finding is the following: *in order for the effects of IL-2 to be mediated through $GABA_A$ receptors and produce suppression of defensive rage, the region in question requires the absence of the ε subunit of the $GABA_A$ receptor. This apparently is the case in the medial hypothalamus. In the presence of this receptor subunit, IL-2 will interact with other receptors, such as the NK_1 receptor, which has excitatory properties and which is the case in the PAG, thus producing facilitation of defensive rage rather than suppression.*

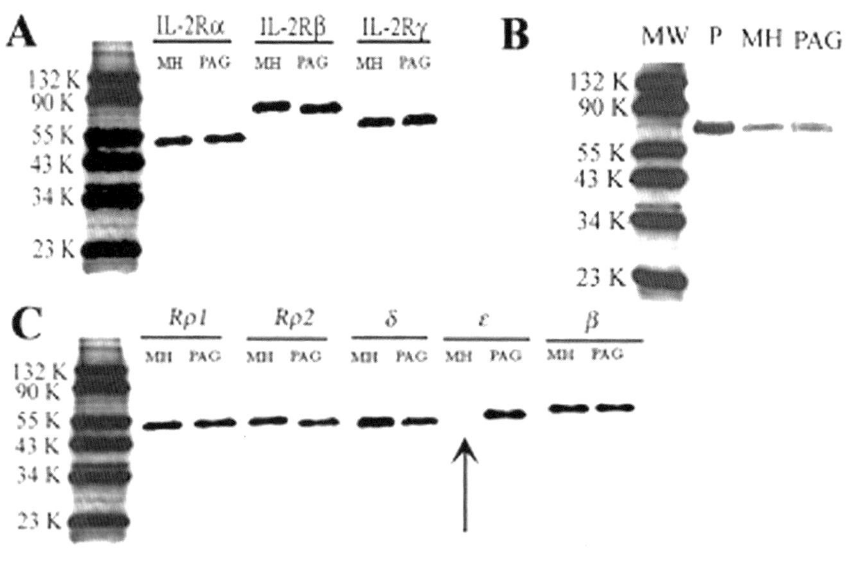

Fig. 13 Analysis of the receptor subunits for interleukin (IL)-2 (A) and natural killer (NK)-1(B) revealed no differences between medial hypothalamus and periaqueductal gray (PAG), the two regions involved in defensive rage in cat while (C) analysis of the receptor subunit for $GABA_A$ revealed that ε chain of the $GABA_A$ receptor was absent in medial hypothalamus. MH, medial hypothalamus; P, Positive control; MW, Molecular weight marker

6 Peripheral Nervous System

6.1 Preliminary Studies Utilizing LPS

The studies described above represent an analysis of the effects of activation of cytokine receptors within key regions of the CNS that mediate defensive rage behavior. In contrast, we have no knowledge of what role, if any, cytokines may play in altering aggressive behavior when they are released in the body. In the preliminary studies summarized below, this approach has been utilized through the application of the bacterial endotoxin LPS.

Administration of LPS caused an elevation in body temperature and a depression in behavioral functions, including defensive rage behavior. In the analysis of these functional deficits, the precise time course of events is of critical importance. Specifically, at 60 min, post-injection, there was a marked increase in the latency for defensive rage, while there was little change in body temperature. Moreover, since LPS had no effect upon head turning responses elicited by midbrain stimulation (Fig. 14), this time period provided a window by which one could examine the effects of LPS

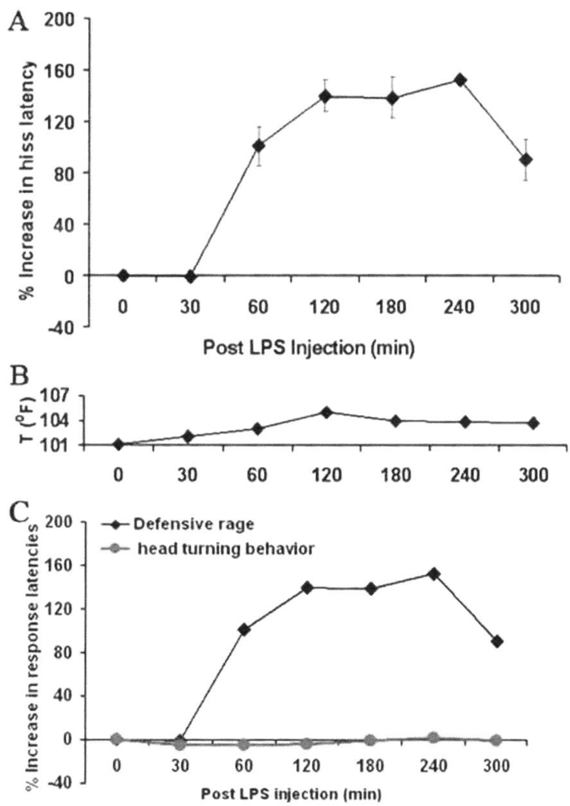

Fig. 14 (A) lipopolysaccharide (LPS) administration suppresses periaqueductal gray (PAG) elicited defensive rage behavior, beginning 60 min, post-injection; (B) Elevation of rectal body temperature to 105°F at 2 h, post-injection; (C) Comparison of LPS administration upon defensive rage behavior and head turning elicited from similar regions of the dorsal PAG in separate cats. Note that response latencies for head turning were not affected by LPS administration ($N=2$). (Unpublished observations)

treatment in the absence of other non-specific effects on motor functions that may have been induced at later periods by LPS. *The strategy present here is that if one could identify and isolate the powerful factor(s) that suppress defensive rage behavior, then such a finding might constitute a major contribution to our understanding of the mechanisms which inhibit defensive rage behavior.*

Inasmuch as LPS induces the release of a cascade of cytokines such as IL-1, IL-6, and TNF-α, it is possible that the suppressive effects of LPS were manifest through the actions of any of these proinflammatory cytokines or, in fact, other substances released by these cytokines. In attempting to assess the peripheral cytokines affective in suppressing defensive rage, the following findings have been noted. Pretreatment with an IL-1 antibody failed to alter the suppressive effects of LPS over a 5-h testing period ($N=2$ cats). In contrast, pretreatment with a TNF-α antibody completely blocked the suppressive effects of LPS as well as its associated fever ($N=3$). The complete blockade of the suppressive effects of LPS manifest by the TNF-α antibody suggests that it is the release of TNF-α that is primarily responsible for the "sickness behavior" and suppression of defensive rage observed following LPS treatment.

An equally important if not more critical question concerns the central mechanisms involved in LPS-induced suppression of defensive rage behavior and induction of sickness behavior. In order to address this question, we hypothesized that it was likely that any central mechanism controlling this process would require involvement of the medial hypothalamus. Therefore, in the first experiment that addressed this hypothesis, a TNF-α antibody (100 ng/0.5 μl) was microinjected into the medial hypothalamus prior to peripheral administration of LPS ($N=2$) in which defensive rage was elicited by stimulation from the PAG. The results indicated that TNF-α antibody treatment had no effect upon LPS-induced suppression of defensive rage behavior.

In a related experiment, we sought to determine whether the suppressive effects of LPS upon defensive rage might be due to the passing of LPS directly into the CNS. To test this possibility, LPS was microinjected directly into the medial hypothalamus and was shown to have little or no effect upon PAG elicited defensive rage behavior over 2 h following microinjections, but did cause some suppression at the 3–4 h period ($N=2$). However, the key point here is that, since peripheral administration of LPS induced suppression of defensive rage as early as 60 min post-injection, it may be concluded that it is highly unlikely that the LPS were manifest through its entrance into the brain through the blood–brain barrier.

Inasmuch as our laboratory has previously shown that 5-HT_{1A} receptors in the medial hypothalamus play a potent role in suppressing defensive rage behavior, it was reasonable to propose the hypothesis that perhaps, these receptors in the medial hypothalamus mediate the suppressive effects of LPS upon defensive rage behavior.

This hypothesis was tested using the same paradigm as described above for the analysis of TNF-α. Administration of the 5-HT_{1A} receptor antagonist into the medial hypothalamus, p-MPPI (12 nmol/0.5 μl, $N=3$), 5 min prior to LPS administration completely blocked the suppressive effects of LPS including significant increases in body temperature characteristic of sickness behavior associated with LPS treatment, in two cats tested (Fig. 15). In fact, visual observation of the cats did not

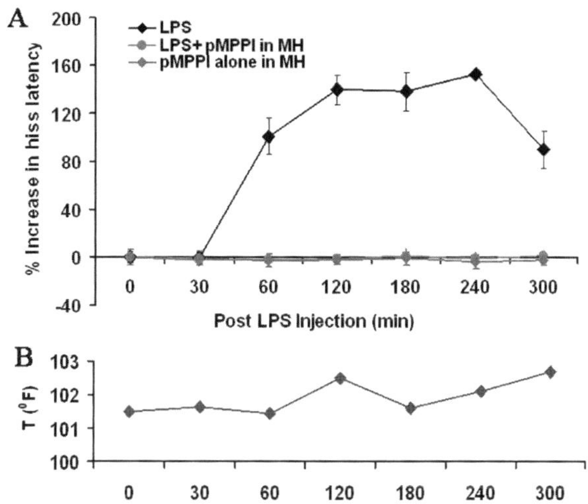

Fig. 15 Microinjections of the 5-hydroxytryptamineb $(5\text{-HT})_{1A}$ receptor antagonist, pMPPI, into a site within the medial hypothalamus from which defensive rage prior to lipopolysaccharide (LPS) delivery completely blocks: (A) the suppressive effects of LPS upon periaqueductal gray (PAG) elicited defensive rage behavior, and (B) elevation in rectal body temperature ($N=2$). (Unpublished observations)

reveal any differences in their overall behavior in comparison to their behavior prior to LPS treatment (Bhatt et al., 2007). *This finding provides evidence that 5-HT$_{1A}$ receptors in the medial hypothalamus mediate LPS-induced suppression of defensive rage behavior.*

Since there is evidence that prostaglandin expression is mediated through serotonin (Linhart et al., 2003), we sought to determine whether blockade of prostaglandin (PG)E$_2$ receptors in the medial hypothalamus would affect the suppressive effects of LPS upon defensive rage behavior. Using the same paradigm as described above for the study of 5-HT$_{1A}$ receptors, the PGE$_2$ receptor antagonist, SC19220 (500 ng/0.5 µl) was microinjected into the medial hypothalamus 5 min prior to LPS administration. The results were similar to those described above for 5-HT$_{1A}$ receptor blockade, namely that infusion of SC19220 blocked the suppressive effects of LPS administration upon defensive rage and upon elevation of body temperature ($N=2$). *These results thus suggest that PGE$_2$ is an important molecule in LPS-induced suppression of defensive rage behavior and in the regulation of sickness behavior.*

7 Summary and Conclusions

The results of the studies described above provide evidence linking immune functions with aggressive behavior. The more recent findings concerning the roles of proinflammatory cytokines released peripherally and the central cytokine effects upon defensive rage behavior were identified, compared, and summarized below as well as indicated in Fig. 16.

1. Activation of IL-I receptors in the medial hypothalamus and PAG potentiates defensive rage behavior in the cat. Facilitation of defensive rage is mediated through 5-HT$_2$ receptors in the medial hypothalamus and PAG.

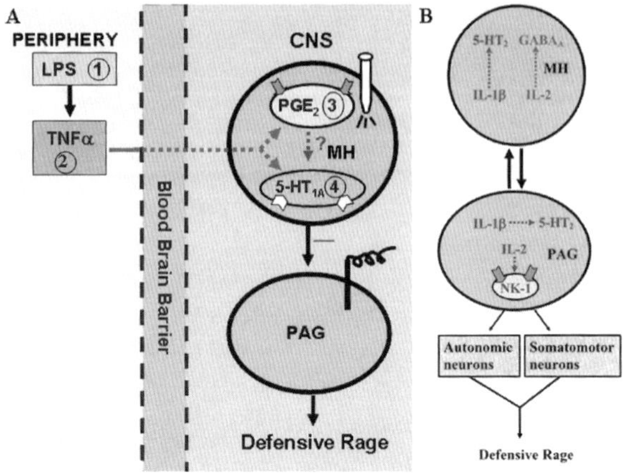

Fig. 16 (A) Model depicting mechanisms by which lipopolysaccharide (LPS) powerfully suppresses defensive rage behavior. The model indicates that LPS causes the release of tumor necrosis factor (TNF)α, which acts upon prostaglandin (PG)E_2 and 5-hydroxytryptamine ($5\text{-}HT_{1A}$) receptors in medial hypothalamus (MH) through unknown pathways, inducing sickness behavior and suppression of defensive rage. (B) Schematic diagram depicting the relationship between the medial hypothalamus and periaqueductal gray (PAG) with respect to the expression of defensive rage behavior. In both the medial hypothalamus and PAG, activation of interleukin (IL)-1β receptors potentiates defensive rage behavior via a mechanism involving $5\text{-}HT_2$ receptors. In contrast, the effects of IL-2 upon defensive rage behavior are dependant upon its locus along the medial hypothalamic-PAG axis. Specifically, IL-2 suppresses defensive rage in the medial hypothalamus by acting through GABA receptors and facilitates this response in the PAG by acting through NK_1 receptors

2. Activation of IL-2 receptors in the medial hypothalamus and PAG differentially affect defensive rage behavior. In the medial hypothalamus, IL-2 receptors suppress defensive rage and this effect is mediated through $GABA_A$ receptors; in contrast, in the PAG, IL-2 receptors facilitate the occurrence of defensive rage behavior and such effects are mediated through substance P NK_1 receptors.
3. Concerning peripheral mechanisms, LPS administration induces the release of a cascade of proinflammatory cytokines. Among the cytokines released, TNF-α appears to play a significant role in the induction of the suppressive effects of LPS upon defensive rage and in sickness behavior in the cat.
4. With respect to the central mechanisms regulating LPS induced suppression of defensive rage and sickness behavior, serotonin $5\text{-}HT_{1A}$ and PGE_2 receptors in the medial hypothalamus appears to play key roles in controlling these processes. What remains unanswered is the precise sequence and linkage by which LPS acts through $5\text{-}HT_{1A}$ and PGE_2 receptors in the medial hypothalamus to induce its suppressive effects upon defensive rage and activate sickness behavior.

Acknowledgements This research is supported by NIH (NINDS) grant NS 07941–36.

References

Araujo DM, Lapchak PA, Collier B, and Quirion R (1989) Localization of interleukin-2 immunoreactivity and interleukin-2 receptors in the rat brain: interaction with the cholinergic system. Brain Res 498:257–266.

Avitsur R, Stark JL, and Sheridan JF (2001) Social stress induces glucocorticoid resistance in subordinate animals. Horm Behav 39:247–257.

Bhatt S, Bhatt R, Zalcman S, and Siegel A (2007) Peripheral administration of lipopolysaccharide (LPS) suppresses midbrain periaqueductal gray (PAG) elicited defensive rage behavior in cats: role of 5-HT_{1A} receptors. Soc Neurosci Abstr 531.18.

Bhatt S, Bhatt R, Zalcman SS, and Siegel A (2008) Role of IL-1 beta and 5-HT2 receptors in midbrain periaqueductal gray (PAG) in potentiating defensive rage behavior in cat. Brain Behav Immun 22:224–233.

Bhatt S, Gregg TR, and Siegel A (2003) NK1 receptors in the medial hypothalamus potentiate defensive rage behavior elicited from the midbrain periaqueductal gray of the cat. Brain Res 966:54–64.

Bhatt S and Siegel A (2006) Potentiating role of interleukin 2 (IL-2) receptors in the midbrain periaqueductal gray (PAG) upon defensive rage behavior in the cat: Role of neurokinin NK1 receptors. Behav Brain Res 167:251–260.

Bhatt S, Zalcman S, Hassanain M, and Siegel A (2005) Cytokine modulation of defensive rage behavior in the cat: Role of GABA(A) and interleukin-2 receptors in the medial hypothalamus. Neuroscience 133:17–28.

Boccia ML, Laudenslager ML, Broussard CL, and Hijazi AS (1992) Immune responses following competitive water tests in two species of macaques. Brain Behav Immun 6:201–213.

Capitanio JP, Mendoza SP, Lerche NW, and Mason WA (1998) Social stress results in altered glucocorticoid regulation and shorter survival in simian acquired immune deficiency syndrome. Proc Natl Acad Sci U S A 95:4714–4719.

Capuron L, Ravaud A, Miller AH, and Dantzer R (2004) Baseline mood and psychosocial characteristics of patients developing depressive symptoms during interleukin-2 and/or interferon-alpha cancer therapy. Brain Behav Immun 18:205–213.

Cheu JW and Siegel A (1998) GABA receptor mediated suppression of defensive rage behavior elicited from the medial hypothalamus of the cat: role of the lateral hypothalamus. Brain Res 783:293–304.

Christensen AJ, Edwards DL, Wiebe JS, Benotsch EG, McKelvey L, Andrews M, and Lubaroff DM (1996) Effect of verbal self-disclosure on natural killer cell activity: moderating influence of cynical hostility. Psychosom Med 58:150–155.

Cohen S, Line S, Manuck SB, Rabin BS, Heise ER, and Kaplan JR (1997) Chronic social stress, social status, and susceptibility to upper respiratory infections in nonhuman primates. Psychosom Med 59:213–221.

Devoino L, Alperina E, Kudryavtseva N, and Popova N (1993) Immune responses in male mice with aggressive and submissive behavior patterns: strain differences. Brain Behav Immun 7:91–96.

Dreau D, Sonnenfeld G, Fowler N, Morton DS, and Lyte M (1999) Effects of social conflict on immune responses and *E. coli* growth within closed chambers in mice. Physiol Behav 67:133–140.

Dunn AJ (1992) Endotoxin-induced activation of cerebral catecholamine and serotonin metabolism: comparison with interleukin-1. J Pharmacol Exp Ther 261:964–969.

Eizenberg O, Faber-Elman A, Lotan M, and Schwartz M (1995) Interleukin-2 transcripts in human and rodent brains: possible expression by astrocytes. J Neurochem 64:1928–1936.

Fuchs SAG, Edinger HM, and Siegel A (1985a) The organization of the hypothalamic pathways mediating affective defense behavior in the cat. Brain Res 330:77–92.

Fuchs SAG, Edinger HM, and Siegel A (1985b) The role of the anterior hypothalamus in affective defense behavior elicited from the ventromedial hypothalamus of the cat. Brain Res 330:93–108.

Gasparotto OC, Ignacio ZM, Lin K, and Goncalves S (2002) The effect of different psychological profiles and timings of stress exposure on humoral immune response. Physiol Behav 76:321–326.

Gemma C, Imeri L, De Simoni MG, and Mancia M (1997) Interleukin-1 induces changes in sleep, brain temperature, and serotonergic metabolism. Am J Physiol 272:R601–R606.

Gemma C, Imeri L, and Opp MR (2003) Serotonergic activation stimulates the pituitary-adrenal axis and alters interleukin-1 mRNA expression in rat brain. Psychoneuroendocrinology 28:875–884.

Granger DA, Booth A, and Johnson DR (2000) Human aggression and enumerative measures of immunity. Psychosom Med 62:583–590.

Gregg TR and Siegel A (2003) Differential effects of NK1 receptors in the midbrain periaqueductal gray upon defensive rage and predatory attack in the cat. Brain Res 994:55–66.

Gryazeva NI, Shurlygina AV, Verbitskaya LV, Mel'nikova EV, Kudryavtseva NN, and Trufakin VA (2001) Changes in various measures of immune status in mice subject to chronic social conflict. Neurosci Behav Physiol 31:75–81.

Gust DA, Gordon TP, Wilson ME, Ahmed-Ansari A, Brodie AR, and McClure HM (1991) Formation of a new social group of unfamiliar female rhesus monkeys affects the immune and pituitary adrenocortical systems. Brain Behav Immun 5:296–307.

Hanisch UK and Quirion R (1996) Interleukin-2 as a neuroregulatory cytokine. Brain Res Rev 21:246–284.

Hardy CA, Quay J, Livnat S, and Ader R (1990) Altered T-lymphocyte response following aggressive encounters in mice. Physiol Behav 47:1245–1251.

Hassanain M, Bhatt S, and Siegel A (2003a) Differential modulation of feline defensive rage behavior in the medial hypothalamus by 5-HT1A and 5-HT2 receptors. Brain Res 981:201–209.

Hassanain M, Bhatt S, Zalcman S, and Siegel A (2005) Potentiating role of interleukin-1 beta (IL-1 beta) and IL-1 beta type 1 receptors in the medial hypothalamus in defensive rage behavior in the cat. Brain Res 1048:1–11.

Hassanain M, Zalcman S, Bhatt S, and Siegel A (2003b) Interleukin-1 beta in the hypothalamus potentiates feline defensive rage: Role of serotonin-2 receptors. Neuroscience 120:227–233.

Hessing MJ, Scheepens CJ, Schouten WG, Tielen MJ, and Wiepkema PR (1994) Social rank and disease susceptibility in pigs. Vet Immunol Immunopathol 43:373–387.

Imeri L, Mancia M, and Opp MR (1999) Blockade of 5-hydroxytryptamine (serotonin)-2 receptors alters interleukin-1-induced changes in rat sleep. Neuroscience 92:745–749.

Irnaten M, Walwyn WM, Wang J, Venkatesan P, Evans C, Chang KS, Andresen MC, Hales TG, and Mendelowitz D (2002) Pentobarbital enhances GABAergic neurotransmission to cardiac parasympathetic neurons, which is prevented by expression of $GABA_A$ epsilon subunit. Anesthesiology 97:717–724.

Kavelaars A, Heijnen CJ, Tennekes R, Bruggink JE, and Koolhaas JM (1999) Individual behavioral characteristics of wild-type rats predict susceptibility to experimental autoimmune encephalomyelitis. Brain Behav Immun 13:279–286.

Kiecolt-Glaser JK, Glaser R, Cacioppo JT, MacCallum RC, Snydersmith M, Kim C, and Malarkey WB (1997) Marital conflict in older adults: endocrinological and immunological correlates. Psychosom Med 59:339–349.

Kraus MR, Schafer A, Faller H, Csef H, and Scheurlen M (2003). Psychiatric symptoms in patients with chronic hepatitis C receiving interferon alfa-2b therapy. J Clin Psychiatry 64:708–714.

Laviano A, Cangiano C, Fava A, Muscaritoli M, Mulieri G, and Rossi FF (1999) Peripherally injected IL-1 induces anorexia and increases brain tryptophan concentrations. Adv Exp Med Biol 467:105–108.

Leger L, Charnay M, Hof PR, Bouras C, and Cespuglio R (2001) Anatomical distribution of serotonin-containing neurons and axons in the central nervous system of the cat. J Comp Neurol 433:157–182.

Leyhausen P (1979) Cat Behavior: The Predatory and Social Behavior of Domestic and Wild Cats. Garland STPM Press, New York.

Linhart O, Obreja O, and Kress M (2003) The inflammatory mediators serotonin, prostaglandin E2 and bradykinin evoke calcium influx in rat sensory neurons. Neuroscience 118:69–74.

Luo B, Cheu JW, and Siegel A (1998). Cholecystokinin B receptors in the periaqueductal gray potentiate defensive rage behavior elicited from the medial hypothalamus of the cat. Brain Res 796:27–37.

McHutchison JG, Gordon SC, Schiff ER, Shiffman ML, Lee WM, Rustgi VK, Goodman ZD, Ling MH, Cort S, and Albrecht JK (1998) Interferon alfa-2b alone or in combination with ribavirin as initial treatment for chronic hepatitis C. Hepatitis Interventional Therapy Group. N Engl J Med 339:1485–1492.

Merali Z, Lacosta S, and Anisman H (1997). Effects of interleukin-1β and mild stress on alterations of norepinephrine, dopamine and serotonin neurotransmission: a regional microdialysis study. Brain Res 761:225–235.

Petitto JM and Huang Z (1994) Molecular cloning of a partial cDNA of the interleukin-2 receptor-beta in normal mouse brain: in situ localization in the hippocampus and expression by neuroblastoma cells. Brain Res 650:140–145.

Petitto JM, Lysle DT, Gariepy JL, Clubb PH, Cairns RB, and Lewis MH (1993) Genetic differences in social behavior: relation to natural killer cell function and susceptibility to tumor development. Neuropsychopharmacology 8:35–43.

Petitto JM, Lysle DT, Gariepy JL, and Lewis MH (1994) Association of genetic differences in social behavior and cellular immune responsiveness: effects of social experience. Brain Behav Immun 8:111–122.

Sapolsky RM and Spencer EM (1997) Insulin-like growth factor I is suppressed in socially subordinate male baboons. Am J Physiol 273:R1346–R1351.

Schubert K, Shaikh MB, and Siegel A (1996) NMDA receptors in the midbrain periaqueductal gray mediate hypothalamically evoked hissing behavior in the cat. Brain Res 726:80–90.

Shaikh MB, Barrett JA, and Siegel A (1987) The pathways mediating affective defense and quiet biting attack behavior from the midbrain central gray of the cat: an autoradiographic study. Brain Res 437:9–25.

Shaikh MB, De Lanerolle NC, and Siegel A (1997) Serotonin 5-HT_{1A} and 5-$HT_{2/1C}$ receptors in the midbrain periaqueductal gray differentially modulate defensive rage behavior elicited from the medial hypothalamus of the cat. Brain Res 765:198–207.

Siegel A (2005) The Neurobiology of Aggression and Rage. CRC Press, Boca Raton, FL.

Siegel A, Roeling TAP, Gregg TR, and Kruk MR (1999) Neuropharmacology of brain-stimulation-evoked aggression. Neurosci Biobehav Rev 23:359–389.

Song C, Merali Z, and Anisman H (1999) Variations of nucleus accumbens dopamine and serotonin following systemic interleukin-1, interleukin-2 or interleukin-6 treatment. Neuroscience 88:823–836.

Stark JL, Avitsur R, Padgett DA, Campbell KA, Beck FM, and Sheridan JF (2001) Social stress induces glucocorticoid resistance in macrophages. Am J Physiol Regul Integr Comp Physiol 280:R1799–R1805.

Stefanski V (1998) Social stress in loser rats: opposite immunological effects in submissive and subdominant males. Physiol Behav 63:605–613.

Stefanski V and Ben-Eliyahu S (1996) Social confrontation and tumor metastasis in rats: Defeat and [beta]-adrenergic mechanisms. Physiol Behav 60:277–282.

Stefanski V and Engler H (1998) Effects of acute and chronic social stress on blood cellular immunity in rats. Physiol Behav 64:733–741.

Suarez EC, Lewis JG, Krishnan RR, and Young KH (2004) Enhanced expression of cytokines and chemokines by blood monocytes to in vitro lipopolysaccharide stimulation are associated with hostility and severity of depressive symptoms in healthy women. Psychoneuroendocrinology 29:1119–1128.

Zalcman S, Green-Johnson JM, Murray L, Nance DM, Dyck D, Anisman H, and Greenberg AH (1994) Cytokine-specific central monoamine alterations induced by interleukin-1, -2 and -6. Brain Res 643:40–49.

Zubareva OE, Krasnova IN, Abdurasulova IN, Bluthe RM, Dantzer R, and Klimenko VM (2001) Effects of serotonin synthesis blockade on interleukin-1 beta action in the brain of rats. Brain Res 915:244–247.

Neurochemical and Behavioral Changes Induced by Interleukin-2 and Soluble Interleukin-2 Receptors

Steven S. Zalcman, Randall T. Woodruff, Ruchika Mohla, and Allan Siegel

Abstract It has become abundantly clear that in addition to regulating immune function, interleukin (IL)-2 potently modulates brain neurochemical activity and behavior. Central targets of IL-2 include mesolimbic and mesostriatal dopaminergic processes, hippocampal cholinergic activity, and monoaminergic and neuroendocrine activity in hypothalamus, and brainstem. Of further significance, IL-2 influences related behaviors and responses, including (but not limited to) exploratory activity, stereotypic behaviors, hedonic responses, and aggressive behavior. IL-2 may induce behavior-activating or depressive-like effects, depending on the stressfulness of the test conditions, chronicity of its administration, and brain region into which it is microinjected, among other factors. Neurochemical and behavioral effects of IL-2 differ in many respects from those characteristics of proinflammatory cytokines that induce the classic symptoms of sickness behavior. Clinically, abnormal production and levels of IL-2 and soluble IL-2 receptors (sIL-2R) are evident in various psychiatric disorders, including schizophrenia, mania, and depression, among other disorders. Increased levels of sIL-2Rα are further evident in subgroups of patients displaying abnormal motor activity, suggesting that it might act as an etiological agent. Consistent with this idea, we have discovered in mice that injections of purified forms of the sIL-2Rα or β subunit induce abnormal and subunit-specific increases in motor activity and stereotypic behaviors. These findings further suggest that sIL-2Rs act as novel messengers between the brain and immune system. More research is needed to shed light on the mechanisms underlying IL-2's potent modulation of brain neurotransmitter activity and behavior, as well as its role in psychiatric disorders.

Keywords IL-2 · Cytokines · Dopamine · Norepinephrine · Serotonin · HPA-axis · Striatum · Neurotransmitters · Locomotion · Stereotypy · Reward · Schizophrenia · Mania

---

S.S. Zalcman (✉)
Department of Psychiatry, UMDNJ – New Jersey Medical School, Behavioral Health Sciences Building, Room F-1559, 183 South Orange Avenue, Newark, NJ 07103, USA
e-mail: zalcmass@umdnj.edu

Supported by a grant from NIH to SSZ (R01 MH 074689)

A. Siegel, S.S. Zalcman (eds.), *The Neuroimmunological Basis of Behavior and Mental Disorders*, DOI 10.1007/978-0-387-84851-8_13,

1 Introduction

Over the past three decades, it has been established that the brain and immune system interact in a bi-directional manner. There are two prominent and mutually inclusive models of such interactions: (1) a model of a neuroimmune feedback loops regulating an ongoing orchestrated immune response (Besedovsky and del Rey, 1996; Nance and Sanders, 2007), and (2) the sickness behavior model describing a series of adaptive changes in mood and motivation that occur following infection (Hart, 1988; Dantzer et al., 1999). In parallel, studies have shown that pathogens and immune-derived elements (notably cytokines) may influence brain activity, potently modulate neurotransmitter activity, and alter behavior (Banks, 2005; Dantzer et al., 1999). Such work dispelled the long-held notion that the brain is completely immune privileged. Of further importance, it has become clear that immune-derived substances are also endogenously found in brain, and act as potent neuromodulators even in the absence of an ongoing immune response (Maier et al., 2001).

Landmark studies of the behavioral consequences of immunological challenge have focused on adaptive depressive-like behavior and changes in motivational status following exposure to infectious agents and certain inflammatory cytokines (e.g., interleukin (IL)-1, TNFα; Dantzer et al., 1999). In examining the neurochemical consequences of inflammatory cytokines, we found that cytokine-specific alterations of central monoamine activity and behavior are induced by IL-1, IL-2, and IL-6 (Zalcman et al., 1994a, b, 1998). *Of unique interest was the appearance of behavior activating effects following a single injection of IL-2.*

In this chapter, we will focus on neurochemical and behavioral effects of IL-2, and on its suggested role in various psychiatric disorders. We will also discuss data that we have recently collected indicating that soluble IL-2 receptors induce marked behavior activating effects, and thus, may act as novel immune to brain messengers (Fig. 1).

2 Interleukin-2

IL-2 is a glycoprotein with a molecular weight of approximately 15–18 kDa. Following antigen-specific activation, T helper (Th)1 cells produce proinflammatory cytokines, including IL-2, IL-12, interferon (IFN)-γ, and Lymphotoxin. These cytokines are primarily involved in cell-mediated immune responses. IL-2 plays a pivotal role in the expansion, differentiation, and survival of antigen-selected T cell clones. Of further importance, IL-2 regulates B cells, natural killer (NK) cells, and regulatory T cells (Smith, 1988; Klebb et al., 1996). IL-2 is not uniquely found nor are its biological effects restricted to the peripheral immune system. IL-2 mRNA transcripts, IL-2-like immunoreactivity, and genes for IL-2 receptor subunits have been detected in many brain regions including the striatum, cortex, hippocampus, and septal area (Hanisch and Quirion, 1995). In vitro studies have further shown that IL-2 promotes the survival of neurons in cortical, striatal, and septal neurons (Awatsuji et al., 1993).

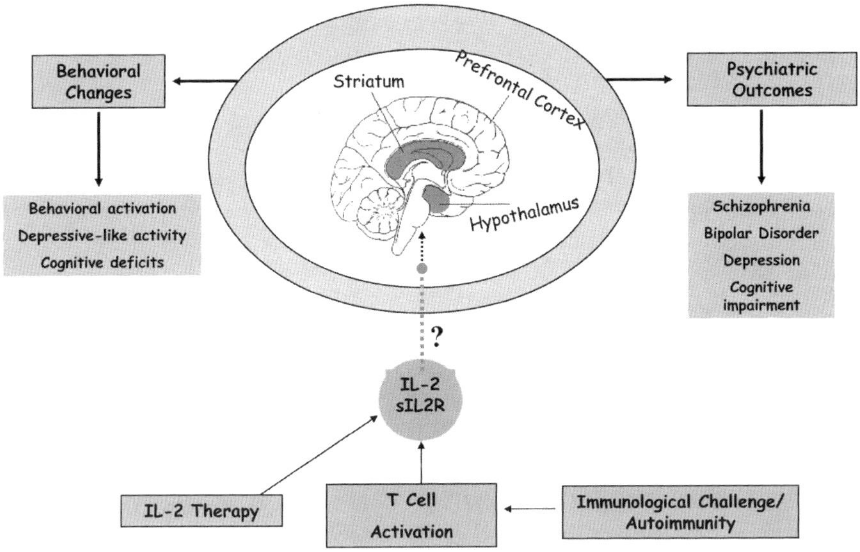

Fig. 1 Highly schematic model depicting interleukin (IL)-2-induced alterations of brain activity and behavior. Circulating levels of IL-2 and soluble IL-2 receptors (sIL-2R) may be increased following T cell activation or in patients receiving IL-2 therapy. By an unknown mechanism, IL-2/sIL-2R potently modulate activity in various brain regions (such as the striatum, prefrontal cortex, and hypothalamus). Region-specific neurochemical alterations lead to a range of behavioral changes and psychiatric outcomes

3 Interleukin-2-Induced Alterations of Central Neurochemical Activity

Zalcman et al. (1994a) showed that IL-1, IL-2, and IL-6 induce cytokine-specific alterations of central monoamine activity. With regard to IL-2, studies conducted over the past two decades have focused on this cytokine's modulation of central monoamine activity and acetylcholine release. These studies are summarized and discussed here.

3.1 Dopamine

The mesolimbic and mesostriatal systems were among the first neural systems to be identified as targets of IL-2. In this section, we review data showing that IL-2 is a potent modulator of dopaminergic activity in these systems.

In vitro application of IL-2 to rat mesencephalic slices influences K+- and *N*-methyl-D-aspartate (NMDA)-evoked [3H] dopamine release (Alonso et al., 1993). IL-2 also modulates spontaneous [3H] dopamine release in preloaded neonatal rat striatal slices (Lapchak, 1992), and endogenous veratrine-evoked dopamine release in striatal slices of adult mice (Petitto et al., 1997). IL-2's effects are biphasic in that dopamine release is increased by relatively low doses of IL-2 and inhibited at higher doses (although there are between region differences in the doses required to affect dopamine release). In addition to modulating dopamine release, IL-2 is a potent modulator of membrane conductance in dopamine neurons (Ye et al., 2001). Specifically, relatively low doses inhibit NMDA- and kainate-activated currents in dopamine neurons freshly isolated from the ventral tegmental area (VTA) of neonatal rats (Ye et al., 2001, 2005). Relatively high doses of IL-2 potentiate NMDA-activated current. Inasmuch as IL-2 increases the survival of dopamine neurons during development (Awatsuji et al., 1993), it was suggested the inhibitory effect of IL-2 on NMDA-activated currents could prevent the development of excitotoxicity in mesolimbic neurons (Ye et al., 2001). The fact that IL-2 modulates NMDA- and kainate-activated currents in dopamine neurons in the VTA also raises the possibility that it plays a role in mediating the effects of excitatory glutamate inputs (e.g., from the prefrontal cortex) into this region.

IL-2 also influences central dopamine activity in vivo. For example, Zalcman et al. (1994a) showed in mice that a peripheral injection of IL-2 induces an increase in dopamine utilization in the medial prefrontal cortex (Zalcman et al., 1994a). Similar increases are induced in the rat hypothalamus and hippocampus (Connor et al., 1998). A single peripheral injection of IL-2 also reduces extracellular dopamine levels in the nucleus accumbens (Anisman et al., 1996; Song et al., 1999), an effect possibly related to increased activity in the medial prefrontal cortex.

Based on these findings, one can make the following conclusions: the effects of IL-2 on brain dopamine activity are (1) dose-dependent and (2) site-specific. In view of these findings, more research is needed to identify the specific properties governing this complex relationship.

The effects of IL-2 on dopaminergic activity in the mesostriatal system are related to the chronicity of IL-2 treatment. Hassanain et al. (2006) showed in mice that repeated injections of IL-2 (one injection a day for 5 days) provoked marked increases in c-fos expression within subterritories of the caudate nucleus, including the central, dorsolateral, and ventromedial aspects. This activation occurred coincident with an increase in the expression of stereotypic behaviors, an effect not seen with single injections of IL-2. In contrast with this effect, Lacosta et al. (2000) reported that 7 days of IL-2 treatment (but not a single injection) resulted in a decrease in dopamine turnover within the caudate nucleus and substantia nigra in mice. The reasons for the between study disparities remain to be determined. However, differences in dose, duration of treatment, strain of mouse, and test conditions, among other factors, could have contributed. Chronic IL-2 treatment also affects dopamine receptor binding in mesocorticolimbic and striatal structures. Reductions in D1 receptor binding in frontoparietal cortex and D2 binding in the nucleus accumbens were evident in rats receiving intracerebroventricular (i.c.v) infusions of IL-2 for up to 14 days (Hanisch et al., 1997). However, it should be noted that the regimen of IL-2 treatment used was also shown to have neurotoxic effects. It remains to be determined whether IL-2-induced changes in receptor expression are uniquely associated with neurotoxicity, or whether they are related to IL-2-induced alterations of dopamine release, a cytokine–receptor interaction, or to other mechanisms.

It is important to note that the pathway(s) by which peripheral IL-2 comes to affect brain dopamine activity (or that of other neurotransmitters) remain to be identified. Regarding possible transport across the blood–brain barrier (BBB), very small amounts of injected IL-2 enter the brain (Banks et al., 2004). Indeed, a saturable brain to blood efflux system exists by which IL-2 is transported in the brain to blood direction. Of further significance, the net amount of IL-2 that enters the brain is not influenced by the chronicity of its administration. Thus, differences in the behavioral consequences of single and repeated injections of IL-2 do not appear to be related to BBB transport.

3.2 Norepinephrine and Serotonin

In addition to modulating central dopaminergic activity, IL-2 influences norepinephrine and serotonin activity. Lapchak and Araujo (1993) showed that IL-2 modulates K+ evoked norepinephrine release in the hypothalamus. In mice, a single peripheral injection of IL-2 induces a marked increase in the accumulation of hypothalamic MHPG, a metabolite of norepinephrine, suggesting a cytokine-induced increase in utilization of the transmitter (Zalcman et al., 1994a; Lacosta et al., 2000). Unit activity in the hypothalamus is similarly increased by IL-2 (Bartholomew and Hoffman, 1993). The effects of IL-2 on norepinephrine activity are related to the chronicity of cytokine treatment in that repeated but not single injections result in an increase in norepinephrine utilization within the median eminence, arcuate nucleus, hippocampus, central amygdala, and medial prefrontal cortex (Lacosta et al., 2000). Repeated i.c.v. administration of IL-2 likewise increases brain norepinephrine and serotonin levels in olfactory bulbectomized rats (Song and Leonard, 1995). A single i.c.v injection of a relatively high dose of IL-2 in rats also increases extracellular levels of serotonin and its metabolite, 5-hydroxyindoleacetic acid in the hippocampus (Pauli et al., 1998). This effect is blocked by an IL-1 receptor antagonist, suggesting that an IL-1 receptor mechanism mediates IL-2-induced alterations of serotonin release. Serotonin levels in the hippocampus and prefrontal cortex of mice are also affected by IL-2 treatment (Lacosta et al., 2000).

3.3 Acetylcholine

It has been suggested that IL-2 is the most potent substance known to modulate brain acetylcholine (ACh) release (Hanisch and Quirion, 1995). In a series of studies, it was discovered that IL-2 modulates hippocampal and frontocortical K+ evoked, but not basal ACh release (Araujo et al., 1989; Hanisch et al., 1993). This effect occurs via an IL-2Rα/Tac-mediated mechanism. Low doses of IL-2 stimulate while relatively high doses inhibit ACh release. IL-2 and CD4(+)CD25(−) T-cells inhibit hippocampal short- and long-term potentiation (Tancredi et al., 1990; Lewitus et al., 2007), suggesting a role for this cytokine in synaptic transmission and neuronal communication within the hippocampus. In vivo, a chronic IL-2 treatment regimen alters the binding levels of cholinergic M1 and M2 receptors, although this effect could be related to the fact that the treatment regimen had neurotoxic effects (Hanisch et al., 1997). It is important to note that the relationship between IL-2 and cholinergic activity is cytokine specific in that IL-1, IL-4, and IL-6 among other cytokines are not effective in altering ACh release (Hanisch and Quirion, 1995).

3.4 Neuroendocrine Activity

In vitro studies have established that IL-2 influences the release of adrenocorticotropin (ACTH), corticotropin releasing hormone (CRH), met-enkephalin, β-endorphin, leutinizing hormone, follicle-stimulating hormone, and thyrotropic hormone, among other hormones in the hypothalamus and pituitary (Hanisch and Quirion, 1995). IL-2's modulatory effects are potent; indeed, it is more potent than CRH in stimulating ACTH release (Raber et al., 1995). In vivo studies have also shown that IL-2 influences hypothalamo-pituitary-adrenocortical (HPA)-axis activity. McCann and colleagues showed that plasma concentrations of ACTH are increased in rats receiving single injections of IL-2 into the third ventricle (McCann et al., 2000). Of further importance, the effects of peripherally injected IL-2 on HPA axis activity are related to the stressor environment in which it is injected (Zalcman et al., 1994a, b) and on the chronicity of its administration (Hanisch et al., 1997; Song and Leonard, 1995).

4 IL-2-Induced Behavioral Alterations

As discussed in the previous section, neural targets of IL-2 include mesolimbic and mesostriatal dopaminergic processes, as well as activity in the hypothalamus and brain stem. Accordingly, a variety of investigators have focused attention on the effects of IL-2 on behaviors and responses associated with alterations in these regions, including novelty-induced exploratory activity, stereotypic behaviors, and hedonic responses. Moreover, in light of the fact that IL-2 is a powerful modulator of hippocampal ACh release, investigators have evaluated the effects of IL-2 on performance in cognitive tasks. Of further significance is the link between IL-2 and aggression, among other behaviors. In this section, we will review studies using animal models to examine the behavioral consequences of IL-2 administration.

4.1 Motor Activity

4.1.1 Exploratory Locomotion

A single peripheral or i.c.v injection of IL-2 results in an increase in exploration and locomotion (Nistico and De Sarro, 1991; Petitto et al., 1997; Zalcman et al., 1998;

Zalcman, 2001). IL-2 is potent in stimulating exploratory locomotion, and effects are evident over a relatively wide dose range (Zalcman, 2001). IL-2-treated mice also exhibit marked increases in the investigation of a novel stimulus (Zalcman et al., 1998). Behavioral effects of IL-2 are dependent upon the novelty of the test conditions in that IL-2 treatment does not appreciably alter exploration and locomotion in mice that were preexposed to the test cage or following home cage injections (Zalcman et al., 2001). These findings also imply that IL-2 does not induce a restricted motor response. Inasmuch as acutely administered IL-2 does not alter responding in the elevated plus maze, IL-2's potentiation of novelty-induced activity is not related to effects on anxiety-like behavior or fearfulness (Petitto et al., 1997; Connor et al., 1998). It should be mentioned that centrally, IL-2's effects on activity and general arousal are related to the site of its injection since soporific effects are induced following injections into the locus coeruleus (De Sarro et al., 1990).

Little is known about the mechanisms underlying the behavior-activating effects of IL-2. There is evidence, however, that pretreating mice with naloxone attenuates IL-2's effects on exploratory activity, suggesting that brain opioid receptors play an important role (Nistico and De Sarro, 1991).

In addition to potentiating novelty-induced exploratory activity, IL-2 further augments the behavior activating effects of drugs that target brain dopaminergic or cholinergic processes. For example, pretreating mice with IL-2 potentiates the locomotor-stimulating effects of GBR 12909, a highly selective dopamine uptake inhibitor (Zalcman, 2001). IL-2 pretreatment likewise potentiates behavior-activating effects of scopolamine, a cholinergic M1 receptor antagonist (Bianchi and Panerai, 1993).

In contrast with studies showing that single injections of IL-2 potentiate novelty-induced exploratory activity, other investigators either found that IL-2 inhibits or fails to alter such activity. For example, i.c.v administration of IL-2 resulted in hypoactivity and signs of sickness behavior in rats (Pauli et al., 1998). Relatively high doses of IL-2 were required to induce these effects, however. Moreover, the time spent engaged in locomotion and approach to a novel stimulus were found to be unaffected by a single peripheral injection of IL-2 (Bianchi and Panerai, 1993; Lacosta et al., 1999). It is important to note that in the Lacosta et al. (1999) study, locomotion was measured in an open field that had in its center a cylindrical opaque plastic container. Thus, it is possible that the seemingly disparate findings across studies relate to differences in the test arenas in which behavior was measured (i.e., open field with or without a novel object, or an elevated plus maze). It should also be considered that differences in measurements of locomotor activity (distance traveled vs. time spent engaged in locomotor activity) could have contributed. Along the same lines, apparent discrepancies may be explained by the way in which exploratory activity was defined and measured. In particular, in view of the fact that IL-2-treated mice tended to explore limited portions of the test cage, Zalcman et al. (1998) evaluated the effects of IL-2 on "ambulatory" and "non-ambulatory" exploration. It was found that non-ambulatory exploration was increased while ambulatory exploration was decreased relative to controls.

It is important to note that an apparent discrepancy in the direction of an IL-2-induced behavioral change is not limited to novelty-induced exploration. As will be discussed in Section 4.4 of this chapter (see also "Cytokines and Aggressive Behavior" Chapter in this book by Siegel et al.), IL-2 may suppress or facilitate feline defensive rage behavior depending on the brain region into which it is microinjected.

4.1.2 Stereotypic Behaviors

Single injections of IL-2 influence the expression of stereotypic motor behaviors, which comprise a group of species-specific adaptive behaviors that help the individual cope with environmental conditions (Bardo et al., 1996). In rodents, the exploration of novelty involves a variety of stereotyped motor behaviors, including sniffing, digging, gnawing, biting, rearing, and climbing behavior, among other behaviors. However, abnormal repetition of such behaviors is associated with certain psychiatric and neurological disorders and with abnormal changes in neuronal activation patterns in the nigrostriatal system (Saka and Graybiel, 2003). It is thus of unique interest that the potentiating effects of peripherally administered IL-2 on novelty-induced exploratory activity are associated with an increased expression of rearing, sniffing and digging behavior (Zalcman et al., 1998; Zalcman, 2001). Moreover, microinjections of IL-2 into nigrostriatal structures (substantia nigra pars compacta or caudate nucleus) result in ipsilateral turning and circling behavior (Nistico and DeSarro, 1991). Asymmetric body posture is also seen in these mice. Parenthetically, a link between IL-2 and ataxia is also evident in transgenic mice carrying the IL-2 gene (Katsuki et al., 1989). In a study by Magid et al. (2005), a single peripheral injection of IL-2 potentiated the expression of repetitive grooming behavior associated with dopamine D1 receptor stimulation (Magid et al., 2005). Curiously, hypolocomotion associated with D2/3 receptor stimulation was exaggerated by IL-2 treatment. Based on this study, the following conclusions can be made: (1) the nature of IL-2's effects on activity vary with the neurotransmitter receptors stimulated in temporal congruity with IL-2 administration; (2) IL-2-dopamine receptor interactions are subtype specific; and (3) the modulatory effects of IL-2 cannot be predicted based on the cytokine's behavioral effects alone.

The effects of IL-2 on stereotypic behavior are related to the chronicity of IL-2 administration. Zalcman (2002) showed that repeated, but not single injections of IL-2 increase the expression of stereotypical climbing activity. These effects are mediated by dopamine D1 and D2 receptor mechanisms. A chronic treatment regimen of IL-2 (a daily injection for 5 days) was also shown to increase the expression of repetitive rearing/sniffing behavior (Hassanain et al., 2006). It should be noted that although locomotion was persistently activated throughout the treatment regimen, it was not progressively increased by repeated injections of IL-2 (Zalcman, 2001). Consistent with this finding, Song and Leonard (1995) found that increases in horizontal and vertical activity evident in olfactory bulbectomized rats were unaffected by repeated injections of IL-2. However, these investigators also showed that increased grooming and defecation were attenuated by repeated IL-2 treatment. It remains to be determined whether IL-2 reversed these effects through actions on relevant neural systems or by affecting emotional behavior. Lacosta et al. (1999), moreover, found that modest decreases in exploration and investigation of a novel stimulus were induced by a 7-day treatment IL-2 regimen, but not by single injections of IL-2. As discussed earlier, various factors could have contributed to differences across studies in the direction of IL-2's effects, including doses used, duration of treatment, and test conditions.

It is important to note that chronic IL-2 treatment induces a long-lasting change in sensitivity to behavior-activating effects induced by psychostimulant challenge. Zalcman (2001) showed that increases in rearing behavior following an injection of GBR 12909 were significantly augmented in mice with a history of IL-2 treatment. This finding implies that IL-2 changed the sensitivity of brain dopaminergic neurons to subsequent stimulation.

4.2 Reward

Inasmuch as IL-2 is a potent modulator of neural activity in the mesolimbic and mesostriatal systems, it is reasonable to hypothesize that it influences reward processes. There are several studies that provide support for this hypothesis. For example, using a drug discrimination paradigm, Ho et al. (1994) showed that the ability of rats to discriminate between amphetamine and saline (and thus the amount of reinforcement received) was potentiated by IL-2 pretreatment. This effect was blocked when IL-2 was co-administered with naloxone. IL-2 also enhanced discrimination between saline and the opioid agonist ethylketocyclazocine. A link between IL-2 and opiates was also provided by Gu et al. (2005), who showed that IL-2 or pcDNA3-IL-2 attenuates morphine withdrawal syndrome.

Dopamine release and membrane conductance in the VTA and nucleus accumbens are potently modulated by IL-2. It may thus be predicted that IL-2 also modulates hedonic responses associated with stimulation of these regions. In support of this idea, IL-2 induced a modest reduction in rates of responding for intracranial self-stimulation (ICSS) from the medial forebrain bundle (Anisman et al., 1996). Of particular interest, responding for ICSS was significantly reduced in animals tested 1–7 days later (in the absence of IL-2), suggesting that IL-2's effects on responding for rewarding brain stimulation are long lasting. I.c.v administration of IL-2 likewise resulted in marked decreases in rates of responding for ICSS elicited from the ventral tegmental area (VTA; Hebb et al., 1998; Miguelez et al., 2004). In the study by Miguelez et al., IL-2's effects were also long lasting in that the increased thresholds were sustained over a subsequent one-month period of testing. In contrast with IL-2-induced alterations in rewarding brain stimulation,

IFNα had no effect (Kentner et al., 2007). Paralleling findings with IL-2, rates of responding for ICSS elicited from the nucleus accumbens are significantly reduced at the time of the peak antibody response to a T cell-dependent antigen (Zacharko et al., 1997).

4.3 Cognition

A consistent feature of chronic IL-2 immunotherapy, which is used to treat cancer, acquired immune deficiency syndrome (AIDS), and hepatitis C, is cognitive impairment (Meyers and Valentine, 1995; Rothermundt et al., 2001). In rodents, IL-2 treatment induces performance deficits in cognitive tasks. For example, IL-2 potentiates the amnesic effects of a cholinergic M1 receptor antagonist (Bianchi and Panerai, 1993). Paralleling these findings, chronic IL-2 treatment induces performance deficits in the Morris Water Maze (Hanisch et al., 1997; Lacosta et al., 1999), suggesting that IL-2 influences spatial memory and hippocampal activity. Deletion of the IL-2 gene results in similar deficits and in abnormal development of the hippocampus (Petitto et al., 1999). Based on these findings, Petitto and colleagues made an intriguing suggestion, namely that IL-2 is required for the development of normal cognitive function (Petitto et al., 1999; see "Interleukin-2 and Septohippocampal Neurons: Neurodevelopment and Autoimmunity" Chapter in this book). Kipnis et al. (2004), further suggested that peripheral T cells are required for normal cognitive function since cognitive deficits evident in mice lacking mature T cells are reversed by T cell restoration.

4.4 Aggression

A link between IL-2 and aggressive behavior has been made in several species. For example, Petitto et al. (1994) showed that the production of Th1 cytokines, including IL-2 and IFNγ, are elevated in mice bred for high aggression. In patient populations, increases in hostility/aggression and irritability develop during a chronic regimen of IL-2 therapy (Capuron et al., 2004).

In a series of studies, we characterized the effects of IL-2 on feline aggression, and identified neuroanatomical substrates and underlying neurotransmitter receptor mechanisms (see "Cytokines and Aggressive Behavior" Chapter in this book by Siegel et al.; Zalcman and Siegel, 2006). We found that microinjections of IL-2 into the medial hypothalamus potently inhibit periaqueductal gray (PAG)-elicited defensive rage behavior (Bhatt et al., 2005). This effect is mediated by IL-2-receptor and $GABA_A$-receptor mechanisms. In light of the reciprocal feedback loop between the medial hypothalamus and PAG, it was also determined whether microinjections of IL-2 into the PAG modulate defensive rage behavior (Bhatt and Siegel, 2006). Results showed that IL-2 in the PAG facilitates defensive rage, and that this occurs through IL-2-receptor and NK_1 receptor mechanisms. In contrast with its effects in the medial hypothalamus, a $GABA_A$ receptor antagonist did not block IL-2's effects in the PAG. Subsequent analyses revealed that $GABA_A$ receptors mediate IL-2's effects on defensive rage behavior only when the epsilon subunit of the $GABA_A$ receptor is absent.

These studies underscore an intriguing feature of the relationship between IL-2 and behavior: IL-2 may facilitate or inhibit a given behavior depending on the brain region into which it is microinjected. Of further importance, IL-2 interacts with neurotransmitter systems in highly selective manners. This profile is cytokine-specific: in contrast with IL-2's effects, IL-1β suppresses defensive rage through a 5-HT_2-receptor mechanism whether it is microinjected into the medial hypothalamus or PAG (Hassanain et al., 2003, 2005; Bhatt et al., 2008).

4.5 Anxiety

Behavioral responses in the elevated plus maze, which is used as an animal model of anxiety, are unaffected by single peripheral injections of IL-2 (Petitto et al., 1997; Connor et al., 1998). However, repeated i.c.v injections of IL-2 reverse anxiety-like behavior otherwise seen in olfactory bulbectomized rats (Song and Leonard, 1995). The finding that rats exhibiting low anxiety had increased levels of IL-2 mRNA in the prefrontal cortex (Pawlak et al., 2005) would seem to be consistent with the latter effect of injected IL-2. In contrast with these findings, however, increases in IL-2 mRNA in the ventral striatum are evident in high anxiety rats, suggesting that the relationship between IL-2 and anxiety-like behavior is region-specific (Pawlak et al., 2003).

4.6 Feeding and Anorexia

Inflammatory cytokines are known to influence appetitive behavior and induce anorectic effects. For example, peripheral or central administration of IL-1β, IL-6, or TNFα induces marked reductions in the consumption of standard laboratory chow and palatable substances (Dantzer et al., 1999; Dunn, 2001; Plata-Salaman, 2001; Merali et al., 2003; Asarian and Langhans, 2005). Such effects are evident under pathological (e.g., after endotoxin challenge) and physiological conditions. A link between inflammatory cytokines and anorexia has also been made (e.g., Geary et al., 2004; Lennie, 2004). There is evidence that IL-2 influences feeding and induce anorectic effects. For example, weight gain otherwise seen over the course of a 30-day observation period is reduced in rats treated with IL-2 prior to the observation period (Miguelez et al., 2004). IL-2 treatment also results in a decrease in consumption of a palatable substance (Sudom et al., 2004). Moreover, there is evidence for a link between genetic obesity and physiological effects of IL-2. In particular, i.c.v administration of IL-2 induces a moderate increase in body temperature in lean but not obese rats (Plata-Salaman, 2001). A reduction in food consumption and subsequent weight loss is evident in mice deficient in the IL-2 gene (Gaetke et al., 2002). It appears that this effect is related to abnormal leptin responses. The extent to which this reflects an interactive effect between IL-2 and leptin in regulating feeding behavior or whether effects are indirectly related to deletion of the IL-2 gene remain to be determined.

4.7 *Relevance of Behavior Activating Effects of Interleukin-2 to Sickness Behavior*

Systemic injections of endotoxin, or certain inflammatory cytokines (notably IL-1β) induce an increased expression of depressive-like behaviors that comprise the classic symptoms of sickness behavior (Dantzer et al., 1999). These behaviors serve adaptive purposes during an immune response. As discussed, IL-2 induces transient increases in exploratory motor activity and stereotypical behaviors. Zalcman et al. (1998) suggested that during an orchestrated immune response, such effects could serve the adaptive role of activating a transiently depressed individual that would be vulnerable to predators. However, abnormal or protracted increases in IL-2 may result in abnormal behavioral responses and psychiatric abnormalities. The fact that various behavioral abnormalities and cognitive deficits develop with repeated injections of IL-2 is consistent with this idea.

5 Interleukin-2 and Psychiatric Disorders

The fact that IL-2 induces a range of behavioral changes, including behavior activating and depressive-like effects, raises the possibility that it is involved in a variety of psychiatric abnormalities. In the present section we will discuss the evidence implicating IL-2 and soluble IL-2 receptors in various psychopathological outcomes.

5.1 *Schizophrenia*

Many investigators have reported that the ability of circulating lymphocytes to produce IL-2 in vitro is reduced in schizophrenic patients (e.g., Villemain et al., 1989; Ganguli et al., 1995; Bessler et al., 1995; Rothermundt et al., 2001; Hinze-Selch and Pollmächer, 2001; Potvin et al., 2008). However, there are disparities in the literature regarding this effect. For example, other investigators observed that IL-2 production is either increased (e.g., Cazzullo et al., 2001), or unaltered in patients (e.g., Wilke et al., 1996). It is tempting to suggest that disparities in the literature could be due to differences in patient characteristics and methodological issues across studies. However, after conducting an exhaustive review of the literature, Hinze-Selch and Pollmächer (2001) concluded that such differences do not readily explain these disparities. Nonetheless, they suggested that greater attention should be focused on diagnostic subgroups, duration of medication, and the role of other confounding variables (e.g., bi-directional interactions between the CNS and immune system and cigarette smoking). The authors also emphasized that there is a need for more hypothesis driven studies.

Regarding IL-2 concentrations, there are reports linking plasma IL-2, dopamine, and symptom expression in schizophrenic patients. In one study, elevated plasma levels of IL-2 and homovanillic acid (HVA), a metabolite of dopamine, were found to coincide with increased symptom expression. Treatment with haloperidol, a typical antipsychotic, attenuated the increases in plasma IL-2 and HVA (Kim et al., 2000). In another study, increased serum IL-2 concentrations were reversed by treatment with haloperidol or risperidone, an atypical antipsychotic (Zhang et al., 2004). Despite such findings, a meta-analysis of 62 cross-sectional studies found no significant effect size for IL-2 levels (Potvin et al., 2008).

It should be mentioned that the main objective of the meta-analysis conducted by Potvin and colleagues was to confirm the hypothesis that an imbalance in T helper cell subsets, with a predominance of Th2 cells, is evident in schizophrenic patients (Schwarz et al., 2001). The analysis did not provide support for this hypothesis. Nonetheless, significant effect sizes were found for serum levels of soluble IL-2 receptors (see later in this section), IL-1 receptor antagonist, and IL-6, a cytokine that modulates brain dopamine and serotonin activity, and that potentiates novelty- and psychostimulant-induced behavioral changes (Zalcman et al., 1994a, 1998, 1999). It was further suggested that serum IL-6 plays a role in the disease process (see also Chapter by Muller and Schwarz in this book; see Schwarz et al., 2001).

In most studies examining the relationship between IL-2 and schizophrenia, immunological analyses understandably involve serum samples. Relatively few investigators have examined the IL-2-schizophrenia relationship using samples obtained from the cerebrospinal fluid (CSF). In one such study, Licinio et al. (1993) found that IL-2 levels in the CSF were elevated in a cohort of neuroleptic-free schizophrenic patients (Licinio et al., 1993), suggesting that IL-2 plays a role in the disease process. Additional support for this notion stems from the finding that CSF levels of IL-2 predict the expression of psychotic symptoms in patients (McCallister et al., 1995). Of further importance, concurrent reductions in CSF levels of IL-2 and expression of psychotic symptoms were (a) evident following treatment with haloperidol and (b) increased following withdrawal from haloperidol (McCallister et al., 1995). Paralleling these findings, the antipsychotic drugs chlorpromazine and flupentixol were found to inhibit IL-2 release from glial cell cultures (Kowalski et al., 2000). In contrast with findings showing a positive relationship between IL-2 levels in the CSF and symptom expression, others failed to confirm such findings (el-Mallakh et al., 1993; Barak et al., 1995; Rapapport et al., 1997).

Soluble IL-2 receptors (sIL-2R) are found in the circulation after being shed from activated T cells (Hornberg et al., 1995). Thus, an increase in sIL-2R is thought to reflect a previous T cell activation. A variety of investigators found that levels of the alpha subunit (sIL-2Rα) are increased in schizophrenic patients (e.g., Ganguli and Rabin, 1989; Rapapport et al., 1997; Gaughran et al., 1998; Akiyama, 1999; Sirota et al., 2005). Mitogen-stimulated increases in sIL-2Rα have also been reported (Kowalski et al., 2000). The latter effect was attenuated by haloperidol treatment. In contrast with these findings, one study found that CSF levels of sIL-2Rα are decreased in patients (Barak et al., 1995).

A meta-analysis revealed significant effect sizes for circulating levels of sIL-2Rα in schizophrenic patients (Potvin et al., 2008). A sub analysis further showed that sIL-2Rα concentrations were altered in European and North American patients receiving antipsychotic medication, suggesting that the increased concentrations were not associated with the disease process per se. However, it should be considered that sIL-2Rα concentrations may vary among subgroups of medicated patients. Rapapport and colleagues provided support for this notion by showing that serum levels of sIL-2Rα are increased in medicated patients with tardive dyskinesia but not in medicated patients without tardive dyskinesia (Rapapport and Lohr, 1994). A link between IL-2 and motor abnormalities was additionally seen by Rapapport et al. (1997) who showed that concurrent increases in serum sIL-2Rα levels and muscle force instability were evident in a subgroup of schizophrenic patients (Rapapport et al., 1997), including those who were neuroleptic-naïve. Parenthetically, these findings are consistent with those showing that IL-2 may induce motor abnormalities in animal models (see Section 4 of this chapter). These findings also underscore the need to identify patient subgroups when considering the complex relationship between IL-2 and schizophrenia.

It is important to consider that many of the findings regarding the relationship between cytokines and schizophrenia are derived from cross-sectional studies. It ought to be considered that failure to demonstrate a link between a cytokine measured at one time point and a psychopathological outcome does not preclude a role for that cytokine in the development or expression of that outcome. Indeed, cytokines are potent and tightly regulated, and often have relatively short half-lives. However, cytokines may also produce relatively long-lasting increases in the sensitivity of neural and behavioral responses to subsequent and seemingly different challenges. For example, as discussed earlier, in rodents IL-2 treatment produces long-lasting increases in sensitivity to psychostimulant challenge. Thus, individuals with a history of abnormal increases in IL-2 could be more vulnerable to the behavioral consequences of a later challenge. It is also relevant here that IL-2 potently modulates neurotransmitter activity in mesocorticolimbic and mesostriatal structures during neurodevelopmental periods. Thus, alterations in IL-2 activity during such periods could have profound long-lasting effects on the development of these systems and in turn, on behaviors associated with these systems. This is of unique interest since schizophrenia is presumed to be a developmental disorder.

In summary, there are links between IL-2/sIL-2Rα and schizophrenia. However, there is controversy regarding the nature of this relationship. Two common findings are that in vitro production of IL-2 is reduced and that circulating sIL-2Rα levels are increased in schizophrenic patients. A meta-analysis of cross-sectional studies concluded that increased levels of serum sIL-2Rα are associated with antipsychotic medications, and not with the disease process per se. It should be noted, however, that the finding of increased serum levels of sIL-2Rα in a subgroup of patients exhibiting motor abnormalities underscores the need to determine whether abnormal expression of IL-2/sIL-2R is associated with a specific subgroup(s) of schizophrenic patients. In CSF, a link between IL-2 levels, dopamine activity, and symptom expression was observed by some investigators but not others. As suggested by Hinze-Selch and Pollmächer (2001), greater attention on patient characteristics and generation of hypothesis driven studies could shed important light on the complex relationship between IL-2 and schizophrenia.

5.2 Depression

Inflammatory cytokines are implicated in the etio-pathogenesis of depressive illness (see Chang et al., "Cytokine-Induced Sickness Behavior and Depression" Chapter in this book; see also Schleifer, "Immunity and Depression: A Clinical Perspective" Chapter in this book; Dantzer et al., 1999; Anisman and Merali, 2002; Yirmiya et al., 2000). Although much of the research has focused on links between IL-1, IL-6, and TNFα and depressive illness, there is evidence implicating IL-2 and sIL-2R. For example, Maes et al. (1995) showed that plasma levels of sIL-2R were increased in a cohort of depressed patients. A positive correlation was also found between sIL-2R and the tension-anxiety subscale and the HAM-D in a cohort of depressed patients (Kagaya et al., 2001). However, a negative correlation was found between IL-2 receptor-mediated blastoformation and the severity of depression in another study (Kanba et al., 1998). Of further interest was the finding that sIL-2R significantly correlated with levels of IL-6 and with the transferrin receptor (TfR). A similar link between sIL-2R and TfR was found in patients suffering from obsessive-compulsive disorder (Maes et al., 1994). It should be mentioned, however, that sIL-2R levels are not typically altered in minor psychiatric disorders (Tsai et al., 2001).

An increase in sIL-2R was also evident in individuals attempting suicide (Nässberger and Träskman-Bendz, 1993), although such increases do not necessarily imply a link between sIL-2R and depression per se. Along the same lines, decreased levels of serum IL-2 were evident in survivors of traumatic events. Reductions in mitogen-induced IL-2 production were also observed in a cohort of anxious patients (Arranz et al., 2007).

Perhaps the most compelling evidence linking IL-2 with depression stems from psychiatric evaluations of patients receiving IL-2 therapy (alone or in combination with IFNα; Capuron et al., 2004). A common side effect is the development of clinical depression, although other psychiatric abnormalities also develop in these patients. In one study, depressive symptoms developed within 3–5 days of onset of IL-2/IFNα therapy coincident with a decrease in serum levels of dipeptidyl peptidase IV, which is an enzyme that catalyzes the cleavage of certain peptides and cytokines (including IL-2) and thus, modulates their production (Maes et al., 2001). Thus, it is possible that the increase in depressive symptoms were mediated through effects on other cytokines.

5.3 Bipolar Disorder

There is evidence for a link between IL-2 and bipolar mania. Maes et al. (1995) reported that sIL-2R levels were increased in manic patients compared with healthy controls. Consistent with this finding, Tsai et al. (2001) found that relative to controls, plasma sIL-2R levels were increased in acute mania but not in remitted patients. In a study conducted by Breunis and colleagues et al., circulating levels of activated T cells and sIL-2R were similarly found to be increased in euthymic, manic, and depressed patients, suggesting that increased T cell activity may be a trait marker in bipolar disorder (Breunis et al., 2003). Of further importance and in line with the findings of Maes et al. and Tsai et al., sIL-2R levels were higher in manic patients than in depressed bipolar patients.

5.4 Psychiatric Abnormalities Associated with IL-2 Therapy

Psychiatric abnormalities are evident in patients receiving IL-2 therapy, which is used to treat hepatitis C, cancer, and AIDS. Indeed, a range of psychopathological outcomes, including major depressive disorder, cognitive impairment, anxiety, schizophrenic-like behavior, and psychosis have been observed in patients receiving chronic administration of IL-2 alone or in combination with IFNα (e.g., Denicoff et al., 1987; Ellison et al., 1990; Pizzi et al., 2002; Capuron et al., 2004; Pizzi et al., 2002; Meyers and Valentine, 1995). A study by Pizzi et al. (2002) exemplifies the range of psychiatric abnormalities associated with IL-2 therapy. In this study, significantly increased scores on the Minnesota Multiphasic Personality Inventory (MMPI) were evident in scales of schizophrenia and psychopathic deviate, conversion hysteria, depression, and psychasthenia (Pizzi et al., 2002). The psychiatric abnormalities associated with IL-2 therapy are thought to be a direct consequence of IL-2 administration since they are not typically evident prior to onset of therapy, and disappear after termination of treatment. The specific psychiatric abnormalities that are expressed in a given patient are likely related to a variety of factors, including doses used, duration of treatment, immunological status, existence of co-morbid illnesses, and genetic factors, among other factors.

6 A Potentially Novel Immune-Brain Messenger: Behavior-Activating Effects Induced by the Soluble IL-2 Receptor

The IL-2 receptor is normally expressed on several cell types, including T cells, B cells, natural killer cells, and monocytes (Fulop et al., 1991). It is composed of three subunits (α, β, and γ). The α subunit confers individuality to IL-2, while the β and γ subunits are shared with other cytokines (notably IL-15; Ellery and Nicholls, 2002). After being shed from activated T cells and released into the circulation, sIL-2R binds IL-2. Increases in sIL-2R are evident in various disorders involving T cell activation, including autoimmune and neoplastic diseases (Suh and Kim, 2008; Bien and Balcerska, 2008). Thus, it has been suggested (e.g., Suh and Kim, 2008) that circulating levels of sIL-2R may be used as diagnostic markers, and perhaps as predictors of disease activity.

In recent years, it has become apparent that soluble cytokine receptors have additional biological functions. For example, the soluble TNFR1 helps to modulate the transport of substances across the blood–brain barrier (Taylor and Pollard, 2007). Such findings raise the possibility that a soluble cytokine receptor may affect brain function, and thus, play a role in the disease process. In view of the findings that serum levels of sIL-2Rα are increased in manic patients and in a subgroup of schizophrenics exhibiting motor abnormalities, we used an animal model to test the hypothesis that administration of a purified form of sIL-2R stimulates motor activity.

In a series of experiments, male young adult Balb/c mice received single subcutaneous injections of saline or a purified form of a given sIL-2R subunit. Immediately thereafter, the mice were individually placed into a test arena (Coulbourn TruScan activity arenas, Coulbourn Inc.) for 2 h. The test sessions were also filmed with a VHS camera and scored by an experienced rater at a later date. As illustrated in Fig. 2, sIL-2Rα induced marked increases in the number of rearing and grooming episodes compared with controls. Locomotion was also increased as a function of sIL-2Rα treatment (data not shown). As seen in Fig. 3, sIL-2Rβ induced significant increases in the number of rearing episodes and stereotypic movements, as well as other measures of motor activity (data not shown). Grooming behavior was reduced in mice treated with sIL-2Rβ. In contrast with findings in sIL-2Rα and sIL-2Rβ-treated mice, an injection of sIL-2Rγ did not appreciably affect these behavioral measures.

In summary, we have discovered that single injections of sIL-2R stimulate motor activity and increase the expression of stereotypic behaviors. Remarkably, sIL-2R-induced behavioral changes occur in a subunit-specific manner. The nature of these behavioral changes implies that sIL-2Rs influence activity in the mesocorticolimbic and mesostriatal systems. Thus, we suggest that sIL-2Rs act as messengers between the immune system and the brain. Inasmuch as sIL-2Rs are increased in certain disorders involving repetitive motor stereotypies, it should be considered that they act as etiological agents in such disorders.

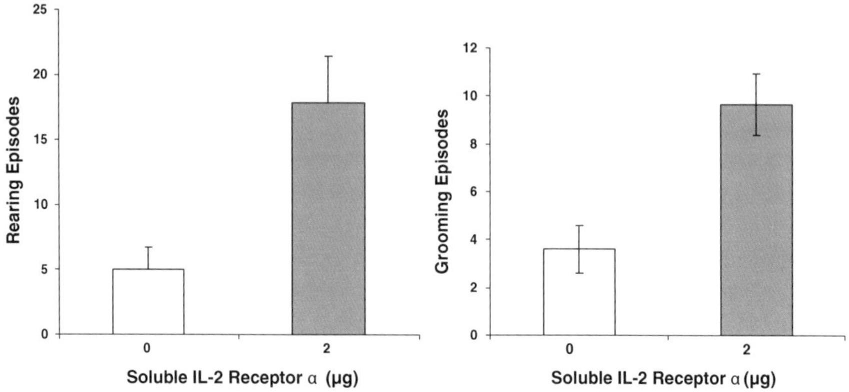

Fig. 2 Effects of single subcutaneous injection of soluble interleukin (sIL)-2Rα on motor activity in Balb/c mice. Compared with controls, sIL-2Rα-treated mice displayed significant increases in the number of rearing episodes (*left panel*) and grooming episodes (*right panel*) (p's<.05)

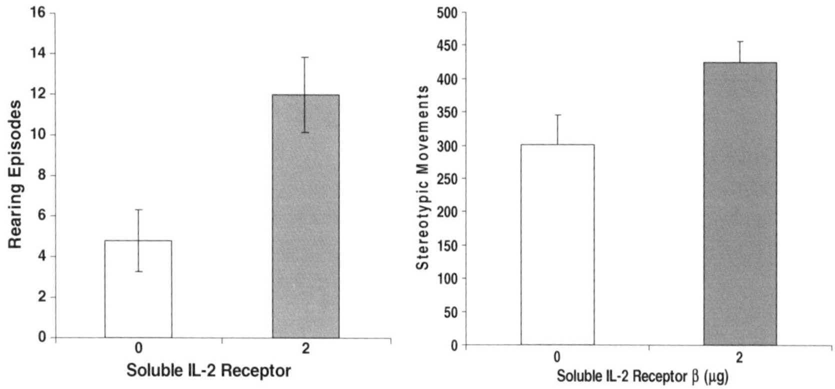

Fig. 3 Effects of single subcutaneous injection of soluble interleukin (sIL-2Rβ) on motor activity in Balb/c mice. Compared with controls, sIL-2Rβ-treated mice displayed significant increases in the number of rearing episodes (*left panel*) and stereotypic movements (*right panel*) (p's < .05)

7 Concluding Remarks

It has become abundantly clear that in addition to regulating cell-mediated responses, IL-2 is a potent modulator of activity in brain regions associated with mood and motivation. Curiously, unlike IL-1, which is the classically and most studied proinflammatory cytokine, IL-2 induces a more restricted range of central neurochemical changes. These effects notably include potent modulation of mesolimbic and mesostriatal dopaminergic processes and hippocampal acetylcholine release. Of further significance, and perhaps not surprising given its neurochemical profile, IL-2 induces a range of behavioral alterations, including behavior-activating and depressive-like effects. Of further importance, behavioral effects of IL-2 are brain-region specific. IL-2 and sIL-2Rα are also implicated in a number of different psychiatric disorders, including schizophrenia, depression, and bipolar disorder, among others. In recent studies in our laboratory, we have discovered that injections of purified forms of the sIL-2R α or β subunit stimulate motor activity and increase expression of stereotypic behaviors in mice. These effects occur in a receptor subunit-specific manner. More research is needed to help shed light on IL-2's complex modulatory effects on neurotransmission and behavior. Alongside clinical studies, such work could help provide us with a greater understanding of IL-2's role in psychiatric disorders and lead to the development of effective first line and/or adjunctive therapies.

References

Akiyama K (1999) Serum levels of soluble IL-2 receptor alpha, IL-6 and IL-1 receptor antagonist in schizophrenia before and during neuroleptic administration. Schizophr Res 37(1):97–106.

Alonso R, Chaudieu I, Diorio J, Krishnamurthy A, Quirion R, Boksa P (1993) Interleukin-2 modulates evoked release of [3H]dopamine hi rat cultured mesencephalic cells. J Neurochem 61:1284–90.

Anisman H, Kokkinidis L, Merali, Z (1996) Interleukin-2 decreases accumbal dopamine efflux and responding for rewarding lateral hypothalamic stimulation. Brain Res 731:1–11.

Anisman H, Merali Z (2002) Cytokines, stress, and depressive illness. Brain Behav Immun 16(5):513–24.

Araujo DM, Lapchak PA, Collier B, Quirion R (1989) Localization of interleukin-2 immunoreactivity and interleukin-2 receptors in the rat brain: interaction with the cholinergic system. Brain Res 498:257–66.

Arranz L, Guayerbas N, De la Fuente M (2007) Impairment of several immune functions in anxious women. J Psychosom Res 62(1):1–8.

Asarian L, Langhans W (2005) Current perspectives on behavioural and cellular mechanisms of illness anorexia. Int Rev Psychiatry 17(6):451–9.

Awatsuji DM, Furukawa Y, Nakajima M, Furukawa S, Hayashi K (1993) Interleukin-2 as a neurotrophic factor for supporting the survival of neurons cultured from various regions of fetal rat brain. J Neurosci Res 35:305–11.

Banks WA (2005) Blood-brain barrier transport of cytokines: a mechanism for neuropathology. Curr Pharm Des 11(8):973–84.

Banks WA, Niehoff ML, Zalcman SS (2004) Permeability of the mouse blood-brain barrier to murine interleukin-2: predominance of a saturable efflux system. Brain Behav Immun 18:434–42.

Barak V, Barak Y, Levine J, Nisman B, Roisman I (1995) Changes in interleukin-1 beta and soluble interleukin-2 receptor levels in CSF and serum of schizophrenic patients. J Basic Clin Physiol Pharmacol 6(1):61–9.

Bardo MT, Donohew RL, Harrington NG (1996) Psychobiology of novelty seeking and drug seeking behavior. Behav Brain Res 77:23–43.

Bartholomew SA, Hoffman SA (1993) Effects of peripheral cytokine injections on multiple unit activity in the anterior hypothalamic area of the mouse. Brain Behav Immun 7(4):301–16.

Besedovsky HO, del Rey, A (1996) Immune-neuro-endocrine interactions: facts and hypotheses. Endoc Rev 17:64–102.

Bessler H, Levental Z, Karp L, Modai I, Djaldetti M, Weizman A (1995) Cytokine production in drug-free and neuroleptic-treated schizophrenic patients. Biol Psychiatry 38(5):297–302.

Bhatt S, Siegel A (2006) Potentiating role of interleukin 2 (IL-2) receptors in the midbrain periaqueductal gray (PAG) upon defensive rage behavior in the cat: role of neurokinin NK(1) receptors. Behav Brain Res 167(2):251–60.

Bhatt S, Bhatt R, Zalcman SS, Siegel A (2008) Role of IL-1 beta and 5-HT2 receptors in midbrain periaqueductal gray (PAG) in potentiating defensive rage behavior in cat. Brain Behav Immun 22(2):224–33.

Bhatt S, Zalcman S, Hassanain M, Siegel A (2005) Cytokine modulation of defensive rage behavior in the cat: role of GABAA and IL-2 receptors in the medial hypothalamus. Neuroscience 133(l):17–28.

Bianchi M, Panerai AE (1993) Interleukin-2 enhances scopolamine-induced amnesia and hyperactivity in the mouse. Neuroreport 4(8):1046–8.

Bien E, Balcerska A (2008) Serum soluble interleukin 2 receptor alpha in human cancer of adults and children: a review. Biomarkers 13(1):1–26.

Breunis MN, Kupka RW, Nolen WA, Suppes T, Denicoff KD, Leverich GS, Post RM, Drexhage HA (2003) High numbers of circulating activated T cells and raised levels of serum IL-2 receptor in bipolar disorder. Biol Psychiatry 53(2):157–65.

Capuron L, Ravaud A, Miller AH, Dantzer R (2004) Baseline mood and psychosocial characteristics of patients developing depressive symptoms during interleukin-2 and/or interferon-alpha cancer therapy. Brain Behav Immun 18:205–13.

Cazzullo CL, Sacchetti E, Galluzzo A, Panariello A, Colombo F, Zagliani A, Clerici M (2001) Cytokine profiles in drug-naive schizophrenic patients. Schizophr Res 47(2–3):293–8.

Connor TJ, Song C, Leonard BE, Merali Z, Anisman H (1998) An assessment of the effects of central interleukin-1 P, -2, -6 and tumor necrosis factor-o, administration on some behavioral, neurochemical, endocrine and immune parameters in the rat. Neuroscience 84:923–55.

Dantzer R, Aubert A, Bluthe RM, Gheusi G, Cremona S, Laye S, Konsman JP, Parnet P, Kelley KW (1999) Mechanisms of the behavioural effects of cytokines. Adv Exp Med Biol 461:83–105.

Denicoff KD, Rubinow DR, Papa MZ, Simpson C, Seipp CA, Lotze MT, Chang AE, Rosenstein D, Rosenberg SA (1987) The neuropsychiatric effects of treatment with interleukin-2 and lymphokine-activated killer cells. Ann Intern Med 107:293–300.

De Sarro D, Masuda Y, Asciotti C, Audino MG, Nistico G (1990) Behavioural and EcOG spectrum changes induced by intracerebral infusion of interferons and interleukin 2 in rats is antagonized by naloxone. Neuropharmacology 29:167–79.

Dunn AJ (2001) Effects of cytokines and infections on brain neurochemistry. In: Psychoneuroimmunology, 3rd Edition, Ader, R., Felten, D.L., Cohen, N. (Eds.), Academic Press, San Diego, pp. 649–686.

el-Mallakh RS, Suddath RL, Wyatt RJ (1993) Interleukin-1 alpha and interleukin-2 in cerebrospinal fluid of schizophrenic subjects. Prog Neuropsychopharmacol Biol Psychiatry 17(3):383–91.

Ellery JM, Nicholls PJ (2002) Possible mechanism for the alpha subunit of the interleukin-2 receptor (CD25) to influence interleukin-2 receptor signal transduction. Immunol Cell Biol 80(4):351–7.

Ellison MD, Krieg RJ, Povlishock JT (1990) Differential central nervous system responses following single and multiple recombinant interleukin-2 infusions. J Neuroimmunol 28(3):249–60.

Fülöp T Jr, Utsuyama M, Hirokawa K (1991) Determination of interleukin 2 receptor number of Con A stimulated human lymphocytes with aging. J Clin Lab Immunol 34(1):31–6.

Gaetke LM, Oz HS, de Villiers WJ, Varilek GW, Frederich RC (2002) The leptin defense against wasting is abolished in the IL-2-deficient mouse model of inflammatory bowel disease. J Nutr 132(5):893–6.

Ganguli R, Brar JS, Chengappa KR, DeLeo M, Yang ZW, Shurin G, Rabin BS (1995) Mitogen-stimulated interleukin-2 production in never-medicated, first-episode schizophrenic patients. The influence of age at onset and negative symptoms. Arch Gen Psychiatry 52(8):668–72.

Ganguli R, Rabin BS (1989) Increased serum interleukin 2 receptor concentration in schizophrenic and brain-damaged subjects. Arch Gen Psychiatry 46(3):292.

Gaughran F, O'Neill E, Cole M, Collins K, Daly RJ, Shanahan F (1998) Increased soluble interleukin 2 receptor levels in schizophrenia. Schizophr Res 29(3):263–7.

Geary N, Asarian L, Sheahan J, Langhans W (2004) Estradiol-mediated increases in the anorexia induced by intraperitoneal injection of bacterial lipopolysaccharide in female rats. Physiol Behav 82:251–61.

Gu J, Yao M, Wang J, Zhou W, Yang G, Liu H, Zhang Z, Liu X (2005) Suppression of morphine withdrawal syndrome by interleukin-2 and its gene. Neuroreport 16(4):387–91.

Hanisch U-K, Neuhaus J, Rowe W, van Rossum D, Moller T, Kettenmann H, Quirion R (1997) Neurotoxic consequences of central long-term administration of interleukin-2 in rats. Neuroscience 79, 799–818.

Hanisch U-K, Quirion R (1995) Interleukin-2 as a neuroregulatory cytokine. Brain Res Rev 21:246–84.

Hanisch UK, Seto D, Quirion R (1993) Modulation of hippocampal acetylcholine release: a potent central action of interleukin-2. J Neurosci 13(8):3368–74.

Hart BL (1988) Biological basis of the behavior of sick animals. Neurosci Biobehav Rev 12(2):123–37.

Hassanain M, Bhatt S, Zalcman S, Siegel A (2005) Potentiating role of interleukin-1β (IL-1β) and IL-1β type 1 receptors in the medial hypothalamus in defensive rage behavior in the cat. Brain Res 1048(1–2):1–11.

Hassanain M, Rossi-George A, Bobbin MD, Zalcman SS (2006). Repeated Administration of Interleukin-2 Induces c-fos Expression in Highly Specific Areas within the Striatum and Septal Area. *Psychoneuroimmunology Research Society Meeting Abstracts*, Miami, FL.

Hassanain M, Zalcman S, Bhatt S, Siegel A (2003). Interleukin-1β in the hypothalamus potentiates feline defensive rage: role of serotonin-2 receptors. Neuroscience 120:227–33.

Hebb AL, Zacharko RM, Anisman H (1998) Self-stimulation from the mesencephalon following intraventricular interleukin-2 administration. Brain Res Bull 45(6):549–56.

Hinze-Selch D, Pollmächer T (2001) In vitro cytokine secretion in individuals with schizophrenia: results, confounding factors, and implications for further research. Brain Behav Immun 15(4):282–318.

Ho BT, Lu JG, Huo YY, Fan SH, Meyers CA, Tansey LW, Payne R, Levin VA (1994) Neurochemical basis of interleukin 2-modified discrimination behaviour. Cytokine 6(4):365–7.

Hornberg M, Arolt V, Wilke I, Kruse A, Kirchner H (1995) Production of interferons and lymphokines in leukocyte cultures of patients with schizophrenia. Schizophr Res 15(3):237–42.

Kagaya A, Kugaya A, Takebayashi M, Fukue-Saeki M, Saeki T, Yamawaki S, Uchitomi Y (2001) Plasma concentrations of interleukin-1beta, interleukin-6, soluble interleukin-2 receptor and tumor necrosis factor alpha of depressed patients in Japan. Neuropsychobiology 43(2):59–62.

Kanba S, Manki H, Shintani F, Ohno Y, Yagi G, Asai M (1998) Aberrant interleukin-2 receptor-mediated blastoformation of peripheral blood lymphocytes in a severe major depressive episode. Psychol Med 28(2):481–4.

Katsuki M, Kimura M, Ohta M, Otani H, Tanaka O, Yamamoto T, Nozawa-Kimura S, Yokoyama M, Nomura T, Habu S (1989) Lymphocyte infiltration into cerebellum in transgenic mice carrying human IL-2 gene. Int Immunol 1(2):214–8.

Kentner AC, James JS, Miguelez M, Bielajew C (2007) Investigating the hedonic effects of interferon-alpha on female rats using brain-stimulation reward. Behav Brain Res 177(1):90 9.

Kim YK, Kim L, Lee MS (2000) Relationships between interleukins, neurotransmitters and psychopathology in drug-free male schizophrenics. Schizophr Res 44:165–75.

Kipnis J, Cohen H, Cardon M, Ziv Y, Schwartz M (2004) T cell deficiency leads to cognitive dysfunction: implications for therapeutic vaccination for schizophrenia and other psychiatric conditions. Proc Natl Acad Sci USA 101:8180–5.

Klebb G, Autenrieth I, Haber B, Gillert E, Sadlack B, Smith KA, Horak I (1996) Interleukin-2 is indispensable for development of immunological self-tolerance. Clin Immunol Immunopathol 81:282–6.

Kowalski J, Blada P, Kucia K, Lawniczek T, Madej A, Belowski D, Herman ZS (2000) In-vitro immunornodulatory effects of haloperidol and perazine in schizophrenia. World J Biol Psychiatry 1(4):190–6.

Lacosta S, Merali Z, Anisman H (1999) Influence of acute and repeated interleukin-2 administration on spatial learning, locomotor activity, exploratory behaviors, and anxiety. Behav Neurosci 113:1030–41.
Lacosta S, Merali Z, Anisman H (2000) Central monoamine activity following acute and repeated systemic interleukin-2 administration. Neuroimmunomodulation 8(2):83–90.
Lapchak PA (1992) A role for interleukin-2 in the regulation of striatal dopaminergic function. Neuroreport 3:165–8.
Lapchak PA, Araujo DM (1993) Interleukin-2 regulates monoamine and opioid peptide release from the hypothalamus. Interleukin-2 regulates monoamine and opioid peptide release from the hypothalamus. Neuroreport . 4(3):303–6.
Lennie TA (2004) Sex differences in severity of inflammation-induced anorexia and weight loss. Biol Res Nurs 5(4):255–64.
Lewitus GM, Zhu J, Xiong H, Hallworth R, Kipnis J (2007) CD4(+)CD25(−) effector T-cells inhibit hippocampal long-term potentiation in vitro. Eur J Neurosci 26(6):1399–406.
Licinio J, Seibyl JP, Altemus M, Charney DS, Krystal JH (1993) Elevated CSF levels of interleukin-2 in neuroleptic-free schizophrenic patients. Am J Psychiatry 150:1408–10.
Maes M, Bosmans E, Calabrese J, Smith R, Meltzer HY (1995) Interleukin-2 and interleukin-6 in schizophrenia and mania: effects of neuroleptics and mood stabilizers. J Psychiatr Res 29(2):141–52.
Maes M, Capuron L, Ravaud A, Gualde N, Bosmans E, Egyed B, Dantzer R, Neveu PJ (2001) Lowered serum dipeptidyl peptidase IV activity is associated with depressive symptoms and cytokine production in cancer patients receiving interleukin-2-based immunotherapy. Neuropsychopharmacology 24(2):130–40.
Maes M, Meltzer HY, Bosmans E (1994) Psychoimmune investigation in obsessive-compulsive disorder: assays of plasma transferrin, IL-2 and IL-6 receptor, and IL-1 beta and IL-6 concentrations. Neuropsychobiology 30(2–3):57–60.
Magid J, Bobbin MD, Zalcman SS (2005). Interleukin-2 Enhances the Expression of Stereotypic Behavior Associated with Dopamine D_1 Receptor Stimulation. Society for Neuroscience Abstracts; 35th Annual Meeting, Program number 907.8.
Maier SF, Watkins LR, Nance DM (2001) Multiple routes of of action of interleukin-1 on the nervous system. In: Psychoneuroimmunology, Ader, R., Cohen, N., Felten, D.L. (Eds.), Academic Press, San Diego, pp. 563–83.
McCallister CG, Van Kammen DP, Rehn TJ, Miller AL, Gurklis J, Kelley ME, Yao J, Peters JL (1995) Increases in CSF levels of interleukin-2: effects of recurrence of psychosis and medication status. Am J Psychiatry 152:1291–7.
McCann SM, Kimura M, Karanth S, Yu WH, Mastronardi CA, Rettori V (2000) The mechanism of action of cytokines to control the release of hypothalamic and pituitary hormones in infection. Ann N Y Acad Sci 917:4–18.
Merali Z, Brennan K, Brau P, Anisman H (2003) Dissociating anorexia and anhedonia elicited by interleukin-1beta: antidepressant and gender effects on responding for "free chow" and "earned" sucrose intake. Psychopharmacology (Berl). 165:413–8.
Meyers CA, Valentine AD (1995) Neurological and psychiatric adverse effects of immunological therapy. CNS Drugs 3:56–68.
Miguelez M, Lacasse M, Kentner AC, Rizk I, Fouriezos G, Bielajew C (2004) Short- and long-term effects of interleukin-2 on weight, food intake, and hedonic mechanisms in the rat. Behav Brain Res 154(2):311–9.
Nance DM, Sanders VM (2007) Autonomic innervation and regulation of the immune system (1987–2007). Brain Behav Immun 21(6):736–45.
Nässberger L, Träskman-Bendz L (1993) Increased soluble interleukin-2 receptor concentrations in suicide attempters. Acta Psychiatr Scand 88(1):48–52.
Nistico G, De Sarro G (1991) Behavioral and electrocortical spectrum power effects after microinfusion of Lymphokines in several areas of the rat brain. Ann N Y Acad Sci 621:119–34.
Pauli S, Linthorst AC, Reul JM (1998) Tumour necrosis factor-alpha and interleukin-2 differentially affect hippocampal serotonergic neurotransmission, behavioural activity, body temperature and hypothalamic-pituitary-adrenocortical axis activity in the rat. Eur J Neurosci 10(3):868–78.

Pawlak CR, Schwarting RK, Bauhofer A (2003) Relationship between striatal levels of interleukin-2 mRNA and plus-maze behaviour in the rat. Neurosci Lett 341(3):205–8.
Pawlak CR, Schwarting RK, Bauhofer A (2005) Cytokine mRNA levels in brain and peripheral tissues of the rat: relationships with plus-maze behavior. Brain Res Mol Brain Res 137 (1–2):159–65.
Petitto JM, Lysle DT, Gariepy JL, Lewis MH (1994) Association of genetic differences in social behavior and cellular immune responsiveness: effects of social experience. Brain Behav Immun. 8(2):111–22.
Petitto JM, McCarthy DB, Rinker CM, Huang Z, Getty T (1997) Modulation of behavioral and neurochemical measures of forebrain dopamine function in mice by species-specific interleukin-2. J Neuroimmunol 73:183–90.
Petitto JM, McNamara RK, Gendreau PL, Huang Z (1999) Impaired learning and memory and altered hippocampal neurodevelopment resulting form interleukin-2 gene deletion. J Neurosci Res 56:441–6.
Pizzi C, Caraglia M, Cianciulli M, Fabbrocini A, Libroia A, Matano E, Contegiaconuv A, Del Prete S, Abbruzzese A, Martignetti A, Tagliaferri P, Bianco AR (2002) Low-dose recombinant IL-2 induces psychological changes: monitoring by Minnesota Multiphasic Personality Inventory (MMPI). Anticancer Res 22(2A):727–32.
Plata-Salaman CR (2001) Cytokines and feeding. Int J Obes Relat Metab Disord 25 Suppl 5:S48–52.
Pollmächer T, Hinze-Selch D, Mullington J, Holsboer F (1995) Clozapine-induced increase in plasma levels of soluble interleukin-2 receptors. Arch Gen Psychiatry 52(10):877–8.
Potvin S, Stip E, Sepehry AA, Gendron A, Bah R, Kouassi E (2008) Inflammatory cytokine alterations in schizophrenia: a systematic quantitative review. Biol Psychiatry 63(8):801–8.
Raber J, Koob GF, Bloom FE (1995) Interleukin-2 (IL-2) induces corticotropin-releasing factor (CRF) release from the amygdala and involves a nitric oxide-mediated signaling; comparison with the hypothalamic response. J Pharmacol Exp Ther 272(2):815–24.
Rapapport MH, Caliguiri MP, Lohr JB (1997) An association between increased soluble interleukin-2 receptors and a disturbance in motor muscle force in schizophrenic patients. Prog Neuropsychopharmacol Biol Psychiatry 21:817–27.
Rapapport MH, Lohr JB (1994) Serum soluble interleukin-2 receptors in neuroleptic naive schizophrenic patients and in medicated schizophrenic subjects with and without tardive dyskinesia. Acta Psychiatr Scand 90:311–5.
Rothermundt M, Arolt V, Bayer TA (2001) Review of immunological and immunopathological findings in schizophrenia. Brain Behav Immun 15:319–39.
Saka E, Graybiel AM (2003) Pathophysiology of Tourette's syndrome: striatal pathways revisited. Brain Dev 25 Suppl 1:815–9.
Schwarz MJ, Chiang S, Muller N, Ackenheil M (2001) T-helper-1 and T-helper-2 responses in psychiatric disorders. Brain Behav Immun 15:340–70.
Sirota P, Meiman M, Herschko R, Bessler H (2005) Effect of neuroleptic administration on serum levels of soluble IL-2 receptor-alpha and IL-1 receptor antagonist in schizophrenic patients. Psychiatry Res 134(2):151–9.
Smith KA (1988) A design approach to the structural analysis of interleukin 2. Science 240:1169–76.
Song C, Leonard BL (1995) Interleukin-2-induced changes in behavioural, neurotransmitter, and immunological parameters in the olfactory bulbectomized rat. Neuroimmunomodulation 2(5):263–73.
Song C, Merali Z, Anisman H (1999) Variations of nucleus accumbens dopamine or serotonin following systemic interleukin-1, interleukin-2 or interleukin-6 treatment. Neuroscience 88:823–36.
Sudom K, Turrin NP, Hayley S, Anisman H (2004) Influence of chronic interleukin-2 infusion and stressors on sickness behaviors and neurochemical change in mice. Neuroimmunomodulation 11(5):341–50.
Suh CH, Kim HA (2008) Cytokines and their receptors as biomarkers of systemic lupus erythematosus. Expert Rev Mol Diagn 8(2):189–98.

Tancredi V, Zona C, Velotti F, Eusebi F, Santoni A (1990) Interleukin-2 suppresses established long-term potentiation and inhibits its induction in the rat hippocampus. Brain Res 525(1):149–51.

Taylor JM, Pollard JD (2007) Soluble TNFR1 inhibits the development of experimental autoimmune neuritis by modulating blood-nerve-barrier permeability and inflammation. J Neuroimmunol 183(1–2):118–24.

Tsai SY, Yang YY, Kuo CJ, Chen CC, Leu SJ (2001) Effects of symptomatic severity on elevation of plasma soluble interleukin-2 receptor in bipolar mania. J Affect Disord 64(2–3):185–93.

Villemain F, Chatenoud L, Galinowski A, Homo-Delarche F, Ginestet D, Loo H, Zarifian E, Bach JF (1989) Aberrant T cell-mediated immunity in untreated schizophrenic patients: deficient interleukin-2 production. Am J Psychiatry 146(5):609–16.

Wilke I, Arolt V, Rothermundt M, Weitzsch C, Hornberg M, Kirchner H (1996) Investigations of cytokine production in whole blood cultures of paranoid and residual schizophrenic patients. Eur Arch Psychiatry Clin Neurosci 246(5):279–84.

Ye JH, Tao L, Zalcman SS (2001) Interleukin-2 modulates *N*-methyl-D-aspartate receptors of native mesolimbic neurons. Brain Res 894:241–8.

Ye JH, Zalcman SS, Tao L (2005) Kainate-activated currents in the ventral tegmental area of neonatal rats are modulated by interleukin-2. Brain Res 1049(2):227–33.

Yirmiya R, Pollak Y, Morag M, Reichenberg A, Barak O, Avitsur R, Shavit Y, Ovadia H, Weidenfeld J, Morag A, Newman ME, Pollmächer T (2000) Illness, cytokines, and depression. Ann N Y Acad Sci 917:478–87.

Zacharko RM, Zalcman S, Macneil G, Andrews M, Mendella PD, Anisman H (1997) Differential effects of immunologic challenge on self-stimulation from the nucleus accumbens and the substantia nigra. Pharmacol Biochem Behav 58(4):881–6.

Zalcman SS (2001) Interleukin-2 potentiates novelty- and GBR 12909-induced exploratory activity. Brain Res 899:1–9.

Zalcman SS (2002) Interleukin-2-induced increases in climbing behavior: inhibition by dopamine D-l and D-2 receptor antagonists. Brain Res 944:157–64.

Zalcman S, Green-Johnson JM, Brown A-M, Nance DM, Greenberg AH (1994b) Interleukin-2 and uncontrollable stressors interact to alter neuroendocrine and immune activity. Soc Neurosci Abstr 20:948.

Zalcman S, Green-Johnson JM, Murray L, Dyck DG, Nance DM, Anisman H, Greenberg AH (1994a) Cytokine specific central monoamine alterations induced by interleukin-1, -2, and -6. Brain Res 643:40–9.

Zalcman S, Murray L, Dyck DG, Greenberg AH, Nance DM (1998) Interleukin-2 and -6 induce behavioral activating effects in mice. Brain Res 811:111–21.

Zalcman S, Savina L, Wise RA (1999) Interleukin-6 increases sensitivity to the locomotor-stimulating effects of amphetamine in rats. Brain Res 847:276–83.

Zalcman SS, Siegel A (2006) The neurobiology of aggression and rage: role of cytokines. Brain Behav Immun 20(6):507–14.

Zhang XY, Zhou DF, Cao LY, Zhang PY, Wu GY, Shen YC (2004) Changes in serum interleukin-2, -6, and -8 levels before and during treatment with risperidone and haloperidol: relationship to outcome in schizophrenia. J Clin Psychiatr 65(7):940–7.

Part III
Neuroimmunological Basis of Mental Disorders

Immunity and Depression: A Clinical Perspective

Steven J. Schleifer

Abstract Depressive disorders, among the most common of human ailments, are a major public health challenge, including increased risk for somatic diseases. This chapter summarizes the complex observations suggesting an immune component for such medical morbidity as a consequence of two broad categories of immune change: decreased immune functions and increased inflammation. Reciprocal effects of the immune system, especially inflammation, on brain and behavior, suggest that some psychiatric disorders, including depression, may be immunologically mediated in part.

Keywords Depression · Unipolar · Bipolar · Cytokines · NKCA · Mitogens · Inflammation

Depression is among the first and most extensively studied of the clinical psychiatric syndromes in psychoneuroimmunology. Research followed studies that demonstrated immune changes as a consequence of stressful experiences (Herbert and Cohen 1993b). One such stressor, bereavement, is associated with the grieving process which is often behaviorally indistinguishable from depressive disorders. Additionally, depressive disorders are associated with hypothalamo-pituitary–adrenocortical (HPA) axis abnormalities, including hypercortisolism and resistance to feedback inhibition, which would predict a possible depression–immune link (Anisman and Merali 2002). While stimulating much general interest, early observations of decreased immune function in depression were, at times, inconsistent and attempts to develop a coherent theory of depression and immunity were then complicated by emerging evidence that depression is associated with immune activation (Smith 1991). It now appears that several distinct immune changes are associated with depression, and that inconsistencies in the literature may reflect the heterogeneity of depressive disorders as well as the effects of various depression-related behavioral and central nervous

S.J. Schleifer (✉)
Department of Psychiatry, UMDNJ – NJ Medical School, Behavioral Health Science Building, Room F-1430, 183 South Orange Avenue, Newark, NJ 07103, USA
e-mail: schleife@umdnj.edu

A. Siegel, S.S. Zalcman (eds.), *The Neuroimmunological Basis of Behavior and Mental Disorders*, DOI 10.1007/978-0-387-84851-8_14,

system (CNS) states. Finally, a more recent line of investigation has reversed the polarity of the putative depression–immunity causal link, suggesting that the immune system contributes to the pathophysiology of depression. Diseases associated with immune activation and, most provocatively, medically induced immune activation can trigger depressive states. Immune activation has, accordingly, become implicated as a cause of depressive symptoms and disorders.

This chapter will seek to identify the relevant clinical and behavioral dimensions in the relationship between depression and the immune system. It will consider depression as a disorder, as well as associated clinical factors that may also influence immunity. Despite a large and growing literature, however, the number of such factors is large and systematic research on most of the potentially relevant variables is still quite limited.

Several areas are beyond the scope of this chapter. These include the literature on depression and immunity in persons with established medical disorders, including disorders for which the pathophysiology is itself linked to the immune system (e.g., neoplasia, HIV). Generalizations from studies of otherwise healthy persons to those with such illnesses must be made with caution due to potentially important interacting effects (e.g., Ben-Eliyahu et al. 2007). The role of the psychoneuroimmunology of depression in the onset and course of medical diseases (e.g., neoplasia, infections, and inflammatory disorders such as cardiovascular disease) is also beyond the scope of this chapter. Finally, the mechanisms of the depression–immune associations, such as neuroendocrine and catecholamine processes, will not be considered systematically.

1 Depressive States and Altered Immunity

Major Depressive Disorder (MDD) is a clinically defined diagnostic entity, among the most salient of the depressive disorders (American Psychiatric Association 2000). It is manifested by a pervasive mood disturbance for at least several weeks, together with at least several physical and cognitive disturbances such as altered sleep, appetite, concentration, and energy level, as well as impaired functioning in activities of daily living. Unipolar disorders are defined by a natural history including one or many depressive episodes. They are distinct from bipolar disorders, which have manic as well as depressive episodes.

Early studies that examined associations of depression with immunity focused on the readily measurable, mostly nonspecific, immune measures. They found that patients suffering from (unipolar) MDD differed from nondepressed persons on immune measures such as the distribution of lymphocyte subsets, mitogen responses, and NK cell activity (NKCA; Schleifer et al. 1984; Irwin et al. 1987a, b; Evans et al. 1992). Later studies began to find increased levels of proinflammatory cytokines in major depression including IL-1, IL-2, IL-6, tumor necrosis factor (TNF)-α, and soluble receptors for IL-2, IL-6, and the IL-1 receptor antagonist (Anisman and Merali 2002). Meta-analytic reviews on depressive disorders and immunity were published in 1993 and 2001 (Herbert and Cohen 1993a; Zorrilla et al. 2001). Herbert and Cohen (1993a) found 35 studies comparing patients meeting criteria for syndromal depression with nondepressed persons. They found evidence for increased WBCs across the studies, with neutrophilia and lymphopenia, and decreased T cells, helper T cells, and suppressor/cytotoxic T cells. Decreased NK cells and helper/suppressor ratios were also found, although less reliably. The meta-analysis revealed decreased responses to the mitogens PHA, ConA and PWM and decreased NKCA. These associations were sustained in a more restricted, methodologically rigorous subgroup of the studies. Zorrilla et al. (2001), examining the updated literature on unipolar depression and using additional meta-analytic strategies, found similar overall effects. These included neutrophilia and lymphopenia, and decreased circulating NK and T cells. Depression was again associated with decreased mitogen responses and NKCA, and also with decreased neutrophil phagocytosis. It was also now found to be associated with immune activation, including increased CD4+cells, CD4/CD8 ratios, cells with activation markers, inflammatory cytokines (IL-6, sIL-2R), and acute phase plasma proteins. The more restrictive analytic strategy showed significant effects for lymphopenia, neutrophilia, increased CD4/CD8, increased haptoglobin, and decreased mitogen response and NKCA. The two reviews reflect what has become the general consensus: that depressive disorder, at least of the unipolar type, is associated with decrements in certain immune functions as well as increased inflammation.

The relationship between the two domains of immune change in depression has received only limited attention. There is, however, evidence that the two broad categories of effects are at least partly dissociated. Pike and Irwin (2006) found increased IL-6 and decreased NKCA in a sample of male patients with acute major depression, but found no correlation between the NKCA and inflammatory effects. Studies by the same group and others suggested that there may be different mechanisms for the functional and inflammatory effects, with corticosteroid resistance associated with inflammation but not decreased NKCA (Raison and Miller 2003). NKCA effects, in contrast, may be linked to behaviors such as tobacco use (Jung and Irwin 1999).

Several laboratories have noted another dissociation in the immune profile of MDD patients. It would be expected that the number of circulating NK cells would contribute to, and be correlated with, NKCA, and both reduced NKCA and circulating NK cells are found in depression (Zorrilla et al. 2001). This relationship, however, may be dissociated in depressed patients. Evans et al. (1992) noted that the expected correlation of NK cell numbers and function was found among nondepressed controls but not in the depressed subjects. We observed, in young adults with MDD, decreased circulating CD56+ cells and decreased NKCA, but, rather than accounting for the NKCA effect, correcting statistically for circulating NK cells enhanced the NKCA effect (Schleifer et al. 1996). In a study with depressed adolescents, which found decreased NK cells and (unexpectedly) increased NKCA, we found, similar to Evans et al. (1992), no correlation of circulating NK cells with NKCA among the depressed subjects, but the expected correlation among the nondepressed subjects (Schleifer et al. 2002). Moreover, in a study of the contribution of depression to immune change in persons with alcoholism, we found the same NK cell–NKCA dissociation for depression, but not for alcoholism (the expected correlation was found for both alcoholic and nonalcoholic persons; Schleifer et al. 2006). The evidence that altered NKCA in depression is associated with depression-related behaviors such as heavy tobacco use (Jung and Irwin

1999; Schleifer et al. 2006) provides one avenue of exploration for this dissociation. Finally, some studies have found a dissociation between depression effects for mitogens and NKCA, which in our studies appears to be age-dependent (Section 5.1).

2 The Heterogeneity of Depressive States and Immunity

A major research challenge relates to the heterogeneous and at times imprecise diagnostic status of clinical depression. The diagnostic entity is itself defined empirically, relying on symptom pattern criteria. Considering MDD as a single entity may therefore obscure differences among clinically similar but biologically heterogeneous conditions. Moreover, the syndromal phenotype often overlaps with affective states that are not considered "disorders," such as "normal grief." While many studies work with an implied qualitative difference between depressed and nondepressed persons, it is unclear which, if any, effects are linked to "MDD" per se. Immune effects may relate to core aspects of the depressive syndrome (MDD), but also to subgroups of depressive disorders marked by specific symptoms patterns, to individual symptoms within the syndrome, or to secondary behaviors such as tobacco and alcohol use.

Considering these clinical issues, two general models associating depression and immunity may be considered: (1) A syndromal model in which a biologically distinct syndrome of MDD (possibly mimicked by other phenotypes, however) is associated with a specific pattern of immune changes. The syndrome may represent a chronic and persistent configuration of the CNS, with the periodic overlay of acute depressive episodes. Both the underlying and the acute states may have immune correlates. (2) A symptom model in which the immune status of persons with MDD reflects the aggregate immune effects of individual symptoms (e.g., moods, increased sleep, decreased sleep), with no syndromal effect per se. Behaviors frequently expressed in depressed persons, such as alcohol use, may also contribute to the immune effects. Even a syndromal immune effect, moreover, would likely be modulated by effects of individual symptoms. It is also important to recognize that some clinical factors may serve as confounders of research designs. For example, a study of cytokine levels in a large number of psychiatric patients upon hospital admission found few immune differences related to diagnosis when effects for age, BMI, gender, tobacco use, recent infections, and medications (clozapine, lithium, benzodiazepines) were statistically controlled (Haack et al. 1999). In some cases, this may be consistent with a symptom model but, in others, such uncontrolled factors may not represent depression-related effects at all.

3 Immunity and the Core Syndrome of Depression

Studies, such as those considered in the meta-analyses, assessed immunity in persons meeting MDD criteria. They are consistent with an association of the core syndrome of MDD with immune change, but do not rule out the alternative model relating to the symptoms of the syndrome. There are also "depression" studies that are more ambiguous diagnostically, in which depression is measured as an aggregate of the MDD symptoms using quantitative rating scales (e.g., Beck Depression Inventory). High scores on these measures are likely to be correlated with depressive syndromes and the scales are also useful for assessing clinical severity and the course of the disorders. Such studies may also access community (vs. clinical) samples more readily. Findings in research using scaled measures have tended to be consistent with the studies of diagnosed MDD. Inflammatory markers, for example, were increased in "depressed" subgroups of community dwelling persons in several studies (Penninx et al. 2003; Suarez 2004). In contrast, a recent study found decreased inflammatory cytokine production by peripheral blood mononuclear cell (PBMC)s among midlife women with depressive symptoms (Cyranowski et al. 2007). While possibly reflecting sample differences, these differing findings may also relate to biological differences in measures of circulating cytokines vs. in vitro stimulated cytokine production (Anisman and Merali 2002; Cyranowski et al. 2007).

A recent report provides less ambiguous evidence for a link between syndromal affective disorders and at least one aspect of immunity. Padmos et al. (2008), assessing monocytes from persons with bipolar disorder (manic, depressed, or euthymic), found aberrant expression of mRNAs for genes associated with inflammation. Increased expression was also found in the subjects' offspring, both in those positive for mood disorders and in several who first manifested the disorder later in the study. Associated inflammatory proteins, however, were not increased. The investigators speculated that the dissociation of protein and mRNA expression reflects a tenuous compensatory process which is subject to disruption by stress. The stress effect may itself involve acute elevations of inflammatory cytokines, triggering a depressive or manic episode. Besides genetic–environmental interactions, there is evidence for an interaction between early life environmental factors, depression, and immunity. In a recent study, major depression in young adults was associated with increased inflammation (CRP) only in those with a history of childhood maltreatment (Danese et al. 2007).

4 Immunity and the Symptoms of Depression

Many component symptoms of MDD may be associated with immune change, although most have been studied to only a limited extent, whether in persons with MDD or in those without mood disorders.

4.1 Dysphoric Mood

Depressed or other disturbed moods have not been assessed extensively in isolation. An exploratory study using a composite measure of mood disturbance (POMS) in otherwise healthy adults found some changes in the distribution of T cells and NKT cells (Lutgendorf et al. 2005). Another study, with otherwise healthy students, found that the POMS predicted decreased delayed cutaneous hypersensitivity to primary antigen exposure, but only in those with specific HLA genotypes (Smith et al. 2005). Anxiety, also a common symptom in depression, may have effects opposing those of depression. While not studying mood in isolation, Andreoli et al. (1992) found that the presence of an anxiety disorder was associated with higher PHA responses than those of depressed patients without concurrent anxiety disorder. More recently, attention has focused on anger and hostility as symptoms of depression that are risk factors for cardiovascular disease (Frasure-Smith et al. 2007; Miller et al. 2002). Hostility has been associated with increased inflammatory cytokines in community and stressed subjects (Suarez 2004; Graham et al. 2006). One study reported that hostility in otherwise healthy men predicted increased IL-6 only in the presence of depressive symptoms (Suarez 2003). Another found that negative mood states elicited by recalling an anger-inducing experience were associated with increased cytokine secretion in those with higher insulin resistance (IR), the opposite effect occurring for lower-IR subjects (Suarez et al. 2006; see also below, Miller et al. 2003). Finally, reciprocal effects of immune processes on the CNS influence hostility. Cytokines, such as IL-1 and IL-2, modulate CNS pathways associated with aggression (Zalcman and Siegel 2006).

Other emotional dimensions associated with depression warrant attention. Anhedonia, diminished interest or pleasure, may have different biological correlates than depressed mood. Another construct, loneliness, has been associated with altered circulating NK cells (Steptoe et al. 2004). Loneliness may be related to lower levels of social support, an important modulator of the clinical course of depression, and was linked to changes in circulating NK cells, at least in older MDD subjects (Bouhuys et al. 2004). Positive emotions have also received some attention. An interaction between stressors and the trait of optimism has been associated with immune processes, but appears unrelated to depressed mood (Segerstrom 2005). Similarly, increases in inflammatory cytokines following experimentally induced viral infection were associated with diminished positive mood but no change in negative mood (Janicki-Deverts et al. 2007) and positive (but not negative) emotional style was associated with diminished IL-6 (but also IL-10) production (Prather et al. 2007).

4.2 Decreased or Increased Weight/Appetite

This symptom dimension is receiving increasing attention, with evidence that obesity and depression may interact to increase inflammatory markers such as CRP (Ladwig et al. 2003). Obesity in depression may lead to increased IL-6 and CRP both directly and as a consequence of leptin secretion (Miller et al. 2003). Reciprocally, appetite change and weight gain may induce cytokine effects that contribute to the depressive state (Andreasson et al. 2007). Potential correlations between patterns of weight change and subtypes of depression have also been reported (typical vs. atypical depression associated with weight loss vs. weight gain; Andreasson et al. 2007).

4.3 Decreased or Increased Sleep

Sleep patterns have long been recognized as important modulators of immunity (Moldofsky et al. 1989; Irwin 2002a). In a community sample, increased sleep and decreased fatigue were associated with increased NKCA (Shakhar et al. 2007). Another study found that disturbed sleep impaired subsequent NK cell mobilization in response to a mild stressor (Wright et al. 2007). Supporting the symptom model, a controlled sleep study with MDD patients found that elevated nocturnal IL-6 and adhesion molecules (sICAM) were largely attributable to the patients' sleep disturbance (Motivala et al. 2005). The behavioral dimension of fatigue warrants independent consideration since, besides being a consequence of sleep deprivation, it can reflect inflammation (e.g., sickness behavior) and is a component of complex states such as chronic fatigue syndrome (CFS), consideration of which is beyond the scope of this chapter.

4.4 Motor Activity

Depression can be associated with dramatic changes in physical behavior, including marked agitation and restlessness and its opposite, severe motor retardation or inhibition. As with other polar symptoms in depression (e.g., altered sleep or appetite), these motor states do not represent distinct syndromal entities, except that they tend to occur differentially in bipolar vs. unipolar disorders and in atypical vs. typical depression (see below). An early study found reduced β-adrenergic responsiveness of lymphocytes in depressed patients who were agitated, but not in those with motor retardation (Mann et al. 1985). In studies of motor activity, diminished physical activity was associated with reduced lymphocyte proliferation (Miller et al. 1999) and some (but not other) exercise interventions ameliorated the elevation of inflammatory cytokines (Kohut et al. 2006).

4.5 Suicidality

Suicidal ideation and behavior are potentially catastrophic concomitants of depressive disorders. It is of interest, however, that risk factors for and pathophysiologic aspects of suicidality, including serotonergic changes (associated with impulsivity), can be distinguished from the depressive states themselves. A series of postmortem brain studies found suicide to be associated with microgliosis, irrespective of whether the clinical diagnosis was depression or schizophrenia (Steiner et al. 2008).

Systematic study of peripheral as well as CNS immune processes in relation to suicidality is indicated.

4.6 Other Symptoms

Feelings of worthlessness and guilt are common in depression. One study found increased TNF-α activity in association with induced shame, but not guilt, in otherwise healthy subjects (Dickerson et al. 2004). Cognitive deficits in depression, such as diminished concentration and indecisiveness, may reflect specific CNS deficits with potential immune correlates.

5 Modulators of Immunity in Depression

5.1 Age

One meta-analytic review suggested that immune changes in depression are more prominent in older individuals (Herbert and Cohen 1993a). Our laboratory has assessed otherwise healthy untreated persons with depressive disorders across a wide age range. The studies suggested that decreased functional measures, such as mitogen response and NKCA, are more readily found in older depressed adults (Schleifer et al. 1989). A collaborative study, using similar methodology, found a correlation between increasing age and decreasing mitogen responses (Andreoli et al. 1993). Our studies further suggested that some aspects of immunity are relatively increased in depressed younger adults compared with nondepressed persons (Schleifer et al. 1989). In subsequent studies of depressed young adults and prepubertal children, a dissociation was found between NKCA, which was decreased, and mitogen responses, which were either no different from matched controls or increased (Schleifer et al. 1996; Bartlett et al. 1995). A somewhat different pattern of dissociation was found in a study of MDD adult women who showed generally decreased mitogen responses but age-related NKCA effects: NKCA decreased among older depressed women, increased among the younger subjects (Miller et al. 1999). We found a pattern similar to that of Miller et al. (1999) in still another study focusing on depressed adolescents who, unlike other groups studied in our laboratory, showed increased NKCA but possibly decreased mitogen responses (Schleifer et al. 2002).

Multiple factors could account for the various patterns of age-related findings: interactions among the aging process, depression, and immunity (e.g., neuroendocrine or catecholamine effects); symptom or syndrome profiles that vary as a function of age (e.g., the pathophysiologically distinct "vascular depression" described in older adults (Alexopoulos 2006)); and "secondary" behaviors (e.g., substance use) varying with age. Calendar age, moreover, is likely to be correlated with illness duration, which is associated with increasing clinical severity and episode frequency (Kessing 2008). In a multivariate study, age of onset of depression predicted decreased circulating NK cells and NKCA (Frank et al. 2002). Another study of MDD found increased IL-1β in late-life depression that was unrelated to age of onset but was related to duration of the current episode (Thomas et al. 2005).

5.2 Sex

At least some immune effects in depression appear to differ by sex. While our studies of lymphocyte proliferation and NKCA in major depression did not find differences (Schleifer et al. 1989), Evans et al. (1992) found that circulating NK cells and NKCA were reduced in male but not female depressed subjects. Another study suggested that elevated CRP is associated with depression in males but not females (Liukkonen et al. 2006). A population-based study similarly found that a history of depressive disorder was associated with increased CRP in men, but not women (Danner et al. 2003). In a study including only men, increased IL-6 was associated with concurrent hostility and depressive symptoms (Suarez 2003), linking depression, inflammation, and cardiovascular disease (Miller and Manji 2006). A recent large study suggested that depressive symptoms in women approaching menopause are not associated with inflammatory markers (CRP), however health risk was found to be a consequence of increased coagulability (Matthews et al. 2007). Ravindran et al. (1998) found evidence for a complex interaction among subtypes of depression (typical vs. atypical, major depression vs. dysthymia) and sex in relation to treatment-sensitive alterations in circulating NK cells.

6 Depressive Disorder Variants

6.1 Bipolar Disorder

Immunity in patients with bipolar disorder is of interest, being clinically distinct from the more common unipolar depressive disorders, and responding to different pharmacologic interventions. This syndrome has the potential to demonstrate state-related phenomena since the same person experiences mania and depression with dramatic differences in motor behavior, sleep, and appetite. Studies of bipolar disorder can be especially difficult in mania due to the severity of the behavioral disturbance. Nevertheless, there is now an expanding literature in this field. Effects in patients with bipolar disorder, whether depressed, manic or euthymic, are often comparable to those in unipolar depression, at least with respect to increased inflammatory cytokines (Breunis et al. 2003; O'Brien et al. 2006). There has been a long standing interest in increased autoimmune antibodies, especially thyroid-related, in bipolar disorder (Padmos et al. 2004; Vonk et al. 2007). Autoimmune thyroiditis has been postulated as an etiologic factor for this disorder (Vonk et al. 2007).

Some studies found state-dependent immune effects in bipolar disorder, although the effects are variable. Kim et al. (2007) found the manic phase associated with increased monocyte-derived proinflammatory cytokines (but not IL2 or IFN-γ), which decreased with treatment; levels of anti-inflammatory cytokines were low throughout. The same group found increased IL-12 in unipolar, but not bipolar, depression (Kim et al. 2002). Another recent report suggested that bipolar patients have increased IL-6 and TNF-α in the depressed phase, and increased TNF-α and IL-4 in the manic phase (Ortiz-Domínguez et al. 2007). Others reported lower IFN-γ in acute mania and in the remitted euthymic state, but no changes in IL-2, IL-4 or IL-10 in mania or remission (Liu et al. 2004). Evidence for an inflammatory state (increased adhesion molecules) has also been found in brain tissue from patients with bipolar disorder, and to a lesser extent, unipolar disorder (Thomas et al. 2004). As already described, increased mRNA expression for inflammatory genes independent of clinical state has been found in bipolar patients and family members (Padmos et al. 2008). These studies together provide evidence for increased inflammation in bipolar disorder that is at least as strong as that for unipolar depression. Less is known concerning functional immunity in bipolar disorder.

6.2 Melancholia

The term melancholia is often used to describe a subgroup of patients with more severe major depression, a dimension that has itself been associated with functional immune changes (e.g., Schleifer et al. 1989). Melancholia may either be a variant of major depression (marked by profound anhedonia and diurnal variability), or refer more nonspecifically to severe depressive states. One report suggested that melancholia is associated with more pronounced increases in CD4+cells and possibly inflammatory cytokines compared with other depressive states (Schlatter et al. 2004). In contrast, another group reported that mitogen induced IL-1β production was increased in non-melancholic, but not melancholic, depressed patients (Kaestner et al. 2005). Dose-related (i.e., of severity) interactions of neuroendocrine and immune effects or possible syndromal differences have been postulated (e.g., Torres et al. 2005; Salicrú et al. 2007).

6.3 Typical vs. Atypical Depression

Atypical depressive disorders, as noted, are of interest since the typical motor and neurovegetative symptoms are often reversed (e.g., increased sleep and weight). Nevertheless, it is likely, in contrast to bipolar depression, that atypical depression is not fundamentally different from typical depression. Anisman et al. found increased circulating IL-1β levels in patients with atypical, but not typical, depression (Griffiths et al. 1996); IL-1β production by stimulated lymphocytes, however, did not differ between depressed persons with typical and atypical features (Anisman et al. 1999a). An observed increase in circulating NK cells was less dramatic in patients with atypical profiles (Ravindran et al. 1998).

6.4 Other Depressive Disorders

Recently, clinical attention has been directed to depressive states that are similar to major depression, but less severe. It appears that such disorders are often similar to major depression in their morbidity and pharmacologic responsivity (Lyness et al. 2006), and might therefore be variants of the same syndrome. Dysthymic disorder, a diagnostic entity (American Psychiatric Association 2000) marked by long-term depression, but fewer symptoms than major depression, has been associated with larger increases in mitogen-induced IL-1β production than in major depression itself (Anisman et al. 1999a, b). In a study of older persons with depression, however, increased IL-1β was found in major depression compared with persons with no history of depression, while those with a history of major depression but who were currently "subsyndromal" had intermediate levels (Thomas et al. 2005).

7 Paradepressive States

Immune change in states sharing features with depression, such as bereavement, provided a rationale for studies of immunity in depressive disorders. Other clinical states with behavioral similarities to depressive disorders may also warrant consideration.

7.1 Bereavement

Grief is a ubiquitous response to loss. It has variable behavioral manifestations, in part culturally determined, that include almost all the symptoms of major depression. It is often only distinguished from that disorder by the social context. Despite early interest, the number of immune studies remains modest. Decreased mitogen responses (Bartrop et al. 1977; Schleifer et al. 1983; Gerra et al. 2003), NKCA (Irwin et al. 1987a; Gerra et al. 2003), and antibody responses to influenza vaccination (Phillips et al. 2006) have been reported in otherwise healthy bereaved individuals, especially during the first months after the loss. Not all studies have noted immune effects, however (Beem et al. 1999). Altered mitogen and NKCA in the bereaved may be more pronounced in those with more prominent or persistent depressive symptoms (Gerra et al. 2003; Zisook et al. 1994). Little is known concerning inflammatory processes in this classic paradepressive state.

7.2 Vital Exhaustion

This putative syndrome, of much interest to clinicians and investigators in Europe, overlaps with depressive disorders with respect to fatigue and irritability. It has been associated with increased inflammatory markers and, possibly, resistance to glucocorticoid suppression (Wirtz et al. 2003). In one study, vital exhaustion, but not depressive symptoms per se, was associated with increased IL-6 and CRP in women with coronary artery disease (Janszky et al. 2005).

7.3 Burnout

This controversial behavioral syndrome, marked primarily by emotional and physical exhaustion, was associated with increased leukocyte adhesiveness in one study (Lerman et al. 1999).

7.4 Sickness Behavior

This syndrome, classically a consequence of infectious illnesses, manifests some features of major depression and has elicited considerable research interest, initially in animal models (Dantzer and Kelley 2007). It is receiving clinical attention as therapeutic strategies exposing medically ill persons to cytokines induce sickness behavior with high frequency (Capuron and Miller 2004). A subgroup of cytokine treated persons, moreover, proceed to develop what appear to be major depressive episodes (Capuron and Miller 2004; Irwin and Miller 2007). This important clinical phenomenon, discussed further below (Section 8), may provide a tool for dissecting the bidirectional behavioral correlates of inflammatory vs. functional immune changes in depression.

7.5 Other Syndromes

Another group of syndromes, to which both depressive symptoms and immune changes are relevant, are identified by somatic symptoms not readily accounted by traditional medical diagnostic categories. These disorders, such as chronic fatigue syndrome and fibromyalgia, will not be considered here. They can be controversial, may represent pathophysiologically heterogeneous disorders, but may provide models for assessing interactions among depression, immunity, and physical health.

8 State-Related Effects

8.1 Effects of Treatment, Clinical Improvement, and Remission

Immune studies in depressive disorders have mostly considered persons with acute episodes of depression or have measured immunity in relation to the intensity of current symptomatology. As suggested by recent genetic studies (e.g., Padmos et al. 2008), some immune changes in depression may be linked to an underlying syndromal diathesis for depressive (or manic) episodes in which the phenotype is periodically expressed (analogous to diabetes mellitus or hypertension). Immune changes may be found between episodes as well as prior to the initial episode. There has also been a modest literature concerning whether the immune changes observed during the acute episode persist following clinical improvement or remission.

In unipolar as well as bipolar disorders, methodologic constraints limit the generalizability of longitudinal studies since most of those who present for treatment receive antidepressant or other medications that themselves can alter immune processes (e.g., Castanon et al. 2002; Frank et al. 1999). Several studies suggest that some immune changes in acute depression revert to more normative states with successful treatment. Irwin et al. (1992) found that baseline decreases in NKCA reverted with clinical improvement. Another study, which had found increased NK cell numbers in an acutely depressed sample, reported decreases to control levels with antidepressant treatment (Ravindran et al. 1995). A longitudinal study we conducted with young adult depressed persons (who showed elevated mitogen response) found decreases toward control levels with treatment that were not associated with antidepressant blood levels (Schleifer et al. 1999). A recent report noted that the interactions between the HPA axis and TNF-α levels changed when patients treated for depression entered remission (Himmerich et al. 2006). Both the genetic and several longitudinal studies suggest overall, however, that the inflammatory state in depression persists even after clinical improvement (Maes 1999) and in remission (Kling et al. 2007).

8.2 Immune Effects of Cytokine Treatments

The use of cytokines, such as IFN-α and IL-2, to treat medical disorders such as neoplasia, hepatitis C, and HIV has resulted in clinically significant psychiatric morbidity. These treatments induce neurologic and behavioral changes with high frequency, becoming limiting factors in their use. Systemic exposure to inflammatory cytokines is associated with a variety of behavioral and psychiatric consequences, including especially depression (Capuron and Miller 2004; Irwin and Miller 2007), while treatments that block cytokines, such as TNF, can improve mood (Tyring et al. 2006). Studies have suggested that the behavioral effects of cytokines are heterogeneous, with at least two types of depressive presentations. These include: (1) an almost universal rapid-onset cluster of the largely somatic, "sickness behavior" depressive symptoms (fatigue, sleep disturbance, pain, gastrointestinal symptoms, and psychomotor retardation); and (2) a syndrome with additional mood and cognitive abnormalities that closely resembles, or is identical to, the full syndrome of MDD. The latter alone responds to antidepressant treatment and it is more likely to occur in patients with baseline depression (Capuron et al. 2004; Beratis et al. 2005). Induction of the full syndrome may represent the interaction of elevated inflammatory cytokines with constitutionally sensitive immunocytes in at-risk persons. The cytokines may then trigger a secondary response in which compensatory CNS processes are overwhelmed, leading to the full depressive syndrome.

9 Concluding Comments

Depressive disorders are among the most common of human ailments. They are a recognized major public health challenge, if only as a consequence of the morbidity and mortality of suicide, but also due to increased risk for somatic diseases. This chapter has summarized immunologic observations which suggest that two categories of immune change in depression may account for medical morbidity, decreased immune function, and increased inflammation. The former may be associated with increased morbidity from infection and neoplasia, the latter with increased morbidity from disorders of inflammation and autoimmunity, including atherosclerotic cardiovascular disorders (Irwin 2002b). Reciprocal effects of the immune system, especially inflammation, on brain and behavior, suggest that the pathophysiology of psychiatric disorders, including depression, may be immunologically mediated. Immunologic interventions may ultimately play a role in the prevention and treatment of depressive disorders.

References

Alexopoulos GA (2006) The vascular depression hypothesis: 10 years later. Biol Psychiatry 60:1304–5

American Psychiatric Association (2000) Diagnostic and statistical manual of mental disorders, 4th edition, text revision (DSM-IV-TR). American Psychiatric Association, Washington, DC

Andreasson A, Arborelius L, Erlanson-Albertsson C, Lekander M (2007) A putative role for cytokines in the impaired appetite in depression. Brain Behav Immun 21:147–52

Andreoli A, Keller SE, Rabaeus M, Zaugg L, Garrone G, Taban C (1992) Immunity, major depression, and panic disorder comorbidity. Biol Psychiatry 31:896–908

Andreoli AV, Keller SE, Rabaeus M, Marin P, Bartlett JA, Taban C (1993) Depression and immunity: age, severity, and clinical course. Brain Behav Immun 7:279–92

Anisman H, Ravindran AV, Griffiths J, Merali Z (1999a) Endocrine and cytokine correlates of major depression and dysthymia with typical or atypical features. Mol Psychiatry 4:182–8

Anisman H, Ravindran AV, Griffiths J, Merali Z (1999b) Interleukin-1ß production in dysthymia before and after pharmacotherapy. Biol Psychiatry 46:1649–55

Anisman H, Merali Z (2002) Cytokines, stress, and depressive illness. Brain Behav Immun 16:513–24

Bartlett J, Schleifer S, Demetrikopoulos M, Keller S (1995) Immune differences in children with and without depression. Biol Psychiatry 38:771–4

Bartrop RW, Luckhurst E, Lazarus L, Kiloh LG, Penny R (1977) Depressed lymphocyte function after bereavement. Lancet 1(8016):834–6

Beem EE, Hooijkaas H, Cleiren MH, Schut HA, Garssen B, Croon MA, Jabaaij L, Goodkin K, Wind H, de Vries MJ (1999) The immunological and psychological effects of bereavement: does grief counseling really make a difference? A pilot study. Psychiatry Res 85:81–93

Ben-Eliyahu S, Page GG, Schleifer SJ (2007) Stress, NK cells, and cancer: still a promissory note. Brain Behav Immun 21:881–7

Beratis S, Katrivanou A, Georgiou S, Monastirli A, Pasmatzi E, Gourzis P, Tsambaos D (2005) Major depression and risk of depressive symptomatology associated with short-term and low-dose interferon-alpha treatment. J Psychosom Res 58:15–8

Bouhuys AL, Flentge F, Oldehinkel AJ, van den Berg MD (2004) Potential psychosocial mechanisms linking depression to immune function in elderly subjects. Psychiatry Res 127:237–45

Breunis MN, Kupka RW, Nolen WA, Suppes T, Denicoff KD, Leverich GS, Post RM, Drexhage HA (2003) High numbers of circulating activated T cells and raised levels of serum IL-2 receptor in bipolar disorder. Biol Psychiatry 53:157–65
Capuron L, Miller AH (2004) Cytokines and psychopathology: lessons from interferon-alpha. Biol Psychiatry 56:819–24.
Capuron L, Ravaud A, Miller AH, Dantzer R (2004) Baseline mood and psychosocial characteristics of patients developing depressive symptoms during interleukin-2 and/or interferon-alpha cancer therapy. Brain Behav Immun 18:205–13
Castanon N, Leonard BE, Neveu PJ, Yirmiya R (2002) Effects of antidepressants on cytokine production and actions. Brain Behav Immun 16:569–74
Cyranowski JM, Marsland AL, Bromberger JT, Whiteside TL, Chang Y, Matthews KA (2007) Depressive symptoms and production of proinflammatory cytokines by peripheral blood mononuclear cells stimulated in vitro. Brain Behav Immun 21:229–37
Danese A, Pariante CM, Caspi A, Taylor A, Poult R (2007) Childhood maltreatment predicts adult inflammation in a life-course study. PNAS 104:1319–24
Danner M, Kasl SV, Abramson JL, Vaccarino V (2003) Association between depression and elevated C-reactive protein. Psychosom Med 65:347–56
Dantzer R, Kelley KW (2007) Twenty years of research on cytokine-induced sickness behavior. Brain Behav Immun 21:153–60
Dickerson SS, Kemeny ME, Aziz N, Kim KH, Fahey JL (2004) Immunological effects of induced shame and guilt. Psychosom Med 66:124–31
Evans DL, Folds JD, Petitto JM, Golden RN, Pedersen CA, Corrigan M, Gilmore JH, Silva SG, Quade D, Ozer H (1992) Circulating natural killer cell phenotypes in men and women with major depression. Relation to cytotoxic activity and severity of depression. Arch Gen Psychiatry 49:388–95
Frank MG, Hendricks SE, Johnson DR, Wieselerd JL, Burkeb HJ (1999) Antidepressants augment natural killer cell activity: in vivo and in vitro. Neuropsychobiology 39:18–24
Frank MG, Wieseler Frank JL, Hendricks SE, Burke WJ, Johnson DR (2002) Age at onset of major depressive disorder predicts reductions in NK cell number and activity. J Affect Disord 71:159–67
Frasure-Smith N, Lespérance F, Irwin MR, Sauvé C, Lespérance J, Théroux P (2007) Depression, C-reactive protein and two-year major adverse cardiac events in men after acute coronary syndromes. Biol Psychiatry 62:302–8
Gerra G, Monti D, Panerai AE, Sacerdote P, Anderlini R, Avanzini P, Zaimovic A, Brambilla F, Franceschi C (2003) Long-term immune-endocrine effects of bereavement: relationships with anxiety levels and mood. Psychiatry Res 121:145–58
Graham JE, Robles TF, Kiecolt-Glaser JK, Malarkey WB, Bissell MG, Glaser R (2006) Hostility and pain are related to inflammation in older adults. Brain Behav Immun 20:389–400
Griffiths J, Ravindran AV, Merali Z, Anisman H (1996) Immune and behavioral correlates of typical and atypical depression. Soc Neurosci Abstr 22:1350
Haack M, Hinze-Selch D, Fenzel T, Kraus T, Kuhn M, Schuld A, Pollmacher T (1999) Plasma levels of cytokines and soluble cytokine receptors in psychiatric patients upon hospital admission: effects of confounding factors and diagnosis. J Psychiatr Res 33:407–18
Herbert TB, Cohen S (1993a) Depression and immunity: a meta-analytic review. Psychol Bull 113:472–86
Herbert TB, Cohen S (1993b). Stress and immunity in humans: a meta-analytic review. Psychosom Med 55:364–79
Himmerich H, Binder EB, Künzel HE, Schuld A, Lucae S, Uhr M, Pollmächer T, Holsboer F, Ising M (2006) Successful antidepressant therapy restores the disturbed interplay between TNF-alpha system and HPA axis. Biol Psychiatry 60:882–8
Irwin M, Daniels M, Smith TL, Bloom E, Weiner H (1987a) Impaired natural killer cell activity during bereavement. Brain Behav Immun 1:98–104
Irwin M, Smith TL, Gillin JC (1987b) Low natural killer cytotoxicity in major depression. Life Sci 41:2127–33
Irwin M, Lacher U, Caldwell C (1992) Depression and reduced natural killer cytotoxicity: a longitudinal study of depressed patients and control subjects. Psychol Med 22:1045–50

Irwin M (2002a) Effects of sleep and sleep loss on immunity and cytokines. Brain Behav Immun 16:503–12
Irwin M (2002b) Psychoneuroimmunology of depression: clinical implications. Brain Behav Immun 16:1–16
Irwin MR, Miller AH (2007) Depressive disorders and immunity: 20 years of progress and discovery. Brain Behav Immun 21:374–83
Janicki-Deverts D, Cohen S, Doyle WJ, Turner RB, Treanor JJ (2007) Infection-induced proinflammatory cytokines are associated with decreases in positive affect, but not increases in negative affect. Brain Behav Immun 21:301–7
Janszky I, Lekander M, Blom M, Georgiades A, Ahnve S (2005) Self-rated health and vital exhaustion, but not depression, is related to inflammation in women with coronary heart disease. Brain Behav Immun 19:555–63
Jung W, Irwin M (1999) Reduction of natural killer cytotoxic activity in major depression: interaction between depression and cigarette smoking. Psychosom Med 61:263–70
Kaestner F, Hettich M, Peters M, Sibrowski W, Hetzel G, Ponath G, Arolt V, Cassens U, Rothermundt M (2005) Different activation patterns of proinflammatory cytokines in melancholic and nonmelancholic major depression are associated with HPA axis activity. J Affect Disord 87:305–11
Kessing LV (2008) Severity of depressive episodes during the course of depressive disorder. Br J Psychiatry 192:290–3
Kim YK, Suh IB, Kim H, Han CS, Lim CS, Choi SH, Licinio J (2002) The plasma levels of interleukin-12 in schizophrenia, major depression, and bipolar mania: effects of psychotropic drugs. Mol Psychiatry 7:1107–14
Kim YK, Jung HG, Myint AM, Kim H, Park SH (2007) Imbalance between pro-inflammatory and anti-inflammatory cytokines in bipolar disorder. J Affect Disord 104:91–5
Kling MA, Alesci S, Csako G, Costello R, Luckenbaugh DA, Bonne O, Duncko R, Drevets WC, Manji HK, Charney DS, Gold PW, Neumeister A (2007) Sustained low-grade pro-inflammatory state in unmedicated, remitted women with major depressive disorder as evidenced by elevated serum levels of the acute phase proteins C-reactive protein and serum amyloid A. Biol Psychiatry 62:309–13
Kohut ML, McCann DA, Russell DW, Konopka DN, Cunnick JE, Franke WD, Castillo MC, Reighard AE, Vanderah E (2006) Aerobic exercise, but not flexibility/resistance exercise, reduces serum IL-18, CRP, and IL-6 independent of [beta]-blockers, BMI, and psychosocial factors in older adults. Brain Behav Immun 20:201–9
Ladwig KH, Marten-Mittag B, Löwel H, Döring A, Koenig W (2003) Influence of depressive mood on the association of CRP and obesity in 3205 middle aged healthy men. Brain Behav Immun 17:268–75
Lerman Y, Melamed S, Shragin Y, Kushnir T, Rotgoltz Y, Shirom A, Aronson M. (1999) Association Between Burnout at Work and Leukocyte Adhesiveness/Aggregation Psychosomatic Medicine 61:828–833.
Liu HC, Yang YY, Chou YM, Chen KP, Shen WW, Leu SJ (2004) Immunologic variables in acute mania of bipolar disorder. J Neuroimmunol 150:116–22
Liukkonen T, Silvennoinen-Kassinen S, Jokelainen J, Räsänen P, Leinonen M, Meyer-Rochow VB, Timonen M (2006) The association between C-reactive protein levels and depression: results from the northern Finland 1966 birth cohort study. Biol Psychiatry 60:825–30
Lutgendorf SK, Moore M, Bradley S, Shelton B, Lutz C (2005) Distress and natural killer receptors on lymphocytes. Brain Behav Immun 19:185–94
Lyness JM, Heo M, Datto CJ, Ten Have TR, Katz IR, Drayer R, Reynolds CF 3rd, Alexopoulos GS, Bruce ML (2006) Outcomes of minor and subsyndromal depression among elderly patients in primary care settings. Ann Intern Med 144:496–504
Maes M (1999) Major depression and activation of the inflammatory response system. Adv Exp Med Biol 461:25–46
Mann JJ, Brown RP, Halper JP, Sweeney JA, Kocsis JH, Stokes PE, Bilezikian JP (1985) Reduced sensitivity of lymphocyte beta-adrenergic receptors in patients with endogenous depression and psychomotor agitation. N Engl J Med 313:715–20

Matthews KA, Schott LL, Bromberger J, Cyranowski J, Everson-Rose SA, Sowers MF (2007) Associations between depressive symptoms and inflammatory/hemostatic markers in women during the menopausal transition. Psychosom Med 69:124–30
Miller GE, Cohen S, Herbert TB (1999) Pathways linking major depression and immunity in ambulatory female patients. Psychosom Med 61:850–60
Miller GE, Stetler CA, Carney RM, Freedland KE, Banks WA (2002) Clinical depression and inflammatory risk markers for coronary heart disease. Am J Cardiol 90:1279–83
Miller GE, Freedland KE, Carney RM, Stetler CA, Banks WA (2003) Pathways linking depression, adiposity, and inflammatory markers in healthy young adults. Brain Behav Immun 17:276–85
Miller AH, Manji HK (2006) On redefining the role of the immune system in psychiatric disease. Biol Psychiatry 60:796–8
Moldofsky H, Lue FA, Davidson JR, Gorczynski R (1989) Effects of sleep deprivation on human immune functions. FASEB J 3:1972–7
Motivala SJ, Sarfatti A, Olmos L, Irwin MR (2005) Inflammatory markers and sleep disturbance in major depression. Psychosom Med 67: 187–94
O'Brien SM, Scully P, Scott LV, Dinan TG (2006) Cytokine profiles in bipolar affective disorder: focus on acutely ill patients. J Affect Disord 90:263–7
Ortiz-Domínguez A, Hernández ME, Berlanga C, Gutiérrez-Mora D, Moreno J, Heinze G, Pavón L (2007) Immune variations in bipolar disorder: phasic differences. Bipolar Disord 9:596–602
Padmos RC, Bekris L, Knijff EM, Tiemeier H, Kupka RW, Cohen D, Nolen WA, Lernmark A, Drexhage HA (2004) A high prevalence of organ-specific autoimmunity in patients with bipolar disorder. Biol Psychiatry 56:476–82
Padmos RC, Hillegers MHJ, Knijff EM, Vonk R, Bouvy A, Staal FJT, de Ridder D, Kupka RW, Nolen WA, Drexhage HA (2008) A discriminating messenger RNA signature for bipolar disorder formed by an aberrant expression of inflammatory genes in monocytes. Arch Gen Psychiatry 65:395–407
Penninx BWJH, Kritchevsky SB, Yaffe K, NewmanAB, Simonsick EM, Rubin S, Ferrucci L, Harris T, Pahor M (2003) Inflammatory markers and depressed mood in older persons: results from the health, aging and body composition study. Biol Psychiatry 54:566–72
Phillips AC, Carroll D, Burns VE, Ring C, Macleod J, Drayson M (2006) Bereavement and marriage are associated with antibody response to influenza vaccination in the elderly. Brain Behav Immun 20:279–89
Pike JL, Irwin MR (2006) Dissociation of inflammatory markers and natural killer cell activity in major depressive disorder. Brain Behav Immun 20:169–74.
Prather AA, Marsland AL, Muldoon MF, Manuck SB (2007) Positive affective style covaries with stimulated IL-6 and IL-10 production in a middle-aged community sample. Brain Behav Immun 21:1033–7
Raison C, Miller AH (2003) When not enough is too much: the role of insufficient glucocorticoid signaling in the pathophysiology of stress-related disorders. Am J Psychiatry 160:1554–65
Ravindran AV, Griffiths J, Merali Z, Anisman H (1995) Lymphocyte subsets in major depression and dysthymia: modification by antidepressant treatment. Psychosom Med 57:555–63
Ravindran AV, Griffiths J, Merali Z, Anisman H (1998) Circulating lymphocyte subsets in major depression and dysthymia with typical or atypical features. Psychosom Med 60:283–9
Salicrú AN, Sams CF, Marshall GD (2007) Cooperative effects of corticosteroids and catecholamines upon immune deviation of the type-1/type-2 cytokine balance in favor of type-2 expression in human peripheral blood mononuclear cells. Brain Behav Immun 21:913–20
Schlatter J, Ortuño F, Cervera-Enguix S (2004) Lymphocyte subsets and lymphokine production in patients with melancholic versus nonmelancholic depression. Psychiatry Res 128:259–65
Schleifer SJ, Keller SE, Camerino MS, Thornton J, Stein M (1983) Suppression of lymphocyte stimulation following bereavement. JAMA 250:374–377
Schleifer SJ, Keller SE, Meyerson AT, Raskin MJ, Davis KL, Stein M (1984) Lymphocyte function in major depressive disorder. Arch Gen Psychiatry 41:484–6
Schleifer SJ, Keller SE, Bond RN, Cohen J, Stein M (1989) Depression and immunity: role of age, sex, and severity. Arch Gen Psychiatry 46:81–7

Schleifer SJ, Keller SE, Bartlett JA Eckholdt HM, Delaney BR (1996) Immunity in young adults with major depressive disorder. Am J Psychiatry 153:477–82
Schleifer SJ, Keller SE, Bartlett JA (1999) Depression and immunity: clinical factors and therapeutic course. Psychiatry Res 85:63–9
Schleifer SJ, Bartlett JA, Keller SE Eckholdt HM, Shiflett SC, Delaney BR (2002) Immunity in adolescents with major depression. J Am Acad Child Adolesc Psychiatry 41:1054–60
Schleifer SJ, Keller SE, Czaja S (2006) Major depression and immunity in alcohol dependent persons. Brain Behav Immun 20:80–91
Segerstrom SC (2005) Optimism and immunity: Do positive thoughts always lead to positive effects? Brain Behav Immun 19:195–200
Shakhar K, Valdimarsdottir HB, Guevarra JS, Bovbjerg DH (2007) Sleep, fatigue, and NK cell activity in healthy volunteers: significant relationships revealed by within subject analyses. Brain Behav Immun 21:180–4
Smith RS (1991) The macrophage theory of depression. Med Hypotheses 35:298–306
Smith A, Vollmer-Conna U, Geczy A, Dunckley H, Bennett B, Hickie I, Lloyd A (2005) Does genotype mask the relationship between psychological factors and immune function? Brain Behav Immun 19:147–52
Steiner J, Bielau H, Brisch R, Danos P, Ullrich O, Mawrin C, Bernstein HG, Bogerts (2008) Immunological aspects in the neurobiology of suicide: elevated microglial density in schizophrenia and depression is associated with suicide. J Psychiatr Res 42:151–7
Steptoe A, Owen N, Kunz-Ebrecht SR, Brydon L (2004) Loneliness and neuroendocrine, cardiovascular, and inflammatory stress responses in middle-aged men and women. Psychoneuroendocrinology 29:593–611
Suarez EC (2003) Joint effect of hostility and severity of depressive symptoms on plasma interleukin-6 concentration. Psychosom Med 65: 523–27
Suarez EC (2004) C-reactive protein is associated with psychological risk factors of cardiovascular disease in apparently healthy adults. Psychosom Med 66:684–91
Suarez EC, Boyle SH, Lewis JG, Hall RP, Young KH (2006) Increases in stimulated secretion of proinflammatory cytokines by blood monocytes following arousal of negative affect: the role of insulin resistance as moderator. Brain Behav Immun 20:331–8
Thomas AJ, Davis S, Ferrier IN, Kalaria RN, O'Brien JT (2004) Elevation of cell adhesion molecule immunoreactivity in the anterior cingulate cortex in bipolar disorder. Biol Psychiatry 55:652–5
Thomas AJ, Davis S, Morris C, Jackson E, Harrison R, O'Brien JT (2005) Increase in interleukin-1beta in late-life depression. Am J Psychiatry 162:175–7
Torres KC, Antonelli LR, Souza AL, Teixeira MM, Dutra WO, Gollob KJ (2005) Norepinephrine, dopamine and dexamethasone modulate discrete leukocyte subpopulations and cytokine profiles from human PBMC. J Neuroimmunol 166:144–57
Tyring S, Gottlieb A, Papp K, Gordon K, Leonardi C, Wang A, Lalla D, Woolley M, Jahreis A, Zitnik R, Cella D, Krishnan R (2006) Etanercept and clinical outcomes, fatigue, and depression in psoriasis: double-blind placebo-controlled randomised phase III trial. Lancet 367(9504):29–35
Vonk R, van der Schot AC, Kahn RS, Nolen WA, Drexhage HA (2007) Is autoimmune thyroiditis part of the genetic vulnerability (or an endophenotype) for bipolar disorder? Biol Psychiatry 62:135–40
Wirtz PH, von Känel R, Schnorpfeil P, Ehlert U, Frey K, Fischer JE (2003) Reduced glucocorticoid sensitivity of monocyte interleukin-6 production in male industrial employees who are vitally exhausted. Psychosom Med 65:672–8
Wright CE, Erblich J, Valdimarsdottir HB, Bovbjerg DH (2007) Poor sleep the night before an experimental stressor predicts reduced NK cell mobilization and slowed recovery in healthy women. Brain Behav Immun 21:358–63
Zalcman SS, Siegel A (2006) The neurobiology of aggression and rage: role of cytokines. Brain Behav Immun 20:507–14

Zisook S, Shuchter SR, Irwin M, Darko DF, Sledge P, Resovsky K (1994) Bereavement, depression, and immune function. Psychiatr Res 52:1–10
Zorrilla EP, Luborsky L, McKay JR, Rosenthal R, Houldin A, Tax A, McCorkle R, Seligman DA, Schmidt K (2001) The relationship of depression and stressors to immunological assays: a meta-analytic review. Brain Behav Immun 15:199–226.

Cytokines, Immunity and Schizophrenia with Emphasis on Underlying Neurochemical Mechanisms

Norbert Müller and Markus J. Schwarz

Abstract This overview presents a hypothesis to bridge the gap between psychoneuroimmunological findings and recent results from pharmacological, neurochemical and genetic studies in schizophrenia. In schizophrenia, a glutamatergic hypofunction is discussed to be crucially involved in dopaminergic dysfunction. This view is supported by findings of the neuregulin and dysbindin genes, which have functional impact on the glutamatergic system. Glutamatergic hypofunction is mediated by NMDA (*N*-methyl-D-aspartate) receptor antagonism. The only endogenous NMDA receptor antagonist identified up to now is kynurenic acid (KYN-A). KYN-A also blocks the nicotinergic acetylcholine receptor, i.e. increased KYN-A levels can explain psychotic symptoms and cognitive deterioration. KYN-A levels are described to be higher in the CSF and in critical CNS regions of schizophrenics.

Another line of evidence suggests that the immune system in schizophrenic patients is characterized by an imbalance between the type-1 and the type-2 immune responses with a partial inhibition of the type-1 response, while the type-2 response is relatively over-activated. This immune constellation is associated with the inhibition of the enzyme indoleamine 2,3-dioxygenase (IDO), because type-2 cytokines are potent inhibitors of IDO. Due to the inhibition of IDO, tryptophan is predominantly metabolized by tryptophan 2,3-dioxygenase (TDO), which is located in astrocytes, but not in microglia cells. As indicated by increased levels of S100B, astrocytes are activated in schizophrenia. On the other hand, the kynurenine metabolism in astrocytes is restricted to the dead-end arm of KYN-A production. Accordingly, an increased TDO activity and an accumulation of KYN-A in the CNS of schizophrenics have been described. Thus, the immune-mediated glutamatergic–dopaminergic dysregulation may lead to the clinical symptoms of schizophrenia. Therapeutic consequences, e.g. the use of anti-inflammatory cyclooxygenase-2 inhibitors, which are also able to directly decrease KYN-A, are discussed.

N. Müller (✉)
Department of Psychiatry and Psychotherapy, Ludwig-Maximilians-University Munich, Hospital for Psychiatry and Psychotherapy, Nussbaumstrasse 7, D-80336 Munich-Germany
e-mail: norbert.mueller@med.uni-muenchen.de

A. Siegel, S.S. Zalcman (eds.), *The Neuroimmunological Basis of Behavior and Mental Disorders*, DOI 10.1007/978-0-387-84851-8_15,

Keywords Schizophrenia · Immune system · Glutamate · NMDA receptor · COX-2, PGE2

Abbreviations CNS, central nervous system; COX, cyclooxygenase; CSF, cerebrospinal fluid; HLA, human leukocyte antigen; 3-HK, 3-hydroxykynurenine; ICAM-1, intercellular adhesion molecule-1; IDO, indoleamine 2,3-dioxygenase; Ig, Immunoglobulin; IFN, interferon; IL, interleukin; KMO, kynurenine-monoxygenase; KYN, kynurenine; KYN-A, kynurenic acid; NMDA, *N*-methyl-D-aspartate; PCP, Phencyclidine; PGE2, prostaglandin E2; TDO, tryptophan 2,3-dioxygenase; Th1/2, T helper cell type 1 or 2, respectively

1 Dopaminergic – Glutamatergic Imbalance in Schizophrenia

There is no doubt that a disturbance in the dopaminergic neurotransmission plays a key role in the pathogenesis of schizophrenia (Carlsson, 1988; Jentsch and Roth, 1999). This view is based on the evidence that most drugs ameliorating psychotic symptoms act as dopamine receptor blockers, in particular D2 receptor blockers. However, despite the fact that only a part of the patients respond to antipsychotic drugs and the long-term outcome of antipsychotic treatment is unsatisfactory in many cases, attempts to explain the disease solely in terms of dopaminergic dysfunction leave many aspects of schizophrenia unsolved.

The glutamate hypothesis of schizophrenia postulates an equilibrium between inhibiting dopaminergic and glutamatergic neurons; the model of a cortico-striato-thalamo-cortical control loop integrates the glutamate hypothesis with neuroanatomical aspects on the pathophysiology of schizophrenia (Carlsson et al., 2001). A hypofunction of the glutamatergic cortico-striatal pathway—i.e. an inhibition of the inhibitory GABA neurons by hyperactivated dopaminergic receptors as well as reduction of the glutamatergic input—is associated with opening of the thalamic filter, which leads to an uncontrolled flow of sensory information to the cortex and to psychotic symptoms (Carlsson, 2006).

Dopamine release can be challenged by amphetamine, which can be blocked by NMDA-receptor antagonists. In an animal experiment, the treatment with NMDA (*N*-methyl-D-aspartate) receptor antagonists leads to a marked, dose-dependent increase of amphetamine-induced dopamine release (Miller and Abercrombie, 1996). This observation has been confirmed in humans: Blocking of the NMDA-receptor by ketamine induced a significant enhancement of amphetamine-induced dopamine release in healthy controls (Kegeles et al., 2000). The magnitude of the increase was comparable to the exaggerated response patients with schizophrenia had to amphetamine alone (Laruelle et al., 2005); in schizophrenics, the amphetamine-induced dopamine release is much higher compared to healthy controls (Laruelle et al., 1996). This observation is in accordance with the view that the abnormally elevated dopamine release revealed by the amphetamine challenge in schizophrenia results from a disruption of glutamatergic neuronal systems regulating dopaminergic activity (Laruelle et al., 2005).

2 NMDA-Receptor Hypofunction and Schizophrenia

Hypofunction of the glutamatergic neurotransmitter system as a causal mechanism in schizophrenia was first proposed due to the observation of low concentrations of glutamate in the cerebrospinal fluid (CSF) of schizophrenic patients (Kim et al., 1980).

Phencyclidine (PCP), ketamine, and MK-801 all block the NMDA receptor complex and are associated with schizophrenia-like symptoms through hypofunction of the glutamatergic neurotransmission (Krystal et al., 1994; Olney and Farber, 1995). Other substances, acting as NMDA antagonists, but not at the PCP site, have psychotogenic properties, too (CPP, CPP-ene, CGS 19755; 8). NMDA receptor hypofunction can explain schizophrenic positive and negative symptoms; cognitive deterioration and structural brain changes can be a consequence of NMDA receptor dysfunction (Olney and Farber, 1995). Recently, the first in vivo evidence for a NMDA receptor deficit has been reported in medication-free schizophrenic patients (Pilowsky et al., 2006).

Decreased plasma levels of the NMDA co-agonist glycine in schizophrenics and a correlation of glycine levels with schizophrenic negative symptoms were found (Sumiyoshi et al., 2004). Baseline glycine levels predicted the treatment outcome of clozapine in negative symptoms (Sumiyoshi et al., 2005). Clinical investigations targeted the glycine co-agonist site of the NMDA receptor by administering the amino acids glycine or D-serine, or a glycine pro-drug such as milacemide (Tamminga et al., 1992). Some of these studies have yielded positive results, particularly against the schizophrenic deficit syndrome (Heresco-Levy et al., 1999).

3 Genetics of the Immune System and of the NMDA System

Schizophrenia is a complex genetic disorder. Given the hereditary component and the role of an inflammatory/immunological process in schizophrenia, immunologically relevant genes may shape up as susceptibility genes for schizophrenia.

Genetic factors influence acquiring infectious diseases, both with respect to susceptibility (Cook and Hill, 2001) and to resistance to infection (Hill, 1996). Mechanisms for genetically mediated responses to infection occur through genetic variations in immune mediators such as cytokines and HLA genes. The HLA region on chromosome 6 is located within or very near a region which has a high susceptibility risk for schizophrenia (Schwab et al., 2000). So far, associations of certain HLA-loci with schizophrenia were described (Laumbacher et al., 2003), but replication in larger, independent samples are still lacking. Studies of genetic polymorphisms of the pro-inflammatory cytokine TNF-alpha – also located in the HLA region of chromosome 6 – show divergent results, the difference in the outcome possibly being related to ethnic differences between the samples (Riedel et al., 2002; Meira-Lima et al., 2003). An analysis of polymorphisms of the type-I cytokine IL-2 and the type-II cytokine IL-4 revealed a possible genetic base for this imbalance (Schwarz et al., 2005). Several other cytokine genes and components of the immune system have been studied without conclusive results. Methodological problems of samples and diagnoses, the small genetic load of every individual gene, the pleiotropic function of the immune components, and their marked functional compensatory abilities may explain weak genetic associations.

There is consensus that genetic variations of dysbindin, located also on chromosome 6p22 (Numakawa et al., 2004) and neuregulin-1, located at chromosome 8p (Williams et al., 2003) are associated with an increased risk for schizophrenia. Both genes have been identified in large studies over the last years. Although the functions are not yet fully elucidated and at least the neuregulin-1 has multiple functions, interestingly, both genes code for proteins which are involved in the glutamatergic neurotransmission (Collier and Li, 2003). Neuregulin-1 regulates the NMDA receptor expression/presence in glutamatergic synaptic vesicles, and dysbindin – located in glutamatergic neurons – is reduced in schizophrenia (Talbot et al., 2004). These recent findings support the view that the glutamatergic neurotransmission plays a key role in schizophrenia.

4 Neurodevelopmental Aspects of Inflammation and NMDA Receptor Dysfunction

The discovery of environmental risk factors for schizophrenia, acting before, during, and shortly after birth has been central for the neurodevelopmental hypothesis of schizophrenia (Murray and Lewis, 1987). Genetic and environmental risk factors interact during the crucial phase of development of the central nervous system (CNS) causing subtle abnormalities, which leave the individual vulnerable to psychosis in later life. Established risk factors are obstetric complications, prenatal or postnatal infections. Cytokines, mediators of the immune response, are growth factors of the nervous system and of glial cells, therefore crucial for the development of the CNS. Obstetric complications such as hypoxia and injury of the CNS are associated with a change in cytokine release in the CNS (Dean and Murray, 2005).

On the other hand, it has been argued that the effect of obstetric complications might be mediated by glutamatergic excitotoxic damage in the foetal/neonatal brain (Fearon et al., 2000). This view is supported by an animal model showing that glutamatergic damage is not associated with functional impairment in early life, but regularly manifests itself during early adulthood (Farber et al., 1995). The sensitization of the CNS to glutamatergic toxicity seems to be a process of maturation, which becomes symptomatic during adulthood. Accordingly, the occurrence of psychotic symptoms following the use of the NMDA receptor antagonist ketamine in humans is age dependent, with psychotic symptoms occurring rarely, if ever, in prepubertal children, but manifesting in nearly 50% of young to middle-aged adults (Marshall and Longnecker, 1990).

Sensitization studies with pro-inflammatory cytokines show that an increased production of pro-inflammatory cytokines such as IL-1 during an infectious or inflammatory process in the perinatal period can induce long-lasting, probablypermanent, alterations in the CNS neurotransmitter systems. These results indicate that an increased production of IL-1 during an infectious or inflammatory process in the perinatal period can induce long-lasting, probably permanent, alterations in the central (and peripheral) neurotransmitter systems (Kabiersch et al., 1998).

5 Inflammation and Schizophrenia

The role of infection in the aetiology of schizophrenia has gained more attention during the last years (Rapaport and Müller, 2001; Müller, 2004a). Infection during pregnancy in mothers of offsprings later developing schizophrenia has been repeatedly described (Brown et al., 2004; Buka et al., 2001; Westergaard et al., 1999) and is discussed as an explanation for the seasonality of schizophrenic births between December and May (Torrey et al., 1997).

Results of a Finnish epidemiological study showed that infection of the CNS in childhood increases the risk of becoming psychotic later on 5-fold (Koponen et al., 2004). A 5-fold increased risk for developing psychoses later on was also observed in Brazil after a (bacterial) meningitis epidemic (Gattaz et al., 2004). Taking into account a sensitization process, an infection during early childhood is in accordance with the assumption that an infection-triggered disturbance in brain development might play a key role in the aetiology of schizophrenia.

On the other hand, a persistent (chronic) infection as aetiological factor in schizophrenia is discussed since many years. Signs of inflammation were observed in schizophrenic brains (Körschenhausen et al., 1996) and the term 'mild localized chronic encephalitis' was proposed (Bechter et al., 2003). Following the hypothesis of an ongoing immune process, it would be expected that the brain volume reductions, regularly found in schizophrenia, are not only due to a neurodevelopmental disturbance (Jakob and Beckmann, 1986), but also directly preceding the first episode of schizophrenia and being further progressive. In fact, the Edinburgh High Risk Study, recently showed that a marked reduction of the inferior temporal gyrus over the time preceded the first onset of schizophrenia (Job et al., 2006) and the progressive loss of brain volume has repeatedly been demonstrated (e.g. Chakos et al., 2005).

Due to the characteristics of infectious agents, there are difficulties in proving a localized infection in the brain. A virus or other intracellular infectious agent may be silently hidden in cells of the lymphoid or the nervous system and exacerbate under certain conditions such as stress. The estimation of serum antibody titres is a method with limited sensitivity for a localized mild infectious process in the CNS. Nevertheless, antibody titres against viruses have been examined in the sera of schizophrenic patients for many years (Yolken and Torrey, 1995). The results, however, were inconsistent beneath others due to the fact that interfering factors were not controlled. Antibody levels are associated with the medication state; this finding partly explains the earlier controversial results (Leweke et al., 2004).

There is major evidence supporting pre- or perinatal exposure to infection as a risk factor for developing schizophrenia, with main focal points on the influenza, rubella, measles and herpes simplex viruses (Pearce, 2001). Moreover, viral infections during childhood (Koponen et al., 2004) and even preceding the onset of the illness have been associated with schizophrenia (Leweke et al., 2004). The levels of the pro-inflammatory cytokine IL-8 were increased during the second trimenon of pregnancy in the serum of those mothers, whose offsprings developed schizophrenia later, i.e. increased IL-8 levels were associated with an increased risk for schizophrenia in the offspring (Brown et al., 2004).

Recent research points out that not one single pathogen but the immune response of the mother is related to the increased risk for schizophrenia (Zuckerman and Weiner, 2005).

6 Polarized Type-1 and Type-2 Immune Responses

The cellular arm of the adaptive immune system is mainly activated by T-helper-1 (TH-1) cytokines like Interleukin-2 (IL-2) and Interferon-g (IFN-g). Since not only T-helper cells (CD4+ cells) but also monocytes/macrophages (M1) and other cell types produce these cytokines, the immune response is called type-1 immune response. The humoral arm of the adaptive immune system is mainly activated via the type-2 immune response. T-helper-2 cells (TH-2) or monocytes/macrophages (M2) produce IL-4, IL-10 and IL-13 (Mills et al., 2000). Other pro-inflammatory cytokines such as tumour necrosis factor-α (TNF-α) and IL-6 are primarily secreted from monocytes and macrophages. While TNF-α is a ubiquitious cytokine, mainly activating the cellular immune response, IL-6 activates the type-2 response including the antibody production.

The type-1 system promotes the cell-mediated immune responses against intracellular pathogens, whereas the type-2 response helps B-cell maturation and promotes the humoral immune responses against extracellular pathogens. Type-1 and type-2 cytokines antagonize each other in promoting their own type of response, while suppressing the immune response of the other.

7 Reduced Type-1 Immune Response in Schizophrenia

A well established finding in schizophrenia is the decreased in vitro production of IL-2 and IFN-g (Wilke et al., 1996; Müller et al., 2000) indicating a blunted production of type-1 cytokines. Additionally, decreased levels of IFN-g in the peripheral blood of schizophrenic patients have been described by several groups (Avgustin et al., 2005; Chiang et al., 2004). Although some authors argued that the blunted in vitro production of IL-2 and IFN-γ may indicate an exhaustion of the immune cells due to enhanced in vivo production of these cytokines (Rothermundt et al., 2001), several other immunological data point to a reduced activation of the cellular immune system in schizophrenia. The decreased response of lymphocytes after stimulation with specific antigens reflect a reduced capacity for a type-1 immune response in schizophrenia, as well (Müller et al., 1991). Decreased levels of neopterin, a product of activated monocytes/macrophages (Sperner-Unterweger et al., 1999), as well as decreased levels of soluble (s)ICAM-1 represent an under-activation of the type-1 immune system (Schwarz et al., 2000).

A blunted response of the skin to different antigens in schizophrenia was observed before the era of antipsychotics (Molholm, 1942). A study using a skin test of the cellular immune response (Multitest Merieux) in unmedicated schizophrenic patients also showed a decreased reaction (Riedel et al., 2007).

8 Increased Type-2 Immune Response in Schizophrenia

Several reports described increased serum IL-6 levels in schizophrenia. IL-6 serum levels might be especially high in patients with an unfavourable course of the disease (Müller et al., 2000). IL-6 is a product of activated monocytes and of the activation of the type-2 immune response. Moreover, several other signs of activation of the type-2 immune response are described in schizophrenia, including the increased production of IgE (Schwarz et al., 2001) and an increase of IL-10 serum levels (Cazzullo et al., 1998). IL-10 levels in the CSF are related to the severity of the psychosis (van Kammen et al., 1997).

A lot of studies described an increased antibody production – reflecting type-2 activation – in schizophrenic patients, those observations leading to the discussion of an autoimmune origin of schizophrenia (Ganguli et al., 1987). Although findings have repeatedly shown that about 20–35% of schizophrenic patients show features of an autoimmune process (Müller and Ackenheil, 1998), the role of actual or former therapy with neuroleptics may not have been taken enough into consideration. An increase of IgG – antibodies are mainly IgG antibodies – in the CSF has been described especially in patients with predominant negative symptoms (Müller and Ackenheil, 1995). Increased antibodies against heat shock protein 60 are one of the recent interesting findings in schizophrenia, because it may reflect a mechanism of loss of neuronal protection (Schwarz et al., 1999; Kilidireas et al., 1992).

The key cytokine for the type-2 immune response is IL-4. Increased levels of IL-4 in the CSF of juvenile schizophrenic patients have recently been reported (Mittleman et al., 1997). The CSF findings point out that the increased type-2 response in schizophrenia is not only a phenomenon observed in the peripheral immune system.

Given the heterogeneity of the schizophrenic syndrome, the heterogeneity and high variability of the immune system, methodological problems in the determination of immune variables, in particular of circulating cytokines, and the multiplicity of interfering variables, it is not astonishing that the results of immunological studies in schizophrenia in part also show diverging results (Rapaport and Müller, 2001; Kim et al., 2002). The crucial role of the interfering antipsychotic medication for the outcome of immune variables, which is not regarded in several studies, is discussed elsewhere (Müller et al., 2000; Schuld et al., 2004; Schwarz et al., 2001).

9 Anti-psychotic Drugs Rebalance the Type-1/Type-2 Imbalance

In vitro studies show that the blunted IFN-g production becomes normalized after therapy with neuroleptics (Wilke et al., 1996). An increase of $CD4^{+}CD45RO^{+}$ cells ('memory cells') – one of the main sources of IFN-g production – during anti-psychotic therapy with neuroleptics was observed by different groups (Müller et al., 1997b). Additionally, an increase of soluble IL-2 receptors (sIL-2R) – the increase reflects an increase of activated, IL-2 bearing T-cells – during anti-psychotic treatment was described (Müller et al., 1997a). Reduced sICAM-1 levels show a significant increase during short-term anti-psychotic therapy (Schwarz et al., 2000) and the leucocyte function antigen-1 (LFA-1), the ligand of ICAM-1, shows a significantly increased expression during anti-psychotic therapy (Müller et al., 1999). Moreover, the blunted reaction to vaccination with *salmonella typhi* was not observed in patients medicated with anti-psychotics (Ozek et al., 1971). Recently, an elevation of IL-18 serum levels was described in medicated schizophrenics (Tanaka et al., 2000). Since IL-18 plays a pivotal role in the type-1 immune response, this finding is consistent with other descriptions of type-1 activation during antipsychotic treatment.

Regarding the type-2 response, several studies point out that anti-psychotic therapy is accompanied by a functional decrease of the IL-6 system (Maes et al., 1997; Müller et al., 2000).

However, there are also several studies, indicating an anti-inflammatory effect of antipsychotic drugs (for review see Drzyzga et al., 2006).

10 Type I – Type II Immune Response Imbalance in Schizophrenia Promote the Production of the Endogenous NMDA Receptor Antagonist Kynurenic Acid

The only known naturally occurring NMDA receptor antagonist in the human CNS is KYN-A (Stone, 1993). KYN-A is one of the three neuroactive intermediate products of the kynurenine pathway. Kynurenine (KYN) is the primary major degradation product of tryptophan (TRP). While the excitatory KYN metabolites 3-hydroxykynurenine (3HK) and quinolinic acid are synthesized from KYN en route to NAD, kynurenic acid (KYN-A) is formed in a dead end side arm of the pathway (Schwarcz and Pellicciari, 2002).

KYN-A acts both, as a blocker of the glycine co-agonist site of the NMDA receptor (Kessler et al., 1989) and as a noncompetitive inhibitor of the $\alpha 7$ nicotinic acetylcholine receptor (Hilmas et al., 2001).

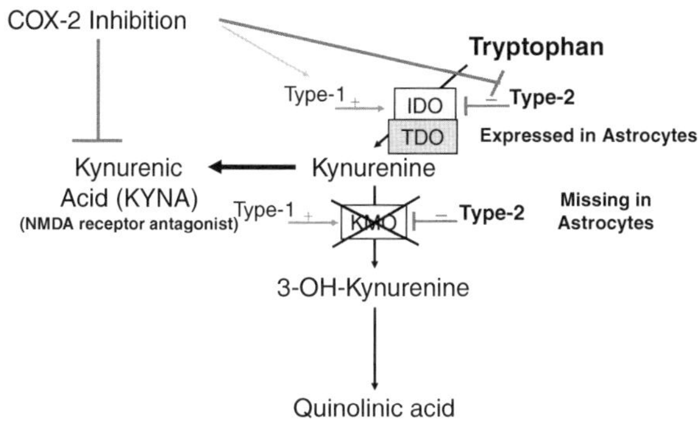

Fig. 1 Metabolization pathways form tryptophan/kynurenine to kynurenic acid (KYNA) and quinolinic acid. Indoleamine 2,3-dioxygenase (IDO) and kynurenine-monoxygenase (KMO) are activated by type-1 cytokines and inhibited by type-2 cytokines. Tryptophan 2,3-dioxygenase (TDO) is expressed in astrocytes, KMO is missing in astrocytes. KYN-A decreases during cyclooxygenase (COX)-2 inhibition

The production of KYN-A is regulated by indoleamine 2,3-dioxygenase (IDO) and tryptophan 2,3-dioxygenase (TDO). Both enzymes catalyze the first step in the pathway, the degradation from tryptophan to kynurenine. Type-1 cytokines, such as IFN-g and IL-2 stimulate the activity of IDO, while type-2 cytokines like IL-4 and IL-10 are IDO inhibitors (Grohmann et al., 2003; Fig. 1).

There is a mutual inhibitory effect of TDO and IDO: a decrease in TDO activity occurs concomitantly with IDO induction, resulting in a coordinate shift in the site (and cell types) of tryptophan degradation (Takikawa et al., 1986). While it has been known for a long time that IDO is expressed in different types of CNS cells, TDO was thought to be restricted to liver tissue for many years. It is known today, however, that TDO is also expressed in CNS cells including neurons and astrocytes (Miller et al., 2004).

The type-2 or Th-2 shift in schizophrenia may result in two functional consequences: The expression of IDO, normally increased by type-1 cytokines, in particular INF-g, is down-regulated, while TDO is up-regulated. Thus, the type-1/type-2 imbalance is associated with the IDO/TDO imbalance. Second, the enzymatic imbalance between IDO and TDO is related to the activation of astrocytes, which in turn are involved in the type-1/type-2 related astrocyte/microglial imbalance (Aloisi et al., 2000).

The functional overweight of astrocytes leads to further accumulation of the end product KYN-A.

Indeed, a study referring to the expression of IDO and TDO in schizophrenia showed exactly these results. An increased expression of TDO compared to IDO was observed in schizophrenic patients and the increased TDO expression was found, as expected, in astrocytes, not in microglial cells (Miller et al., 2004, 2006).

11 Astrocytes, Microglia, and Type-1/Type-2 Response

The cellular sources for the polarized immune response in the CNS are astrocytes and microglia cells. Microglial cells, deriving from peripheral macrophages, secrete preferably type-1 cytokines such as IL-12, while astrocytes inhibit the production of IL-12 and ICAM-1 and secrete the type-2 cytokine IL-10 (Xiao and Link, 1999; Aloisi et al., 2000).

Therefore, the type-1 / type-2 imbalance in the CNS seems to be represented by the imbalance in the activation of microglial cells and astrocytes, although it has to be taken into consideration that the production of cytokines by astrocytes and microglial cells depends on the activation conditions and partly on environment conditions. The view of an over-activation of astrocytes in schizophrenia is supported by the finding of increased levels of S100B – a marker of astrocyte activation of the medication state (Rothermundt et al., 2004a, b). Microglia activation, however, was only found in a small percentage of schizophrenics and is discussed to be a medication effect (Bayer et al., 1999). A type-1 immune activation as an effect of neuroleptic treatment has been observed repeatedly.

12 Cellular Source of Kynurenic Acid in the CNS

Astrocytes play a key role in the production of kynurenic acid in the CNS because astrocytes are the main source of KYN-A (Heyes et al., 1997). The cellular localization of the kynurenine metabolism is primarily in macrophages and microglial cells, but also in astrocytes (Speciale and Schwarcz, 1993; Kiss et al., 2003).

Interestingly, kynurenine 3-monooxygenase (KMO), a critical enzyme in the kynurenine metabolism, is absent in human astrocytes (Guillemin et al., 2001). Accordingly, it has been described that astrocytes cannot produce the intermediate 3HK but are able to produce large amounts of KYN and KYN-A (Guillemin et al., 2001). This supports the observation that inhibition of KMO leads to an increased KYN-A production in the CNS (Chiarugi et al., 1996). The complete metabolism of kynurenine to quinolinic acid is observed only in microglial cells, not in astrocytes. Due to the lack of KMO, KYN-A accumulates in astrocytes.

A second key player in the metabolizing of 3-HK is monocytic cells infiltrating the CNS. Monocytes are responsible for the conversion of astrocytic produced KYN to quinolinic acid (Guillemin et al., 2001). However, the low levels of sICAM-1 (ICAM-1 is the molecule that mainly mediates the penetration of monocytes and lymphocytes into the CNS) in the serum and in the CSF of nonmedicated schizophrenic patients (Schwarz et al., 2000) and the increase of adhesion molecules during antipsychotic therapy indicate that the penetration of monocytes may be reduced in nonmedicated schizophrenic patients (Müller et al., 1999).

13 The Possible Role of Kynurenic Acid in Schizophrenia

The accumulation of KYN-A may lead to schizophrenic symptoms (Erhardt et al., 2003). Accordingly, increased levels of KYN-A have been observed in the CSF of schizophrenic patients (Erhardt et al., 2001a). Since most of the patients in this study were drug-naive first-episode patients, this increase could not be caused by antipsychotic treatment. At any rate, chronic drug treatment with antipsychotics does not result in an increase, but rather in a decrease of KYN-A (Ceresoli-Borroni et al., 2006).

An investigation of CNS tissue specimens in different cortical regions revealed increased KYN-A levels in schizophrenics compared to a control sample, particularly in the prefrontal cortex (Schwarcz et al., 2001). In the amygdala, a small but not significant increase of KYN-A in medicated schizophrenics was observed (Miller et al., 2006). The prefrontal cortex is an area involved in the pathophysiology of schizophrenia (Andreasen et al., 1992).

In recent years, drugs acting as elevators of endogenous KYN-A in the CNS have been identified. One of these substances is PNU 156561A, an inhibitor of KMO. This substance enables studies of the effects of increased endogenous KYN-A levels in animals (Speciale et al., 1996). The effects were similar to the effects observed after administration of MK-801 or PCP: In particular, dopaminergic neurons in the midbrain showed an increased activity (Erhardt et al., 2001b).

Clozapine, however, has modulating, in higher doses inhibitory effects on the activity of dopaminergic neurons in the midbrain, which is mediated by the glycine site of the NMDA receptor (Schwieler et al., 2004). This inhibitory effect of clozapine may account for its beneficial effects in ameliorating symptoms of schizophrenia.

Besides the effects on the NMDA-receptor, KYN-A is also a potent antagonist of the $\alpha7$ nicotinic acetylcholine receptors (Hilmas et al., 2001). This antagonism is associated with cognitive impairment. Compared to other schizophrenic symptoms, cognitive decline is a basic disturbance in schizophrenia (Huber, 1983). The effect on acetylcholine receptors is effective already at lower concentrations of KYN-A compared to the antagonism to the NMDA receptor; the affinity of KYN-A to the $\alpha7$ nicotinic acetylcholine receptor is about twice as high compared to the NMDA receptor (Hilmas et al., 2001). This finding indicates that the impairment of cognitive functions is induced by lower concentrations of KYN-A, while psychotic symptoms appear only at higher concentrations of KYN-A. This view fits with the earlier onset of cognitive disturbance in schizophrenia compared to the acute psychotic symptoms.

It was suggested that increased intracerebral KYN-A levels should be related to an enhanced dopaminergic neurotransmission. A recent study published by Robert Schwarcz's group, however, demonstrated that in the striatum, KYN-A significantly *inhibits* dopamine release (Wu et al., 2007). This effect is mediated through the $\alpha7$ nicotinic ACh receptor antagonism, while the NMDA receptor was obviously not involved in this effect. The authors proved this effect for both acute and chronic (knockout mice) modulation of KYN-A levels. Since this effect has specifically been demonstrated for the striatum, it remains to be investigated if KYN-A has the same effect in other brain regions.

14 COX-2 Inhibitors Inhibit the Production of Kynurenic Acid, Rebalance the Type-1/Type-2 Immune Response, and Have Therapeutic Effects in Early Stages of Schizophrenia

Additional to the above described immunological mechanism, selective cyclooxygenase-2 (COX-2) inhibitors reduce the KYN-A levels by a prostaglandin-mediated mechanism (Schwieler et al., 2005). COX inhibition provokes differential effects on the kynurenine metabolism: while COX-1 inhibitors increase the levels of KYN-A, COX-2 inhibitors decrease them. Therefore, psychotic symptoms and cognitive dysfunctions, observed during therapy with COX-1 inhibitors, were assigned to the COX-1-mediated increase of KYN-A (Schwieler et al., 2005). The balance between KYN-A and quinolinic acid – as the balance between the type-1 and type-2 immune responses – seems to be crucial not only in schizophrenia, as indicated by a reduced KYN-A/quinolinic acid ratio in Huntington's disease (Stoy et al., 2005; Guidetti et al., 2004).

Recently, prostaglandin E2 (PGE2) has been shown to enhance the production of type-2 cytokines such as IL-4, IL-5, IL-6 and IL-10; PGE2 also drastically inhibits the production of the type-1 cytokines IFN-γ, IL-2 and IL-12 (Stolina et al., 2000). Therefore, inhibition of PGE2 synthesis is hypothesized to be beneficial in the treatment of disorders with dysregulated T-helper cell responses (Harris et al., 2002).

One class of modern drugs is well known to induce a shift from the type-1 like to a type-2 dominated immune response: the selective COX-2 inhibitors. Several studies demonstrated the type-2 inducing effect of PGE2 – the major product of COX-2, while inhibition of COX-2 is accompanied by inhibition of type-2 cytokines and induction of type-1 cytokines (Pyeon et al., 2000; Stolina et al., 2000). PGE2 levels in schizophrenia are not well studied; increased levels of PGE2, however, have been described (Kaiya et al., 1989). PGE2 induces the production of IL-6, a cytokine which is consistently described to be increased in schizophrenia. COX-2 inhibition seems to balance the type-1/type-2 immune response by inhibition of IL-6, PGE2, and by stimulating the type-1 immune response (Litherland et al., 1999). Moreover, an increased COX-2 expression was found in schizophrenia (Das and Khan, 1998), although controversial data have recently been published (Yokota et al., 2004). Therefore COX-2 inhibition seems to be a promising approach in the therapy of schizophrenia.

In a prospective, randomized, double-blind study of therapy with the COX-2 inhibitor celecoxib add-on to risperidone in acute exacerbation of schizophrenia, a therapeutic effect of celecoxib was observed (Müller et al., 2002). Immunologically, an increase of the type-1 immune response was found in the celecoxib treatment group (Müller et al., 2004c). The clinical effect of COX-2 inhibition was especially pronounced regarding cognition in schizophrenia (Müller et al., 2005). The finding of a clinical advantage of COX-2 inhibition, however, could not be replicated in a second study. Further analysis of the data revealed that the outcome depends on the duration of the disease (Müller et al., 2004c). The efficacy of therapy with a COX-2 inhibitor seems most pronounced in the first years of the schizophrenic disease process. This observation is in accordance with results from animal studies showing that the effects of COX-2 inhibition on cytokines, hormones, and particularly on behavioural symptoms are dependent on the duration of the preceding changes and the time-point of application of the COX-2 inhibitor (Casolini et al., 2002). Thus, a point of no return for therapeutic effects regarding the pathological changes during an inflammatory process has to be postulated.

Regarding the role of the inflammatory process in schizophrenia and possibly other psychiatric disorders, anti-inflammatory therapy should be taken into the focus for further research (Müller et al., 2004b); COX-2 inhibition is one option among others. Therapeutic research, however, has to consider different levels and different mechanisms for therapeutic targets in the neuroimmune system and the dopaminergic–glutamatergic neurotransmission circuits including the kynurenine pathway of the tryptophan metabolism.

References

Aloisi F, Ria F, Adorini L (2000) Regulation of T-cell responses by CNS antigen-presenting cells: different roles for microglia and astrocytes. Immunol Today 21: 141–147

Andreasen NC, Rezai K, Alliger R, Swayze VW, Flaum M, Kirchner P, et al. (1992) Hypofrontality in neuroleptic-naive patients and in patients with chronic schizophrenia. Assessment with xenon 133 single-photon emission computed tomography and the Tower of London. Arch Gen Psychiatry 49: 943–958

Avgustin B, Wraber B, Tavcar R (2005) Increased Th1 and Th2 immune reactivity with relative Th2 dominance in patients with acute exacerbation of schizophrenia. Croat Med J 46: 268–274

Bayer TA, Buslei R, Havas L, Falkai P (1999) Evidence for activation of microglia in patients with psychiatric illnesses. Neurosci Lett 271: 126–128

Bechter K, Schreiner V, Herzog S, Breitinger N, Wollinsky KH, Brinkmeier H, et al. (2003) CSF filtration as experimental therapy in therapyresistant psychoses in borna disease virus-seropositive patients. Psychiatr Prax 30: 216–220

Brown AS, Begg MD, Gravenstein S, Schaefer CA, Wyatt RJ, Bresnahan M, et al. (2004) Serologic evidence of prenatal influenza in the etiology of schizophrenia. Arch Gen Psychiatry 61: 774–780

Buka SL, Tsuang MT, Torrey EF, Klebanoff MA, Bernstein D, Yolken RH (2001) Maternal infections and subsequent psychosis among offspring. Arch Gen Psychiatry 58: 1032–1037

Carlsson A (1988) The current status of the dopamine hypothesis of schizophrenia. Neuropsychopharmacology 1: 179–186

Carlsson A, Waters N, Holm-Waters S, Tedroff J, Nilsson M, Carlsson ML (2001) Interactions between monoamines, glutamate, and GABA in schizophrenia: new evidence. Annu Rev Pharmacol Toxicol 41: 237–260

Carlsson A (2006) The neurochemical circuitry of schizophrenia. Pharmacopsychiatry 39(Suppl 1): S10–S14

Casolini P, Catalani A, Zuena AR, Angelucci L (2002) Inhibition of COX-2 reduces the age-dependent increase of hippocampal inflammatory markers, corticosterone secretion, and behavioral impairments in the rat. J Neurosci Res 68: 337–343

Cazzullo CL, Scarone S, Grassi B, Vismara C, Trabattoni D, Clerici M, et al. (1998) Cytokines production in chronic schizophrenia patients with or without paranoid behaviour. Prog Neuropsychopharmacol Biol Psychiatry 22: 947–957

Ceresoli-Borroni G, Rassoulpour A, Wu HQ, Guidetti P, Schwarcz R (2006) Chronic neuroleptic treatment reduces endogenous kynurenic acid levels in rat brain. J Neural Transm 113(10): 1355–1365

Chakos MH, Schobel SA, Gu H, Gerig G, Bradford D, Charles C, et al. (2005) Duration of illness and treatment effects on hippocampal volume in male patients with schizophrenia. Br J Psychiatry 186: 26–31

Chiang SSW, Riedel M, Müller N, Ackenheil M, Gruber R, Schwarz MJ (2004) Th2-shift in schizophrenia: primary findings from whole blood in vitro stimulation. Psychiatry Online

Chiarugi A, Carpenedo R, Moroni F (1996) Kynurenine disposition in blood and brain of mice: effects of selective inhibitors of kynurenine hydroxylase and of kynureninase. J Neurochem 67: 692–698

Collier DA, Li T (2003) The genetics of schizophrenia: glutamate not dopamine? Eur J Pharmacol 480: 177–184

Cook GS, Hill DR (2001) Genetics of susceptibility to human infectious disease. Nat Rev Genet 2: 967–977

Das I, Khan NS (1998) Increased arachidonic acid induced platelet chemiluminescence indicates cyclooxygenase overactivity in schizophrenic subjects. Prostaglandins Leukot Essent Fatty Acids 58: 165–168

Dean K, Murray RM (2005) Environmental risk factors for psychosis. Dialogues Clin Neurosci 7: 69–80

Drzyzga L, Obuchowicz E, Marcinowska A, Herman ZS (2006) Cytokines in schizophrenia and the effects of antipsychotic drugs. Brain Behav Immun 20: 532–545

Erhardt S, Blennow K, Nordin C, Skogh E, Lindstrom LH, Engberg G (2001a) Kynurenic acid levels are elevated in the cerebrospinal fluid of patients with schizophrenia. Neurosci Lett 313: 96–98

Erhardt S, Oberg H, Mathe JM, Engberg G (2001b) Pharmacological elevation of endogenous kynurenic acid levels activates nigral dopamine neurons. Amino Acids 20: 353–362

Erhardt S, Schwieler L, Engberg G (2003) Kynurenic acid and schizophrenia. Adv Exp Med Biol 527: 155–165

Farber NB, Wozniak DF, Price MT, Labruyere J, Huss J, St PH, et al. (1995) Age-specific neurotoxicity in the rat associated with NMDA receptor blockade: potential relevance to schizophrenia? Biol Psychiatry 38: 788–796

Fearon P, Cotter P, Murray RM (2000) Is the association between obstretic complications and schizophrenia mediated by glutaminergic excitotoxic damage of the foetal/neonatal brain? In: Revely M, Deacon B, eds. Psychopharmacology of Schizophrenia. London: Chapman and Hall 21–40

Ganguli R, Rabin BS, Kelly RH, Lyte M, Ragu U (1987) Clinical and laboratory evidence of autoimmunity in acute schizophrenia. Ann N Y Acad Sci 496: 676–685

Gattaz WF, Abrahao AL, Foccacia R (2004) Childhood meningitis, brain maturation and the risk of psychosis. Eur Arch Psychiatry Clin Neurosci 254: 23–26

Grohmann U, Fallarino F, Puccetti P (2003) Tolerance, DCs and tryptophan: much ado about IDO. Trends Immunol 24: 242–248

Guidetti P, Luthi-Carter RE, Augood SJ, Schwarcz R (2004) Neostriatal and cortical quinolinate levels are increased in early grade Huntington's disease. Neurobiol Dis 17: 455–461

Guillemin GJ, Kerr SJ, Smythe GA, Smith DG, Kapoor V, Armati PJ, et al. (2001) Kynurenine pathway metabolism in human astrocytes: a paradox for neuronal protection. J Neurochem 78: 842–853

Harris SG, Padilla J, Koumas L, Ray D, Phipps RP (2002) Prostaglandins as modulators of immunity. Trends Immunol 23: 144–150

Heresco-Levy U, Javitt DC, Ermilov M, Mordel C, Silipo G, Lichtenstein M (1999) Efficacy of high-dose glycine in the treatment of enduring negative symptoms of schizophrenia. Arch Gen Psychiatry 56: 29–36

Heyes MP, Chen CY, Major EO, Saito K (1997) Different kynurenine pathway enzymes limit quinolinic acid formation by various human cell types. Biochem J 326: 351–356

Hill AV (1996) Genetics of infectious disease resistance. Curr Opin Genet Dev 6: 348–353

Hilmas C, Pereira EF, Alkondon M, Rassoulpour A, Schwarcz R, Albuquerque EX (2001) The brain metabolite kynurenic acid inhibits alpha7 nicotinic receptor activity and increases non-alpha7 nicotinic receptor expression: physiopathological implications. J Neurosci 21: 7463–7473

Huber G (1983) Das Konzept substratnaher Basissymptome und seine Bedeutung für Theorie und THerapie schizophrener Erkrankungen [The concept of substrate-close basic symptoms and its significance for the theory and therapy of schizophrenic diseases]. Nervenarzt 54: 23–32

Jakob H, Beckmann H (1986) Prenatal developmental disturbances in the limbic allocortex in schizophrenics. J Neural Transm 65: 303–326

Jentsch JD, Roth RH (1999) The neuropsychopharmacology of phencyclidine: from NMDA receptor hypofunction to the dopamine hypothesis of schizophrenia. Neuropsychopharmacology 20: 201–225

Job DE, Whalley HC, McIntosh AM, Owens DG, Johnstone EC, Lawrie SM (2006) Grey matter changes can improve the prediction of schizophrenia in subjects at high risk. BMC Med 4: 29

Kabiersch A, Furukawa H, del RA, Besedovsky HO (1998) Administration of interleukin-1 at birth affects dopaminergic neurons in adult mice. Ann N Y Acad Sci 840: 123–127

Kaiya H, Uematsu M, Ofuji M, Nishida A, Takeuchi K, Nozaki M, et al. (1989) Elevated plasma prostaglandin E2 levels in schizophrenia. J Neural Transm 77: 39–46

Kegeles LS, bi-Dargham A, Zea-Ponce Y, Rodenhiser-Hill J, Mann JJ, Van Heertum RL, et al. (2000) Modulation of amphetamine-induced striatal dopamine release by ketamine in humans: implications for schizophrenia. Biol Psychiatry 48: 627–640

Kessler M, Terramani T, Lynch G, Baudry M (1989) A glycine site associated with *N*-methyl-D-aspartic acid receptors: characterization and identification of a new class of antagonists. J Neurochem 52: 1319–1328

Kilidireas K, Latov N, Strauss DH, Gorig AD, Hashim GA, Gorman JM, et al. (1992) Antibodies to the human 60 kDa heat-shock protein in patients with schizophrenia. Lancet 340: 569–572

Kim JS, Kornhuber HH, Schmid-Burgk W, Holzmuller B (1980) Low cerebrospinal fluid glutamate in schizophrenic patients and a new hypothesis on schizophrenia. Neurosci Lett 20: 379–382

Kim YK, Suh IB, Kim H, Han CS, Lim CS, Choi SH, et al. (2002) The plasma levels of interleukin-12 in schizophrenia, major depression, and bipolar mania: effects of psychotropic drugs. Mol Psychiatry 7: 1107–1114

Kiss C, Ceresoli-Borroni G, Guidetti P, Zielke CL, Zielke HR, Schwarcz R (2003) Kynurenate production by cultured human astrocytes. J Neural Transm 110: 1–14

Koponen H, Rantakallio P, Veijola J, Jones P, Jokelainen J, Isohanni M (2004) Childhood central nervous system infections and risk for schizophrenia. Eur Arch Psychiatry Clin Neurosci 254: 9–13

Körschenhausen DA, Hampel HJ, Ackenheil M, Penning R, Müller N (1996) Fibrin degradation products in post mortem brain tissue of schizophrenics: a possible marker for underlying inflammatory processes. Schizophr Res 19: 103–109

Krystal JH, Karper LP, Seibyl JP, Freeman GK, Delaney R, Bremner JD, et al. (1994) Subanesthetic effects of the noncompetitive NMDA antagonist, ketamine, in humans. Psychotomimetic, perceptual, cognitive, and neuroendocrine responses. Arch Gen Psychiatry 51: 199–214

Laruelle M, Abi-Dargham A, van Dyck CH, Gil R, D'Souza CD, Erdos J, et al. (1996) Single photon emission computerized tomography imaging of amphetamine-induced dopamine release in drug-free schizophrenic subjects. Proc Natl Acad Sci U S A 93: 9235–9240

Laruelle M, Frankle WG, Narendran R, Kegeles LS, bi-Dargham A (2005) Mechanism of action of antipsychotic drugs: from dopamine D(2) receptor antagonism to glutamate NMDA facilitation. Clin Ther 27(Suppl A): S16–S24

Laumbacher B, Müller N, Bondy B, Schlesinger B, Gu S, Fellerhoff B, et al. (2003) Significant frequency deviation of the class I polymorphism HLA-A10 in schizophrenic patients. J Med Genet 40: 217–219

Leweke FM, Gerth CW, Koethe D, Klosterkotter J, Ruslanova I, Krivogorsky B, et al. (2004) Antibodies to infectious agents in individuals with recent onset schizophrenia. Eur Arch Psychiatry Clin Neurosci 254: 4–8

Litherland SA, Xie XT, Hutson AD, Wasserfall C, Whittaker DS, She JX, et al. (1999) Aberrant prostaglandin synthase 2 expression defines an antigen-presenting cell defect for insulin-dependent diabetes mellitus. J Clin Invest 104: 515–523

Maes M, Bosmans E, Kenis G, De Jong R, Smith RS, Meltzer HY (1997) In vivo immunomodulatory effects of clozapine in schizophrenia. Schizophr Res 26: 221–225

Marshall BE, Longnecker DE (1990) General anesthetics. In: Goodman LS, Gilman A, Rall TW, et al., eds. The pharmacological basis of therapeutics. Elmsford, NY: Pergamon Press 285–310

Meira-Lima IV, Pereira AC, Mota GF, Floriano M, Araujo F, Mansur AJ, et al. (2003) Analysis of a polymorphism in the promoter region of the tumor necrosis factor alpha gene in schizophrenia and bipolar disorder: further support for an association with schizophrenia. Mol Psychiatry 8: 718–720

Miller DW, Abercrombie ED (1996) Effects of MK-801 on spontaneous and amphetamine-stimulated dopamine release in striatum measured with in vivo microdialysis in awake rats. Brain Res Bull 40: 57–62

Miller CL, Llenos IC, Dulay JR, Barillo MM, Yolken RH, Weis S (2004) Expression of the kynurenine pathway enzyme tryptophan 2,3-dioxygenase is increased in the frontal cortex of individuals with schizophrenia. Neurobiol Dis 15: 618–629

Miller CL, Llenos IC, Dulay JR, Weis S (2006) Upregulation of the initiating step of the kynurenine pathway in postmortem anterior cingulate cortex from individuals with schizophrenia and bipolar disorder. Brain Res 1073–1074: 25–37

Mills CD, Kincaid K, Alt JM, Heilman MJ, Hill AM (2000) M-1/M-2 macrophages and the Th1/Th2 paradigm. J Immunol 164: 6166–6173

Mittleman BB, Castellanos FX, Jacobsen LK, Rapoport JL, Swedo SE, Shearer GM (1997) Cerebrospinal fluid cytokines in pediatric neuropsychiatric disease. J Immunol 159: 2994–2999

Molholm HB (1942) Hyposensitivity to foreign protein in schizophrenic patients. Psychiatr Quarterly 16: 565–571

Müller N, Ackenheil M, Hofschuster E, Mempel W, Eckstein R (1991) Cellular immunity in schizophrenic patients before and during neuroleptic treatment. Psychiatry Res 37: 147–160

Müller N, Ackenheil M (1995) Immunoglobulin and albumin content of cerebrospinal fluid in schizophrenic patients: relationship to negative symptomatology. Schizophr Res 14: 223–228

Müller N, Empl M, Riedel M, Schwarz M, Ackenheil M (1997a) Neuroleptic treatment increases soluble IL-2 receptors and decreases soluble IL-6 receptors in schizophrenia. Eur Arch Psychiatry Clin Neurosci 247: 308–313

Müller N, Riedel M, Schwarz MJ, et al. (1997b) Immunomodulatory effects of neuroleptics to the cytokine system and the cellular immune system in schizophrenia. In: Wieselmann G, ed. Current Update in Psychoimmunology. Wien, NY: Springer 57–67

Müller N, Ackenheil M (1998) Psychoneuroimmunology and the cytokine action in the CNS: implications for psychiatric disorders. Prog Neuropsychopharmacol Biol Psychiatry 22: 1–33

Müller N, Riedel M, Hadjamu M, Schwarz MJ, Ackenheil M, Gruber R (1999) Increase in expression of adhesion molecule receptors on T helper cells during antipsychotic treatment and relationship to blood-brain barrier permeability in schizophrenia. Am J Psychiatry 156: 634–636

Müller N, Riedel M, Ackenheil M, Schwarz MJ (2000) Cellular and humoral immune system in schizophrenia: a conceptual re-evaluation. World J Biol Psychiatry 1: 173–179

Müller N, Riedel M, Scheppach C, Brandstätter B, Sokullu S, Krampe K, et al. (2002) Beneficial antipsychotic effects of celecoxib add-on therapy compared to risperidone alone in schizophrenia. Am J Psychiatry 159: 1029–1034

Müller N (2004a) Immunological and infectious aspects of schizophrenia. Eur Arch Psychiatry Clin Neurosci 254: 1–3

Müller N, Riedel M, Schwarz MJ (2004b) Psychotropic effects of COX-2 inhibitors – a possible new approach for the treatment of psychiatric disorders. Pharmacopsychiatry 37: 266–269

Müller N, Ulmschneider M, Scheppach C, Schwarz MJ, Ackenheil M, Möller HJ, et al. (2004c) COX-2 inhibition as a treatment approach in schizophrenia: immunological considerations and clinical effects of celecoxib add-on therapy. Eur Arch Psychiatry Clin Neurosci 254: 14–22
Müller N, Riedel M, Schwarz MJ, Engel RR (2005) Clinical effects of COX-2 inhibitors on cognition in schizophrenia. Eur Arch Psychiatry Clin Neurosci 255: 149–151
Murray RM, Lewis SW (1987) Is schizophrenia a neurodevelopmental disorder? Br Med J (Clin Res Ed) 295: 681–682
Numakawa T, Yagasaki Y, Ishimoto T, Okada T, Suzuki T, Iwata N, et al. (2004) Evidence of novel neuronal functions of dysbindin, a susceptibility gene for schizophrenia. Hum Mol Genet 13: 2699–2708
Olney JW, Farber NB (1995) Glutamate receptor dysfunction and schizophrenia. Arch Gen Psychiatry 52: 998–1007
Ozek M, Toreci K, Akkok I, Guvener Z (1971) Influence of therapy on antibody-formation. Psychopharmacologia 21: 401–412
Pearce BD (2001) Schizophrenia and viral infection during neurodevelopment: a focus on mechanisms. Mol Psychiatry 6: 634–646
Pilowsky LS, Bressan RA, Stone JM, Erlandsson K, Mulligan RS, Krystal JH, et al. (2006) First in vivo evidence of an NMDA receptor deficit in medication-free schizophrenic patients. Mol Psychiatry 11: 118–119
Pyeon D, Diaz FJ, Splitter GA (2000) Prostaglandin E(2) increases bovine leukemia virus tax and pol mRNA levels via cyclooxygenase 2: regulation by interleukin-2, interleukin-10, and bovine leukemia virus. J Virol 74: 5740–5745
Rapaport MH, Müller N (2001) Immunological states associated with schizophrenia. In: Ader A, Felten DL, Cohnen N, eds. Psychoneuroimmunology, vol 2, 3rd ed. San Diego, CA: Academic Press 373–382
Riedel M, Krönig H, Schwarz MJ, Engel RR, Kuhn KU, Sikorski C, et al. (2002) No association between the G308A polymorphism of the tumor necrosis factor-alpha gene and schizophrenia. Eur Arch Psychiatry Clin Neurosci 252: 232–234
Riedel M, Spellmann I, Schwarz MJ, Strassnig M, Sikorski C, Möller HJ, et al. (2007) Decreased T cellular immune response in schizophrenic patients. J Psychiatr Res 41:3–7
Rothermundt M, Arolt V, Bayer TA (2001) Review of immunological and immunopathological findings in schizophrenia. Brain Behav Immun 15: 319–339
Rothermundt M, Falkai P, Ponath G, Abel S, Burkle H, Diedrich M, et al. (2004a) Glial cell dysfunction in schizophrenia indicated by increased S100B in the CSF. Mol Psychiatry 9: 897–899
Rothermundt M, Ponath G, Arolt V (2004b) S100B in schizophrenic psychosis. Int Rev Neurobiol 59: 445–470
Schuld A, Hinze-Selch D, Pollmacher T (2004) Zytokinnetzwerke bei Patienten mit Schizophrenie und ihre Bedeutung für die Pathophysiologie der Erkrankung [Cytokine network in patients with schizophrenia and its significance for the pathophysiology of the illness]. Nervenarzt 75: 215–226
Schwab SG, Hallmayer J, Albus M, Lerer B, Eckstein GN, Borrmann M, et al. (2000) A genome-wide autosomal screen for schizophrenia susceptibility loci in 71 families with affected siblings: support for loci on chromosome 10p and 6. Mol Psychiatry 5: 638–649
Schwarcz R, Rassoulpour A, Wu HQ, Medoff D, Tamminga CA, Roberts RC (2001) Increased cortical kynurenate content in schizophrenia. Biol Psychiatry 50: 521–530
Schwarcz R, Pellicciari R (2002) Manipulation of brain kynurenines: glial targets, neuronal effects, and clinical opportunities. J Pharmacol Exp Ther 303: 1–10
Schwarz MJ, Riedel M, Gruber R, Ackenheil M, Müller N (1999) Antibodies to heat shock proteins in schizophrenic patients: implications for the mechanism of the disease. Am J Psychiatry 156: 1103–1104
Schwarz MJ, Riedel M, Ackenheil M, Müller N (2000) Decreased levels of soluble intercellular adhesion molecule-1 (sICAM-1) in unmedicated and medicated schizophrenic patients. Biol Psychiatry 47: 29–33

Schwarz MJ, Chiang S, Müller N, Ackenheil M (2001) T-helper-1 and T-helper-2 responses in psychiatric disorders. Brain Behav Immun 15: 340–370
Schwarz MJ, Kronig H, Riedel M, Dehning S, Douhet A, Spellmann I, et al. (2005) IL-2 and IL-4 polymorphisms as candidate genes in schizophrenia. Eur Arch Psychiatry Clin Neurosci 256(2): 72–76
Schwieler L, Engberg G, Erhardt S (2004) Clozapine modulates midbrain dopamine neuron firing via interaction with the NMDA receptor complex. Synapse 52: 114–122
Schwieler L, Erhardt S, Erhardt C, Engberg G (2005) Prostaglandin-mediated control of rat brain kynurenic acid synthesis – opposite actions by COX-1 and COX-2 isoforms. J Neural Transm 112: 863–872
Speciale C, Schwarcz R (1993) On the production and disposition of quinolinic acid in rat brain and liver slices. J Neurochem 60: 212–218
Speciale C, Wu HQ, Cini M, Marconi M, Varasi M, Schwarcz R (1996) (R,S)-3,4-dichlorobenzoylalanine (FCE 28833A) causes a large and persistent increase in brain kynurenic acid levels in rats. Eur J Pharmacol 315: 263–267
Sperner-Unterweger B, Miller C, Holzner B, et al. (1999) Measurement of neopterin, kynurenine and tryptophan in sera of schizophrenic patients. In: Müller N, ed. Psychiatry, Psychoimmunology, and Viruses. Wien, NY: Springer 115–119
Stolina M, Sharma S, Lin Y, Dohadwala M, Gardner B, Luo J, et al. (2000) Specific inhibition of cyclooxygenase 2 restores antitumor reactivity by altering the balance of IL-10 and IL-12 synthesis. J Immunol 164: 361–370
Stone TW (1993) Neuropharmacology of quinolinic and kynurenic acids. Pharmacol Rev 45: 309–379
Stoy N, Mackay GM, Forrest CM, Christofides J, Egerton M, Stone TW, et al. (2005) Tryptophan metabolism and oxidative stress in patients with Huntington's disease. J Neurochem 93: 611–623
Sumiyoshi T, Anil AE, Jin D, Jayathilake K, Lee M, Meltzer HY (2004) Plasma glycine and serine levels in schizophrenia compared to normal controls and major depression: relation to negative symptoms. Int J Neuropsychopharmacol 7: 1–8
Sumiyoshi T, Jin D, Jayathilake K, Lee M, Meltzer HY (2005) Prediction of the ability of clozapine to treat negative symptoms from plasma glycine and serine levels in schizophrenia. Int J Neuropsychopharmacol 8: 451–455
Takikawa O, Yoshida R, Kido R, Hayaishi O (1986) Tryptophan degradation in mice initiated by indoleamine 2,3-dioxygenase. J Biol Chem 261: 3648–3653
Talbot K, Eidem WL, Tinsley CL, Benson MA, Thompson EW, Smith RJ, et al. (2004) Dysbindin-1 is reduced in intrinsic, glutamatergic terminals of the hippocampal formation in schizophrenia. J Clin Invest 113: 1353–1363
Tamminga CA, Cascella N, Fakouhl TD, et al. (1992) Enhancement of NMDA-mediated transmission in schizophrenia: effects of milacemide. In: Meltzer HY, ed. Novel Antipsychotic Drugs. New York: Raven Press 171–177
Tanaka KF, Shintani F, Fujii Y, Yagi G, Asai M (2000) Serum interleukin-18 levels are elevated in schizophrenia. Psychiatry Res 96: 75–80
Torrey EF, Miller J, Rawlings R, Yolken RH (1997) Seasonality of births in schizophrenia and bipolar disorder: a review of the literature. Schizophr Res 28: 1–38
van Kammen DP, McAllister-Sistilli CG, Kelley ME (1997) Relationship between immune and behavioral measures in schizophrenia. In: Wieselmann G, ed. Current Update in Psychoimmunology. Wien, NY: Springer 51–55
Westergaard T, Mortensen PB, Pedersen CB, Wohlfahrt J, Melbye M (1999) Exposure to prenatal and childhood infections and the risk of schizophrenia: suggestions from a study of sibship characteristics and influenza prevalence. Arch Gen Psychiatry 56: 993–998
Wilke I, Arolt V, Rothermundt M, Weitzsch C, Hornberg M, Kirchner H (1996) Investigations of cytokine production in whole blood cultures of paranoid and residual schizophrenic patients. Eur Arch Psychiatry Clin Neurosci 246: 279–284

Williams NM, Preece A, Spurlock G, Norton N, Williams HJ, Zammit S, et al. (2003) Support for genetic variation in neuregulin 1 and susceptibility to schizophrenia. Mol Psychiatry 8: 485–487
Wu HQ, Rassoulpour A, Schwarcz R (2007) Kynurenic acid leads, dopamine follows: a new case of volume transmission in the brain? J Neural Transm 114:33–41
Xiao BG, Link H (1999) Is there a balance between microglia and astrocytes in regulating Th1/Th2-cell responses and neuropathologies? Immunol Today 20: 477–479
Yokota O, Terada S, Ishihara T, Nakashima H, Kugo A, Ujike H, et al. (2004) Neuronal expression of cyclooxygenase-2, a pro-inflammatory protein, in the hippocampus of patients with schizophrenia. Prog Neuropsychopharmacol Biol Psychiatry 28: 715–721
Yolken RH, Torrey EF (1995) Viruses, schizophrenia, and bipolar disorder. Clin Microbiol Rev 8: 131–145
Zuckerman L, Weiner I (2005) Maternal immune activation leads to behavioral and pharmacological changes in the adult offspring. J Psychiatr Res 39: 311–323

Immunobiological and Neural Substrates of Cancer-Related Neurocognitive Deficits

Martin Klein

Abstract In addition to neurocognitive deficits, cancer patients may also experience fatigue, mood disorders, sleep disturbance, and sexual dysfunction, all of which are factors that negatively affect both neurocognitive functioning and health-related quality of life. The exact mechanisms responsible for neurocognitive and behavioral symptoms in cancer patients are not well understood, challenging clinicians to develop effective treatments.

Side effects related to systemic chemotherapy and radiotherapy have been extensively investigated in clinical trials, but the effect of cancer treatment on neurocognitive function is one of the less well-characterized experiences of the patient. Apart from the ability of chemotherapy to enter the brain across the blood-brain barrier, DNA damage and shortened telomere length resulting from chemotherapy, problems with neural repair such as the presence of apolipoprotein E4 alleles, and decreased neurotransmitter activity as potential mechanisms underlying neurocognitive deficits in cancer patients, problems with cytokine regulation might provide an important explanations of compromised neurocognitive functioning.

This chapter will discuss how changes in levels of cytokines are associated with changes in neurocognitive functioning in cancer patients.

Keywords Cancer · Cytokines · Neurocognitive functioning · Radiotherapy · Chemotherapy · Mood

1 Introduction

Although a moderate increase in survival rates is seen in distinct patient groups (Gloeckler Ries et al. 2003), cancer remains one of the most difficult life-threatening diseases to treat, a consequence of factors which include limitations of animal models, tumor diversity, drug resistance, and side effects of therapy. Moreover, due

M. Klein (✉)
Department of Medical Psychology – D349, VU University Medical Center, Van der Boechorststraat 7, 1081 BT Amsterdam, The Netherlands
e-mail: m.klein@vumc.nl

A. Siegel, S.S. Zalcman (eds.), *The Neuroimmunological Basis of Behavior and Mental Disorders*, DOI 10.1007/978-0-387-84851-8_16,

to the aging populations in most countries, the incidence of cancer is increasing, and it has been estimated that in 20 years time there will be 20 million new cancer patients worldwide each year (Sikora et al. 2004). An optimistic view, however, is that in the coming decades great strides will be made in the treatment of cancer using a combination of modern surgical techniques, radiation, chemotherapy treatment, and smart drugs custom-designed to block the growth of cancer cells thus rendering cancer becoming considered not as a fatal but as a chronic disease.

Up till then, however, significant numbers of cancer patients will be confronted with reductions in neurocognitive functioning caused by the tumor, by local therapies such as surgery and radiotherapy, systemic therapies such as chemotherapy and endocrine therapy, and biological agents such as biological response-modifiers and molecular-targeted therapies, or, not in the least, by depressed mood as a psychological consequence of the often dismal prognosis. More likely, a combination of these factors will contribute to neurocognitive dysfunction.

Many side effects related to systemic therapies and radiotherapy have been extensively investigated in clinical trials, but the effect of cancer treatment on neurocognitive function is one of the less well-characterized experiences of the patient. Apart from the ability of chemotherapy to enter the brain across the blood–brain barrier, DNA damage and shortened telomere length resulting from chemotherapy, problems with neural repair such as the presence of apolipoprotein E4 alleles, and decreased neurotransmitter activity as potential mechanisms underlying neurocognitive deficits in cancer patients (Ahles and Saykin 2007), problems with cytokine regulation might provide an important explanans of compromised neurocognitive functioning.

This chapter will discuss how changes in levels of cytokines are associated with changes in neurocognitive functioning in cancer patients.

There are a number of factors that increase the likelihood that cancer patients will exhibit activation of inflammatory pathways. Surgery, chemotherapy, and radiation are all associated with significant tissue damage and destruction, which in turn is related to activation of innate immune responses. Also, chemotherapeutic agents and gamma-irradiation are capable of directly inducing nuclear factor κB and its downstream proinflammatory gene products (Aggarwal et al. 2006). Moreover, receiving a diagnosis of cancer and battling with chronic uncertainties regarding treatment, recurrence, and mortality is one of the greatest stressors imaginable. Given the impact of stress on inflammatory responses, the confluence of the physical and psychological challenges inherent in having and being treated for cancer place the cancer patient at high risk for the development of inflammation-induced behavioral alterations.

2 Immune System and Cancer

The immune system serves as one of the primary defenses against cancer. When normal tissue becomes a tumor or cancerous tissue, new antigens develop on their surface. These antigens send a signal to immune cells such as the cytotoxic T lymphocytes, natural killer (NK) cells, and macrophages, which in turn directly may kill the tumor cells or release substances like cytokines that ultimately may bring about tumor cell death. Thus, under normal circumstances, the immune system provides continued surveillance and eliminates cells that become cancers. However, tumors may survive by hiding or disguising their tumor antigens, or by producing substances that allow suppressor T cells to proliferate. A number of studies have supported the association between inflammation and an increased risk of developing several types of cancer, including bladder, cervical, and ovarian (Balkwill and Mantovani 2001, de Visser et al. 2006). In addition, cytokines have been shown to be increased in cancer patients before treatment. In patients with advanced breast cancer, interleukin (IL)-6 and tumor necrosis factor-α were increased compared with healthy controls (Tsavaris et al. 2002).

2.1 Cancer Immunotherapeutic Agents

Researchers have been working on stimulating immune cells in cancer patients with substances broadly classified as biological response modifiers. Biological response modifiers are basically used alone or as adjuvants to cancer chemotherapeutic agents. Biological response modifiers can act passively by enhancing the immunologic response to tumor cells or actively by altering the differentiation/growth of tumor cells. Active immunotherapy with cytokines, which can also be classified as biological response modifiers, such as interferons (IFNs) and ILs, is a form of nonspecific active immune stimulation. During the 1990s the number of cytokines identified increased enormously and the functions associated with them are of immense potential in diagnostics and immune therapy. Some of the key cytokines that have proven therapeutic value in cancer are IL2, IFNγ, and IL12. The IFNs have been tested as therapies for many hematological and solid neoplasms and have demonstrated therapeutic benefits in various cancers. Moreover, IL2 has already gained FDA approval for the treatment of renal cell carcinoma and metastatic melanoma. Success has been achieved in the area of immunotherapy, especially in the area of passive immunotherapy using monoclonal antibodies. Other strategies, such as the use of antiangiogenic agents, matrix metalloprotease inhibitors, tyrosine kinase inhibitors, and tumor vaccines, have also been met with some success.

Antibodies are a key component of the adaptive immune response, playing a central role in both the recognition of foreign antigens and the stimulation of an immune response to them. It is not surprising therefore that many immunotherapeutic approaches involve the use of antibodies. The advent of monoclonal antibody technology has made it possible to raise antibodies against specific antigens such as the unusual antigens that are presented on the surfaces of tumors.

The development and testing of second-generation immunotherapies are well under way. While antibodies targeted to disease-causing antigens can be effective under certain circumstances, in many cases, their efficacy may be limited by other factors. In the case of tumors, the microenvironment is immunosuppressive, allowing even those tumors that present unusual antigens to survive and flourish in spite of the immune response generated by the cancer patient, against his or her own tumor tissue. Certain members of a group of cytokines, such as IL2 play a key role in modulating the immune response, and have been tried in conjunction with antibodies in order to generate an even more devastating immune response against the tumor. While the therapeutic administration of such cytokines may cause systemic inflammation, resulting in serious side effects and toxicity, a new generation of chimeric molecules consisting of an immune-stimulatory cytokine attached to an antibody that targets the cytokine's activity to tumors, are able to generate a very effective yet localized immune response against the tumor tissue, destroying the cancer-causing cells without the unwanted side effects.

One of the major adverse effects of cancer chemotherapy is immunosuppression, which leads to many opportunistic infections; so hematopoietic factors, such as colony-stimulating factor (CSF), have been utilized to increase the immune response. Hematopoietic agents such as granulocyte macrophage CSF (sargramostim) and granulocyte CSF (filgrastim) have been used to increase immunity.

2.2 Cytokines Within the Central Nervous System (CNS)

The concept of the brain as an immunologically isolated organ separated entirely from circulating immune cells by the blood–brain barrier has been modified since various evidence has shown many functional interactions between neural, immune, and neuroendocrine systems (Felten and Felten 1991, Cserr and Knopf 1992, Ader et al. 1995, McGeer and McGeer 1995).

In particular, many molecules and pathways associated with immune and inflammatory responses have been identified in the CNS, and are activated in various pathophysiological conditions. Among these molecules, pro- and anti-inflammatory cytokines mediate a functional cross talk between neurons, glia, and cells of the immune system.

Cytokines have been implicated as mediators and modulators of various physiological CNS functions, including effects on the hypothalamo-pituitary-adrenal (HPA) axis, fever responses, and somnogenic effects (Schobitz et al. 1994, Hopkins and Rothwell). Cytokines also affect various neurotransmitter systems (De Simoni and Imeri 1998, Rothwell and Hopkins 1995), and the expression of neuropeptides and neurotrophic factors in several forebrain areas (Lapchak and Araujo 1993, Scarborough et al. 1989, Spranger et al. 1990). Deregulation of cytokine activity has been shown to play a role in neurotoxicity and diverse forms of neurodegenerative disorders, including Alzheimer's disease, vascular dementia, multiple sclerosis, and Parkinson's disease (Tonelli et al. 2005). In particular, electrophysiological findings have shown that relatively low concentrations of IL1β, IL6, and tumor necrosis factor-α (TNFα). inhibit long-term potentiation (Katsuki et al. 1990, Bellinger et al. 1993, Cunningham et al. 1996) and affect excitatory synaptic transmission (Coogan and O'Connor 1997, D'Arcangelo et al. 1997, Zeise et al. 1997, Wang et al. 2000) and ionic currents (Wang et al. 2000, Plata-Salaman and Ffrench-Mullen 1992).

Besides infiltration or indirect signaling from the periphery, cytokines (including IFNα, IFNγ, IL1, IL2, IL6, and TNFα), as well as their receptors, have been reported to be constitutively produced in the CNS itself, mainly by astrocytes and microglia (McGeer and McGeer 1995, Breder et al. 1988, Freidin et al. 1992, Breder et al. 1993). Cytokine production has been found to occur in both subcortical and cortical brain areas, including the periventricular regions, hypothalamus, hippocampus, cerebellum, forebrain regions, basal ganglia, and brain stem nuclei (Breder et al. 1988, Kronfol and Remick 2000, Anisman and Merali 2002). IL1 release within the brain has most clearly been demonstrated in the hypothalamus and hippocampus (Breder et al. 1988, Maier and Watkins 1998).

Communication between peripheral cytokines outside the CNS and cytokines in the brain and spinal fluid takes place through various mechanisms, including direct penetration of the brain by peripheral cytokines across the blood–brain barrier via active transport mechanisms or indirectly via vagal nerve stimulation by peripheral cytokines stimulating the release of central cytokines (Wilson et al. 2002, Kelley et al. 2003).

Although the exact functional role of constitutive central cytokine production remains to be elucidated, centrally synthesized proinflammatory cytokines, including IL1, IL6, TNFα, and IFNγ, are thought to contribute to neuronal development and plasticity, synaptogenesis, and tissue repair (Aloisi et al. 1995, Munoz-Fernandez and Fresno 1998, Beattie et al. 2002). In contrast, proinflammatory cytokines may also have adverse effects on brain cell function, as IL1 has been found to promote neuronal damage after central insults (Anisman and Merali 2002). Next to constitutive expression, cytokine production by glial cells may also be triggered by antigenic challenges (such as viral infection), local or peripheral inflammatory responses, and brain injury (Cserr and Knopf 1992).

3 Cytokines, Neuronal Pathways, and Neurocognitive Functioning

Cytokines have been associated with effects on monoaminergic neurotransmission, i.e., serotonin, noradrenaline, and dopamine which all are neurotransmitters important for normal neurocognitive functioning (Wilson et al. 2002, Linthorst et al. 1995, Merali et al. 1997, Pauli et al. 1998, Song et al. 1999, Lacosta et al. 2000). The profile of neurochemical changes, which varies among different brain regions, is cytokine-specific (Song et al. 1999).

Neurocognitive changes associated with cytokine deregulation are most likely associated with both direct and indirect mechanisms. Neuronal damage secondary to cytokine exposure can be caused by various mechanisms, including excitotoxic glutamate receptor-mediated damage and oxidative stress (Wilson et al. 2002). Consistent with their involvement as mediators of bidirectional communication between the CNS and the peripheral immune system, cytokines play a key role in the HPA axis activation seen in stress and depression. Indirect means by which peripheral or central cytokine deregulation could affect cognition include impaired sleep regulation, micronutrient deficiency induced by appetite suppression, and an array of endocrine interactions (Wilson et al. 2002).

The effects of proinflammatory cytokines on brain function have been investigated primarily for IL1. It has been found that IL1 may stimulate the production of other cytokines (such as IL6 and TNFα) by astrocytes and microglia, hence promoting inflammatory processes in the brain (Maier and Watkins 1998).

Systemic administration of IL1 in rats has been observed to increase dopamine, noradrenaline, and serotonin activities in the hypothalamic nuclei, controlling body temperature, hunger, thirst, fatigue, anger, and circadian cycles; the nucleus accumbens, playing an important role in reward, laughter, pleasure, addiction and fear (Schwienbacher et al. 2004)., and the limbic system, including the hippocampus, supporting a variety of functions including emotion, behavior and long-term memory storage (Merali et al. 1997, Song et al. 1999, Anisman and Merali 1999, Dunn et al. 1999).

Increases in hippocampal extracellular serotonin concentrations have also been found following intracerebroventricular administration of IL1β in rats (Linthorst et al. 1995). Furthermore, IL1 may be considered as one of the key mediators of the neurochemical alterations that occur in response to an immunological challenge: the increase in hippocampal noradrenaline and serotonin concentrations following intraperitoneal administration of the bacterial endotoxin lipopolysaccharide was significantly attenuated by intracerebroventricular pre-treatment with IL1 receptor antagonist (IL1RA; Linthorst et al. 1995, Linthorst and Reul 1998). Moreover, the stimulating effects of intraperitoneal lipopolysaccharide on hippocampal serotonin neurotransmission can be mimicked by intracerebroventricular application of IL1β (Linthorst et al. 1995). Patients with acute myelogenous leukemia or myelodysplastic syndrome were found to have highly increased levels of the circulating cytokines IL1, IL1RA, IL6, IL8, and TNFα when compared with normal controls. Higher IL6 levels in these patients were associated with poorer executive function, whereas, surprisingly, higher IL8 levels were associated with better memory performance (Meyers et al. 2005). Furthermore, higher levels of IL6 in these patients were associated with poorer performance on measures of executive function before treatment.

The results obtained by studies investigating the neurochemical effects of IL2 appear rather heterogeneous: while acute systemic IL2 administration in rats only caused an increase in noradrenaline activity in the paraventricular nucleus of the hypothalamus, repeated systemic administration of IL2 also stimulated noradrenaline neurotransmission in the median eminence, the hippocampus, the central amygdala, which performs a primary role in the formation and storage of memories associated with emotional events, and the prefrontal cortex, whose basic activity is considered to be the orchestration of thoughts and actions in accordance with internal goals (i.e., executive function; Lacosta et al. 2000).

In contrast, repeated IL2 administration decreased noradrenaline turnover in the locus coeruleus as well as interstitial dopamine concentrations in the caudate nucleus and substantia nigra (Lacosta et al. 2000). In a study performed by Song et al. (1999) systemic IL2 administration attenuated dopamine turnover in the nucleus accumbens, but did not affect serotonin activity in this specific brain area. In addition, whereas intracerebroventricular administration of TNFα in rats did not affect hippocampal serotonin neurotransmission, central IL2 administration has been reported to increase extracellular serotonin concentrations in the hippocampus (Pauli et al. 1998). The stimulating effects of IL2 on serotonin activity may be mediated by endogenous IL1, as intracerebroventricular pre-treatment with IL1RA inhibited the IL2-induced alterations in hippocampal extracellular serotonin concentrations (Pauli et al. 1998). Symptoms in cancer patients who have received immunotherapy with IL2 or IFNα include depression, weakness, fatigue, and neurocognitive disruption (Trask et al. 2000). The incidence of late psychiatric side effects, which include symptoms of depression, anxiety, and occasional suicidal ideation, range from 0 to 70% in cancer patients undergoing IL2 or IFNα treatment. However, depression associated with IFNα treatment may partially be reduced by psychostimulants including methylphenidate (Camacho and Ng 2006).

Follow-up studies into the effects of IFNα and IL2 on neurocognitive performance have shown deterioration specifically in information processing speed, executive function, spatial ability, and reaction time that could not primarily be attributed to depressed mood (Scheibel et al. 2004, Capuron et al. 2001). Patients treated with IL2 alone had impaired spatial working memory and lower accuracy of planning abilities. In contrast, patients treated with IFNα alone did not show any impairment in performance accuracy, but showed longer latencies in reaction time tests (Capuron et al. 2001).

Systemic administration of IL6 has been found to increase serotonin neurotransmission in the hippocampus, nucleus accumbens, and frontal cortex (Song et al. 1999, Dunn et al. 1999, Muller and Ackenheil 1998) and to decrease interstitial dopamine concentrations in the nucleus accumbens (Song et al. 1999). An intravenous injection of Escherichia coli endotoxin (0.2 ng/kg) in healthy controls had no effect on body temperature, cortisol levels, blood pressure or heart rate, but circulating levels of TNFα and IL6 increased 2- and 7-fold, respectively. Declarative memory performance in these subjects was inversely correlated with increased cytokine levels suggesting very low-dose endotoxemia to affect neurocognitive functioning independently of physical stress symptoms, and activity of the HPA axis. Another study in healthy subjects also injected with Escherichia coli endotoxin showed that a postexposure increase in IL6, and upregulation of the normally rare, stress-induced AChE-R variant, was associated with improved postexposure working memory performance, indicating that limited cholinergic reactions may be beneficial for cognition (Ofek et al. 2007).

Neuroendocrine effects of peripheral and central proinflammatory cytokines, such as IL1 and IL6, involve activation of the HPA axis, which is reflected in increased plasma concentrations of corticotrophin-releasing hormone, vasopressin

(VP), ACTH, and corticosteroids (Pauli et al. 1998, Sapolsky et al. 1987, Harbuz et al. 1992, Xu et al. 1999, Brebner et al. 2000). An increase in the synthesis of corticotrophin-releasing hormone and VP (which act synergistically to trigger pituitary ACTH release) by hypothalamic neurons may occur through stimulation by nerve fibers projecting from the nucleus tractus solitarius (which receives input from the vagus nerve concerning peripheral IL1 signaling; Maier and Watkins 1998). corticotrophin-releasing hormone production may also be stimulated by central cytokines (particularly IL1) in an indirect manner, via their effects on central monoaminergic neurotransmission and, consequently, the secretory profile of hypothalamic neurons (Dunn et al. 1999, Xu et al. 1999, Maes et al. 1995, Crane et al. 2003). A small number of studies indicate that standard-dose chemotherapy is associated with increases in cytokine levels. Specifically, paclitaxel and the almost identical docetaxel have been associated with increased levels of IL6, IL8, and IL10 (Tsavaris et al. 2002, Wang et al. 2006, Penson et al. 2000, Pusztai et al. 2004). However, data on the association between chemotherapy use, cytokine levels, and neurocognitive functioning in long-term surviving patients is lacking. A number of studies among breast cancer survivors suggest that survivors who experience persistent fatigue 2 years or more after chemotherapy have been shown to have elevations in proinflammatory cytokines including the IL1RA, soluble TNF receptor type II, neopterin, IL6, and TNFα (Bower et al. 2002, Collado-Hidalgo et al. 2006).

Although proinflammatory cytokines are potent activators of the HPA axis, it should be noted that, under physiological circumstances, cytokines only seem to be able to stimulate HPA axis activity to a certain extent, as the activity of this neuroendocrine system appears to be tightly regulated through an inhibitory feedback mechanism. This negative feedback mechanism, which is inherent to the HPA axis, ensures that high concentrations of corticosteroids, which may result from cytokine-induced increases in HPA axis activity, attenuate further stimulation of the HPA axis (Sapolsky et al. 1985). Excessive stimulation of CS receptors in the hypothalamus and pituitary gland following increased CS secretion is considered to cause a decreased synthesis of corticotrophin-releasing hormone and ACTH, thereby limiting the extent to which the HPA axis can be stimulated (Sapolsky et al. 1985).

4 Cytokines and Mood

The presence of compromised mood or depression can be associated with some degree of impairment in neurocognitive function (or in the case of most brain-diseased patients, with incremental impairment). It is unclear, however, which aspects of neurocognitive performance are most vulnerable to depression in mood. There are also conceptual problems such as whether depression should be regarded as a nonneurological factor potentially affecting neurocognitive performance or as a complicating cerebral dysfunction, perhaps one primarily involving derangement of right hemisphere mechanisms (Kronfol et al. 1978, Flor-Henry 1979).

Given the well-known role of psychological stress in the development of a wide variety of behavioral disorders (Kendler et al. 1999), it is intriguing to note that stress can activate inflammatory cytokines and their signaling pathways both in the periphery and in the brain (Bierhaus et al. 2003, Pace et al. 2006, Frank et al. 2007, Johnson et al. 2005). Also, data in rats indicate that stress can activate microglia in the brain and increase their sensitivity to immunologic stimuli (i.e., lipopolysaccharide (LPS); Frank et al. 2007). Of note, stress-induced IL1 in the brain has been shown to significantly reduce the expression of brain-derived neurotrophic factor (BDNF), which is believed to play a pivotal role in neuronal growth and development, learning, synaptic plasticity, and, ultimately, behavioral disorders (Barrientos et al. 2004, Duman and Monteggia 2006).

The effects of stress on brain inflammatory pathways are believed to be mediated by activation of the sympathetic nervous system and the release of catecholamines which bind to alpha and beta adrenergic receptors on relevant cells (Bierhaus et al. 2003, Johnson et al. 2005). Interestingly, data suggest that the parasympathetic nervous system via the release of acetylcholine and subsequent activation of the alpha 7 subunit of the nicotinic acetylcholine receptor can inhibit inflammatory signaling pathways (Pavlov and Tracey 2005), suggesting that sympathetic and parasympathetic pathways have an opposing influence on inflammatory responses during stress.

Apart from affecting the functionality of the HPA axis, proinflammatory cytokines can also have direct effects on mood. The role of these cytokines is to mobilize whole body responses, including behavioral changes that help to fight against infection. One of the ways to accomplish this is to conserve energy normally expended on activity not relevant to the fight against infection. As a result these cytokines induce a behavioral syndrome associated with withdrawal from social interaction, reduced sexual activity and appetite, increased slow-wave sleep, decreased locomotor activity, and reduced aggressive behavior (Dantzer 2004, Kent et al. 1992). Healthy people given IL2 and TNFα develop sickness behavior that is associated with depressed mood, increased somatic concern, neurocognitive impairment, fatigue, and difficulties with motivation and flexible thinking (Maier and Watkins 1998, Kelley et al. 2003, Cleeland et al. 2003). These symptoms occur very quickly after cytokine administration and vanish soon after discontinuation of treatment, suggesting that these cytokines play a causal role in induction of the symptoms. However, the dose of cytokines given in these experiments was very large, although they do demonstrate the principle that a link exists between cytokines and mood.

Interestingly, many of the behavioral changes associated with sickness behavior (Dantzer 2004, Kent et al. 1992) are also observed following stressor exposure (Short and Maier 1993, Milligan et al. 1998). These similarities have led some investigators to postulate that the neural circuitry underlying the behavioral effects of stressor exposure and immune challenge may also be similar. Many of these behavioral changes observed following immune stimulation are mediated by central production of the proinflammatory cytokine IL1. Central administration of IL1 produces fever, hyperalgesia (Watkins et al. 1994), induces slow-wave sleep (Opp and

Krueger 1991), reduces food and water intake (Kent et al. 1996), alters peripheral immune function (Sullivan et al. 1997), increases plasma ACTH and glucocorticoids (Dunn et al. 1999), reduces social interaction (Kent et al. 1992), and decreases some measures of anxiety (Montkowski et al. 1997). Many of the behavioral changes produced by intracerebroventricular administration of IL1 can be blocked or attenuated by prior intracerebroventricular administration of IL1RA; Opp and Krueger 1991, Kent et al. 1996). Thus, central production of IL1 appears to be a critical component of host defense against peripheral infection and subsequent recovery. Besides its role in mediating sickness behaviors, central production of IL1 has also emerged as an important mediator of behavioral and neuroendocrine responses to stress. Shintani et al. (1995) have shown that central injection of IL1 produced a robust activation of the HPA axis and increased hypothalamic monoamine turnover. These changes are typically considered the hallmarks of stressor exposure. Importantly, IL1RA has been shown to block the HPA and monoamine response to immobilization stress (Shintani et al. 1995). Central IL1 has also been implicated in mediating the behavioral consequences of inescapable tail shock since the enhancement of fear conditioning and interference with escape learning produced by this shock experience can also be blocked by intracerebroventricular administration of IL1RA (Maier and Watkins 1995). Likewise, α-MSH administered intracerebroventricular blocked all of the acute phase like changes that have been observed following inescapable tail shock exposure (Milligan et al. 1998). When coupled with the demonstration that exposure to psychological stressors can increase IL1 production in specific brain regions (Nguyen et al. 2000), it can be concluded that stress-induced production of IL1 may be critically involved in long-term behavioral and physiological adjustments that are produced by stressor exposure. This is not to say that all stressors induce central production of IL1, or that IL1 mediates all effects of stressors. Indeed, there are some stressors, such as exposure to predators that do not affect brain cytokine levels at all (Plata-Salaman et al. 2000). As a result, the critical determinant(s) for the observation of stress-induced increases in brain IL1 remains elusive and demands further study. These efforts must begin by determination of which stressors cause increases in central cytokine production, and the role that these cytokines play in mediating subsequent behavioral and physiological consequences of stressor exposure.

References

Ader R, Cohen N, Felten D. Psychoneuroimmunology: interactions between the nervous system and the immune system. Lancet. Jan 14 1995;345(8942):99–103.

Aggarwal BB, Shishodia S, Sandur SK, Pandey MK, Sethi G. Inflammation and cancer: how hot is the link? Biochem Pharmacol. Nov 30 2006;72(11):1605–1621.

Ahles TA, Saykin AJ. Candidate mechanisms for chemotherapy-induced cognitive changes. Nat Rev Cancer. Mar 2007;7(3):192–201.

Aloisi F, Borsellino G, Care A, et al. Cytokine regulation of astrocyte function: in-vitro studies using cells from the human brain. Int J Dev Neurosci. Jun–Jul 1995;13(3–4):265–274.

Anisman H, Merali Z. Anhedonic and anxiogenic effects of cytokine exposure. Adv Exp Med Biol. 1999;461:199–233.

Anisman H, Merali Z. Cytokines, stress, and depressive illness. Brain Behav Immun. Oct 2002;16(5):513–524.

Balkwill F, Mantovani A. Inflammation and cancer: back to Virchow? Lancet. Feb 17 2001;357(9255):539–545.

Barrientos RM, Sprunger DB, Campeau S, Watkins LR, Rudy JW, Maier SF. BDNF mRNA expression in rat hippocampus following contextual learning is blocked by intrahippocampal IL-1beta administration. J Neuroimmunol. Oct 2004;155(1–2):119–126.

Beattie EC, Stellwagen D, Morishita W, et al. Control of synaptic strength by glial TNFalpha. Science. Mar 22 2002;295(5563):2282–2285.

Bellinger FP, Madamba S, Siggins GR. Interleukin 1 beta inhibits synaptic strength and long-term potentiation in the rat CA1 hippocampus. Brain Res. Nov 19 1993;628(1–2):227–234.

Bierhaus A, Wolf J, Andrassy M, et al. A mechanism converting psychosocial stress into mononuclear cell activation. Proc Natl Acad Sci USA. Feb 18 2003;100(4):1920–1925.

Bower JE, Ganz PA, Aziz N, Fahey JL. Fatigue and proinflammatory cytokine activity in breast cancer survivors. Psychosom Med. Jul–Aug 2002;64(4):604–611.

Brebner K, Hayley S, Zacharko R, Merali Z, Anisman H. Synergistic effects of interleukin-1beta, interleukin-6, and tumor necrosis factor-alpha: central monoamine, corticosterone, and behavioral variations. Neuropsychopharmacology. Jun 2000;22(6):566–580.

Breder CD, Dinarello CA, Saper CB. Interleukin-1 immunoreactive innervation of the human hypothalamus. Science. Apr 15 1988;240(4850):321–324.

Breder CD, Tsujimoto M, Terano Y, Scott DW, Saper CB. Distribution and characterization of tumor necrosis factor-alpha-like immunoreactivity in the murine central nervous system. J Comp Neurol. Nov 22 1993;337(4):543–567.

Camacho A, Ng B. Methylphenidate for alpha-interferon induced depression. J Psychopharmacol (Oxf). Sep 2006;20(5):687–689.

Capuron L, Ravaud A, Dantzer R. Timing and specificity of the cognitive changes induced by interleukin-2 and interferon-alpha treatments in cancer patients. Psychosom Med. May–Jun 2001;63(3):376–386.

Cleeland CS, Bennett GJ, Dantzer R, et al. Are the symptoms of cancer and cancer treatment due to a shared biologic mechanism? A cytokine-immunologic model of cancer symptoms. Cancer. Jun 1 2003;97(11):2919–2925.

Collado-Hidalgo A, Bower JE, Ganz PA, Cole SW, Irwin MR. Inflammatory biomarkers for persistent fatigue in breast cancer survivors. Clin Cancer Res. May 1 2006;12(9):2759–2766.

Coogan A, O'Connor JJ. Inhibition of NMDA receptor-mediated synaptic transmission in the rat dentate gyrus in vitro by IL-1 beta. Neuroreport. Jul 7 1997;8(9–10):2107–2110.

Crane JW, Buller KM, Day TA. Evidence that the bed nucleus of the stria terminalis contributes to the modulation of hypophysiotropic corticotropin-releasing factor cell responses to systemic interleukin-1beta. J Comp Neurol. Dec 8 2003;467(2):232–242.

Cserr HF, Knopf PM. Cervical lymphatics, the blood-brain barrier and the immunoreactivity of the brain: a new view. Immunol Today. Dec 1992;13(12):507–512.

Cunningham AJ, Murray CA, O'Neill LA, Lynch MA, O'Connor JJ. Interleukin-1 beta (IL-1 beta) and tumour necrosis factor (TNF) inhibit long-term potentiation in the rat dentate gyrus in vitro. Neurosci Lett. Jan 12 1996;203(1):17–20.

Dantzer R. Cytokine-induced sickness behaviour: a neuroimmune response to activation of innate immunity. Eur J Pharmacol. Oct 1 2004;500(1–3):399–411.

D'Arcangelo G, Dodt HU, Zieglgansberger W. Reduction of excitation by interleukin-1 beta in rat neocortical slices visualized using infrared-darkfield videomicroscopy. Neuroreport. May 27 1997;8(8):2079–2083.

De Simoni MG, Imeri L. Cytokine-neurotransmitter interactions in the brain. Biol Signals. Jan 1998;7(1):33–44.

de Visser KE, Eichten A, Coussens LM. Paradoxical roles of the immune system during cancer development. Nat Rev Cancer. Jan 2006;6(1):24–37.

Duman RS, Monteggia LM. A neurotrophic model for stress-related mood disorders. Biol Psychiatry. Jun 15 2006;59(12):1116–1127.

Dunn AJ, Wang J, Ando T. Effects of cytokines on cerebral neurotransmission. Comparison with the effects of stress. Adv Exp Med Biol. 1999;461:117–127.

Felten SY, Felten DL. Innervation of Lymphoid Tissue. 2nd ed. San Diego, CA: Academic Press; 1991.

Flor-Henry P. On certain aspects of the localization of the cerebral systems regulating and determining emotion. Biol Psychiatry. 1979;133(677–698).

Frank MG, Baratta MV, Sprunger DB, Watkins LR, Maier SF. Microglia serve as a neuroimmune substrate for stress-induced potentiation of CNS pro-inflammatory cytokine responses. Brain Behav Immun. Jan 2007;21(1):47–59.

Freidin M, Bennett MV, Kessler JA. Cultured sympathetic neurons synthesize and release the cytokine interleukin 1 beta. Proc Natl Acad Sci USA. Nov 1 1992;89(21):10440–10443.

Gloeckler Ries LA, Reichman ME, Lewis DR, Hankey BF, Edwards BK. Cancer survival and incidence from the Surveillance, Epidemiology, and End Results (SEER) program. Oncologist. 2003;8(6):541–552.

Harbuz MS, Stephanou A, Sarlis N, Lightman SL. The effects of recombinant human interleukin (IL)-1 alpha, IL-1 beta or IL-6 on hypothalamo-pituitary-adrenal axis activation. J Endocrinol. Jun 1992;133(3):349–355.

Hopkins SJ, Rothwell NJ. Cytokines and the nervous system. I: Expression and recognition. Trends Neurosci. Feb 1995;18(2):83–88.

Johnson JD, Campisi J, Sharkey CM, et al. Catecholamines mediate stress-induced increases in peripheral and central inflammatory cytokines. Neuroscience. 2005;135(4):1295–1307.

Katsuki H, Nakai S, Hirai Y, Akaji K, Kiso Y, Satoh M. Interleukin-1 beta inhibits long-term potentiation in the CA3 region of mouse hippocampal slices. Eur J Pharmacol. Jun 8 1990;181(3):323–326.

Kelley KW, Bluthe RM, Dantzer R, et al. Cytokine-induced sickness behavior. Brain Behav Immun. Feb 2003;17 Suppl 1:S112–118.

Kendler KS, Karkowski LM, Prescott CA. Causal relationship between stressful life events and the onset of major depression. Am J Psychiatry. Jun 1999;156(6):837–841.

Kent S, Bluthe RM, Kelley KW, Dantzer R. Sickness behavior as a new target for drug development. Trends Pharmacol Sci. Jan 1992;13(1):24–28.

Kent S, Bret-Dibat JL, Kelley KW, Dantzer R. Mechanisms of sickness-induced decreases in food-motivated behavior. Neurosci Biobehav Rev. 1996;20(1):171–175.

Kronfol Z, Remick DG. Cytokines and the brain: implications for clinical psychiatry. Am J Psychiatry. May 2000;157(5):683–694.

Kronfol Z, Hamsher K, Digre K, Waziri R. Depression and hemisphere functions: changes associated with unilateral ECT. Br J Psychiatry. 1978;132:560–567.

Lacosta S, Merali Z, Anisman H. Central monoamine activity following acute and repeated systemic interleukin-2 administration. Neuroimmunomodulation. 2000;8(2):83–90.

Lapchak PA, Araujo DM. Interleukin-2 regulates monoamine and opioid peptide release from the hypothalamus. Neuroreport. Mar 1993;4(3):303–306.

Linthorst AC, Flachskamm C, Muller-Preuss P, Holsboer F, Reul JM. Effect of bacterial endotoxin and interleukin-1 beta on hippocampal serotonergic neurotransmission, behavioral activity, and free corticosterone levels: an in vivo microdialysis study. J Neurosci. Apr 1995;15(4):2920–2934.

Linthorst AC, Reul JM. Brain neurotransmission during peripheral inflammation. Ann NY Acad Sci. May 1 1998;840:139–152.

Maes M, Meltzer HY, Bosmans E, et al. Increased plasma concentrations of interleukin-6, soluble interleukin-6, soluble interleukin-2 and transferrin receptor in major depression. J Affect Disord. Aug 18 1995;34(4):301–309.

Maier SF, Watkins LR. Intracerebroventricular interleukin-1 receptor antagonist blocks the enhancement of fear conditioning and interference with escape produced by inescapable shock. Brain Res. Oct 16 1995;695(2):279–282.

Maier SF, Watkins LR. Cytokines for psychologists: implications of bidirectional immune-to-brain communication for understanding behavior, mood, and cognition. Psychol Rev. Jan 1998;105(1):83–107.

McGeer PL, McGeer EG. The inflammatory response system of brain: implications for therapy of Alzheimer and other neurodegenerative diseases. Brain Res Brain Res Rev. Sep 1995;21(2):195–218.

Merali Z, Lacosta S, Anisman H. Effects of interleukin-1beta and mild stress on alterations of norepinephrine, dopamine and serotonin neurotransmission: a regional microdialysis study. Brain Res. Jul 4 1997;761(2):225–235.

Meyers CA, Albitar M, Estey E. Cognitive impairment, fatigue, and cytokine levels in patients with acute myelogenous leukemia or myelodysplastic syndrome. Cancer. Aug 15 2005;104(4):788–793.

Milligan ED, Nguyen KT, Deak T, et al. The long term acute phase-like responses that follow acute stressor exposure are blocked by alpha-melanocyte stimulating hormone. Brain Res Bull. Nov 9 1998;810(1–2):48–58.

Montkowski A, Landgraf R, Yassouridis A, Holsboer F, Schobitz B. Central administration of IL-1 reduces anxiety and induces sickness behaviour in rats. Pharmacol Biochem Behav. Oct 1997;58(2):329–336.

Muller N, Ackenheil M. Psychoneuroimmunology and the cytokine action in the CNS: implications for psychiatric disorders. Prog Neuropsychopharmacol Biol Psychiatry. Jan 1998;22(1):1–33.

Munoz-Fernandez MA, Fresno M. The role of tumour necrosis factor, interleukin 6, interferon-gamma and inducible nitric oxide synthase in the development and pathology of the nervous system. Prog Neurobiol. Oct 1998;56(3):307–340.

Nguyen KT, Deak T, Will MJ, et al. Timecourse and corticosterone sensitivity of the brain, pituitary, and serum interleukin-1beta protein response to acute stress. Brain Res. Mar 24 2000;859(2):193–201.

Ofek K, Krabbe KS, Evron T, et al. Cholinergic status modulations in human volunteers under acute inflammation. J Mol Med. 2007;85(11):1239–1251.

Opp MR, Krueger JM. Interleukin 1-receptor antagonist blocks interleukin 1-induced sleep and fever. Am J Physiol. Feb 1991;260(2 Pt 2):R453–457.

Pace TW, Mletzko TC, Alagbe O, et al. Increased stress-induced inflammatory responses in male patients with major depression and increased early life stress. Am J Psychiatry. Sep 2006;163(9):1630–1633.

Pauli S, Linthorst AC, Reul JM. Tumour necrosis factor-alpha and interleukin-2 differentially affect hippocampal serotonergic neurotransmission, behavioural activity, body temperature and hypothalamic-pituitary-adrenocortical axis activity in the rat. Eur J Neurosci. Mar 1998;10(3):868–878.

Pavlov VA, Tracey KJ. The cholinergic anti-inflammatory pathway. Brain Behav Immun. Nov 2005;19(6):493–499.

Penson RT, Kronish K, Duan Z, et al. Cytokines IL-1beta, IL-2, IL-6, IL-8, MCP-1, GM-CSF and TNFalpha in patients with epithelial ovarian cancer and their relationship to treatment with paclitaxel. Int J Gynecol Cancer. Jan 2000;10(1):33–41.

Plata-Salaman CR, Ffrench-Mullen JM. Interleukin-1 beta depresses calcium currents in CA1 hippocampal neurons at pathophysiological concentrations. Brain Res Bull. Aug 1992;29(2):221–223.

Plata-Salaman CR, Ilyin SE, Turrin NP, et al. Neither acute nor chronic exposure to a naturalistic (predator) stressor influences the interleukin-1beta system, tumor necrosis factor-alpha, transforming growth factor-beta1, and neuropeptide mRNAs in specific brain regions. Brain Res Bull. Jan 15 2000;51(2):187–193.

Pusztai L, Mendoza TR, Reuben JM, et al. Changes in plasma levels of inflammatory cytokines in response to paclitaxel chemotherapy. Cytokine. Feb 7 2004;25(3):94–102.

Rothwell NJ, Hopkins SJ. Cytokines and the nervous system II: Actions and mechanisms of action. Trends Neurosci. Mar 1995;18(3):130–136.

Sapolsky RM, Meaney MJ, McEwen BS. The development of the glucocorticoid receptor system in the rat limbic brain. III. Negative-feedback regulation. Brain Res. Feb 1985;350(1–2):169–173.

Sapolsky R, Rivier C, Yamamoto G, Plotsky P, Vale W. Interleukin-1 stimulates the secretion of hypothalamic corticotropin-releasing factor. Science. Oct 23 1987;238(4826):522–524.

Scarborough DE, Lee SL, Dinarello CA, Reichlin S. Interleukin-1 beta stimulates somatostatin biosynthesis in primary cultures of fetal rat brain. Endocrinology. Jan 1989;124(1):549–551.

Scheibel RS, Valentine AD, O'Brien S, Meyers CA. Cognitive dysfunction and depression during treatment with interferon-alpha and chemotherapy. J Neuropsychiatry Clin Neurosci. Spring 2004;16(2):185–191.

Schobitz B, De Kloet ER, Holsboer F. Gene expression and function of interleukin 1, interleukin 6 and tumor necrosis factor in the brain. Prog Neurobiol. Nov 1994;44(4):397–432.

Schwienbacher I, Fendt M, Richardson R, Schnitzler HU. Temporary inactivation of the nucleus accumbens disrupts acquisition and expression of fear-potentiated startle in rats. Brain Res. 2004;1027(1–2):87–93.

Shintani F, Nakaki T, Kanba S, et al. Involvement of interleukin-1 in immobilization stress-induced increase in plasma adrenocorticotropic hormone and in release of hypothalamic monoamines in the rat. J Neurosci. Mar 1995;15(3 Pt 1):1961–1970.

Short KR, Maier SF. Stressor controllability, social interaction, and benzodiazepine systems. Pharmacol Biochem Behav. Aug 1993;45(4):827–835.

Sikora K, Timbs O. Cancer 2025: introduction. Expert Rev Anticancer Ther. Jun 2004;4 Suppl 1:S11–12.

Song C, Merali Z, Anisman H. Variations of nucleus accumbens dopamine and serotonin following systemic interleukin-1, interleukin-2 or interleukin-6 treatment. Neuroscience. 1999;88(3):823–836.

Spranger M, Lindholm D, Bandtlow C, et al. Regulation of Nerve Growth Factor (NGF) Synthesis in the Rat Central Nervous System: Comparison between the Effects of Interleukin-1 and Various Growth Factors in Astrocyte Cultures and in vivo. Eur J Neurosci. Jan 1990;2(1):69–76.

Sullivan GM, Canfield SM, Lederman S, Xiao E, Ferin M, Wardlaw SL. Intracerebroventricular injection of interleukin-1 suppresses peripheral lymphocyte function in the primate. Neuroimmunomodulation. Jan–Feb 1997;4(1):12–18.

Tsavaris N, Kosmas C, Vadiaka M, Kanelopoulos P, Boulamatsis D. Immune changes in patients with advanced breast cancer undergoing chemotherapy with taxanes. Br J Cancer. Jul 1 2002;87(1):21–27.

Tonelli LH, Postolache TT, Sternberg EM. Inflammatory genes and neural activity: involvement of immune genes in synaptic function and behavior. Front Biosci. Jan 1 2005;10:675–680.

Trask PC, Esper P, Riba M, Redman B. Psychiatric side effects of interferon therapy: prevalence, proposed mechanisms, and future directions. J Clin Oncol. Jun 2000;18(11):2316–2326.

Wang TH, Chan YH, Chen CW, et al. Paclitaxel (Taxol) upregulates expression of functional interleukin-6 in human ovarian cancer cells through multiple signaling pathways. Oncogene. 2006;25(35):4857–4866.

Wang S, Cheng Q, Malik S, Yang J. Interleukin-1beta inhibits gamma-aminobutyric acid type A (GABA(A)) receptor current in cultured hippocampal neurons. J Pharmacol Exp Ther. Feb 2000;292(2):497–504.

Watkins LR, Wiertelak EP, Goehler LE, Smith KP, Martin D, Maier SF. Characterization of cytokine-induced hyperalgesia. Brain Res. Aug 15 1994;654(1):15–26.

Wilson CJ, Finch CE, Cohen HJ. Cytokines and cognition – the case for a head-to-toe inflammatory paradigm. J Am Geriatr Soc. Dec 2002;50(12):2041–2056.

Xu Y, Day TA, Buller KM. The central amygdala modulates hypothalamic-pituitary-adrenal axis responses to systemic interleukin-1beta administration. Neuroscience. 1999;94(1):175–183.

Zeise ML, Espinoza J, Morales P, Nalli A. Interleukin-1beta does not increase synaptic inhibition in hippocampal CA3 pyramidal and dentate gyrus granule cells of the rat in vitro. Brain Res. Sep 12 1997;768(1–2):341–344.

Autoimmunity and Brain Dysfunction

Steven A. Hoffman and Boris Sakic

Abstract This chapter summarizes current knowledge on the relationship between systemic autoimmune/inflammatory processes, brain pathology, and mental (dys) function. Several lines of evidence are presented, as well as the molecular mechanisms instrumental in the initiation and maintenance of complex pathogenic circuits. Although the emphasis is on systemic lupus erythematosus and its associated neuropsychiatric manifestations, other disorders relevant to the immunological theory of mental disorders are also reviewed. Together with relevant immunopathology, the associated permissive conditions (e.g., breached blood-brain barrier) are discussed in the context of an upregulation in brain-reactive autoantibodies, neuroactive cytokines, complement components, and immune complexes. The gaps in our present knowledge and future directions are outlined at the end of the chapter.

Keywords Autoimmunity · Behavior · Mental function · Neuropsychiatric manifestations · Brain pathology · Lupus · Schizoprenia · Autism · Multiple sclerosis · Brain-reactive autoantibodies · Blood-brain barrier · Neuroactive cytokines · Complement · Immune complexes · Homeostatic metasystem

1 Introduction

From a functional point of view, behavior can be considered a multidimensional output of complex physiologic responses to an ever-changing external environment. Defense reactions (also known as "fight or flight" behavioral responses) enable an individual to maintain homeostasis, prolong one's own survival, and expand the species. Physical and psychological stressors continuously affect three major domains of behavior. A wide variety of factors (ranging from radiation, temperature, noise, and light, to threats from predators, pathogens, hunger, etc.) induce a relatively limited

S.A. Hoffman (✉)
Arizona State University, College of Liberal Arts and Sciences, School of Life Sciences, Neuroimmunology Labs, PO Box 874501, Tempe, AZ 85287–4501, USA
e-mail: SteveHoffman@asu.edu

A. Siegel, S.S. Zalcman (eds.), *The Neuroimmunological Basis of Behavior and Mental Disorders*, DOI 10.1007/978-0-387-84851-8_17,

number of responses in locomotion, emotional reactivity, and learning/memory. When certain stressors become excessive, prolonged, or when they combine with other factors, the body's coping mechanisms may break down and result in signs and symptoms, which, in evolutionary context, reflect maladaptive behaviors. Common manifestations of aberrant central nervous system (CNS) function in humans include seizures, anxiety, depression, psychosis, and cognitive deficits that differ in severity and duration. The concept on etiological heterogeneity of mental disorders first appeared in papers of Emil Kraepelin, one of the founders of scientific psychiatry. In particular, in 1920 he wrote: "Sifting through the manifestations of disease for insight into the underlying disease process is well-nigh impossible if we only deal with the general ways in which human beings respond to various harmful insults. In these circumstances we must be very wary of claiming that a particular disorder is characteristic of one and only one particular disease process" (Kraepelin, 1992). Such a multifaceted approach to our understanding of mental illness etiology continues to dominate modern biological psychiatry and neuroscience (Caspi and Moffitt, 2006).

This book chapter reviews accumulating experimental and clinical evidence that proposes systemic autoimmunity as one of the principal pathogenic mechanisms in some forms of mental illness. Together with tumor formation, infections, and inflammatory responses, autoimmunity can be regarded as a powerful physiological stressor, which can severely damage brain structure, or alter normal brain processes by nondestructive, functional interference. As discussed below, the "autoimmune theory" does not exclude interactions with genetic, psychological, and other environmental factors. It merely emphasizes the importance of the neuro-immuno-endocrine network (or homeostatic metasystem) in maintenance of physical and mental health (Fig. 1). One may expect that a deeper appreciation and understanding of this complex network in general, and autoimmunity in particular, would lead to more effective prevention of common mental illnesses, including the development of novel pharmacological approaches.

2 Autoimmunity and CNS Dysfunction

Systemic autoimmunity is a life-threatening pathological condition characterized by an imbalanced cytokine network, hyperactivity of immune cells, excessive production of self-reactive antibodies, immune complex deposition, and widespread inflammation of various tissues and organs. Causes of autoimmunity are not fully understood yet, but genetic and environmental factors (e.g., sunlight, exposure to chemicals, bacterial, and viral pathogens) are proposed to play key roles in triggering and maintaining chronic autoimmune reactions (Rhodes and Vyse, 2007).

Experimental and clinical studies are presently largely focused on identification of candidate genes and environmental risk factors that underlie the pathogenesis of common mental disorders, such as schizophrenia, major depression, Alzheimer's disease, and autism. The autoimmune theory does not exclude, yet rather complements new conceptual approaches in studying multifactorial etiologies and complex pathogenic

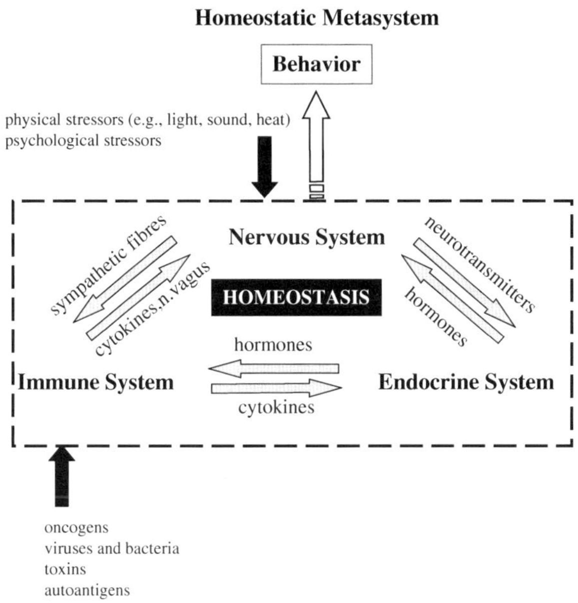

Fig. 1 The Homeostatic Metasystem involves a complex network of immune, nervous, and endocrine systems. Diverse responses to external and internal stressors followed by inter-systemic interactions via humoral and neural pathways continuously influence brain plasticity and mental function. This is ultimately expressed by dynamic changes in behavior that, within the environmental context, can be adaptive or maladaptive. Note: The interacting agents may also include components of the complement system, autoantibodies, chemokines, other peptide messengers, and T cells.

circuits (Fig. 2). In particular, pluripotency of genes (pleiotropism) provides a theoretical basis for the notion that aberrant gene expression/protein synthesis associated with disturbed neurotransmission (or neuronal morphology) can also be linked to a functional lesion in the immune system. Immunologic imbalances in subgroups of psychiatric patients and immunomodulatory effects of common anti-psychotics corroborate this possibility (Muller, 1995; Muller and Ackenheil, 1995; Muller and Schwarz, 2006a; Kronfol and Remick, 2000; Hayley and Anisman, 2005). Along the same line, imprinting of genetic material into the host and/or molecular mimicry (Nahmias et al., 2006; Buka et al., 2007) may underlie epigenetic phenomena induced by exposure to various pathogens, and super antigens. If so, it is important to recognize that such gene–environment interactions may have a common denominator, which is organ-specific or systemic autoimmune response. A significant amount of evidence has been collected about brain involvement in various immunologic disorders (Ainiala et al., 2001; Brey et al., 2002; Carr et al., 1978; Johnson and Richardson, 1968; Kim et al., 1982; Kumar et al., 1988, 1989; Monastero et al., 2001; Ramos and Mandybur, 1975; Sofat et al., 2006; Steiner and Gelbloom, 1959; Wekking et al., 1991; Wekking, et al., 1993; Zandman-Goddard et al., 2007). Well-documented neurologic and psychiatric (NP) manifestations in systemic lupus erythematosus, as

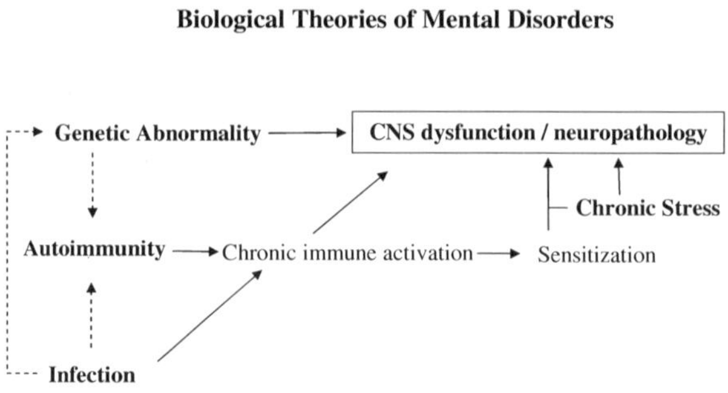

Fig. 2 The etiologies of mental disorders are likely multifactorial. Genetic, environmental, and immune factors likely play a synergistic role in triggering, maintenance, and severity of behavioral deficits. Autoimmune disease may result from either inherited genetic lesion, or from exposure to environmental factors (including external pathogens) which are able to induce DNA imprinting and molecular mimicry. Consequent autoimmune/inflammatory responses can affect central nervous system (CNS) functioning directly, via circulating cells and soluble immune factors crossing the blood–brain barrier, or by enhancing detrimental effects of exogenous stressors via cytokine-induced sensitization of the pituitary-adrenal axis.

well as autoimmune and psychiatric manifestations in multiple sclerosis support the hypothesis that autoimmunity causes brain dysfunction. However, despite the significant comorbidity of autoimmune and psychiatric manifestations, the nature of clinical evidence is still largely correlational. The sections below are listed in the order that they reflect the existing empirical knowledge on causal relationships between autoimmunity and brain pathology/aberrant behavior.

2.1 Aberrant Behavior in Autoimmune Animals

Most of the convincing empirical evidence for the causal relationship between autoimmunity and behavioral dysfunction comes from animal models of the autoimmune disease. Inbred strains of NZB, NZB/W, BXSB, and MRL mice spontaneously develop lupus-like manifestations, ranging from a disturbed cytokine network and increased autoantibody production, to fatal immune complex-mediated glomerulonephritis (Theofilopoulos, 1992; Sakic et al., 1997a). Besides systemic autoimmunity and inflammation, a large percentage of these animals develop various brain abnormalities and a constellation of behavioral deficits. The lack of an adequate (genetically similar) control group and a high incidence of inherited brain anomalies in NZB and BXSB strains (Sherman et al., 1987, 1990a, b), however, precludes a direct linkage between immunopathogenic mechanisms on one side, and brain damage and aberrant behavior on the other (Lal et al., 1988; Forster et al., 1988a, b; Schrott et al., 1993, 1992; Schrott and Crnic, 1996a). Conversely, the MRL strain (developed in The Jackson Laboratory in the late 70s) consists of the MRL/MpJ-Fas^{lpr}/J (MRL/lpr) substrain and the congenic (control) MRL/MpJ (MRL/Mp) substrain. Most importantly, lack of inherited brain abnormalities, rapid onset of lupus-like disease, and emergence of behavioral deficits during the onset of spontaneous lupus-like disease rendered the MRL model an adequate preparation in studying the link between systemic autoimmunity and CNS involvement (Sakic et al., 1997b; Szechtman et al., 1997; Ballok, 2007).

The MRL/lpr and the congenic MRL/Mp mice are comparable in many respects (appearance, size, and reproductive age), except in the onset of systemic autoimmune disease. Due to the presence of the lymphoproliferative (lpr) gene on chromosome 19 and a subsequent deficit in apoptotic Fas receptor expression (Singer et al., 1994) MRL/lpr mice develop an accelerated form of the chronic lupus-like condition (Theofilopoulos, 1992). This murine form of SLE is accompanied by a constellation of behavioral deficits, operationally labeled "autoimmunity-associated behavioral syndrome" (Sakic et al., 1997b). At the onset of serological manifestations the deficits are most consistently noted in tasks reflective of emotional reactivity and affective behavior (Szechtman et al., 1997), while at advanced stages of lupus-like disease learning/memory deficits may emerge (Sakic et al., 1992; Hess et al., 1993). In particular, during an early imbalance in the cytokine network and the hyperproduction of autoantibodies young (4–5 weeks of age) MRL/lpr mice develop a progressive anxious/depressive-like behavioral state, as indicated by increased thigmotaxic behavior (Sakic et al., 1993b, a, 1992), impaired exploration of novel objects and spaces, excessive floating in the forced swim test (Sakic et al., 1994), reduced responsiveness to a palatable stimulus (Sakic et al., 1996a), and reduced isolation-induced inter-male fighting (Sakic et al., 1998a). Impaired cognitive flexibility was inferred from response perseveration in spatial learning tasks, such as the Morris water maze and the spontaneous alternation task (Sakic et al., 1992; Ballok et al., 2004b). Besides the deficits systemically measured in behavioral tasks, a substantial proportion (30–40%) of diseased MRL/lpr mice exhibit aberrant social behavior. Signs of social withdrawal and discomfort are evidenced by the tendency of diseased animals to reduce contact duration with other cagemates (ms in preparation).

The fact that several early deficits in motivated behavior (e.g., blunted responsiveness to palatable food, immobility in the forced swim test, reduced aggressiveness, etc.) appear before signs of severe peripheral symptomatology (Sakic et al., 1992, 1996b; Brey et al., 1995) is consistent with clinical reports on depressive episodes before SLE is even diagnosed (van Dam et al., 1994). They also suggest that impaired performances are not an epiphenomenon produced by systemic organ involvement, but rather a consequence of direct neuro-immune interactions. Brain inflammation and behavioral deficits have been reported in other murine models of SLE (Forster et al., 1988a; Denenberg et al., 1991; Vogelweid et al., 1994; Schrott and Crnic, 1994; Schrott et al., 1994; Schrott and Crnic, 1996a). Compared to other disease models, however, the knowledge obtained from the MRL model encompasses genetic, immunological, neuropathological, and behavioral aspects, thus providing a broad support for further studies on gene-autoimmune-CNS interactions.

The causative role of autoimmunity and inflammation in the pathogenesis of aberrant behavior has been supported by studies employing the immunosuppressive drug cyclophosphamide (CY), which (in addition to normalized neuronal morphology) improved behavioral performance in MRL/lpr mice. More specifically, sustained administration of CY significantly attenuated anxiety- and depressive-like behaviors, as indicated by restored novel object exploration, increased responsiveness to a sweet palatable solution, and reduced floating in the forced swim test (Sakic et al., 1996b, a; Farrell et al., 1997).

2.2 Neuropsychiatric Manifestations in SLE

Systemic lupus erythematosus (SLE) is a chronic autoimmune/inflammatory disease with a broad spectrum of clinical presentations, thus justifying its colloquial term the "great imitator." This heterogeneity is also evident from NP manifestations, which mimic a variety of classical CNS disorders (ACR Ad Hoc Committee, 1999). Although NP manifestations have been known for more than 100 years (Kaposi, 1872) and have been studied extensively since, their etiology and pathogenesis remain largely unknown (Johnson and Richardson, 1968; Carr et al., 1978; Carbotte et al., 1986; Bluestein et al., 1986; Wekking et al., 1991; Wekking, 1993; Denburg et al., 1993; Navarrete and Brey, 2000; Zandman-Goddard et al., 2007). They occur in up to 75% of the patients and have been proposed to reflect a more severe form of the disease (Scolding and Joseph, 2002; Hanly et al., 2005). Overt symptomatology involves diffuse manifestations (acute confusional state, psychosis, anxiety, depression, cognitive deficits), focal abnormalities (seizures, cerebrovascular disease, chorea, and myelopathy, transverse myelitis, demyelinating syndrome and aseptic meningitis, headaches), and various deficits in the peripheral nervous system (polyneuropathies and mononeuropathies, autonomic disorders, plexopathy, myasthenia gravis-like manifestations).

No reliable diagnostic standards of CNS involvement are available yet, but neuroimaging, neuropsychological testing, and soluble markers are helpful in contemporary diagnosis (Borchers et al., 2005). For example, despite the evidence that SLE patients do not differ from controls in the number of correct responses to a memory task, their style of obtaining these responses is different and less effective, scoring more poorly on correct consecutive responses and processing more slowly (Shucard et al., 2004). Although detection of autoantibodies in serum could eventually be important in the diagnosis of CNS damage (Jennekens and Kater, 2002), increased intrathecal synthesis (as revealed by an elevated IgG index and oligoclonal banding) in patients with CNS dysfunction (McLean et al., 1995; Hirohata et al., 1985; Winfield et al., 1983) and antigen-specific autoantibodies in the cerebrospinal fluid (CSF; Yoshio et al., 2005) seem better predictors of NP manifestations (Greenwood et al., 2002). Also, neuroactive cytokines, such as IL-1 and IL-6, are frequently reported in CSF samples from patients with neuropsychiatric lupus (Baraczka et al., 2004; Tsai et al., 1994; Alcocer-Varela et al., 1992). Along the line of immune messengers, soluble forms of the chemokine CX3CL1 in CSF from patients with active NP-SLE suggest an active pathological process in the brain parenchyma (Kasama et al., 2008). Similarly, elevated production of serum matrix metalloproteinase MMP-9 is proposed to be associated with small-vessel cerebral vasculopathy, which may increase the risk of cerebral ischemic events and cognitive dysfunction (Ainiala et al., 2004). Given the diverse psychiatric symptomatology, this heterogeneity in potentially CNS compromising immune factors and mechanisms is not surprising.

Directly linking autoimmunity to CNS damage is often confounded by several factors, such as kidney damage, infections, and steroid therapy. Accumulation of toxic metabolites (due to glomerulonephritis), opportunistic bacterial infections, and steroid-induced psychosis are likely secondary contributors to overall deterioration of brain functioning when SLE becomes a chronic condition (Hanly et al., 2005). On the other hand, these factors are insufficient to account for an early CNS involvement (i.e., before lupus affects peripheral organs), when no infections can be detected, and/or when steroids are not used in therapy. Despite these complex pathogenic mechanisms and notorious heterogeneity in clinical presentation, an accumulating body of evidence suggests that a set of humoral immune factors play key roles in the induction of psychiatric manifestations via structural and functional brain damage.

2.3 Neuropsychiatric Manifestations in MS

Multiple sclerosis (MS) is a common neurological disorder characterized by inflammation and degeneration of myelin sheath in brain and spinal cord. Symptoms of MS mainly include visual deficits, muscle weakness, disturbed coordination and balance, and various neurological sensations. The current scientific consensus is that autoimmune attack to myelin basic protein underlies pathogenesis of MS. It is proposed that a constellation of pathogenic mechanisms, rather than a single one, is involved in the CNS damage. They include proteolysis, axonal injury, apoptosis, T cell activation and extravasation, cytokine and chemokine activation, complement activation, and epitope spreading (Scarisbrick, 2008).

Although neurological symptoms are the hallmark of MS, many clinicians would agree that various disturbances in emotional and cognitive functioning could be seen in a significant number of their patients (Schiffer and Hoffman, 1991). Epidemiological studies estimate that up to 50% of MS patients develop major depressive disorder and various cognitive impairments during the course of their illness (Ghaffar and Feinstein, 2007). The prevalence of other manifestations, such as bipolar affective disorder, anxiety, and pathological laughing and crying, are also increased in comparison to the general population (Feinstein et al., 1997; Sa, 2007; Beiske et al., 2008). Although further supporting evidence is required from studies with large cohorts, the incidence of MS-related psychosis seems to follow a similar pattern (Patten et al., 2005; Harel et al., 2007).

Depressive episodes are more common during relapses and may exacerbate fatigue and cognitive dysfunction. Moreover, interferon treatment does not seem to be associated with behavioral dysfunctions, which often overlap and may increase the rate of suicidal thoughts (Feinstein et al., 1999; Sa, 2007). It is proposed that diffuse white matter lesions (rather than focal damage) largely accounts for deficits in working, semantic and episodic memory, as well as concept formation and abstract reasoning (Feinstein, 2004).

2.4 Schizophrenia Spectrum Disorders

The link between immunity and schizophrenia spectrum disorders has been proposed for several decades (Heath, 1969; Heath and Krupp, 1967) and the question of whether infectious and/or autoantigens are one of the principal pathogenic factors was the topic of many review articles (McAllister et al., 1991; Rapaport and McAllister, 1991; Kirch, 1993; Gaughran, 2002; Muller and Schwarz, 2006b). Despite accumulating evidence, many inconsistencies and contradictions still exist, likely due to dissimilarities in subpopulation of patients recruited, different methodology employed, and significant differences in sample sizes (Hill and McMillan, 2006).

The possibility that autoimmunity and schizophrenia are related is historically based on clinical observations that endogenous psychoses parallel autoimmune disorders with respect to early onset, genetic vulnerability, and waxing and waning course (Ganguli et al., 1993). Also, a subgroup of patients shows a variety of immunological aberrations reminiscent of chronic autoimmune and inflammatory conditions (Sirota et al., 1995; Rothermundt et al., 1996; Wilke et al., 1996; Naudin et al., 1996; Maes et al., 1997; Monteleone et al., 1997; Arolt et al., 1997; Lin et al., 1998; Muller et al., 1998; Cazzullo et al., 1998). They include high antibody titers to diverse antigens (Ganguli et al., 1993), a distinct cytokine profile in serum and CSF (Schwartz and Silver, 2000; Muller et al., 1997), and altered lymphocyte counts (Nikkila et al., 2001; Maino et al., 2007). Altered pattern in cytokine production suggests that arms of the immune system responsible for cell-mediated (Th1) and antibody-mediated (Th2) immunity are not in balance (Maino et al., 2007). Activation of the cellular arm of the immune system after standard therapy suggests not only neurochemical, but also immunological potency of anti-psychotics (Pollmacher et al., 1996; Muller et al., 1997). The immunological abnormalities, however, seem to be restricted to subpopulations of patients and often associated with an early disease onset and proclivity to negative symptomatology.

Besides inflammatory and autoimmune phenomena, exposure to endogenous retroviruses and parasites are proposed to play major, but rather indirect role in the etiopathology of certain forms of the disease (Torrey and Yolken, 2001; Yolken, 2004). Proposed intruders include herpes simplex virus type 2 (Buka et al., 2007), cytomegalovirus (Torrey et al., 2006), and Toxoplasma gondii (Conejero-Goldberg et al., 2003; Mortensen et al., 2007). As discussed above, infectious mechanisms during gestation do not exclude autoimmune responses because they seem to be associated with increased production of antigen-specific antibodies and emergence of psychiatric manifestations in adulthood (Buka et al., 2001; Leweke et al., 2004; Torrey et al., 2007). Further experimental research is a mandatory step in elucidating the complex interactions among genes, pathogens, and immune system in the etiology of schizophrenia spectrum disorders.

2.5 Autism Spectrum Disorders

Despite extensive experimental and clinical studies, the etiologies of autism spectrum disorders remain largely unknown. For many years genetic and environmental insults have been proposed as predominant determinants (Gillberg, 1999). More recently, chronic activation of the immune system emerges as an important factor to interfere with critical phases of brain development and maturation (van Gent et al., 1997; Trottier et al., 1999). The complex relationship between etiological factors can be illustrated by the evidence that the genome of autistic patients shows more frequent major histocompatibility complex haplotypes, including an association with the C4b null allele (encoding for the fourth component of complement), thus suggesting a genetically determined imbalance in immune system function (Daniels et al., 1995; Warren and Singh, 1996; Warren et al., 1996).

Cases of autism have been linked to prenatal exposure to rubella, cytomegalovirus, varicella zoster, syphilis, and toxoplasmosis (Deykin and MacMahon, 1979). Additionally, 9% of children with congenital rubella are autistic (Chess, 1977). In an animal model of autism-like disorder, perinatal exposure to neurotropic Borna disease virus results in pro-inflammatory changes in the brain, abnormal righting reflexes, hyperactivity, impaired exploration, stereotypic behaviors, disrupted architecture in hippocampus and cerebellum, and reduction in granule and Purkinje cell numbers (Hornig et al., 1999) impairment. These findings suggest that exposure of developing brain to viral infections results in behavioral and morphological abnormalities, likely stemming from the damage of the CNS, yet devoid of functional blood–brain barrier (BBB; Rodier, 1994). The role of the immune system can be supported by the evidence that autistic children show compromised cellular and humoral immune activity (van Gent et al., 1997). In particular, reduced numbers of the CD4+ helper cells, which are essential to the function of both humoral and cell-mediated immunity, are common in autism (Warren et al., 1986; Yonk et al., 1990). In 40–50% of cases, existing T cell populations do not proliferate at normal levels when stimulated by the T cell mitogens, phytohemmaglutinin, or conconavalin A. Similarly, T cell and B cell response to pokeweed mitogen can be impaired (Warren et al., 1986). Together with decreased activity of natural killer cells, which play a role in destruction of virus/bacteria-infected cells (Warren et al., 1987), the above changes may contribute to an increased susceptibility to viral pathogens, both prenatally and during early childhood.

In addition to the viral theory, an accumulating body of evidence suggests increased autoimmune activity in autistic patients. It is important to note that the viral and the autoimmune theory are not mutually exclusive, because the presence of an exogenous pathogen can induce an immune response that may spread to tissue-specific autoantigens in genetically predisposed individuals, eventually determining progression to disease (Mamula, 1998; McCluskey et al., 1998). Bystander activation and molecular mimicry between viral and self antigens could, in some instances, initiate autoimmunity (Tomer and Davies, 1995; Gianani and Sarvetnick, 1996; Horwitz and Sarvetnick, 1999). Indeed, autistic children or their family members are at a higher risk of developing certain autoimmune diseases (Money et al., 1971; Comi et al., 1999) and similarly to people suffering from juvenile rheumatoid arthritis and lupus, many patients show a reduction in the T cell suppressor–inducer subset (Warren et al., 1990), impaired T-cell activation (Warren et al., 1995), and increased cytokine levels in plasma (Singh, 1996). This forms the basis for the body's failure to properly downregulate immune cell activity, resulting in the attack on the body's own tissues, including the brain (Zimmerman et al., 1993). The serum from autistic patients often contains antibodies to the surface antigens of several neural tissues, including the myelin basic proteins, as well as antibodies for the neuron-axon filament proteins (Singh et al., 1997; van Gent et al., 1997). The CSF of approximately 32% of autistic patients contains antibodies to glial fibrillary acidic proteins (GFAP). The increase in GFAP and anti-GFAP within the autistic brain may cause cyclic astrocyte replication, and contribute to the neuropathology (Ahlsen et al., 1993). Along the same line, unusual and varied responses to immunizations support the notion of aberrant activity of the immune system. Following the first vaccination for the rubella virus, many patients exhibit a decreased immune response with undetectable antibody titers (Stubbs, 1976). After a second immunization however, antibodies can be detected. The immunization of some children may also result in an extreme immune response, including illness and fever. It is possible that these events are related to the development of cross-reactive serum antibodies, which subsequently play a role in sustaining the autoimmune activity (Zimmerman et al., 1993).

3 Putative Pathogenic Mechanisms

There are a variety of mechanisms by which the immune system can affect nervous system functioning and lead to mental disorders. It is well conceived that vascular pathology, neuronal cell loss, or glial cell dysfunction may compromise cytoarchitecture of the CNS. Alternatively, alterations in the immune system may lead to transient, non-destructive changes in brain functioning. There is intensive research on the potential role of autoantibodies, but not nearly as much is known about the role of cytokines, chemokines, or complement components. Importantly, it needs to be kept in mind that a more permeable BBB is a mandatory condition for circulating cells and molecules to get into the brain.

3.1 Neuropathology

Direct support for the autoimmunity-induced CNS damage can be drawn from the MRL model and evidence that changes in behavior coincide temporally with infiltration of lymphoid cells into the brain (Vogelweid et al., 1991; Farrell et al., 1997), that brains show reduced complexity of pyramidal neurons (Sakic et al., 1998b) and ventricular enlargement (Denenberg et al., 1992), as well as increased expression of mRNA for proinflammatory cytokines (Tomita et al., 2001b, a). Compared to congenic controls, the cortical and hippocampal neurons from MRL/lpr mice are characterized by an age-dependent aberrancy in dendritic morphology, reduced density, and less complex branching (Sakic et al., 1998b). Importantly, the degree of morphological change correlated with the severity of autoimmune disease, as evidenced by more profound spine/neuronal loss in animals with high anti-nuclear antibody titers and severe proteinuria (standard indices of autoimmunity).

The above autoimmunity–CNS relationship was corroborated by observations that immunosuppressive treatment with CY prevents neuronal atrophy and that autoimmune symptoms correlate significantly with dendritic spine loss (Sakic et al., 2000a). Consistent with reduced neuronal density (Sakic et al., 1998b), increased density of TUNEL+ cells with neuron-like morphology, their periventricular distribution and their partial colocalization with CD4/CD8+ cells, mutually pointed to excessive neuronal death and infiltration of T-cells in the subventricular regions of MRL/lpr brains (Sakic et al., 2000b). The same study revealed the presence of CD4+ and CD8+ cells in the ventricular lumen, suggesting that circulating lymphocytes had infiltrated the brain via drainage into the CSF. Neuron-like cytoplasmic staining in some cells, and limited overlapping between apoptotic and CD3, CD4, and CD8-positive cells suggested that 70% of dying cells are not T-lymphocytes. A facilitatory factor for leukocyte entry into the brain parenchyma (Farrell et al., 1997) is likely an increased expression of cell adhesion molecules in areas around the ventricles (Sakic et al., 2000b; Zameer and Hoffman, 2003). Moreover, it was recently documented that CSF samples from NP-SLE patients and MRL/lpr mice are neurotoxic to hippocampal neurons (Maric et al., 2001; DeGiorgio et al., 2001). The latter study revealed that astrocytes are not affected and that serum from MRL/lpr mice does not impair survival of cultured cells. Significant correlations between neurodegenerative/neuroinflammatory and behavioral performances in MRL/lpr mice suggest a close relationship between structural brain damage and functional impairments. For example, deficits in spatial learning/memory emerged concomitant with hippocampal damage (Ballok et al., 2004b), aberrant performance in the sucrose preference test (i.e., impaired motivated behavior) coincided with lesions of the nucleus accumbens (Anderson et al., 2006), and hypoactivity (i.e., decreased locomotor capacity) accompanied the appearance of degeneration in the substantia nigra (Ballok et al., 2004a).

Cytokine-producing microglia appear to play an important role in excitotoxic damage and in an injured or immunologically challenged brain (Gwag et al., 1997). They readily activate by transforming from a ramified, resting state into amoeboid cells to express MHC molecules (Lawson et al., 1990), as observed in brains of diseased MRL-lpr mice (McIntyre et al., 1990). Inflammatory responses are then perpetuated by both cyclooxygenase (COX) and nitric oxide (NO). In lupus mice, however, COX inhibition was not effective in ameliorating CNS disease (Ballok et al., 2006), thus suggesting non-prostaglandin-dependent mechanisms. Conversely, cell appearance at the ultrastructural level (Ballok et al., 2006) and significant increases in brain glutamine, glutamate, and lactate concentrations (Alexander et al., 2005b) further supports the notion of excitotoxic cell death in this model of CNS-SLE. In the context of functional closeness between dopamine and glutamate systems (Mora et al., 2007), it was not surprising to reveal that CSF anti-DNA antibodies cross-reacted with central NMDA receptors, producing neuronal apoptosis in mouse hippocampus (DeGiorgio et al., 2001). Consistent with clinical data (Omdal et al., 2005), a similar IgG-mediated neurodegenerative process is proposed to occur in MRL-lpr brains (Sidor et al., 2005; Sakic et al., 2005). The cause–effect relationship was corroborated by the evidence that CY abolishes infiltration of immunocytes into the choroid plexus of MRL-lpr mice (Farrell et al., 1997), prevents atrophy of dendritic spines (Sakic et al., 2000a), and neuronal death (Ballok et al., 2004a). Since CY affects a broad population of cells, terminal factors in the etiology of brain damage remain unknown. Considering that both anti-CD4 treatment (O'Sullivan et al., 1995) and complement inhibitor (Alexander et al., 2005a, 2007) ameliorated CNS disease manifestations in MRL-lpr mice, one may assume that multiple autoimmune and inflammatory mechanisms are involved in induction and maintenance of brain pathology.

Although underlying mechanisms of aberrant behavior and neurological deficits are not yet fully elucidated (Hess et al., 1993; Brey et al., 1997), recent pharmacological evidence supports a link between central dopaminergic circuit damage and AABS. Dopamine catabolism in the limbic system (implicated in reward, movement, and cognitive processes) seems profoundly altered in diseased MRL-lpr animals (Sakic et al., 2002). Pharmacological probing revealed that chronic administration of the selective D2/D3agonist quinpirole induces self-injurious behavior (Sakic et al., 2002; Chun et al., 2007) and stereotyped rotation following acute injection with D1/D2 dopamine agonist apomorphine (Ballok et al., 2004a). Along the same line, acute injection with the indirect dopamine agonist D-amphetamine failed to increase response rate to palatable sucrose solutions (Anderson et al., 2006). Taken together, these results point to damage of central dopaminergic circuits in nigrostriatal, mesolimbic, and mesocortical pathways.

It is well proposed that the buildup of dopamine and its metabolites are associated with increased neuronal vulnerability and neurotoxicity (Blum et al., 2001). If sustained dysfunction in dopamine catabolism occurs over a prolonged period, one may assume that accumulated metabolites compromise normal function and survival of dopaminergic neurons in MRL-lpr brains. Moreover, defective ubiquitin-dependent proteolysis is a cytopathogenic process that may result in neurodegeneration (Layfield et al., 2005). Indeed, the ubiquitin-proteasome system appears to be altered in MRL-lpr brains (Ballok et al., 2004a) and may be related to increased production of anti-ubiquitin autoantibodies in autoimmune mice (Elouaai et al., 1994). Although the Fas (Fas/Apo-1/CD95) surface receptor is not expressed in brains of MRL-lpr mice (Park et al., 1998), an apoptotic mode of neuronal demise may still be mediated by mechanisms such as TNF-α or granzyme B receptors. Considering extensive lymphocytosis of CD8 cytotoxic T cells (known to release granzyme B) into the ventricles and brain parenchyma (Ballok and Sakic, 2002; Ballok et al., 2004b) and TNF-α in CSF and brains of MRL-lpr mice (Ballok et al., 2004a; Ma et al., 2006), both mechanisms may be operational in activating terminal caspase cascades. Given that neurons have high and selective vulnerability to

T cells (Giuliani et al., 2003), activated T cells may accumulate in the CNS and mediate Fas-independent, but contact-dependent neuronal cytotoxicity. Transmission electron microscopy revealed condensation of cytoplasm and organelles, but no classical marks of apoptosis or necrosis were seen (Ballok et al., 2006). Despite this morphological complexity, these results pointed to a severe metabolic perturbation and an excitotoxic-like damage of central neurons during chronic lupus-like disease.

In CNS-SLE functional abnormalities include hypoperfusion (Huang et al., 2002; Handa et al., 2003; Lopez-Longo et al., 2003) and regional metabolic abnormalities (Volkow et al., 1988; Sibbitt and Sibbitt, 1993; Brooks et al., 1997; Komatsu et al., 1999; Yoshida et al., 2007). Brain atrophy is, however, the most frequent structural abnormality (Omdal et al., 1989; Waterloo et al., 1999), likely reflecting widespread neuronal and glial damage (Sibbitt et al., 1994; Bosma et al., 2000; Trysberg et al., 2003; Steens et al., 2004; Appenzeller et al., 2007).

3.2 Breached Blood–Brain Barrier

A breached BBB seems to be a critical initial factor in allowing diverse circulating factors to enter the brain (Hoffman and Harbeck, 1989; McLean et al., 1995; Abbott et al., 2003). There is evidence that the BBB is damaged in a substantial number of NP-SLE patients (as evidenced by an increased albumin quotient) and that autoantibodies can be synthesized within the brain (intrathecally), as suggested by an increased immunoglobulin index (Abbott et al., 2003; Nishimura et al., 2008). In one clinical study an elevated CSF IgG index (associated with intrathecal synthesis) was found in 42% of patients with SLE (Winfield et al., 1983). Q albumin (a measure of BBB permeability) was elevated in 5 of 6 patients with diffuse, major CNS injury, but not in 12 of 13 who had seizure, psychosis, or cranial neuropathy. The authors speculated that BBB impairment was secondary to CNS disease, but that there may be an ongoing humoral immune response within the CNS (associated with IFN-α synthesis in some cases). Another group (McLean et al., 1995) reported BBB breakdown in 30% of patients with SLE and that clinical relapses were associated with worsening of the BBB, while steroid treatment improved barrier function. It is still not clear, however, which mechanism is predominant in SLE, whether brain-reactive autoantibodies (BRAA) passively diffuse from the peripheral blood or are synthesized by leukocytes which, when activated, can enter the CNS (Hickey et al., 1997). One may assume that both mechanisms are operational, but whether one or the other predominates under certain CNS conditions remains to be determined. Similarly, perivascular leakage of IgG and CD4+ cells in choroid plexus of MRL/lpr mice suggested a breached BBB and CNS inflammation at the onset of lupus-like disease (Vogelweid et al., 1991; Farrell et al., 1997; Sidor et al., 2005). The lymphocytic infiltrates were predominantly CD4+, although other T-cells and B-cells were detected (Ma et al., 2006; James et al., 2006). The late onset and low autoimmune, congenic MRL/mp mice did not show either leakage or infiltrates. There is evidence for both the hypothesis that autoantibodies are being secreted in situ, as well as the alternative, that they are crossing the BBB from the periphery. In one study (Zameer and Hoffman, 2004), B and T cells were found in the brains of MRL/lpr. There were more T cell infiltrates, compared to B cells. Most of the cells were associated with the ventricles, but some were also in periventricular tissue. These lymphocytic infiltrates are consistent with the hypothesis that autoantibodies are being secreted in the brain of MRL/lpr mice. The diseased BXSB mice, however, did not show evidence of B or T cells in brain. Since these mice did show Ig binding in their brain, it suggests that autoantibody is coming from the periphery and crossing the BBB. Apparently, there are different mechanisms by which the autoantibodies can get into brain during autoimmune disease. The importance of a breached BBB in the etiology of NP-SLE has been shown in an animal model where active immunization with NR2 antigen and formation of anti-NR2 receptor antibodies led to learning deficits, but only when permeability of the BBB was increased by systemic administration of LPS (Kowal et al., 2004).

Further evidence for the role of the BBB is expression of adhesion molecules in the brains of autoimmune mice. ICAM-1 and VCAM-1 molecules showed increased expression in brains of aged MRL/lpr mice, but not in non-autoimmune controls or 1 month old lupus-prone mice (Zameer and Hoffman, 2003). This expression, associated with endothelial cells, provides a potential mechanism by which B and T cells migrate into the brain and secrete antibodies and/or cytokines. A study in MS and SLE patients provided evidence that adhesion molecules are activated in the human form of these autoimmune diseases, showing elevated levels of soluble ICAM-1 and VCAM-1 in CSF and serum (Baraczka et al., 2001). Intrathecal synthesis of sVCAM-1 was present in both MS and SLE, while sICAM-1 was primarily present in MS, and soluble L-selectin was present in SLE. The data suggested BBB damage in systemic and organ-specific autoimmune diseases of the CNS.

An earlier study on acute immune complex disease is consistent with the notion of the BBB being involved in the pathogenesis of behavioral manifestations. The disease was induced by injecting rabbits with bovine serum albumin (BSA), which led to immune complex deposition in the choroid plexus and increased BBB permeability (Harbeck et al., 1979). An unchanged CSF IgG to albumin ratio and an increased CSF to serum albumin ratio suggested an increase in choroid plexus permeability to serum proteins. This was accompanied by a significant correlation between levels of CSF albumin and IgG with immune complex deposits in the choroid plexus. A follow-up study (Hoffman et al., 1983) confirmed and extended these findings using a double isotope technique. An index of barrier permeability was elevated in the CSF at days 12 and 15, the peak of acute immune complex disease.

As shown in aged MRL/lpr mice, a disrupted BBB allows large molecules and cells to infiltrate into the brain and likely induce CNS damage (Farrell et al., 1997; Sidor et al., 2005). The close associations between serological measures and brain atrophy further supports this notion (Sakic et al., 2000b; Ballok et al., 2006). In addition, associated with changes in BBB permeability, one may assume that alterations in CSF electrolyte composition and pH are able to affect normal behavioral performance.

3.3 Immunopathology

Neuropsychiatric manifestations of SLE are likely mediated by a variety of mechanisms, involving autoantibodies (against neural antigens, phospholipids and ribosomes), cytokines, vascular lesions, and thrombosis (Denburg et al., 1995; Bruns and Meyer, 2006). Although a variety of autoantibodies are believed to have different mechanisms of action, no studies have revealed up to this point, a useful set of diagnostic markers (Alosachie et al., 1998; Zandman-Goddard et al., 2007). Some researchers believe that the autoantibodies primarily play a role in neural cell injury, including defective clearance of apoptotic cells (Scolding and Joseph, 2002). Another hypothesis links communications between the peripheral immune system and the brain, assuming that activated astrocytes, microglia, and possibly neurons, lead to overproduction of cytokines in situ, with resultant tissue injury (Tomita et al., 2004). Indeed, IL-6 has been shown within the CNS of lupus patients and is proposed to play a role in the pathogenesis of neuropsychiatric manifestations (Tsai et al., 1994; Hirohata and Miyamoto, 1990; Gruol and Nelson, 1997). Also, mechanisms of inflammation and neurodegeneration undoubtedly contribute to the overall brain damage (Ferro, 1998; Fieschi et al., 1998; Trysberg and Tarkowski, 2004). As discussed further below, other synergistic or independent mechanisms are also likely to be operational.

3.3.1 Brain-Reactive Autoantibodies (BRAA)

In 1939 the German physician Lechman-Facius was the first to propose the link between autoimmunity and mental illness. He described brain-reactive antibodies (BRA, immunoglobulins/proteins reactive to neuronal antigens) in sera and CSF from psychiatric patients (Lehmann-Facius, 1939). This discovery, however, did not attract significant attention until the 1960s, when Fessel and Solomon published a series of articles on "macroglobulins" or "anti-brain factors" in psychotic patients (Solomon et al., 1969, 1966; Fessel, 1962a, b; Fessel and Hirata-Hibi, 1963). Despite the fact that the pathogenic potential of autoantibodies is presently documented in disorders such as myasthenia gravis, multiple sclerosis, Lambert-Eaton's and Guillain-Barre syndromes (Owens, 2003; Vincent et al., 2003; Henneberg et al., 1991), the notion that BRAA could induce some forms of mental illness did not yet receive appreciation by a wide scientific community (Jankovic, 1985). This was largely due to controversial clinical reports, lack of knowledge on BBB permeability, and models with acceptable construct validity. Renewed enthusiasm for the link between BRAA and psychiatric manifestations (Schott et al., 2003; Tanaka et al., 2003) coincided with the evidence that even in health the CNS is consistently surveyed by a small population of circulating leukocytes (Keane and Hickey, 1997; Hickey, 1999, 2001). If so, one may assume that brain homeostasis will be significantly disrupted if antibody-producing immune cells (activated in chronic autoimmune, inflammatory, and neoplastic disorders) gain access through the BBB.

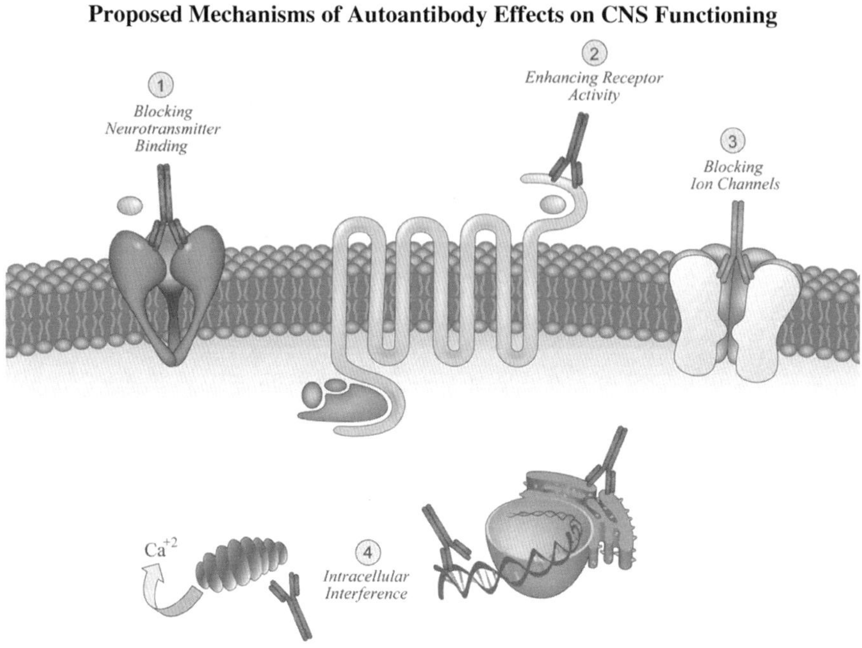

Fig. 3 Proposed mechanisms of brain-reactive autoantibody effect on brain structure and mental functioning: (1) Binding to surface neurotransmitter receptor blocks normal binding, induces receptor internalization, or prevents neurotransmitter release at the synapse. (2) Binding to neurotransmitter receptors mimics the ligand and increases receptor activity. (3) Binding to ion channels alters cell osmolarity and transmembrane pontential. (4) Active passage through the plasma membrane (endocytosis) alters Ca^{+2} metabolism, DNA synthesis, activity of organelles, and secondary messenger systems.

Although diverse mechanisms of anti-neuronal antibody binding may alter normal mental functioning (Fig. 3), the extent to which they mediate neuropsychiatric manifestations is not well understood. This is particularly true for those antibodies that are not cytopathic, but rather transiently affect cellular functioning. Murine models of SLE were extremely helpful in identifying the specificity and biochemical nature of BRAA (Alexander and Quigg, 2007; Hoffman et al., 1988b). Some of this work found reactivity to dissociated murine cerebellar cells (Harbeck et al., 1978b; Hoffman et al., 1978a), as well as other cells of the CNS (Hoffman et al., 1987; Diederichsen and Pyndt, 1970), and neuroblastoma cell lines (Bluestein, 1978, 1979; Hoffman et al., 1988a). These antibodies also react with lymphocyte antigens (Bluestein and Zvaifler, 1976; Harbeck et al., 1978a; Narendran and Hoffman, 1988; Parker et al., 1974; Shirai and Mellors, 1972, 1971) and some subsets react with human brain antigens (Maag and Hoffman, 1993).

In order to define the potential role of these antibodies in the pathogenesis of CNS manifestations they must be characterized. Along these lines the levels of reactivity and chronological appearance of BRAA in autoimmune and non-autoimmune strains of mice were examined. Since it is known that there is a diversity of autoantibodies in SLE, e.g., to DNA, ribonucleoprotein, etc., it was reasonable to expect that there existed autoantibodies which reacted with neurons and which could interfere with normal brain functioning. Bluestein and Zvaifler (1976) showed that thymocytotoxic autoantibody reactivity could be absorbed by brain, indicating the existence of BRAA. Our initial research on the subject showed that BRAA did indeed exist in autoimmune mice (Harbeck et al., 1978b; Hoffman et al., 1978a). These studies were the first to show that not only were there antibodies reactive with thymus, whose activity could be absorbed by brain, but there were independent autoantibodies reactive with neurons from brain. It also showed that there are autoantibodies that react with neuronal cell surface antigens. Later studies confirmed and extended these findings. Using indirect immunofluorescence techniques, it was found that the autoimmune mice (NZB/W, NZB, BXSB, MRL/lpr), in general, at any given age, have higher levels of autoantibody reactivity toward dissociated cerebellar cells than low autoimmune (MRL/Mp) and nonautoimmune (BDF1) mice (Hoffman et al., 1987). There was also an increase in autoantibody level with age. The MRL/lpr and BXSB mice showed the highest reactivity. In order to get an initial characterization of the specificity and diversity of the autoantibodies, sera from the above mice were absorbed against nuclei to eliminate anti-nuclear antibodies and used to immunofluorescently stain sections of mouse brain. The results showed generalized staining of cells of the brain, including hippocampus and cerebellum. This indicated heterogeneity of brain reactivity in systemic autoimmune disease.

These studies set the foundations for the autoantibody hypothesis, showing that BRAA exist, they occur in higher levels in autoimmune than non-autoimmune animals, and they react with cell surface components. It can be hypothesized that there are a variety of BRAA, but that only certain subsets of these are important in mediating the neuropsychiatric manifestations. A diversity of BRAA has been shown by one-dimensional gel electrophoresis and immunoblotting (Narendran and Hoffman, 1989). It was found that there was a much greater diversity of BRAA than originally expected. This diversity exists both within and between strains. The BXSB and MRL/lpr autoimmune strains displayed the greatest diversity of BRAA. Although there is diversity, it was also found that there are dominant autoantibodies, i.e., autoantibodies that occur repeatedly in different mice, both within and between strains. This study also confirmed another part of the hypothesis; that autoantibodies react with integral membrane proteins of brain cells. If autoantibodies mediate some of the CNS manifestations seen in SLE, then it is most likely that they will interfere with cell functioning by binding to integral membrane molecules of the plasma membrane. It has also been speculated, however, that antibodies can get inside cells to mediate alterations of function (Ruiz-Argüelles et al., 2003).

BRAA do not just occur in mice; it is well known that they also occur in SLE patients with neuropsychiatric involvement. It has been proposed for many years that these BRAA play a principal role in mediating different types of psychiatric manifestations (Hoffman et al., 1988b; Harbeck et al., 1978b; Harris and Hughes, 1985; Bluestein and Woods, 1982; Bresnihan et al., 1979; Denburg et al., 1988; Hanly et al., 1994; How et al., 1985; Klein et al., 1991; Lapteva et al., 2006; Long et al., 1990; Quismorio and Friou, 1972; Schiffer and Hoffman, 1991; Toh and Mackay, 1981; Williams et al., 1981; Wilson et al., 1979; Winfield et al., 1978; Zandman-Goddard et al., 2007). The antibodies reactive with brain found in CNS-SLE patients appear to be similar to those found in the murine models. The studies reporting the existence of these autoantibodies also provide evidence that these BRAA mediate, at least some of, the neuropsychiatric manifestations seen in SLE. This is based on correlations between neuropsychiatric involvement and the existence of BRAA. There are, however, discrepancies in the data, such as conflicting reports and weak correlations. There are several explanations for these discrepancies, one of which is that only some of the BRAA have clinical significance. Another is that additional factors (e.g., altered BBB permeability) must be present in order for the psychiatric manifestations to be displayed. Due to these discrepancies BRAA are generally not used as a diagnostic criterion for CNS involvement in SLE (Greenwood et al., 2002).

There have been several different classes of BRAA in the sera of SLE patients. Most of these have been identified on the basis of their reactivities to cells or tissues, including reactivity against several neuroblastoma and glioblastoma cell lines (Bluestein, 1978, 1979; Danon and Garty, 1986; How et al., 1985; Toh and Mackay, 1981; Wilson et al., 1979). There are autoantibodies against lymphocytes capable of being adsorbed by brain (Williams et al., 1981; Bluestein et al., 1981; Bluestein and Zvaifler, 1976; Temesvari et al., 1983; Winfield et al., 1978). Autoantibody reactivity to CNS neurons (Bresnihan et al., 1979; Golombek et al., 1986), neuronal cytoplasm (Quismorio and Friou, 1972), and neuronal receptors (DeGiorgio et al., 2001; Lapteva et al., 2006; Tanaka et al., 2003) has also been reported. BRAA that have been more specifically characterized are against synapsin I (Gitlits et al., 2001), asialo GM1 (Hirano et al., 1980), beta-endorphin (Froelich et al., personal communications) and neurofilament (Kurki et al., 1986). Anti-ribosomal P protein autoantibodies have been well studied and shown to correlate with disease activity in lupus, including neuropsychiatric involvement (Kiss and Shoenfeld, 2007; Toubi and Shoenfeld, 2007).

Anti-phospholipid antibodies have also been well studied and proposed to play a role in mediating CNS involvement in autoimmune diseases (Hogan et al., 1988; Inzelberg and Korczyn, 1989; Uhlig and Dernick, 1989; Denburg et al., 1997; Afeltra et al., 2003; Toubi et al., 1995). These are most likely to play a role in cerebrovascular episodes (Cronin et al., 1988), due to arterial and venous thrombotic vascular occlusions, but a broader range of neuropsychiatric manifestations has also been attributed to this subclass of antibodies (Liberato and Levy, 2007). They have also been reported in children, but with less association with neurological manifestations (Avcin and Silverman, 2007). There is also the suggestion that anti-phospholipid antibodies correlate with cerebral infarcts, but do not directly mediate neuronal dysfunction (Fields et al., 1990). It is common reasoning that vascular damage could mediate the neuropsychiatric manifestations, but while cerebral infarcts have been related to function, measures of cerebral blood do not correlate with neuropsychological involvement (Waterloo et al., 2001).

Using 1D gel electrophoresis and immunoblots, we have found that there are autoantibodies in the sera of patients with rheumatoid arthritis (RA) that bind to integral membrane proteins of mouse brain (Maag and Hoffman, 1993). Thus, it is likely that some of the autoantibodies in murine sera will have reactivity similar to what is found in humans. Even more interesting was the finding that these autoantibodies correlated with disease activity, viz., rheumatoid factor and joint swelling, and cognitive coping strategies in RA patients. There was also a trend toward a correlation between BRAA and depression and daily mood scores. Evidence for a subset of pathogenic autoantibodies that might mediate changes in CNS functioning was also found. These correlations do not imply a causal connection, they do, however, warrant further studies designed to determine the connection between the immune system and CNS functioning.

For this purpose, the murine models are very useful, their neuropathology has been characterized (Alexander et al., 1983; Rudick and Eskin, 1983), and many neurobehavioral studies have been done (as discussed in Section 2.1). Recent articles report anti-DNA antibodies cross-reacting with NR2 glutamate receptor (DeGiorgio et al., 2001; Lapteva et al., 2006). This group has shown that these anti-glutamate receptor antibodies (derived either experimentally or from lupus patients) could affect behavior, if the BBB is opened (Huerta et al., 2006; Kowal et al., 2006). Another interesting study done by Lawrence and colleagues (2007), produced monoclonal autoantibodies from an autoimmune strain of mouse (NZM), derived from NZB/W mice. These monoclonal autoantibodies were directed against mouse dynamin-1. When the monoclonal antibody was injected intravenously into non-autoimmune Balb/C mice they developed behavioral deficits, as compared to controls injected with isotype control, and similar to the original NZM mice. This group has followed up on their studies (Mondal et al., 2008) by trying to elucidate the mechanisms involved in the murine CNS involvement. Another study (Katzav et al., 2007) has injected human anti-ribosomal P protein autoantibodies intracerebroventricularly into mice and tested them behaviorally. It was shown that this treatment induced depressive-like behavior, independent of motor or cognitive dysfunction, as compared to the IgG injected controls. The antibodies bound to neurons in limbic and olfactory areas of the brain. These studies show directly that BRAA can alter behavior.

The functional effects of BRAA have also been tested at the cellular level. Using the patch clamp technique, with the murine neuroblastoma cell line (NBP2) to record from, the hypothesis that BRAA can alter normal neuronal functioning (Crimando et al., 1997) was tested. These are neuronal-like cells that are electrically excitable, synthesize neurotransmitters, and form synaptic contacts when differentiated. Being able to successfully record from them allowed a test of the hypothesis that BRAA from autoimmune mice can alter normal potassium or sodium channel functioning. The cells were untreated, i.e., recorded in buffer alone without any pre-treatment, pre-incubated with IgG from normal Balb/c mice, or pre-incubated with purified IgG from pools of sera from the autoimmune BXSB mice. The sodium and potassium currents from the patch clamp recordings of cells treated with IgG from the non-autoimmune mice did not differ from the buffer-treated cells. One of the six pools of IgG from the autoimmune sera showed a clear inhibition of the inward, sodium currents in the whole cell recordings, as compared to the controls. This same pool did not, however, show any effects on the outward, potassium currents. This shows specificity to the effects of the autoimmune sera. The autoantibodies affected the sodium channels, but not the potassium channels. Another pool of BRAA interfered with potassium currents, but not the sodium currents. These results confirm the hypothesis that there are pathogenic autoantibodies that can depress neuronal activity by affecting sodium channel functioning. It also confirms the hypothesis that there are subsets of autoantibodies which are pathogenic and which affect neuronal functioning in diverse ways.

It is known that at least some of the BRAA found in autoimmune mice are directed against integral membrane proteins of brain cells. There have been limited attempts to fully characterize the BRAA. To a large extent, the specificity of these antibodies is not known, nor where they bind in brain (e.g., what cell types bear the reactive antigens). This information is important for determining the role of these autoantibodies in disease pathogenesis. One study (Gitlits et al., 2001) has gone in this direction, identifying synapsin I as a reasonable candidate autoantigen for mediating CNS manifestations. Another study (Tanaka et al., 2003) has shown serum autoantibodies to neurotransmitter receptors in patients with psychiatric disorders, including schizophrenic and mood disorder, and autoimmune diseases. If autoantibodies play a role in altering mental function, then it is likely that there will be a subset of these BRAA that are predominant. If these autoantibodies can be identified and their role in mental functioning determined, then therapeutic techniques for eliminating the mental disorders can be developed. For example, one possibility would be to use B cell vaccination (Correale et al., 2008) to eliminate the pathogenic BRAA and eliminate the corresponding disorders of mental functioning.

Many investigators have used neuroblastoma cell lines, human and murine, in order to assay for BRAA. The advantage of these cell lines is that they provide a homogeneous population of cells. Unfortunately, they contain a heterogeneous population of antigens. In order to compare the antigens in neuroblastoma to those found on normal mouse brain, lactoperoxidase cell surface labeling and gel electrophoretic techniques were used (Hoffman et al., 1988a). The results showed that there were both similarities and differences between the antigens found in normal brain and neuroblastoma cells. This study shows that it is important to go to the molecular level in order to characterize the specificity of BRAA for a better understanding of how they can interfere with normal brain functioning. The neuroblastoma cells could be used for testing the autoantibody hypothesis, but it is important to determine if the antigens of interest are also on cells of the brain.

We have shown that one level of characterization can be done using both two-dimensional and one-dimensional gel electrophoretic techniques, in combination with immunoblotting. Integral membrane proteins were isolated from brain and separated using two-dimensional gel electrophoresis (Narendran and Hoffman, 1988). These proteins were transferred to nitrocellulose and immunoenzymatically stained using serum from an autoimmune mouse. This technique is useful (Hanly and Hong, 1993) as a first level of characterizing the antigens to which BRAA bind. Adsorption techniques have been used in determining the specificity of BRAA (Hoffman and Madsen, 1990). Sera from various mice were adsorbed against brain, liver, kidney, and spleen, or left unabsorbed. These adsorbed and unadsorbed sera were used for immunoenzymatic staining of one-dimensional immunoblots. The data showed that some autoantibodies are specific to brain, while others react with different tissues. Although useful, these techniques are not good enough for fully characterizing the specificity of the BRAA. The actual molecules to which these antibodies bind have to be identified in an organized and systematic fashion. This can be done using a variety of modern proteomic and genomic techniques.

To detect the presence of in situ immunoglobulin (Ig) in the brain of MRL/lpr and BXSB mice and compare that to control strains of the low autoimmune MRL/Mp and the non-autoimmune C57BL/6 mice, laser confocal microscopy was performed on frozen brain sections (Zameer and Hoffman, 2001). There was a dramatic increase in fluorescence in the brains of MRL/lpr and BXSB at 4 months of age. There was little or no Ig detected in the brains of control mice. This increase in the presence of Ig in the autoimmune mouse brain was paralleled by an increase in the serum titers of BRAA and anti-DNA autoantibodies, as determined by ELISA. This study shows that autoantibodies exist in the brain, providing further evidence that BRAA play a role in autoimmunity-induced behavioral dysfunction. This was further confirmed by the presence of elevated levels of IgG and albumin in the CSF of aged MRL/lpr mice, as compared to the control MRL/Mp (Sidor et al., 2005). This indication of Ig in brain and altered BBB permeability correlated with neurodegeneration in periventricular regions (Ballok et al., 2004b). Furthermore, CSF from these mice has been shown to be cytotoxic to neurons (Maric et al., 2001).

Although common in human and animal forms of SLE (Moore, 1997), BRAA have also been reported in CNS disorders which widely differ in clinical manifestations, as has been reviewed elsewhere (Margutti et al., 2006). The evidence is far from being conclusive, but it is hypothesized that BRAA-mediated lesions occur in a subpopulation of patients with schizophrenia spectrum disorder (Schott et al., 2003; Borda et al., 2002; Schwartz and Silver, 2000; Ganguli et al., 1994), epilepsy (Sokol et al., 2004; Callenbach et al., 2003; Vianello et al., 2002), autism spectrum disorder (Singh and Rivas, 2004; Singh and Jensen, 2003; Singh et al., 2002; Shoenfeld and Aron-Maor, 2000; Connolly et al., 1999; Trottier et al., 1999; Silva et al., 2004), paraneoplastic syndromes, cerebral ataxia, Rasmussen's encephalitis, Morovan's syndrome, and pediatric autoimmune neuropsychiatric disorders associated with streptococcal infections, PANDAS (reviewed in Lang et al., 2003). Further understanding the sources of BRAA, their molecular targets, and pathogenic circuits will likely lead to a profound leap in our understanding of mental disease and its bodily connections.

3.3.2 Neuroactive Cytokines

Psychiatric symptoms may antedate SLE diagnosis and excessive autoantibody production (van Dam et al., 1994). This suggests that other immune messengers (e.g., neuroactive cytokines) affect the neuroendocrine axis and alter behavior well before other soluble factors reach the CNS (Anisman et al., 2005). Much of the evidence for the role of cytokines is indirect and has been discussed in other chapters of this book. There is, however, some evidence that is more relevant to neuropsychiatric manifestations in autoimmune disease. For example, it has been shown that cytokine levels (IL-1, IL-6 and IFN-α) are elevated in the CSF of patients with neuropsychiatric SLE (Hirohata and Miyamoto, 1990; Alcocer-Varela et al., 1992; Shiozawa et al., 1992; Katsumata et al., 2007). In one of these studies, IFN-α (Shiozawa et al., 1992) was elevated in the CSF of five of six patients with lupus psychosis (out of 17 patients with neuropsychiatric manifestations) and these levels decreased when the lupus psychosis subsided. In one patient who died during this study IFN-α was detectable in neurons and microglia. The authors concluded that IFN-α might be synthesized in the brain and be the cause of psychoses in patients with SLE. In a small clinical study (Jara et al., 1998) comparing patients with CNS-SLE, SLE (without CNS involvement), neurocysticercosis, and healthy women, it was found that IL-6 and prolactin levels were significantly elevated in the CSF of the patients with CNS-SLE. The authors suggested that these results indicated intrathecal synthesis of these molecules, which may be indicative of immune-neuroendocrine interactions mediating some of the neuropsychiatric manifestations. An interesting, but difficult to interpret study (Kozora et al., 2001) reported that SLE patients without diagnosed neuropsychiatric manifestations displayed impaired learning and poorer attention, compared to patients with rheumatoid arthritis and healthy controls. Hierarchical regression showed that IL-6 and dehydroepiandrosterone sulfate accounted for some of the patients' performance in the learning and memory tasks. There are also conflicting results, with one study not finding a relation between cytokines and neuropsychiatric involvement in lupus (Jonsen et al., 2003).

In a series of behavioral studies with the NZB/W autoimmune mice, it was shown that cytokines were associated with the behavioral changes (Schrott and Crnic, 1996b, a, 1998). These investigators showed that the anxiety behavior of the NZB/W mice, as compared to the control NZW mice, were likely due to the autoimmune disease (Schrott and Crnic, 1996a). When they injected IFN-α into the NZW mice they found behavior similar to the NZB/W. This shows that this cytokine can induce behavioral alterations similar to those occurring during the autoimmune disease. In a separate study they injected soluble IFN-γ receptor into the animals and were able to reverse the behavioral manifestations (Schrott and Crnic, 1998). The rationale is that this receptor would interfere with IFN-γ or other cytokines (such as IL-1 and IFN-α), which could be altering behavior. In other studies, using an adenovirus carrying mRNA for a murine cytokine, Sakic and colleagues showed that prolonged exposure to circulating IL-6 primarily impairs ingestive behavior (likely reflecting enhanced catabolism), while a relatively brief elevation in systemic IFN-γ has also prolonged effects on motivated behavior (Sakic et al., 2001; Kwant and Sakic, 2004). It should be noted, however, that studies with prolonged expression of neuroactive cytokines could not replicate the constellation of behavioral deficits in autoimmune MRL/lpr mice. Further, these data revealed that although IL-6 and IFN-γ have detrimental effects on ingestive and motivated behavior, other factors and mechanisms seem to significantly contribute to the overall behavioral phenomenon. Nevertheless, all of the above studies provide evidence for the role of cytokines in altering behavior during autoimmunity.

It has also been shown (Hoffman et al., 1978b, 1998) that there are behavioral manifestations, alterations of learning and memory, associated with experimental immune complex disease. Immune complex disease is one of the primary pathogenic mechanisms associated with autoimmune diseases like SLE. The effects on behavior are unlikely to be due to autoantibodies, since we used antibodies to BSA as part of the immune complexes formed. Unless there is cross-reactivity between albumin and brain antigens, the autoantibody hypothesis is unlikely to play a role here. Rather, other mechanisms must be invoked. One of these is that cytokines are mediating the behavioral changes. In order to test the idea that elevated cytokine levels in the vascular system could mediate their effects across the BBB and alter neuronal functioning we used electrophysiological techniques. In support of the cytokine hypothesis we have shown that cytokines can alter normal brain functioning (Bartholomew and Hoffman, 1993). Both IL-1 and IL-2, and to some extent IL-6, were able to alter the electrical activity of neurons in the anterior region of the hypothalamus. These cytokines were administered in the periphery, which more closely mimics the effects of an immune response on the CNS. These studies show that immunologic processes can alter normal neuronal functioning, at the level of the electrical activity of neurons, which in turn could lead to behavioral changes. Ziv and colleagues have reviewed literature (Ziv and Schwartz, 2008) on how cytokines can influence memory and learning by affecting neurogenesis in the hippocampus. Overall, there is good evidence that cytokines are playing a role in altering behavior in autoimmune-mediated diseases.

3.3.3 Complement and Immune Complexes

Immune complexes can also modulate the brain and behavior. This is not only consistent with the cytokine hypothesis, but also with the role of the BBB (discussed above), as well as the role of complement. In two separate studies, immune complex disease was induced in rats, with behavioral changes associated with this immune state. Immune complex disease was induced by immunizing with BSA over a period of time to generate anti-BSA antibodies. After serum antibodies were induced, intravenous injections of BSA were given at a dose to induce circulating, pathogenic immune complexes. This was maintained over several months. Behavioral changes in a fear avoidance paradigm were seen after disease developed (Hoffman et al., 1978b), as determined by elevated proteinuria. In a later study (Hoffman et al., 1998) using a different behavioral paradigm (the Lashley maze) changes in learning were observed associated with immune complex disease. Uremia was not present. These studies showed that the immunologic processes associated with immune complex disease could induce behavioral changes.

In order to determine if immune complex formation within the brain could be mediating these behavioral changes, we infused antibodies and antigens separately into the perifornical region of the hypothalamus. Behavioral changes, i.e., depressed water consumption, without concomitant changes in body temperature, were observed with this procedure (Hoffman et al., 1982). This showed that immune complex reactions within the CNS could alter behavior. What is interesting is that there were no signs of choroid plexus damage in this model of immune complex disease (Peress and Tompkins, 1979).

Complement components are likely mediating the above immune complex-mediated behavioral changes as suggested in two studies by Schupf and Williams (Williams and Schupf, 1977; Schupf and Williams, 1987). They also provided evidence that the behavioral changes, subsequent to immune complex formation within the CNS, might be a consequence of neurotransmitter release (Schupf and Williams, 1985). Namely, they showed that the complement components C3a and C5a could affect catecholaminergic systems and alter behavior (Williams et al., 1985; Schupf et al., 1983). Evidence for C3a receptors in the brain was also provided (Williams et al., 1988).

Subsequent to these initial studies, other investigators have shown that there are receptors for C3a on neurons and glia, as well as in the pituitary gland (Francis et al., 2003). Using in situ hybridization and immunohistochemistry one group (Davoust et al., 1999) found C3a receptors to be highly expressed in cortical and hippocampal neurons, and Purkinje cells. Glial cells showed very low levels, while primary cultures of astrocytes and microglia showed elevated expression of C3aR mRNA. During experimental allergic encephalomyelitis there were no changes in receptor expression on neurons, whereas microglia and astrocytes did show an increase. C5a receptor expression does increase on neurons during an inflammatory state (Nataf et al., 1998), although they are not constitutively expressed. In the pituitary, C3a receptors were also found (Francis et al., 2003) and C3a was able to stimulate pituitary cell cultures to release prolactin, growth hormone, and adrenocorticotropin. This suggests that C3a binding would be able to stimulate the hypothalamic-pituitary-adrenal axis and the stress response.

Addressing the functional significance of complement in CNS-SLE, Alexander has done a series of studies showing that blocking of the complement pathways (Alexander et al., 2003, 2005a) leads to an amelioration of the CNS manifestations in the MRL/lpr murine model of lupus. Furthermore, they showed (Alexander et al., 2007) the importance of the alternative pathway, which when blocked through a deficiency of factor B, led to a reduction of the behavioral alterations in the MRL/lpr mice, as well as a reduction in immune complex deposits in the brain. There was also reduced neutrophil infiltration and apoptosis, suggesting BBB involvement (see above).

Human studies also implicate complement components in the pathogenesis of neuropsychiatric manifestations. In patients with CNS-SLE, indices of C3 and C4 showed elevation of these components in the CSF (Jongen et al., 2000). Complement is generally reduced in the sera of patients with active SLE. The elevated C4 in CSF may be indicative of intrathecal synthesis, or compensatory mechanisms following systemic consumption of complement. In other studies (Hopkins et al., 1988; Belmont et al., 1986) it was reported that complement is activated during SLE and the anaphylatoxins, C3a and C5a, in plasma, correlated with flares of CNS disease in patients. The authors speculated that this was related to vascular injury in the pathogenesis of CNS involvement.

In a more recent report, it was speculated that autoantibodies form immune complexes in the CSF, activate the complement system and contribute to neuropsychiatric manifestations (Sasajima et al., 2006). These investigators found antibodies to a glycolytic enzyme (anti-triosephosphate isomerase antibodies; anti-TPI) in sera and CSF of CNS-SLE patients. Interestingly, the anti-TPI index has a high specificity (94.5%) for SLE patients with neuropsychiatric involvement. In immune complexes isolated from the CSF of CNS-SLE patients TPI was found. The C3d index was higher in anti-TPI positive patients, than in negative patients. The authors speculated that anti-TPI forms immune complexes in the CSF of CNS-SLE patients, activating the complement system and mediating neuropsychiatric manifestations. These studies are interesting in light of the older studies discussed at the beginning of this section. Complement components, such as C3a and C5a, could stimulate neurons or the endocrine system, either in a pathologic or physiologic condition, contributing to alterations of mental functioning. An alternative is that they contribute to increased permeability of the BBB.

4 Summary

The recent introduction of the term "spectrum disorder" in the cases of schizophrenia and autism better reflects the biological complexity and behavioral continuum in these pervasive illnesses. Acknowledgment of such diversity underscores the issues about multiple pathogenic pathways. The evidence reviewed in the sections above provides strong empirical support for the possibility that chronic autoimmunity and inflammation are one of the key factors that can compromise brain morphology and function. This may occur either in isolation, or in synergy with other factors, such as aberrant genetic or detrimental environmental influences. On the other hand, brain deficits associated with changes in the immune system range from diverse neurological to diverse psychiatric signs and symptoms. Co-morbidity and the parallel course of immunological and behavioral manifestations in such disorders further preclude us from inferring simplistic pathogenic mechanisms, and point to the close relationship between immunity and mental function. If the immune system indeed plays a causal role, then it remains to be determined whether circulating immune factors affect the CNS directly or via other bodily systems, such as endocrine, cardiovascular, and renal. The accumulated body of evidence leads to an optimistic view that a deeper appreciation of the neuroimmunologic circuitry in the current century will provide missing links in better understanding normal brain function and some forms of classical mental illness.

5 Future Directions

Despite significant evidence on connections between autoimmunity and mental disorders, further work is needed to elucidate "windows of vulnerability" (i.e., when immuno-neuro-endocrine interactions occur), and pathogenic mechanisms that underlie aberrant behavior.

One of the important goals is to systematically characterize the specificity of BRAA. Knowledge of the molecules to which they bind may allow a determination of the functions that can be affected and preventive diagnosis well before any CNS manifestations emerge. Furthermore, if there is a subset of pathogenic BRAA, then appropriate pharmacological or immunological techniques could be developed to treat or prevent neuropsychiatric episodes.

There have to be more studies on the permissive role of the BBB, both in human conditions and animal models. A major conundrum is the intrathecal synthesis of immune mediators versus increased BBB permeability to circulating factors, and their relationship to neuropsychiatric manifestations. If the latter, then the mechanisms leading to alterations in BBB permeability need to be elucidated.

The neuroactive cytokines inducing specific behavioral deficits need to be identified, likely with significant initial input from animal models. This includes answers on how they cross the BBB, or how they mediate some effects without entering brain parenchyma. Likewise, the role of other soluble factors (e.g., complement components, immune complexes) and their effects on the neuroendocrine system should be further investigated.

At this point in time relatively little is known about the neuropathology of immune-mediated attack. There is evidence for neuronal damage and loss, but cytotoxic mechanisms are largely unknown. Importantly, it needs to be determined whether transient, functional alterations (i.e., without overt neuronal damage) mediate some neuropsychiatric manifestations. As can be seen from the sections above, the autoimmune-mediated mechanisms for altering mental functioning are diverse and complex. This complexity, however, opens new doors for future research in ImmunoPsychiatry.

References

Abbott NJ, Mendonca LL, Dolman DE, 2003. The blood-brain barrier in systemic lupus erythematosus. Lupus 12:908–915.

ACR Ad Hoc Committee, 1999. The American College of Rheumatology nomenclature and case definitions for neuropsychiatric lupus syndromes. Arthritis Rheum. 42:599–608.

Afeltra A, Garzia P, Mitterhofer AP, Vadacca M, Galluzzo S, Del PF, Finamore L, Pascucci S, Gasparini M, Lagana B, Caccavo D, Ferri GM, Amoroso A, Francia A, 2003. Neuropsychiatric lupus syndromes: relationship with antiphospholipid antibodies. Neurology 61:108–110.

Ahlsen G, Rosengren L, Belfrage M, Palm A, Haglid K, Hamberger A, Gillberg C, 1993. Glial fibrillary acidic protein in the cerebrospinal fluid of children with autism and other neuropsychiatric disorders. Biol. Psychiatry 33:734–743.

Ainiala H, Hietaharju A, Dastidar P, Loukkola J, Lehtimaki T, Peltola J, Korpela M, Heinonen T, Nikkari ST, 2004. Increased serum matrix metalloproteinase 9 levels in systemic lupus erythematosus patients with neuropsychiatric manifestations and brain magnetic resonance imaging abnormalities. Arthritis Rheum. 50:858–865.

Ainiala H., Loukkola J., Peltola J., Korpela M., Hietaharju A., 2001. The prevalence of neuropsychiatric syndromes in systemic lupus erythematosus. Neurology 57:496–500.

Alcocer-Varela J, Aleman-Hoey D, Alarcon-Segovia D, 1992. Interleukin-1 and interleukin-6 activities are increased in the cerebrospinal fluid of patients with CNS lupus erythematosus and correlate with local late T-cell activation markers. Lupus 1:111–117.

Alexander JJ, Bao L, Jacob A, Kraus DM, Holers VM, Quigg RJ, 2003. Administration of the soluble complement inhibitor, Crry-Ig, reduces inflammation and aquaporin 4 expression in lupus cerebritis. Biochim. Biophys. Acta 1639: 69–176.

Alexander JJ, Jacob A, Bao L, Macdonald RL, Quigg RJ, 2005a. Complement-dependent apoptosis and inflammatory gene changes in murine lupus cerebritis. J. Immunol. 175:8312–8319.

Alexander JJ, Jacob A, Vezina P, Sekine H, Gilkeson GS, Quigg RJ, 2007. Absence of functional alternative complement pathway alleviates lupus cerebritis. Eur. J. Immunol. 37:1691–1701.

Alexander EL, Murphy ED, Roths JB, Alexander GE, 1983. Congenic autoimmune murine models of central nervous system disease in connective tissue disorders. Ann. Neurol. 14:242–248.

Alexander JJ, Quigg RJ, 2007. Systemic lupus erythematosus and the brain: what mice are telling us. Neurochem. Int. 50:5–11.

Alexander JJ, Zwingmann C, Quigg R, 2005b. MRL/lpr mice have alterations in brain metabolism as shown with [(1)H-(13)C] NMR spectroscopy. Neurochem. Int. 47:143–151.

Alosachie IJ, Terryberry JW, Mevorach D, Chapman Y, Lorber M, Torre D, Youinou P, Peter JB, Shoenfeld Y, 1998. Central nervous system (CNS) involvement in SLE. The diagnostic role of antibodies to neuronal antigens. Clin. Rev. Allergy Immunol. 16:275–284.

Anderson KK, Ballok DA, Prasad N, Szechtman H, Sakic B, 2006. Impaired response to amphetamine and neuronal degeneration in the nucleus accumbens of autoimmune MRL-lpr mice. Behav. Brain Res. 166:32–38.

Anisman H, Merali Z, Poulter MO, Hayley S, 2005. Cytokines as a precipitant of depressive illness: animal and human studies. Curr. Pharm. Des. 11:963–972.

Appenzeller S, Bonilha L, Rio PA, Min LL, Costallat LT, Cendes F, 2007. Longitudinal analysis of gray and white matter loss in patients with systemic lupus erythematosus. Neuroimage. 34:694–701.

Arolt V, Weitzsch C, Wilke I, Nolte A, Pinnow M, Rothermundt M, Kirchner H, 1997. Production of interferon-gamma in families with multiple occurrence of schizophrenia. Psychiatry Res. 66:145–152.

Avcin T, Silverman ED, 2007. Antiphospholipid antibodies in pediatric systemic lupus erythematosus and the antiphospholipid syndrome. Lupus 16:627–633.

Ballok DA, 2007. Neuroimmunopathology in a murine model of neuropsychiatric lupus. Brain Res. Rev. 54:67–79.

Ballok DA, Earls AM, Krasnik C, Hoffman SA, Sakic B, 2004a. Autoimmune-induced damage of the midbrain dopaminergic system in lupus-prone mice. J. Neuroimmunol. 152:83–97.

Ballok DA, Ma X, Denburg JA, Arsenault L, Sakic B, 2006. Ibuprofen fails to prevent brain pathology in a model of neuropsychiatric lupus. J. Rheumatol. 33:2199–2213.

Ballok DA, Sakic B. Neurodegeneration in autoimmune mice as revealed by Fluoro Jade B. Society for Neuroscience Abstracts, 93.15. 2002. Ref Type: Abstract

Ballok DA, Woulfe J, Sur M, Cyr M., Sakic B, 2004b. Hippocampal damage in mouse and human forms of systemic autoimmune disease. Hippocampus 14:649–661.

Baraczka K, Nekam K, Pozsonyi T, Jakab L, Szongoth M, Sesztak M, 2001. Concentration of soluble adhesion molecules (sVCAM-1, sICAM-1 and sL-selectin) in the cerebrospinal fluid and serum of patients with multiple sclerosis and systemic lupus erythematosus with central nervous involvement. Neuroimmunomodulation. 9:49–54.

Baraczka K, Nekam K, Pozsonyi T, Szuts I, Ormos G, 2004. Investigation of cytokine (tumor necrosis factor-alpha, interleukin-6, interleukin-10) concentrations in the cerebrospinal fluid of female patients with multiple sclerosis and systemic lupus erythematosus. Eur. J. Neurol. 11:37–42.

Bartholomew SA, Hoffman SA, 1993. Effects of peripheral cytokine injections on multiple unit activity in the anterior hypothalamic area of the mouse. Brain Behav. Immun. 7:301–316.

Beiske AG, Svensson E, Sandanger I, Czujko B, Pedersen ED, Aarseth JH, Myhr KM, 2008. Depression and anxiety amongst multiple sclerosis patients. Eur. J. Neurol. 15:239–245.

Belmont HM, Hopkins P, Edelson HS, Kaplan HB, Ludewig R, Weissmann G, Abramson S, 1986. Complement activation during systemic lupus erythematosus – C3a and C5a anaphylatoxins circulate during exacerbations of disease. Arthritis Rheum. 29:1085–1089.

Bluestein HG, 1978. Neurocytotoxic antibodies in serum of patients with systemic lupus erythematosus. Proc. Natl. Acad. Sci. U. S. A. 75:3965–3969.

Bluestein HG, 1979. Heterogeneous neurocytotoxic antibodies in systemic lupus erythematosus. Clin. Exp. Immunol. 35:210.

Bluestein HG, Pischel KD, Woods VL, Jr., 1986. Immunopathogenesis of the neuropsychiatric manifestations of systemic lupus erythematosus. Springer Semin. Immunopathol. 9:237–249.

Bluestein HG, Williams GW, Steinberg AD, 1981. Cerebrospinal fluid antibodies to neuronal cells: association with neuropsychiatric manifestations of systemic lupus erythematosus. Am. J. Med. 70:240–246.

Bluestein HG, Woods VL, 1982. Antineuronal antibodies in systemic lupus erythematosus. Arthritis Rheum. 25:773–778.

Bluestein HG, Zvaifler NJ, 1976. Brain-reactive lymphocytotoxic antibodies in the serum of patients with systemic lupus erythematosus. J. Clin. Invest. 57:509–516.

Blum D, Torch S, Lambeng N, Nissou M, Benabid AL, Sadoul R, Verna JM, 2001. Molecular pathways involved in the neurotoxicity of 6-OHDA, dopamine and MPTP: contribution to the apoptotic theory in Parkinson's disease. Prog. Neurobiol. 65:135–172.

Borchers AT, Aoki CA, Naguwa SM, Keen CL, Shoenfeld Y, Gershwin ME, 2005. Neuropsychiatric features of systemic lupus erythematosus. Autoimmun. Rev. 4:329–344.

Borda T, Perez RR, Joensen L, Gomez RM, Sterin-Borda L, 2002. Antibodies against cerebral M1 cholinergic muscarinic receptor from schizophrenic patients: molecular interaction. J. Immunol. 168:3667–3674.

Bosma GP, Rood MJ, Zwinderman AH, Huizinga TW, van Buchem MA, 2000. Evidence of central nervous system damage in patients with neuropsychiatric systemic lupus erythematosus, demonstrated by magnetization transfer imaging. Arthritis Rheum. 43:48–54.

Bresnihan B, Hohmeister R, Cutting J, Travers RL, Waldburger M, Black C, Jones T, Hughes GR, 1979. The neuropsychiatric disorder in systemic lupus erythematosus: evidence for both vascular and immune mechanisms. Ann. Rheum. Dis. 38:301–306.

Brey RL, Cote S, Barohn R, Jackson C, Crawley R, Teale JM, 1995. Model for the neuromuscular complications of systemic lupus erythematosus. Lupus 4:209–212.

Brey RL, Holliday SL, Saklad AR, Navarrete MG, Hermosillo-Romo D, Stallworth CL, Valdez CR, Escalante A, del R, I, Gronseth G, Rhine CB, Padilla P, McGlasson D, 2002. Neuropsychiatric syndromes in lupus: prevalence using standardized definitions. Neurology 58:1214–1220.

Brey RL, Sakic B, Szechtman H, Denburg JA, 1997. Animal models for nervous system disease in systemic lupus erythematosus. Ann. N. Y. Acad. Sci. 823:97–106.

Brooks WM, Sabet A, Sibbitt WL, Barker PB, van Zijl PC, Duyn JH, Moonen CT, 1997. Neurochemistry of brain lesions determined by spectroscopic imaging in systemic lupus erythematosus. J. Rheumatol. 24:2323–2329.

Bruns A, Meyer O, 2006. Neuropsychiatric manifestations of systemic lupus erythematosus. Joint Bone Spine 73:639–645.

Buka SL, Cannon TD, Torrey EF, Yolken RH, 2008. Maternal exposure to herpes simplex virus and risk of psychosis among adult offspring. Biol. Psychiatry. 63:809–815.

Buka SL, Tsuang MT, Torrey EF, Klebanoff MA, Bernstein D, Yolken RH, 2001. Maternal infections and subsequent psychosis among offspring. Arch. Gen. Psychiatry 58:1032–1037.

Callenbach PM, Jol-Van Der Zijde CM, Geerts AT, Arts WF, Van Donselaar CA, Peters AC, Stroink H, Brouwer OF, Van Tol MJ, 2003. Immunoglobulins in children with epilepsy: the Dutch Study of Epilepsy in Childhood. Clin. Exp. Immunol. 132:144–151.

Carbotte RM, Denburg SD, Denburg JA, 1986. Prevalence of cognitive impairment in systemic lupus erythematosus. J. Nerv. Ment. Dis. 174:357–364.

Carr RI, Shucard DW, Hoffman SA, Hoffman AW, Bardana EJ, Harbeck RJ, 1978. In Neurochemical and components of schizophrenia, Goldstein, A.L. and Bergsma, D.(Eds), Birth Defects Orig. Artic. Ser. 14: 209–235.

Caspi A, Moffitt TE, 2006. Gene-environment interactions in psychiatry: joining forces with neuroscience. Nat. Rev. Neurosci. 7:583–590.

Cazzullo CL, Scarone S, Grassi B, Vismara C, Trabattoni D, Clerici M, 1998. Cytokines production in chronic schizophrenia patients with or without paranoid behaviour. Prog. Neuropsychopharmacol. Biol. Psychiatry 22:947–957.
Chess S, 1977. Follow-up report on autism in congenital rubella. J. Autism Child Schizophr. 7:69–81.
Chun S, McEvilly R, Foster JA, Sakic B, 2007. Proclivity to self-injurious behavior in MRL-lpr mice: implications for autoimmunity-induced damage in the dopaminergic system. Mol. Psychiatry. 1359–4184.
Comi AM, Zimmerman AW, Frye VH, Law PA, Peeden JN, 1999. Familial clustering of autoimmune disorders and evaluation of medical risk factors in autism. J. Child Neurol. 14:388–394.
Conejero-Goldberg C, Torrey EF, Yolken RH, 2003. Herpesviruses and Toxoplasma gondii in orbital frontal cortex of psychiatric patients. Schizophr. Res. 60:65–69.
Connolly AM, Chez MG, Pestronk A, Arnold ST, Mehta S, Deuel RK, 1999. Serum autoantibodies to brain in Landau-Kleffner variant, autism, and other neurologic disorders. J. Pediatr. 134:607–613.
Correale J, Farez M, Gilmore W, 2008. Vaccines for multiple sclerosis – progress to date. CNS Drugs 22:175–198.
Crimando J, Cooper K, Hoffman SA, 1997. Inhibition of sodium channel currents by antineuronal autoantibody from autoimmune mice. Ann. N. Y. Acad. Sci. 823:303–307.
Cronin ME, Biswas RM, Van der Straeton C, Fleisher TA, Klippel JH, 1988. IgG and IgM anticardiolipin antibodies in patients with lupus with anticardiolipin antibody associated clinical syndromes. J. Rheumatol. 15:795–798.
Daniels WW, Warren RP, Odell JD, Maciulis A, Burger RA, Warren WL, Torres AR, 1995. Increased frequency of the extended or ancestral haplotype B44-SC30-DR4 in autism. Neuropsychobiology 32:120–123.
Danon YL, Garty BZ, 1986. Autoantibodies to neuroblastoma cell surface antigens in neuropsychiatric lupus. Neuropediatrics 17:23–27.
Davoust N, Jones J, Stahel PF, Ames RS, Barnum SR, 1999. Receptor for the C3a anaphylatoxin is expressed by neurons and glial cells. Glia 26:201–211.
DeGiorgio LA, Konstantinov KN, Lee SC, Hardin JA, Volpe BT, Diamond B, 2001. A subset of lupus anti-DNA antibodies cross-reacts with the NR2 glutamate receptor in systemic lupus erythematosus. Nat. Med. 7:1189–1193.
Denburg JA, Carbotte RM, Denburg SD, 1993. Central nervous system lupus. Rheum. Rev. 2:123–132.
Denburg SD, Carbotte RM, Ginsberg JS, Denburg JA, 1997. The relationship of antiphospholipid antibodies to cognitive function in patients with systemic lupus erythematosus. J. Int. Neuropsychol. Soc. 3:377–386.
Denburg JA, Denburg SD, Carbotte RM, Long AA, Hanly JG, 1988. Nervous system involvement in systemic lupus erythematosus. Isr. J. Med. Sci. 24:754–758.
Denburg JA, Denburg SD, Carbotte RM, Sakic B, Szechtman H, 1995. Nervous system lupus: pathogenesis and rationale for therapy. Scand. J. Rheumatol. 12:263–273.
Denenberg VH, Mobraaten LE, Sherman GF, Morrison L, Schrott LM, Waters NS, Rosen GD, Behan PO, Galaburda AM, 1991. Effects of the autoimmune uterine/maternal environment upon cortical ectopias, behavior and autoimmunity. Brain Res. 563:114–122.
Denenberg VH, Sherman GF, Morrison L, Schrott LM, Waters NS, Rosen GD, Behan PO, Galaburda AM, 1992. Behavior, ectopias and immunity in BD/DB reciprocal crosses. Brain Res. 571:323–329.
Deykin EY, MacMahon B, 1979. Viral exposure and autism. Am. J. Epidemiol. 109:628–638.
Diederichsen H, Pyndt IC, 1970. Antibodies against neurons in a patient with systemic lupus erythematosus, cerebral palsy, and epilepsy. Brain 93:407–412.
Elouaai F, Lule J, Benoist H, Appolinaire-Pilipenko S, Atanassov C, Muller S, Fournie GJ, 1994. Autoimmunity to histones, ubiquitin, and ubiquitinated histone H2A in NZB x NZW and MRL-lpr/lpr mice. Anti-histone antibodies are concentrated in glomerular eluates of lupus mice. Nephrol. Dial. Transplant. 9:362–366.

Farrell M, Sakic B, Szechtman H, Denburg JA, 1997. Effect of cyclophosphamide on leucocytic infiltration in the brain of MRL/lpr mice. Lupus 6:268–274.

Feinstein A, 2004. The neuropsychiatry of multiple sclerosis. Can. J. Psychiatry 49:157–163.

Feinstein A, Feinstein K, Gray T, O'Connor P, 1997. Prevalence and neurobehavioral correlates of pathological laughing and crying in multiple sclerosis. Arch. Neurol. 54:1116–1121.

Feinstein A, O'Connor P, Gray T, Feinstein K, 1999. The effects of anxiety on psychiatric morbidity in patients with multiple sclerosis. Mult. Scler. 5:323–326.

Ferro JM, 1998. Vasculitis of the central nervous system. J. Neurol. 245:766–776.

Fessel WJ, 1962a. Autoimmunity and mental illness. A preliminary report. Arch. Gen. Psychiatry 6:320–323.

Fessel WJ, 1962b. Macroglobin elevations in functional mental illness. Nature 193:1005.

Fessel WJ, Hirata-Hibi M, 1963. Abnormal leucocytes in schizophrenia. Arch. Gen. Psychiatry 106:601–613.

Fields RA, Sibbitt WL, Toubbeh H, Bankhurst AD, 1990. Neuropsychiatric lupus erythematosus, cerebral infarctions, and anticardiolipin antibodies. Ann. Rheum. Dis. 49:114–117.

Fieschi C, Rasura M, Anzini A, Beccia M, 1998. Central nervous system vasculitis. J. Neurol. Sci. 153:159–171.

Forster MJ, Popper MD, Retz KC, Lal H, 1988a. Age differences in acquisition and retention of one-way avoidance learning in C57BL/6NNia and autoimmune mice. Behav. Neural Biol. 49:139–151.

Forster MJ, Retz KC, Lal H, 1988b. Learning and memory deficits associated with autoimmunity: significance in aging and Alzheimer's disease. Drug Dev. Res. 15:253–273.

Francis K, Lewis BM, Akatsu H, Monk PN, Cain SA, Scanlon MF, Morgan BP, Ham J, Gasque P, 2003. Complement C3a receptors in the pituitary gland: a novel pathway by which an innate immune molecule releases hormones involved in the control of inflammation. FASEB J. online.

Ganguli R, Brar JS, Chengappa KN, Yang ZW, Nimgaonkar VL, Rabin BS, 1993. Autoimmunity in schizophrenia: a review of recent findings. Ann. Med. 25:489–496.

Ganguli R, Brar JS, Rabin BS, 1994. Immune abnormalities in schizophrenia: evidence for the autoimmune hypothesis. Harv. Rev. Psychiatry 2:70–83.

Gaughran F, 2002. Immunity and schizophrenia: autoimmunity, cytokines, and immune responses. Int. Rev. Neurobiol. 52:275–302.

Ghaffar O, Feinstein A, 2007. The neuropsychiatry of multiple sclerosis: a review of recent developments. Curr. Opin. Psychiatry 20:278–285.

Gianani R, Sarvetnick N, 1996. Viruses, cytokines, antigens, and autoimmunity. Proc. Natl. Acad. Sci. U. S. A. 93:2257–2259.

Gillberg C, 1999. Neurodevelopmental processes and psychological functioning in autism. Dev. Psychopathol. 11:567–587.

Gitlits VM, Sentry JW, Matthew ML, Smith AI, Toh BH, 2001. Synapsin I identified as a novel brain-specific autoantigen. J. Invest. Med. 49:283.

Giuliani F, Goodyer CG, Antel JP, Yong VW, 2003. Vulnerability of human neurons to T cell-mediated cytotoxicity. J. Immunol. 171:368–379.

Golombek SJ, Graus G, Elkon KB, 1986. Autoantibodies in the cerebrospinal fluid of patients with systemic lupus erythematosus. Arthritis Rheum. 29:1090–1097.

Greenwood DL, Gitlits VM, Alderuccio F, Sentry JW, Toh BH, 2002. Autoantibodies in neuropsychiatric lupus. Autoimmunity 35:79–86.

Gruol DL, Nelson TE, 1997. Physiological and pathological roles of interleukin-6 in the central nervous system. Mol. Neurobiol. 15:307–339.

Gwag BJ, Koh JY, Demaro JA, Ying HS, Jacquin M, Choi DW, 1997. Slowly triggered excitotoxicity occurs by necrosis in cortical cultures. Neuroscience 77:393–401.

Handa R, Sahota P, Kumar M, Jagannathan NR, Bal CS, Gulati M, Tripathi BM, Wali JP, 2003. In vivo proton magnetic resonance spectroscopy (MRS) and single photon emission computerized tomography (SPECT) in systemic lupus erythematosus (SLE). Magn Reson. Imaging 21:1033–1037.

Hanly JG, Fisk JD, Eastwood B, 1994. Brain reactive autoantibodies and cognitive impairment in systemic lupus erythematosus. Lupus 3:193–199.
Hanly JG, Fisk JD, McCurdy G, Fougere L, Douglas JA, 2005. Neuropsychiatric syndromes in patients with systemic lupus erythematosus and rheumatoid arthritis. J. Rheumatol. 32:1459–6.
Hanly JG, Hong C, 1993. Antibodies to brain integral membrane proteins in systemic lupus erythematosus. J. Immunol. Methods 161:107–118.
Harbeck R, Hoffman AA, Hoffman SA, 1978a. Distribution in isolated brain components of antigens combining with thymocytotoxins in New Zealand mice. Transplantation 25:161–163.
Harbeck RJ, Hoffman AA, Hoffman SA, Shucard DW, 1979. Cerebrospinal fluid and the choroid plexus during acute immune complex disease. Clin. Immunol. Immunopathol. 13:413–425.
Harbeck RJ, Hoffman AA, Hoffman SA, Shucard DW, Carr RI, 1978b. A naturally occurring antibody in New Zealand mice cytotoxic to dissociated cerebellar cells. Clin. Exp. Immunol. 31:313–320.
Harel Y, Barak Y, Achiron A, 2007. Dysregulation of affect in multiple sclerosis: new phenomenological approach. Psychiatry Clin. Neurosci. 61:94–98.
Harris EN, Hughes GRV, 1985. Cerebral disease in systemic lupus erythematosus. Springer Semin. Immunopathol. 8:251–266.
Hayley S, Anisman H, 2005. Multiple mechanisms of cytokine action in neurodegenerative and psychiatric states: neurochemical and molecular substrates. Curr. Pharm. Des. 11: 947–962.
Heath RG, 1969. Schizophrenia: evidence of a pathologic immune mechanism. Proc. Annu. Meet. Am. Psychopathol. Assoc. 58:234–252.
Heath RG, Krupp IM, 1967. Schizophrenia as an immunologic disorder. I. Demonstration of antibrain globulins by fluorescent antibody techniques. Arch. Gen. Psychiatry 16:1–9.
Henneberg A, Mayle DM, Kornhuber HH, 1991. Antibodies to brain tissue in sera of patients with chronic progressive multiple sclerosis. J. Neuroimmunol. 34:223–227.
Hess DC, Taormina M, Thompson J, Sethi KD, Diamond B, Rao R, Feldman DS, 1993. Cognitive and neurologic deficits in the MRL/lpr mouse: a clinicopathologic study. J. Rheumatol. 20:610–617.
Hickey WF, 1999. Leukocyte traffic in the central nervous system: the participants and their roles. Semin. Immunol. 11:125–137.
Hickey WF, 2001. Basic principles of immunological surveillance of the normal central nervous system. Glia 36:118–124.
Hickey WF, Lassmann S, Cross AH, 1997. Lymphocyte entry and the initiation of inflammation in the central nervous system. In: Keane, R. W., Hickey, W. F. (Eds.), Immunology of the Nervous System. Oxford University Press, New York, pp. 200–225.
Hill PG, McMillan SA, 2006. Anti-tissue transglutaminase antibodies and their role in the investigation of coeliac disease. Ann. Clin. Biochem. 43:105–117.
Hirano T, Hashimoto H, Shiokawa Y, Iwamori M, Nagai Y, Kasi M, Ochiai Y, Okumura K, 1980. Antiglycolipid autoantibody detected in the sera from systemic lupus erythematosus patients. J. Clin. Invest. 66:1440.
Hirohata S, Hirose S, Miyamoto T, 1985. Cerebrospinal fluid IgM, IgA, and IgG indexes in systemic lupus erythematosus. Their use as estimates of central nervous system disease activity. Arch. Intern. Med. 145:1843–1846.
Hirohata S, Miyamoto T, 1990. Elevated levels of interleukin-6 in cerebrospinal fluid from patients with systemic lupus erythematosus and central nervous system involvement. Arthritis Rheum. 33:644–649.
Hoffman SA, Arbogast DN, Day TT, Shucard DW, Harbeck RJ, 1983. Permeability of the blood cerebrospinal fluid barrier during acute immune complex disease. J. Immunol. 130:1695–1698.
Hoffman SA, Arbogast DN, Ford PM, Shucard DW, Harbeck RJ, 1987. Brain-reactive autoantibody levels in the sera of ageing autoimmune mice. Clin. Exp. Immunol. 70:74–83.
Hoffman SA, Ford P, Kubo R, 1988a. Characterization of cell surface antigens on the adrenergic neuroblastoma clone A2(1). Brain Res. 452:358–366.

Hoffman SA, Harbeck RJ, 1989. CNS lupus and the blood-brain barrier. In: Neuwelt, E. A. (Ed.), Implications of the Blood-Brain Barrier and Its Manipulation. Plenum Medical Book Co., New York-London, pp. 469–494.

Hoffman SA, Hoffman AA, Shucard DW, Harbeck RJ, 1978a. Antibodies to dissociated cerebellar cells in New Zealand mice as demonstrated by immunofluorescence. Brain Res. 142:477–486.

Hoffman SA, Madsen CS, 1990. Brain specific autoantibodies in murine models of systemic lupus erythematosus. J. Neuroimmunol. 30:229–237.

Hoffman SA, Narendran A, Shucard DW, Harbeck RJ, 1988b. Autoantibodies, immune complexes, and behavioral disorders: neuropsychiatric involvement in systemic lupus erythematosus. Drug Dev. Res. 15:237–251.

Hoffman SA, Shucard DW, Brodie HA, Reifenrath C, Harbeck RJ, 1982. Suppression of water intake by immune complex formation in the hypothalamus. Implications for systemic lupus erythematosus. J. Neuroimmunol. 2:167–176.

Hoffman SA, Shucard DW, Harbeck RJ, 1998. The immune system can affect learning: chronic immune complex disease in a rat model. J. Neuroimmunol. 86:163–170.

Hoffman SA, Shucard DW, Harbeck RJ, Hoffman AA, 1978b. Chronic immune complex disease: behavioral and immunological correlates. J. Neuropathol. Exp. Neurol. 37:426–436.

Hogan MJ, Brunet DG, Ford PM, Lillicrap D, 1988. Lupus anticoagulant, antiphospholipid antibodies and migraine. Can. J. Neurol. Sci. 15:420–425.

Hopkins P, Belmont HM, Buyon J, Philips M, Weissmann G, Abramson SB, 1988. Increased levels of plasma anaphylatoxins in systemic lupus erythematosus predict flares of the disease and may elicit vascular injury in lupus cerebritis. Arthritis Rheum. 31:632–641.

Hornig M, Weissenbock H, Horscroft N, Lipkin WI, 1999. An infection-based model of neurodevelopmental damage. Proc. Natl. Acad. Sci. U. S. A. 96:12102–12107.

Horwitz MS, Sarvetnick N, 1999. Viruses, host responses, and autoimmunity. Immunol. Rev. 169:241–253.

How A, Dent PB, Liao SK, Denburg JA, 1985. Antineuronal antibodies in neuropsychiatric systemic lupus erythematosus. Arthritis Rheum. 28:789–795.

Huang WS, Chiu PY, Tsai CH, Kao A, Lee CC, 2002. Objective evidence of abnormal regional cerebral blood flow in patients with systemic lupus erythematosus on Tc-99m ECD brain SPECT. Rheumatol. Int. 22:178–181.

Huerta PT, Kowal C, DeGiorgio LA, Volpe BT, Diamond B, 2006. Immunity and behavior: antibodies alter emotion. Proc. Natl. Acad. Sci. U. S. A. 103:678–683.

Inzelberg R, Korczyn A, 1989. Lupus anticoagulant and late onset seizures. Acta Neurol. Scand. 79:114–118.

James WG, Hutchinson P, Bullard DC, Hickey MJ, 2006. Cerebral leucocyte infiltration in lupus-prone MRL/MpJ-fas lpr mice–roles of intercellular adhesion molecule-1 and P-selectin. Clin. Exp. Immunol. 144:299–308.

Jankovic BD, 1985. From immunoneurology to immunopsychiatry: neuromodulating activity of anti-brain antibodies. Int. Rev. Neurobiol. 26:249–314.

Jara LJ, Irigoyen L, Ortiz, MD, Zazueta B, Bravo G, Espinoza LR, 1998. Prolactin and interleukin-6 in neuropsychiatric lupus erythematosus. Clin. Rheumatol. 17:110–114.

Jennekens FG, Kater L, 2002. The central nervous system in systemic lupus erythematosus. Part 2. Pathogenetic mechanisms of clinical syndromes: a literature investigation. Rheumatology (Oxford) 41:619–630.

Johnson RT, Richardson EP, 1968. The neurological manifestations of systemic lupus erythematosus. Medicine 47:337–369.

Jongen PJ, Doesburg WH, Ibrahim-Stappers JL, Lemmens WA, Hommes OR, Lamers KJ, 2000. Cerebrospinal fluid C3 and C4 indexes in immunological disorders of the central nervous system. Acta Neurol. Scand. 101:116–121.

Jonsen A, Bengtsson AA, Nived O, Ryberg B, Truedsson L, Ronnblom L, Alm GV, Sturfelt G, 2003. The heterogeneity of neuropsychiatric systemic lupus erythematosus is reflected in lack of association with cerebrospinal fluid cytokine profiles. Lupus 12:846–850.

Kaposi MK, 1872. Neue Beitrage zur Kenntnis des Lupus erythematosus. Arch. Belg. Dermatol. Syphiligr. 4:36–52.

Kasama T, Odai T, Wakabayashi K, Yajima N, Miwa Y, 2008. Chemokines in systemic lupus erythematosus involving the central nervous system. Front Biosci. 13:2527–2536.

Katsumata Y, Harigai M, Kawaguchi Y, Fukasawa C, Soejima M, Takagi K, Tanak M, Ichida H, Tochimoto A, Kanno T, Nishimura K, Kamatani N, Hara M, 2007. Diagnostic reliability of cerebral spinal fluid tests for acute confusional state (delirium) in patients with systemic lupus erythematosus: interleukin 6 (IL-6), IL-8, interferon-alpha, IgG index, and Q-albumin. J. Rheumatol. 34:2010–2017.

Katzav A, Solodeev I, Brodsky O, Chapman J, Pick CG, Blank M, Zhang W, Reichlin M, Shoenfeld Y, 2007. Induction of autoimmune depression in mice by anti-ribosomal P antibodies via the limbic system. Arthritis Rheum. 56:938–948.

Keane RW, Hickey WF, 1997. Immunology of the Nervous System. Oxford University Press, New York.

Kim RC, Collins GH, Parisi JE, 1982. Rheumatoid nodule formation within the choroid plexus. Report of a second case. Arch. Pathol. Lab. Med. 106:83–84.

Kirch DG, 1993. Infection and autoimmunity as etiologic factors in schizophrenia: a review and reappraisal. Schizophr. Bull. 19:355–370.

Kiss E, Shoenfeld Y, 2007. Are anti-ribosomal P protein antibodies relevant in systemic lupus erythematosus? Clin. Rev. Allergy Immunol. 69:37–46.

Klein R, Richter C, Berg PA, 1991. Antibodies against central nervous system tissue (anti-CNS) detected by ELISA and western blotting: marker antibodies for neuropsychiatric manifestations in connective tissue diseases. Autoimmunity 10:133–144.

Komatsu N, Kodama K, Yamanouchi N, Okada S, Noda S, Nawata Y, Takabayashi K, Iwamoto I, Saito Y, Uchida Y, Ito H, Yoshikawa K, Sato T, 1999. Decreased regional cerebral metabolic rate for glucose in systemic lupus erythematosus patients with psychiatric symptoms. Eur. Neurol. 42:41–48.

Kowal C, DeGiorgio LA, Lee JY, Edgar MA, Huerta PT, Volpe BT, Diamond B, 2006. Human lupus autoantibodies against NMDA receptors mediate cognitive impairment. Proc. Natl. Acad. Sci. U. S. A. 103:19854–19859.

Kowal C, DeGiorgio LA, Nakaoka T, Hetherington H, Huerta PT, Diamond B, Volpe BT, 2004. Cognition and immunity; antibody impairs memory. Immunity. 21:179–188.

Kozora E, Laudenslager M, Lemieux A, West SG, 2001. Inflammatory and hormonal measures predict neuropsychological functioning in systemic lupus erythematosus and rheumatoid arthritis patients. J. Int. Neuropsychol. Soc. 6:745–754.

Kraepelin E, 1992. The manifestations of insanity. Hist. Psychiatry 3:509–526.

Kronfol Z, Remick DG, 2000. Cytokines and the brain: implications for clinical psychiatry. Am. J. Psychiatry 157:683–694.

Kumar M, Cohen D, Eisdorfer C, 1988. Serum IgG brain reactive antibodies in Alzheimer disease and Down syndrome. Alzheimer Dis. Assoc. Disord. 2:50–55.

Kumar M, Resnick L, Loewenstein DA, Berger J, Eisdorfer C, 1989. Brain-reactive antibodies and the AIDS dementia complex. J. Acquir. Immune Defic. Syndr. 2:469–471.

Kurki P, Helve T, Dahl D, Virtanen I, 1986. Neurofilament antibodies in systemic lupus erythematosus. J. Rheumatol. 13:69–73.

Kwant A, Sakic B, 2004. Behavioral effects of infection with interferon-gamma adenovector. Behav. Brain Res. 151:73–82.

Lal H, Forster M, Retz KC, Reisberg B, 1988. Immune dysfunctions: new targets of drug discovery for Alzheimer's disease and other cognitive disorders. Drug Dev. Res. 15.

Lang B, Dale RC, Vincent A, 2003. New autoantibody mediated disorders of the central nervous system. Curr. Opin. Neurol. 16:351–357.

Lapteva L, Nowak M, Yarboro CH, Takada K, Roebuck-Spencer T, Weickert T, Bleiberg J, Rosenstein D, Pao M, Patronas N, Steele S, Manzano M, van d, V, Lipsky PE, Marenco S, Wesley R, Volpe B, Diamond B, Illei GG, 2006. Anti-*N*-methyl-D-aspartate receptor antibodies, cognitive dysfunction, and depression in systemic lupus erythematosus. Arthritis Rheum. 54:2505–2514.

Lawrence DA, Bolivar VJ, Hudson CA, Mondal TK, Pabello NG, 2007. Antibody induction of lupus-like neuropsychiatric manifestations. J Neuroimmunol 182:185–194.
Lawson LJ, Perry VH, Dri P, Gordon S, 1990. Heterogeneity in the distribution and morphology of microglia in the normal adult mouse brain. Neuroscience 39:151–170.
Layfield R, Lowe J, Bedford L, 2005. The ubiquitin-proteasome system and neurodegenerative disorders. Essays Biochem. 41:157–171.
Lehmann-Facius H, 1939. Serologisch-analytische Versuche mit Liquores und Seren von Schizophrenen. Allg. Z. Psychiatrie 110:232–243.
Leweke FM, Gerth CW, Koethe D, Klosterkotter J, Ruslanova I, Krivogorsky B, Torrey EF, Yolken RH, 2004. Antibodies to infectious agents in individuals with recent onset schizophrenia. Eur. Arch. Psychiatry Clin. Neurosci. 254:4–8.
Liberato B, Levy RA, 2007. Antiphospholipid syndrome and cognition. Clin. Rev. Allergy Immunol. 32:188–191.
Lin A, Kenis G, Bignotti S, Tura GB, DeJong R, Bosmans E, Pioli R, Altamura C, Scharpe S, Maes M, 1998. The inflammatory response system in treatment-resistant schizophrenia: increased serum interleukin-6. Schizophr. Res. 32:9–15.
Long AA, Denburg SD, Carbotte RM, Singal DP, Denburg JA, 1990. Serum lymphocytotoxic antibodies and neurocognitive function in systemic lupus erythematosus. Ann. Rheum. Dis. 49:249–253.
Lopez-Longo FJ, Carol N, Almoguera MI, Olazaran J, onso-Farto JC, Ortega A, Monteagudo I, Gonzalez CM, Carreno L, 2003. Cerebral hypoperfusion detected by SPECT in patients with systemic lupus erythematosus is related to clinical activity and cumulative tissue damage. Lupus 12:813–819.
Ma X, Foster J, Sakic B, 2006. Distribution and prevalence of leukocyte phenotypes in brains of lupus-prone mice. J. Neuroimmunol. 179:26–36.
Maag TJ, Hoffman SA, 1993. Anti-brain antibodies in the sera of rheumatoid arthritis patients: relation to disease activity and psychological status. J. Neuroimmunol. 45:37–45.
Maes M, Delange J, Ranjan R, Meltzer HY, Desnyder R, Cooremans W, Scharpe S, 1997. Acute phase proteins in schizophrenia, mania and major depression: modulation by psychotropic drugs. Psychiatry Res. 66:1–11.
Maino K, Gruber R, Riedel M, Seitz N, Schwarz M, Muller N, 2007. T- and B-lymphocytes in patients with schizophrenia in acute psychotic episode and the course of the treatment. Psychiatry Res. 152:173–180.
Mamula MJ, 1998. Epitope spreading: the role of self peptides and autoantigen processing by B lymphocytes. Immunol. Rev. 164:231–239.
Margutti P, Delunardo F, Ortona E, 2006. Autoantibodies associated with psychiatric disorders. Curr. Neurovasc. Res. 3:149–157.
Maric D, Millward JM, Ballok DA, Szechtman H, Barker JL, Denburg JA, Sakic B, 2001. Neurotoxic properties of cerebrospinal fluid from behaviorally impaired autoimmune mice. Brain Res. 920:183–193.
McAllister CG, Rapaport MH, Pickar D, Paul SM, 1991. Autoimmunity and schizophrenia. In: Tamminga, C. A., Schultz, S. C. (Eds.), Advances in Neuropsychiatry and Psychopharmacology, vol. 1. Raven Press, Ltd., New York, pp. 111–118.
McCluskey J, Farris AD, Keech CL, Purcell AW, Rischmueller M, Kinoshita G, Reynolds P, Gordon TP, 1998. Determinant spreading: lessons from animal models and human disease. Immunol. Rev. 164:209–229.
McIntyre KR, Ayer-LeLievre C, Persson H, 1990. Class II major histocompatibility complex (MHC) gene expression in the mouse brain is elevated in the autoimmune MRL/Mp-lpr/lpr strain. J. Neuroimmunol. 28:39–52.
McLean BN, Miller D, Thompson EJ, 1995. Oligoclonal banding of IgG in CSF, blood-brain barrier function, and MRI findings in patients with sarcoidosis, systemic lupus erythematosus, and Behcet's disease involving the nervous system. J. Neurol. Neurosurg. Psychiatry 58:548–554.

Monastero R, Bettini P, Del Zotto E, Cottini E, Tincani A, Balestrieri G, Cattaneo R, Camarda R, Vignolo LA, Padovani A, 2001. Prevalence and pattern of cognitive impairment in systemic lupus erythematosus patients with and without overt neuropsychiatric manifestations. J. Neurol. Sci.: 33–39.

Mondal TK, Saha SK, Miller VM, Seegal RF, Lawrence DA, 2008. Autoantibody-mediated neuroinflammation: Pathogenesis of neuropsychiatric systemic lupus erythematosus in the NZM88 murine model. Br. Behav. Immunity 22: 949–959.

Money J, Bobrow NA, Clarke FC, 1971. Autism and autoimmune disease: a family study. J. Autism Child Schizophr. 1:146–160.

Monteleone P, Fabrazzo M, Tortorella A, Maj M, 1997. Plasma levels of interleukin-6 and tumor necrosis factor alpha in chronic schizophrenia: effects of clozapine treatment. Psychiatry Res. 71:11–17.

Moore PM, 1997. Autoantibodies to nervous system tissue in human and murine systemic lupus erythematosus. Ann. N. Y. Acad. Sci. 823:289–299.

Mora F, Segovia G, del Arco A, 2007. Aging, plasticity and environmental enrichment: Structural changes and neurotransmitter dynamics in several areas of the brain Brain Res. Rev. 55: 78–88.

Mortensen PB, Norgaard-Pedersen B, Waltoft BL, Sorensen TL, Hougaard D, Torrey EF, Yolken RH, 2007. Toxoplasma gondii as a risk factor for early-onset schizophrenia: analysis of filter paper blood samples obtained at birth. Biol. Psychiatry 61:688–693.

Muller N, 1995. Psychoneuroimmunology: implications for the drug treatment of psychiatric disorders. CNS Drugs 4:125–140.

Muller N, Ackenheil M, 1995. Immunoglobulin and albumin content of cerebrospinal fluid in schizophrenic patients: relationship to negative symptomatology. Schizophr. Res. 14:223–228.

Muller N, Empl M, Riedel M, Schwarz M, Ackenheil M, 1997. Neuroleptic treatment increases soluble IL-2 receptors and decreases soluble IL-6 receptors in schizophrenia. Eur. Arch. Psychiatry Clin. Neurosci. 247:308–313.

Muller N, Schlesinger BC, Hadjamu M, Riedel M, Schwarz M, Ackenheil M, Wank R, Gruber R, 1998. Increased frequency of CD8 positive gamma/delta T- lymphocytes (CD8(+) gamma/delta(+)) in unmedicated schizophrenic patients: relation to impairment of the blood-brain barrier and HLA-DPA 02011. Schizophr. Res. 32:69–71.

Muller N, Schwarz M, 2006a. Schizophrenia as an inflammation-mediated dysbalance of glutamatergic neurotransmission. Neurotox. Res. 10:131–148.

Muller N, Schwarz MJ, 2006b. Neuroimmune-endocrine crosstalk in schizophrenia and mood disorders. Expert. Rev. Neurother. 6:1017–1038.

Nahmias AJ, Nahmias SB, Danielsson D, 2006. The possible role of transplacentally-acquired antibodies to infectious agents, with molecular mimicry to nervous system sialic acid epitopes, as causes of neuromental disorders: prevention and vaccine implications. Clin. Dev. Immunol. 13:167–183.

Narendran A, Hoffman SA, 1988. Identification of autoantibody reactive integral brain membrane antigens – a two dimensional analysis. J. Immunol. Methods 114:227–234.

Narendran A, Hoffman SA, 1989. Characterization of brain-reactive autoantibodies in murine models of systemic lupus erythematosus. J. Neuroimmunol. 24:113–123.

Nataf S, Davoust N, Barnum SR, 1998. Kinetics of C5aR expression during experimental allergic encephalomyelitis. J. Neuroimmunol. 91:147–155.

Naudin J, Mege JL, Azorin JM, Dassa D, 1996. Elevated circulating levels of IL-6 in schizophrenia. Schizophr. Res. 20:269–273.

Navarrete MG, Brey RL, 2000. Neuropsychiatric Systemic Lupus Erythematosus. Curr. Treat. Options. Neurol. 2:473–485.

Nikkila HV, Muller K, Ahokas A, Rimon R, Andersson LC, 2001. Increased frequency of activated lymphocytes in the cerebrospinal fluid of patients with acute schizophrenia. Schizophr. Res. 49:99–105.

Nishimura K, Harigai M, Omori M, Sato E, Hara M, 2008. Blood-brain barrier damage as a risk factor for corticosteroid-induced psychiatric disorders in systemic lupus erythematosus. Psychoneuroendocrinology. 33:395–403.

Omdal R, Brokstad K, Waterloo K, Koldingsnes W, Jonsson R, Mellgren SI, 2005. Neuropsychiatric disturbances in SLE are associated with antibodies against NMDA receptors. Eur. J. Neurol. 12:392–398.

Omdal R, Selseth B, Klow NE, Husby G, Mellgren SI, 1989. Clinical neurological, electrophysiological, and cerebral CT scan findings in systemic lupus erythematosus. Scand. J. Rheumatol. 18:283–289.

O'Sullivan FX, Vogelweid CM, Beschwilliford CL, Walker SE, 1995. Differential effects of CD4(+) T cell depletion on inflammatory central nervous system disease, arthritis and sialadenitis in MRL/lpr mice. J. Autoimmun. 8:163–175.

Owens T, 2003. The enigma of multiple sclerosis: inflammation and neurodegeneration cause heterogeneous dysfunction and damage. Curr. Opin. Neurol. 16:259–265.

Park C, Sakamaki K, Tachibana O, Yamashima T, Yamashita J, Yonehara S, 1998. Expression of Fas antigen in the normal mouse brain. Biochem. Biophys. Res. Commun. 252:623–628.

Parker LM, Chused TM, Steinberg AD, 1974. Immunofluorescence studies on thymocytotoxic antibody from New Zealand black mice. J. Immunol. 112:285.

Patten SB, Svenson LW, Metz LM, 2005. Psychotic disorders in MS: population-based evidence of an association. Neurology 65:1123–1125.

Peress NS, Tompkins DC, 1979. Rat CNS in experimental chronic serum sickness: Integrity of the zonulae occludentes of the choroid plexus epithelium and brain endothelium in experimental chronic serum sickness. Neuropathol. Appl. Neurobiol. 5, 279–288.

Pollmacher T, Hinze-Selch D, Mullington J, 1996. Effects of clozapine on plasma cytokine and soluble cytokine receptor levels. J. Clin. Psychopharmacol. 16:403–409.

Quismorio FP, Friou GJ, 1972. Antibodies reactive with neurons in SLE patients with neuropsychiatric manifestations. Int. Arch. Allergy Appl. Immunol. 43:740–748.

Ramos M, Mandybur TI, 1975. Cerebral vasculitis in rheumatoid arthritis. Arch. Neurol. 32:271–275.

Rapaport MH, McAllister CG, 1991. Neuroimmunologic factors in schizophrenia. In: Gorman, J. M., Kertzner, R. M. (Eds.), Psychoimmunology Update. American Psychiatric Press, Washington, DC, pp. 31–54.

Rhodes B, Vyse TJ, 2007. General aspects of the genetics of SLE. Autoimmunity 40: 550–559.

Rodier PM, 1994. Vulnerable periods and processes during central nervous system development. Environ. Health Perspect. 102 Suppl 2:121–124.

Rothermundt M, Arolt V, Weitzsch C, Eckhoff D, Kirchner H, 1996. Production of cytokines in acute schizophrenic psychosis. Biol. Psychiatry 40:1294–1297.

Rudick RA, Eskin TA, 1983. Neuropathological features of a lupus-like disorder in autoimmune mice. Ann. Neurol. 14:325–332.

Ruiz-Argüelles A, Rivadeneyra-Espinoza L, Alarcón-Segovia D, 2003. Antibody penetration into living cells: pathogenic, preventive and immuno-therapeutic implications. Curr. Pharm. Des. 9:1881–1887.

Sa MJ, 2007. Psychological aspects of multiple sclerosis. Clin. Neurol. Neurosurg. (electronic publication).

Sakic B, Denburg JA, Denburg SD, Szechtman H, 1996a. Blunted sensitivity to sucrose in autoimmune MRL-lpr mice: a curve-shift study. Brain Res. Bull. 41:305–311.

Sakic B, Gurunlian L, Denburg SD, 1998a. Reduced aggressiveness and low testosterone levels in autoimmune MRL-lpr males. Physiol. Behav. 63:305–309.

Sakic B, Kirkham DL, Ballok DA, Mwanjewe J, Fearon IM, Macri J, Yu G, Sidor MM, Denburg JA, Szechtman H, Lau J, Ball AK, Doering LC, 2005. Proliferating brain cells are a target of neurotoxic CSF in systemic autoimmune disease. J. Neuroimmunol. 169:68–85.

Sakic B, Kolb B, Whishaw IQ, Gorny G, Szechtman H, Denburg JA, 2000a. Immunosuppression prevents neuronal atrophy in lupus-prone mice: evidence for brain damage induced by autoimmune disease? J. Neuroimmunol. 111:93–101.

Sakic B, Lacosta S, Denburg J, Szechtman H, 2002. Altered neurotransmission in brains of autoimmune mice: pharmacological and neurochemical evidence. J. Neuroimmunol. 129: 84–96.
Sakic B, Maric I, Koeberle PD, Millward JM, Szechtman H, Maric D, Denburg JA, 2000b. Increased TUNEL-staining in brains of autoimmune Fas-deficient mice. J. Neuroimmunol. 104:147–154.
Sakic B, Szechtman H, Braciak TA, Richards CD, Gauldie J, Denburg JA, 1997a. Reduced preference for sucrose in autoimmune mice: a possible role of interleukin-6. Brain Res. Bull. 44:155–165.
Sakic B, Szechtman H, Denburg JA, 1997b. Neurobehavioral alteration in autoimmune mice. Neurosci. Biobehav. Rev. 21:327–340.
Sakic B, Szechtman H, Denburg SD, Carbotte RM, Denburg JA, 1993a. Brain-reactive antibodies and behavior of autoimmune MRL-lpr mice. Physiol. Behav. 54:1025–1029.
Sakic B, Szechtman H, Denburg SD, Carbotte RM, Denburg JA, 1993b. Spatial learning during the course of autoimmune disease in MRL mice. Behav. Brain Res. 54:57–66.
Sakic B, Szechtman H, Denburg JA, Gorny G, Kolb B, Whishaw IQ, 1998b. Progressive atrophy of pyramidal neuron dendrites in autoimmune MRL-lpr mice. J. Neuroimmunol. 87:162–170.
Sakic B, Szechtman H, Gauldie J, Denburg JA, 2001. Behavioral effects of infection with IL-6 adenovector. Brain Behav. Immun. 15:25–42.
Sakic B, Szechtman H, Keffer M, Talangbayan H, Stead R, Denburg JA, 1992. A behavioral profile of autoimmune lupus-prone MRL mice. Brain Behav. Immun. 6:265–285.
Sakic B, Szechtman H, Stead R, Denburg JA, 1996b. Joint pathology and behavioral performance in autoimmune MRL-lpr mice. Physiol. Behav. 60:901–905.
Sakic B, Szechtman H, Talangbayan H, Denburg SD, Carbotte RM, Denburg JA, 1994. Disturbed emotionality in autoimmune MRL-lpr mice. Physiol. Behav. 56:609–617.
Sasajima T, Watanabe H, Sato S, Sato Y, Ohira H, 2006. Anti-triosephosphate isomerase antibodies in cerebrospinal fluid are associated with neuropsychiatric lupus. J. Neuroimmunol. 181:150–156.
Scarisbrick IA, 2008. The multiple sclerosis degradome: enzymatic cascades in development and progression of central nervous system inflammatory disease. Curr. Top. Microbiol. Immunol. 318:133–175.
Schiffer RB, Hoffman SA, 1991. Behavioral sequelae of autoimmune disease. In: Ader, R., Felten, D. L., Cohen, N. (Eds.), Psychoneuroimmunology. Academic Press, Inc., San Diego, CA, pp. 1037–1066.
Schott K, Schaefer JE, Richartz E, Batra A, Eusterschulte B, Klein R, Berg PA, Bartels M, Mann K, Buchkremer G, 2003. Autoantibodies to serotonin in serum of patients with psychiatric disorders. Psychiatry Res. 121:51–57.
Schrott LM, Crnic LS, 1994. Sensitivity to foot shock in autoimmune NZB x NZW F1 hybrid mice. Physiol. Behav. 56:849–853.
Schrott LM, Crnic LS, 1996a. Anxiety behavior, exploratory behavior, and activity in NZB x NZW F1 hybrid mice: role of genotype and autoimmune disease progression. Brain Behav. Immun. 10:260–274.
Schrott LM, Crnic LS, 1996b. The role of performance factors in the active avoidance conditioning deficit in autoimmune mice. Behav. Neurosci. 110:486–491.
Schrott LM, Crnic LS, 1998. Attenuation of behavioral abnormalities in autoimmune mice by chronic soluble interferon-gamma receptor treatment. Brain Behav. Immun. 12:90–106.
Schrott LM, Denenberg VH, Sherman GF, Waters NS, Rosen GD, Galaburda AM, 1992. Environmental enrichment, neocortical ectopias, and behavior in the autoimmune NZB mouse. Dev. Brain Res. 67:85–93.
Schrott LM, Morrison L, Wimer R, Wimer C, Behan PO, Denenberg VH, 1994. Autoimmunity and avoidance learning in NXRF recombinant inbred strains. Brain Behav. Immun. 8:100–110.
Schrott LM, Waters NS, Boehm GW, Sherman GF, Morrison L, Rosen GD, Galaburda AM, Denenberg VH, 1993. Behavior, cortical ectopias, and autoimmunity in BXSB-Yaa and BXSB-Yaa+ mice. Brain Behav. Immun. 7:205–223.

Schupf N, Williams CA, 1985. Effect of immune complex forming reactants on catecholamine-modulated behaviors in the rat hypothalamus. J. Neuroimmunol. 9:13–27.
Schupf N, Williams CA, 1987. Psychopharmacological activity of immune complexes in rat brain is complement dependent. J. Neuroimmunol. 13:293–303.
Schupf N, Williams CA, Hugli TE, Cox J, 1983. Psychopharmacological activity of anaphylatoxin C3a in rat hypothalamus. J. Neuroimmunol. 5:305–316.
Schwartz M, Silver H, 2000. Lymphocytes, autoantibodies and psychosis–coincidence versus etiological factor: an update. Isr. J. Psychiatry Relat. Sci. 37:32–36.
Scolding NJ, Joseph FG, 2002. The neuropathology and pathogenesis of systemic lupus erythematosus. Neuropathol. Appl. Neurobiol. 28:173–189.
Sherman GF, Galaburda AM, Behan PO, Rosen GD, 1987. Neuroanatomical anomalies in autoimmune mice. Acta Neuropathol. (Berl.) 74:239–242.
Sherman GF, Morrison L, Rosen GD, Behan PO, Galaburda AM, 1990a. Brain abnormalities in immune defective mice. Brain Res. 532:25–33.
Sherman GF, Stone JS, Press DM, Rosen GD, Galaburda AM, 1990b. Abnormal architecture and connections disclosed by neurofilament staining in the cerebral cortex of autoimmune mice. Brain Res. 529:202–207.
Shiozawa S, Kuroki Y, Kim M, Hirohata S, Ogino T, 1992. Interferon-alpha in lupus psychosis. Arthritis Rheum. 35:417–422.
Shirai T, Mellors RC, 1971. Natural thymocytotoxic autoantibody and reactive antigen in New Zealand Black and other mice. Proc. Natl. Acad. Sci. U. S. A. 68:1412–1415.
Shirai T, Mellors RC, 1972. Natural cytotoxic autoantibody against thymocytes in NZB mice. Clin. Exp. Immunol. 12:133–152.
Shoenfeld Y, Aron-Maor A, 2000. Vaccination and autoimmunity-'vaccinosis': a dangerous liaison? J. Autoimmun. 14:1–10.
Shucard JL, Parrish J, Shucard DW, McCabe DC, Benedict RH, Ambrus J, Jr., 2004. Working memory and processing speed deficits in systemic lupus erythematosus as measured by the paced auditory serial addition test. J. Int. Neuropsychol. Soc. 10:35–45.
Sibbitt WL, Haseler LJ, Griffey RH, Hart BL, Sibbitt RR, Matwiyoff NA, 1994. Analysis of cerebral structural changes in systemic lupus erythematosus by proton MR spectroscopy. AJNR Am. J. Neuroradiol. 15:923–928.
Sibbitt WL, Sibbitt RR, 1993. Magnetic resonance spectroscopy and positron emission tomography scanning in neuropsychiatric systemic lupus erythematosus. Rheum. Dis. Clin. North Am. 19:851–868.
Sidor MM, Sakic B, Malinowski PM, Ballok DA, Oleschuk CJ, Macri J, 2005. Elevated immunoglobulin levels in the cerebrospinal fluid from lupus-prone mice. J. Neuroimmunol. 165:104–113.
Silva SC, Correia C, Fesel C, Barreto M, Coutinho AM, Marques C, Miguel TS, Ataide A, Bento C, Borges L, Oliveira G, Vicente AM, 2004. Autoantibody repertoires to brain tissue in autism nuclear families. J. Neuroimmunol. 152:176–182.
Singer GG, Carrera AC, Marshakrothstein A, Martineza C, Abbas AK, 1994. Apoptosis, fas and systemic autoimmunity: The MRL-lpr/lpr model. Curr. Opin. Immunol. 6: 913–920.
Singh VK, 1996. Plasma increase of interleukin-12 and interferon-gamma. Pathological significance in autism. J. Neuroimmunol. 66:143–145.
Singh VK, Jensen RL, 2003. Elevated levels of measles antibodies in children with autism. Pediatr. Neurol. 28:292–294.
Singh VK, Lin SX, Newell E, Nelson C, 2002. Abnormal measles-mumps-rubella antibodies and CNS autoimmunity in children with autism. J. Biomed. Sci. 9: 359–364.
Singh VK, Rivas WH, 2004. Prevalence of serum antibodies to caudate nucleus in autistic children. Neurosci. Lett. 355:53–56.
Singh VK, Warren R, Averett R, Ghaziuddin M, 1997. Circulating autoantibodies to neuronal and glial filament proteins in autism. Pediatr. Neurol. 17:88–90.
Sirota P, Schild K, Elizur A, Djaldetti M, Fishman P, 1995. Increased interleukin-1 and interleukin-3 like activity in schizophrenic patients. Prog. Neuropsychopharmacol. Biol. Psychiatry 19:75–83.

Sofat N, Malik O, Higgens C, 2006. Neurological involvement in patients with rheumatic disease. QJM 99:69–79.

Sokol DK, McIntyre JA, Wagenknecht DR, Dropcho EJ, Patel H, Salanova V, da Costa G, 2004. Antiphospholipid and glutamic acid decarboxylase antibodies in patients with focal epilepsy. Neurology 62:517–518.

Solomon GF, Allansmith M, McCellan B, Amkraut A, 1969. Immunoglobulins in psychiatric patients. Arch. Gen. Psychiatry 20:272–277.

Solomon GF, Moos RH, Fessel WJ, Morgan EE, 1966. Globulins and behavior in schizophrenia. Int. J. Neuropsychiatry 2:20–26.

Steens SC, Admiraal-Behloul F, Bosma GP, Steup-Beekman GM, Olofsen H, Le CS, Huizinga TW, van Buchem MA, 2004. Selective gray matter damage in neuropsychiatric lupus. Arthritis Rheum. 50:2877–2881.

Steiner JW, Gelbloom AJ, 1959. Intracranial manifestations in two cases of systemic rheumatoid disease. Arthritis Rheum. 2:537–545.

Stubbs EG, 1976. Autistic children exhibit undetectable hemagglutination-inhibition antibody titers despite previous rubella vaccination. J. Autism Child Schizophr. 6:269–274.

Szechtman H, Sakic B, Denburg JA, 1997. Behaviour of MRL mice: an animal model of disturbed behaviour in systemic autoimmune disease. Lupus 6:223–229.

Tanaka S, Matsunaga H, Kimura M, Tatsumi K, Hidaka Y, Takano T, Uema T, Takeda M, Amino N, 2003. Autoantibodies against four kinds of neurotransmitter receptors in psychiatric disorders. J. Neuroimmunol. 141:155–164.

Temesvari P, Denburg J, Denburg S, Carbotte R, Bensen W, Singal D, 1983. Serum lymphocytotoxic antibodies in neuropsychiatric lupus: A serial study. Clin. Immunol. Immunopathol. 28:243.

Theofilopoulos AN, 1992. Murine models of lupus. In: Lahita, R. G. (Ed.), Systemic Lupus Erythematosus. Churchill Livingstone, New York, pp. 121–194.

Toh BH, Mackay IR, 1981. Autoantibody to a novel neuronal antigen in systemic lupus erythematosus and in normal human sera. Clin. Exp. Immunol. 44:555–559.

Tomer Y, Davies TF, 1995. Infections and autoimmune endocrine disease. Baillieres Best Pract. Res. Clin. Endocrinol. Metab. 9:47–70.

Tomita M, Holman BJ, Santoro TJ, 2001a. Aberrant cytokine gene expression in the hippocampus in murine systemic lupus erythematosus. Neurosci. Lett. 302:129–132.

Tomita M, Holman BJ, Williams LS, Pang KC, Santoro TJ, 2001b. Cerebellar dysfunction is associated with overexpression of proinflammatory cytokine genes in lupus. J. Neurosci. Res. 64:26–33.

Tomita M, Khan RL, Blehm BH, Santoro TJ, 2004. The potential pathogenetic link between peripheral immune activation and the central innate immune response in neuropsychiatric systemic lupus erythematosus. Med. Hypotheses 62:325–335.

Torrey EF, Bartko JJ, Lun ZR, Yolken RH, 2007. Antibodies to Toxoplasma gondii in patients with schizophrenia: a meta-analysis. Schizophr. Bull. 33:729–736.

Torrey EF, Leweke MF, Schwarz MJ, Mueller N, Bachmann S, Schroeder J, Dickerson F, Yolken RH, 2006. Cytomegalovirus and schizophrenia. CNS Drugs 20:879–885.

Torrey EF, Yolken RH, 2001. The schizophrenia-rheumatoid arthritis connection: infectious, immune, or both? Brain Behav. Immun. 15:401–410.

Toubi E, Khamashta MA, Panarra A, Hughes GRV, 1995. Association of antiphospholipid antibodies with central nervous system disease in systemic lupus erythematosus. Am. J. Med. 99:397–401.

Toubi E, Shoenfeld Y, 2007. Clinical and biological aspects of anti-P-ribosomal protein autoantibodies. Autoimmun. Rev. 6:119–125.

Trottier G, Srivastava L, Walker CD, 1999. Etiology of infantile autism: a review of recent advances in genetic and neurobiological research. J. Psychiatry Neurosci. 24:103–115.

Trysberg E, Nylen K, Rosengren LE, Tarkowski A, 2003. Neuronal and astrocytic damage in systemic lupus erythematosus patients with central nervous system involvement. Arthritis Rheum. 48:2881–2887.

Trysberg E, Tarkowski A, 2004. Cerebral inflammation and degeneration in systemic lupus erythematosus. Curr. Opin. Rheumatol. 16:527–533.

Tsai CY, Wu TH, Tsai ST, Chen KH, Thajeb P, Lin WM, Yu HS, Yu CL, 1994. Cerebrospinal fluid interleukin-6, prostaglandin E2 and autoantibodies in patients with neuropsychiatric systemic lupus erythematosus and central nervous system infections. Scand. J. Rheumatol. 23:57–63.

Uhlig H, Dernick R, 1989. Monoclonal autoantibodies derived from multiple sclerosis patients and control persons and their reactivities with antigens of the central nervous system. Autoimmunity 5:87–99.

van Dam AP, Wekking EM, Callewaert JAC, Schipperijn AJM, Oomen HAPC, Dejong J, Swaak AJG, Smeenk RJT, Feltkamp TEW, 1994. Psychiatric symptoms before systemic lupus erythematosus is diagnosed. Rheumatol. Int. 14:57–62.

van Gent T, Heijnen CJ, Treffers PD, 1997. Autism and the immune system. J. Child Psychol. Psychiatry 38:337–349.

Vianello M, Tavolato B, Giometto B, 2002. Glutamic acid decarboxylase autoantibodies and neurological disorders. Neurol. Sci. 23:145–151.

Vincent A, Dalton P, Clover L, Palace J, Lang B, 2003. Antibodies to neuronal targets in neurological and psychiatric diseases. Ann. N. Y. Acad. Sci. 992:48–55.

Vogelweid CM, Johnson GC, Besch-Williford CL, Basler J, Walker SE, 1991. Inflammatory central nervous system disease in lupus-prone MRL/lpr mice: comparative histologic and immunohistochemical findings. J. Neuroimmunol. 35:89–99.

Vogelweid CM, Wright DC, Johnson JC, Hewett JE, Walker SE, 1994. Evaluation of memory, learning ability, and clinical neurologic function in pathogen-free mice with systemic lupus erythematosus. Arthritis Rheum. 37:889–897.

Volkow ND, Warner N, McIntyre R, Valentine A, Kulkarni M, Mullani N, Gould L, 1988. Cerebral involvement in systemic lupus erythematosus. Am. J. Physiol. Imaging 3:91–98.

Warren RP, Foster A, Margaretten NC, 1987. Reduced natural killer cell activity in autism. J. Am. Acad. Child Adolesc. Psychiatry 26:333–335.

Warren RP, Margaretten NC, Pace NC, Foster A, 1986. Immune abnormalities in patients with autism. J. Autism Dev. Disord. 16:189–197.

Warren RP, Odell JD, Warren WL, Burger RA, Maciulis A, Daniels WW, Torres AR, 1996. Strong association of the third hypervariable region of HLA-DR beta 1 with autism. J. Neuroimmunol. 67:97–102.

Warren RP, Singh VK, 1996. Elevated serotonin levels in autism: association with the major histocompatibility complex. Neuropsychobiology 34:72–75.

Warren RP, Yonk J, Burger RW, Odell D, Warren WL, 1995. DR-positive T cells in autism: association with decreased plasma levels of the complement C4B protein. Neuropsychobiology 31:53–57.

Warren RP, Yonk LJ, Burger RA, Cole P, Odell JD, Warren WL, White E, Singh VK, 1990. Deficiency of suppressor-inducer (CD4+CD45RA+) T cells in autism. Immunol. Invest. 19:245–251.

Waterloo K, Omdal R, Jacobsen EA, Klow NE, Husby G, Torbergsen T, Mellgren SI, 1999. Cerebral computed tomography and electroencephalography compared with neuropsychological findings in systemic lupus erythematosus. J. Neurol. 246:706–711.

Waterloo K, Omdal R, Sjoholm H, Koldingsnes W, Jacobsen EA, Sundsfjord JA, Husby G, Mellgren SI, 2001. Neuropsychological dysfunction in systemic lupus erythematosus is not associated with changes in cerebral blood flow. J. Neurol. 248:595–602.

Wekking EM, 1993. Psychiatric symptoms in systemic lupus erythematosus – an update. Psychosom. Med. 55:219–228.

Wekking EM, Nossent JC, van Dam AP, Swaak AJ, 1991. Cognitive and emotional disturbances in systemic lupus erythematosus. Psychother. Psychosom. 55:126–131.

Wilke I, Arolt V, Rothermundt M, Weitzsch C, Hornberg M, Kirchner H, 1996. Investigations of cytokine production in whole blood cultures of paranoid and residual schizophrenic patients. Eur. Arch. Psychiatry Clin. Neurosci. 246:279–284.

Williams GW, Bluestein HG, Steinberg AD, 1981. Brain-reactive lymphocytotoxic antibody in the cerebrospinal fluid of patients with systemic lupus erythematosus: correlation with central nervous system involvement. Clin. Immunol. Immunopathol. 18:126–132.
Williams CA, Schupf N, 1977. Antigen-antibody reactions in rat brain sites induce transient changes in drinking behavior. Science 196:328–331.
Williams CA, Schupf N, Hugli TE, 1985. Anaphylatoxin C5a modulation of an alpha-adrenergic receptor system in the rat hypothalamus. J. Neuroimmunol. 9:29–40.
Williams CA, Schupf N, Reilly CL, Wagner J, 1988. Complement peptides and neuronal dysfunction in the central nervous system – specificity of receptor sites for anapylatoxin C3a activity in the rat hypothalamus. Drug Dev. Res. 15:175–187.
Wilson HA, Winfield JB, Lahita RG, Koffler D, 1979. Autoantibody to a novel neuronal antigen in systemic lupus erythematosus and in normal human sera. Arthritis Rheum. 22:458–462.
Winfield JB, Brunner CM, Koffler D, 1978. Serologic studies in patients with systemic lupus erythematosus and central nervous system dysfunction. Arthritis Rheum. 21:289–294.
Winfield JB, Shaw M, Silverman LM, Eisenberg RA, Wilson HA, III, Koffler D, 1983. Intrathecal IgG synthesis and blood-brain barrier impairment in patients with systemic lupus erythematosus and central nervous system dysfunction. Am. J. Med. 74:837–844.
Yolken R, 2004. Viruses and schizophrenia: a focus on herpes simplex virus. Herpes 11 Suppl 2:83A–88A.
Yonk LJ, Warren RP, Burger RA, Cole P, Odell JD, Warren WL, White E, 1990. CD4+ helper T cell depression in autism. Immunol. Lett. 25:341–345.
Yoshida A, Shishido F, Kato K, Watanabe H, Seino O, 2007. Evaluation of cerebral perfusion in patients with neuropsychiatric systemic lupus erythematosus using 123I-IMP SPECT. Ann. Nucl. Med. 21:151–158.
Yoshio T, Hirata D, Onda K, Nara H, Minota S, 2005. Antiribosomal P protein antibodies in cerebrospinal fluid are associated with neuropsychiatric systemic lupus erythematosus. J. Rheumatol. 32:34–39.
Zameer A, Hoffman SA, 2001. Immunoglobulin binding to brain in autoimmune mice. J. Neuroimmunol. 120:10–18.
Zameer A, Hoffman SA, 2003. Increased ICAM-1 and VCAM-1 expression in the brains of autoimmune mice. J. Neuroimmunol. 142:67–74.
Zameer A, Hoffman SA, 2004. B and T cells in the brains of autoimmune mice. J. Neuroimmunol. 146:133–139.
Zandman-Goddard G, Chapman J, Shoenfeld Y, 2007. Autoantibodies involved in neuropsychiatric SLE and antiphospholipid syndrome. Semin. Arthritis Rheum. 36:297–315.
Zimmerman AW, Frye VH, Potter NT, 1993. Immunological aspects of autism. Int. Pediatr. 8:161–166.
Ziv Y, Schwartz M, 2008. Immune-based regulation of adult neurogenesis: implications for learning and memory. Brain Behav. Immun. 22:167–176.

Viruses and Psychiatric Disorders

Brad D. Pearce

Abstract Over the last two decades it has become clear that most neuropsychiatric disorders have heterogeneous and multifactorial causes, involving genetic and environmental components. Neurotropic viruses are well established in inducing brain damage that can result in a variety of mental alterations. Moreover, the immune response to a pathogen represents a classic example of the interaction between genetic and environmental factors determining individual variation in the characteristics of diverse symptom types, including those involving the nervous system. Nonetheless, the role of viruses in most mental illnesses has not been established, and this chapter discusses the putative role of viral infections in the etiopathogenesis of Alzheimer's disease, attention deficit hyperactivity disorder, autism spectrum disorders, bipolar disorder, chronic fatigue syndrome, major depression, and schizophrenia.

Keywords Alzheimer's disease · Attention deficit disorder · Autism · Chronic fatigue syndrome · Major depression · Gulf war syndrome · Fibromyalgia · Morgellons · Depression · Bipolar disorder · Obsessive-compulsive disorder · Schizophrenia · Neurogenesis · Viral infection

1 Introduction

Viruses have been linked tentatively with a number of psychiatric disorders, including schizophrenia, autism, major depression, bipolar disorder, and chronic fatigue syndrome (CFS). These disorders likely have multiple contributing etiologies, and viruses may be tapping into pathophysiological pathways that are shared by other environmental triggers. Despite burgeoning interest in the relationship between viral infections and psychiatric illnesses, most of the relevant mechanisms remain undefined.

B.D. Pearce (✉)
Department of Psychology, Emory University, 532 North Kilgo Circle, Atlanta, GA 30332, USA
e-mail: bpearce@emory.edu

A. Siegel, S.S. Zalcman (eds.), *The Neuroimmunological Basis of Behavior and Mental Disorders*, DOI 10.1007/978-0-387-84851-8_18,

It is well established that viral infections can cause encephalitis, white matter abnormalities, and neuroteratogenic malformations in humans (Johnson, 1998). It is not the intention of this chapter to cover the extensive literature on these infections, which typically produce focal neurological signs along with neuropsychiatric symptoms. Instead, I will examine the role of infections in the etiopathogenesis of psychiatric disorders of undefined cause. Accordingly, I will not cover the extensive and important literature on AIDS-related psychiatric and neurological morbidity since the initiating agent (human immunodeficiency virus) is known.

The ability of a virus to produce encephalitis with distinct neurological signs does not exclude the same virus as the culprit in subtler neurobehavioral syndromes. For example, congenital cytomegalovirus infection is a known cause of severe neurological abnormalities in infants. However, the manifestations of this infection vary widely in severity, and some children have latent sequela, which include subtle learning deficits and language abnormalities (Williamson, 1994). Likewise, herpes simplex encephalitis is an overt life threatening illness, yet it has been argued that this organism can also produce a smoldering infection in some brain regions without producing hallmark signs of encephalitis (Itzhaki et al., 2004).

2 Historical Foundations

The idea that mental illness could be caused by viruses dates back to the first three decades of the 1900s when the nascent field of virology discovered that several human and animal diseases could be attributed to an elusive and "filterable" agent (a virus) that was much smaller than a bacterium (Yolken and Torrey, 1995; Nathanson, 1997). The pandemic of viral encephalitis lethargica occurring between 1919 and 1928 drew attention to the ability of infectious agents to induce complex neurobehavioral sequela. Several authors noted the similarity between psychotic features experienced by patients as a consequence of encephalitis lethargica, and the clinical presentation of schizophrenia (Lishman, 1987). By 1926, Menninger's study of psychotic symptoms following influenza led him to formulate the first clear viral hypothesis of schizophrenia (Menninger, 1926). Still, the momentum for viral causes of mental illness began to wane as psychodynamic interpretations gained sway in the ensuing decades.

Between 1950 and the 1970s, technical refinements in multiple fields led to a new era in virology. Prior to 1950, even quantifying the amount of an animal virus in a clinical or experimental isolate could be a difficult and imprecise undertaking. With the refinement of the plaque assay for quantifying infectious virus and the use of the electron microscope to visualize virion structure, the number and diversity of animal viruses began to be fully realized (Nathanson, 1997). Viral immunology was also becoming more sophisticated during this period, and new tools facilitated the identification of viral subtypes and revealed specific mechanisms of pathogenesis. Virology was at the forefront of molecular genetics, and these advances, along with improvements in neurochemistry, helped elucidate the interaction of viruses with nervous system tissue.

Despite these advances, there were few publications examining the role of viruses in specific mental disorders. This began to change in mid 1970s following two publications that marshaled evidence supporting a viral etiology of schizophrenia (Torrey and Peterson, 1973, 1976). These articles were inspired in part by the discovery of the curious properties of a "slow virus" that caused Kuru, which is a neurodegenerative disorder that has an incubation period of greater than 10 years and is now known to be caused by prions (proteinaceous infectious particles; Eggenberger, 2007).

Since most common psychiatric illnesses are now recognized to be multifactorial with contributions from genetics and the environment, the role of viruses as one possible cause has continued to gain acceptance in the mainstream scientific community.

3 Properties of Viruses

Viruses are obligate intracellular organisms; i.e., they depend on host cells to replicate their genome and carry-on their infectious cycle. Therefore, viruses have evolved specific proteins on their surfaces that engage receptors on susceptible host cells, allowing viral entry, fusion, and un-coating so that the viral genetic payload can be delivered inside the cell. Several of the eukaryotic surface receptors allowing the selective entry of viruses have been characterized, though many more remain to be discovered (Gosztonyi and Ludwig, 2001).

For some of these host cell receptors, there are genetic polymorphisms, and hence these molecules represent one of the inherited variables conferring individual differences in susceptibility to viral infections. Some viruses recruit multiple cell surface molecules in order to breach the cell membrane. For example, HIV uses CD4 molecules in conjunction with CCR5 chemokine receptors to gain entry to the cytoplasm (Lederman et al., 2006). People who are homozygous for an uncommon deletion mutation of CCR5 show dramatically reduced rates of HIV infection (Lederman et al., 2006).

Because some viruses use neurotransmitter receptors to gain entry into cells, a virus can have a selective tropism for neuronal subtypes and cause discrete reorganization of neuronal circuitry based on the distribution of those receptors (Gosztonyi and Ludwig, 2001). The relative importance of such a mechanism in human diseases has not been determined, and the impact of a neurotropic virus on host circuitry is determined by a variety of other factors such host age, immune responses, and the route by which the infection travels from neuron to neuron (Pearce, 2003; van den Pol, 2006). Still, the utilization of neurotransmitter receptors for viral entry has lead to the possibility that xenobiotic ligands for such receptors (e.g., drugs developed for psychiatric illnesses) may be exploited as antivirals. For example, mirtazapine was recently tried in the treatment of JC polyomavirus – a demyelinating virus in humans that uses the 5HT2A serotonin receptor for cell entry (Elphick et al., 2004; Verma et al., 2007). Indeed, several antipsychotic medications have antiviral properties, mostly through unknown mechanisms (reviewed in Jones-Brando et al., 2003).

After entering the cell, a "successful" virus replicates its genome and eventually directs the cell to make new virions, which themselves are infectious. Virologists have deciphered an impressive array of replication strategies that different virus species employ at this stage. Among these are retroviruses, which deliver double stranded RNA to the cell interior. Subsequently the RNA is reverse transcribed into DNA. The DNA in turn becomes integrated into the chromosomal DNA of the host cell where it becomes a provirus, permanently altering the genotype of the host cell and all of its progeny. Besides human immunodeficiency virus, the retroviruses have been of particular interest in schizophrenia (Yolken et al., 2000).

4 Virus–Host Interactions in Mental Illnesses

Humans intermingle with viruses daily. Many viral infections do not come to clinical attention, and even when symptoms prompt a physician consultation, the responsible virus is not usually isolated or identified. This presents one of many obstacles to establishing the pathophysiological nexus between a specific viral infection and its neuropsychiatric manifestations. The latency between infection and neuro-symptoms can be variable depending on the virus and host. For example, in rabies the lag-time between the human infection and the fatal neuropsychiatric syndrome (agitation, aggression) ranges from 1 week to several years (Hemachudha et al., 2002). Moreover, viruses often cause disease in only a small subset of the people infected; e.g., most people will become infected with herpes simplex type 1 in their lifetime, but the incidence of central nervous system (CNS) manifestations (encephalitis) is very rare (Whitley, 1997). Establishing a virus–disease connection is further complicated in diseases that have multiple causes, which is likely the case for most psychiatric illnesses.

The immune system is necessary to clear viruses from both peripheral and CNS compartments. Some viruses can evade immune recognition, at least to an extent, and consequently establish a tenuous peace with their hosts in the form of a persistent infection (de la Torre and Oldstone, 1996; Ahmed et al., 1997). Indeed, the CNS is particularly conducive to harboring persistent viral infections because its structural features and unique immunological characteristics permit viruses to be camouflaged from immune surveillance (de la Torre and Oldstone, 1996). Furthermore, modern virology has identified hidden viral infections that are undetectable by conventional neuropathological techniques. For example, non-cytolytic viruses can infect neurons without diminishing cell vitality or causing signs of damage (de la Torre and Oldstone, 1996). Despite the lack of readily identifiable pathology, the differentiated or "luxury" function of these neurons can be disrupted by the infection leading to long-term sequela (de la Torre and Oldstone, 1996). This realization has opened new vistas in viral pathogenesis research, and has created a more accepting climate for theoretical paradigms linking infections with neurological and psychiatric diseases. One such paradigm is that regional and neurotransmitter specificity of a CNS virus can occur due to differential distribution of viral receptors in the brain (Gosztonyi and Koprowski, 2001).

Despite the ability of persistent viruses to avoid complete clearance from the body, most viruses trigger an immune response, and commonly it is the immune effector cells and cytokines, rather than the virus itself, that are responsible for most of the tissue damage (Bilzer and Stitz, 1996). The immune response to a virus may entail a direct cell-to-cell attack on infected neural cells, and in some instances the inflammation produced by this response may go on to injure nearby cells that are mere bystanders to the main assault. Even more indirectly, the CNS can suffer the consequences of a viral infection that is localized in the periphery. For example, in parainfectious or postinfectious encephalitis the immune response is incited against a virus localized exclusively in the periphery. In a manner that is not well understood, this immune reaction is misdirected against cells residing within the CNS (Arnason, 1998). This represents just one of a multitude of ways that the delicate dynamics of immune recognition and host responses can go awry, and ultimately end up compromising the integrity and function of the brain.

As described further below, and in other chapters of this book, there are a multitude of pathways by which viruses can influence neurotransmission and disrupt neuronal circuitry. Generally, these mechanisms exhibit considerable regional and neurochemical specificity, as would be expected if they played a causative role in the etiology of psychiatric disorders.

5 Overview of Psychiatric Disorders that Have Been Connected with Viral Infection

5.1 Alzheimer's Disease (AD)

An infectious etiology of AD has been considered for over 25 years (Ball, 1982; Itzhaki et al., 2004; Gerard et al., 2006). Most recent attention has focused on herpes simplex virus 1 (HSV-1), and the intracellular bacterium, Chlamydia pneumoniae (Itzhaki et al., 2004; Gerard et al., 2006).

Experimentally, HSV-1 DNA can be detected frequently by PCR in the temporal lobe and frontal cortex of elderly brains regardless of whether they have AD pathology. Therefore, the presence of HSV-1 sequences in the brain is not a risk factor for AD. However, it has been argued that HSV-1 contributes to AD pathophysiology in a subset of individuals carrying the APOE-ε-4 apolipoprotein allele (Itzhaki et al., 2004). Although the data supporting this claim have not been entirely uniform (Beffert et al., 1998), the concept has theoretical appeal because APOE protein and HSV-1 virions attach to cells using the same cell surface molecules (heparan sulfate proteoglycans; Itzhaki et al., 2004). Accordingly it has been postulated that APOE protein and HSV-1 may compete for attachment to the cell, and that the APOE-ε-4 isoform might perform poorly at overcoming HSV-1 relative to other APOE isoforms (Itzhaki et al., 2004). This could lead to patients carrying this allele to have a different infection pattern within the brain, or permit the infection to enter the brain at an earlier age.

In contrast to HSV-1, Chlamydia pneumoniae has been reported to be more common in brain tissue from AD patients compared to elderly controls (Gerard et al., 2006). Moreover, this organism is capable of persistence in other tissues, and it has been demonstrated in AD brain tissue by several complementary techniques (PCR, immunohistochemistry, culture; Gerard et al., 2006). Still, there have been some failures to replicate, and animal models may be needed to determine if the infection has a role in AD pathophysiology or is just an epiphenomenon of the disease.

There is growing support for the idea that local inflammation in AD brain tissue contributes to molecular pathways involved in AD neuropathology (Mrak and Griffin, 2005; McGeer et al., 2006). An infection in affected brain regions could be a trigger for glia activation that either initiates or exacerbates these neurodegenerative cascades. Since only a small subset of Alzheimer's patients has the autosomal dominant form of the disease, most cases are polygenic with contributing environmental factors (Blennow et al., 2006). Hence it is plausible that glia activation may be a gateway by which different brain injuries such as infections or head trauma can lead to AD lesions. This can be envisioned to involve the activation of self-propagating regulatory loops between immune mediators (e.g., cytokines, complement proteins) and amyloid precursor proteins (Mrak and Griffin, 2005). Amyloid beta (Aβ) is a well described activator of microglia, and these cells are essential in orchestrating local regulatory networks in the brain (Fig. 1), which involve not only proinflammatory cytokines but also immunomodulatory cytokines (transforming

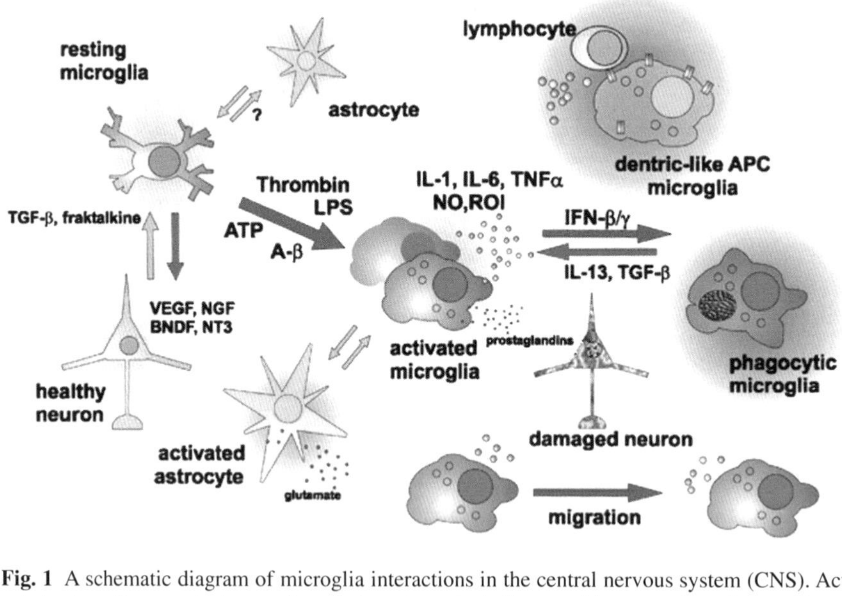

Fig. 1 A schematic diagram of microglia interactions in the central nervous system (CNS). Activated microglia can adopt a number of phenotypes including cytokine-secreting, phagocytic, and antigen-presenting cells. All of these different microglia phenotypes interact with each other as well as with astrocytes via the diffusible inflammatory mediators to produce a coordinated neuroinflammatory response. Reprinted from Garden and Moller (2006)

growth factor beta-TGF-β, IL-13), growth factors (Vascular Endothelial Growth Factor-VEGF, Brain Derived Neurotrophic factor-BDNF, Nerve Growth Factor-NGF, neurotrophin 3-NT3), and free radicals (NO, ROI; Garden and Moller, 2006). The salutary effect of non steroidal anti-inflammatory agents on the incidence of AD is consistent with a role for inflammation. Interestingly the NSAID benefit is reported mainly in patients carrying the APOE-ε-4 allele (Szekely et al., 2008) – the same group of patients that are proposed to be vulnerable to the adverse effects of HSV-1 infection in neural tissue (Itzhaki et al., 2004).

5.2 Attention Deficit Hyperactivity Disorders (ADHD)

This disorder is usually diagnosed in childhood and is characterized by hyperactivity, inattention, and impulsivity. Besides heredity, various prenatal and childhood risk factors have been considered (Biederman and Faraone, 2002). As to the latter, encephalitis lethargica in children was reported to present with a significant subtype in children evidenced by marked personality changes, substantial irritability, restlessness, and hyperkinesia (Vilensky et al., 2007). Subsequently, various congenital or childhood infections have been considered as contributing causes of ADHD, but definitive evidence is lacking (Millichap, 1997; Peterson et al., 2000). Nevertheless, ADHD is heterogeneous, and many of the principles by which infections and immune phenomena can influence behavior need to be considered in light of our growing understanding of ADHD neurobiology.

5.3 Autism

Autism is a neurodevelopmental disorder that presents with impairments in language and social behavior, and is complicated by repetitive movements or other stereotypic behaviors (Pickett and London, 2005; Newschaffer et al., 2007). For research purposes, autism is often considered along with other disorders in the "autism spectrum," such as Asperger's syndrome. Because autism is a life-long condition that is first manifested in early childhood, it is often very taxing on families, which adds further to the frustration of its mysterious origins.

During neurodevelopment, there are probably multiple casual chains that can lead to autism, and viral infections may participate in some cases but not others. Evidence for the link between infection and autism was provided several decades ago in a study of 243 children exposed to congenital rubella (Chess, 1977). The high rate of autism among the exposed children is consistent with prenatal origins. Nevertheless, this association is not disease-specific. Congenital rubella has been linked to other neuropsychiatric disorders (Brown et al., 2001), and other congenital infections have shown some provisional links with autism as well (Libbey et al., 2005). Moreover, non-infectious teratogens such thalidomide and valproate have also been tentatively linked to autism, and obstetric problems such as uterine bleeding also appear to be associated with autism risk (Newschaffer et al., 2007).

Possible mechanisms by which a prenatal infection or other gestational insult could cause autism have been investigated in humans and animal models (Hornig and Lipkin, 2001; Patterson, 2002; Fatemi et al., 2005; Nelson et al., 2006; Zerrate et al., 2007; Zimmerman et al., 2007). Theoretically, a virus could disrupt fetal brain development in a plethora of ways: (i) The virus could directly infect the fetal CNS and influence the survival, division, differentiation, or migration of neural cells. (ii) Infection of the fetus outside the brain could have adverse effects via soluble mediators (e.g., cytokines, hormones) that crossed into the brain, possibly through the immature fetal blood brain barrier. (iii) Infection could induce anomalies in the placenta and maternal–fetal interface that impacted the fetus and developing brain. (iv) Maternal infection that did not spread to the placenta or fetus could have a pernicious influence on the fetal brain via soluble molecules (autoantibodies, cytokines) or metabolic disturbances.

The role of direct infection of the fetal or neonatal brain in autism pathogenesis has been examined mostly in animal models (Fatemi et al., 2002; Patterson, 2002; Pletnikov et al., 2002; Fatemi et al., 2005). One impediment to refining these models is that the neuropathological underpinnings of autism itself are poorly defined (Pickett and London, 2005). Autopsy and neuroimaging studies have implicated abnormalities in the cerebellum, cerebral cortex, and temporal lobe structures (Pickett and London, 2005). The cerebellum has long been a focus of autism research (Courchesne et al., 1988; Pickett and London, 2005), and animal models have examined the neuropathological and behavioral consequences of viruses infecting this brain region (Bautista et al., 1995; Ramirez et al., 1996; Hornig et al., 1999; van den Pol et al., 2002; Bonthius et al., 2007). Specifically, neonatal infection of rats with Borna disease virus, parvovirus, or lymphocytic choriomeningitis virus (LCMV) causes cerebellar hypoplasia. Behavioral studies in rats infected as neonates with Borna disease virus have discovered a number of abnormalities, most notably a decrease in social interactions (Lancaster et al., 2007). Still, there is debate as to whether Borna disease virus infects humans (Durrwald et al., 2007), and while LCMV is a known neuroteratogen of humans (Bonthius et al., 2007), the diverse brain-wide pattern of neuropathology produced by congenital LCMV does not recapitulate the subtle changes seen in autistic brain (Pickett and London, 2005; Bonthius et al., 2007).

A plethora of immunological findings have been reported in autism (van Gent et al., 1997; Korvatska et al., 2002; Pardo et al., 2005). When considered collectively, these data (like many of the biochemical studies in autism) are ostensibly contradictory. There is evidence for both an overactive and underactive immune system in autism spectrum disorders (Plioplys et al., 1994; van Gent et al., 1997; Korvatska et al., 2002; Pardo et al., 2005). Young children who develop autism may have higher rates of some infections, and lower rates of others (Rosen et al., 2007). Within the brain of autistic patients, there are indices of inflammatory activation and also an increase of anti-inflammatory cytokines such as TGF-β1 (Pardo et al., 2005).

At least some of these discrepancies can likely be accounted for by diagnostic heterogeneity and differences in disease stage of the patient populations examined. While many questions remain, the bulk of the data cannot be ignored or discounted, and there is mounting support for the idea that dysregulation of immune responses is involved in some way in the somatic and/or behavioral features of autism (van Gent et al., 1997; Jyonouchi et al., 2005; Pardo et al., 2005; Zimmerman et al., 2007).

To find molecular pathways underlying autism and related disorders, more population-based biomarker research is needed and the information from existing studies needs to be integrated and considered in the context of other disciplines. These studies need to consider that the host response to infection is dynamic, with several waves of immune mediators (including pro- and anti-inflammatory cytokines) acting in networks with feed forward and feedback loops (Biron, 1994). Databases and advances in bioinformatics are helping to untangle these complex regulatory networks, but this has not yet been applied to autism. On somewhat different track, one recent study used a database comprising 1.5 million patient records to examine correlations between diverse diseases including autism (Rzhetsky et al., 2007). Viral infections were among the diseases that correlated significantly with autism.

Heredity plays a substantial role in autism (Newschaffer et al., 2007), which is consistent with a viral etiology because individual susceptibility and response to infections is a classic example of gene–environment interactions. The concept of host susceptibility can be broadened to include non-immune genes expressed exclusively in the brain. For example, such genes could be involved in circuitry plasticity or neuro-repair (Schiefermeier et al., 2000) and be employed by the host as a compensatory mechanism countering the deleterious effects of a viral infection during brain development. Viewed in this manner, infection may be one of many non-optimal conditions that endanger the immature brain, and the failure of the brain to rebound in autism could reflect the crucial genetic component of the disease. There has been little experimental examination of this mechanism.

In concert with the idea of an immunogenetic abnormality in autism, there are reports of an increased rate of autoimmune disorders in family members of autism patients. For example, in a study by Comi et al., mothers of autistic children were 8.8 times more likely to be diagnosed with an autoimmune disorder than mothers of control children (Comi et al., 1999). A variety of autoimmune disorders appeared to be implicated, indicating a generalized predisposition to inflammatory or autoreactive events in autism families. Conversely, another study found that only maternal psoriasis, allergy, and asthma were linked to birth of a child who developed autism, though these mothers have not yet been followed through their major period of risk for many autoimmune disorders (Croen et al., 2005). Data in support of a diathesis toward inflammatory and autoimmune responses in autism also come from studies showing elevations in proinflammatory cytokines in cerebro-spinal fluid of patients with autism spectrum disorders, and increased antibodies that cross react with various brain antigens in the plasma of autism patients and their mothers (Pardo et al., 2005; Zimmerman et al., 2007).

One of the most controversial subjects in autism research is whether vaccines, and mercury-based vaccine preservatives, can engender autistic spectrum disorders (Wakefield, 1999; Vastag, 2001). This possibility is built on the logical framework that measles mumps and rubella (MMR) vaccines are given roughly around the age that autism becomes manifested, and these viruses themselves – as well as some vaccines – have been linked (rarely) with idiosyncratic encephalopathies (Koskiniemi and Vaheri, 1989; Ward et al., 2007). Moreover, we know that stimulation of peripheral immune reactions can have complex neuropsychological outcomes (Raison et al., 2006). In addition, while MMR never contained thimerosal, a number of other childhood vaccines contained this mercury-based preservative. Given that mercury is a known neurotoxin, and that the cumulative amount of thimerosal in some childhood vaccine schedules exceeded Environmental Protection Agency limits (Ball et al., 2001), there was legitimate concern and several studies were launched. However, a series of epidemiological studies have indicated that the MMR vaccine (and thimerosal) are not linked to a significant portion of autism cases (Madsen et al., 2002; Fombonne, 2008; Schechter and Grether, 2008). That does not rule out the possibility of a small number of cases or certain subtypes of autism being caused by one of more of these vaccines, but it strongly indicates that vaccination cannot account for the marked increase in autism rates occurring contemporaneously with childhood multi-vaccination. Moreover, vaccination prevents these childhood viral diseases, some of which themselves can have neuropsychiatric sequela (Koskiniemi and Vaheri, 1989; Dalman et al., 2008).

Nevertheless, it is a valid concept that autism may be caused, at least in some cases, by an unusual immune response to a relatively common exposure from an infectious agent, environmental toxicant, or even iatrogenic event. Thus the negative studies on a vaccine–autism link should not deter investigations of immunogenetic susceptibilities of some children when challenged by an as yet unidentified environmental factor.

While most cases of autism appear to be polygenic and involve gene–environment interactions, there are a few distinct chromosomal defects that are linked with autistic behavior (Weiss et al., 2008). One approach to illuminating autism pathogenesis is to examine genes in this region to uncover the molecular substrates of autism spectrum behavioral phenotypes and neurocircuitry changes. This has lead to an emphasis on brain-specific genes, but gene products that could be viral receptors or determinants of immune system activity should not be ignored. For example, a recent study found that a well-delimited chromosomal defect on chromosome 16 is found in autism patients at a rate 100-fold often than in controls from the general population (Weiss et al., 2008). This data presents good evidence that one or more of the approximately 25 adjacent genes in this defect region participate in molecular pathways entailed in the unknown concatenation of events leading to autistic behavior. Several of the genes in this region are expressed in the immune system, including CD43, which is important for T-lymphocyte infiltration into the brain after viral infection (Onami et al., 2002).

Neuroimmune interactions could hold the key to preventing and treating autism spectrum disorders yet there are large gaps in the knowledge needed to apply such interventions. As recent prevalence estimates of autism septum disorder have climbed to about 60 per 10,000 (Newschaffer et al., 2007), there is a pressing need for more cross-disciplinary study, better definition of diseases subtypes, and open-minded approaches to examine possible pathogenic mechanisms.

5.4 Chronic Fatigue Syndrome, Fibromyalgia, Gulf War Syndrome, Morgellons

These illnesses may share some facets of pathogenesis, or they may have little in common beyond being multisystem syndromes that often include malaise and neurocognitive dysfunction, and for which the research community has not yet succeeded in discovering causative pathways. Nevertheless, there have been some well thought-out hypotheses and experimental progress, which has provided clues to the possible role of infections in these important illnesses.

Chromic Fatigue Syndrome (CFS) is characterized by persistent fatigue that is not relieved by rest and interferes with daily activities (Appel et al., 2007). A number of viral and bacterial infections have been considered as triggers of CFS, but no particular pathogen has been shown unequivocally to be responsible (Appel et al., 2007). There is a substantial literature on immune anomalies and neuroendocrine dysfunction in the syndrome (Appel et al., 2007), but the data have not pinpointed a pathologic mechanism that can account for the majority of patients that meet the current case definition. Nevertheless, accumulating evidence suggests that infections participate in the pathogenesis of CFS, if only in a subset of cases. Indeed, a recent prospective study following the sequela of infection with Epstein-Barr virus, Ross River virus, and Q fever, found that 11% of these postinfective patients met diagnostic criteria for CFS at the 6 month follow-up (Hickie et al., 2006). Thus it is plausible that the critical factor in CFS may be an unusual immunohormonal response to several different infections.

Fibromyalgia shares with CFS the symptoms of fatigue and sleep disturbances, but it also includes widespread pain. There is some evidence that infections (e.g., hepatitis C, HIV) could instigate or perpetuate the disorder (Abeles et al., 2007). The ability of virus-induced cytokines to modulate nocicieptive pathways at multiple levels of the nervous system is consistent with a chronic infection as one possible cause of fibromyalgia (Banks and Watkins, 2006).

Gulf war syndrome is a chronic multisystem illness affecting veterans of the 1991 gulf war. Prominent symptoms include fatigue, polyarthralgia, headache, memory impairment, irritability, and mood disturbances. A follow-up study 10 years after the war found that the syndrome was found in deployed veterans (28.9%) more often than non- deployed veterans (15.8%; Blanchard et al., 2006). A number of possible causes have been debated, including infections (Epstein-Barr virus), vaccines, stress, and environmental agents (Blanchard et al., 2006). Because of the extensive interconnections between psychological variables, immune responses, and infection susceptibility, the role of viruses in the syndrome should not be viewed in isolation from other putative risks factors for gulf war syndrome. Clearly, virus induced cytokines such as interleukin (IL)-1, IL-2, tumor necrosis factor (TNF), and interferons (IFN)-α/γ can have complex effects on behavior. These include most of the psychological symptoms of Gulf war syndrome (fatigue, irritability, anger; Raison et al., 2006; Zalcman and Siegel, 2006; Dantzer et al., 2008). Most of the information on immune changes in gulf war syndrome is controversial, and because the measures have been mainly derived from peripheral blood, it is challenging to discern the possible role of inflammatory activation in the CNS compartment.

Morgellons is a dermopathy in which there is severe pruritis and crawling sensations under the skin (Paquette, 2007). The lesions do not heal well and may contain fibers or specks of undetermined physiochemical composition. The condition is often accompanied by neuropsychiatric symptoms particularly fatigue and memory impairment, and it has some overlapping features with delusion parasitosis (Paquette, 2007). A causative pathogen has not been discovered. An array of viruses, bacteria, and parasites can cause skin lesions in humans; and prions in sheep cause itching and behavioral problems (Eggenberger, 2007). A fruitful direction for research is to investigate the illnesses in view of the shared cellular and molecular communication between the immune system, nervous system, and skin (Misery, 1997).

5.5 Depression, Anxiety, and Bipolar Disorder

The ability of viruses to cause symptoms of depression is not in question – a bout of mononucleosis can cause mood and vegetative symptoms that would overlap with depression considerably. This is not to say that infections are a significant cause of major depression or bipolar illnesses, only that viruses can produce some symptoms of the diseases in humans and animal models (Yolken and Torrey, 1995; Dunn et al., 2005; Dantzer et al., 2008). Viruses induce an inflammatory response, and it is the association between inflammation and depression that forms the core arguments for an infection connection.

Indeed, there are converging lines of evidence to suggest that peripheral induction of proinflammatory cytokines represents one pathway by which the immune system can instigate or exacerbate major depression (Raison et al., 2006; Dantzer et al., 2008). In humans, exogenous administration of cytokines (e.g., as therapy for cancer) can induce a neuropsychological syndrome that mimics many features of major depression (Raison et al., 2006). This syndrome, which includes symptoms of fatigue, anhedonia, social withdrawal, decreased appetite, and sleep disturbances has been referred to as "sickness behavior" (Dunn et al., 2005; Dantzer et al., 2008). This set of neuropsychological and vegetative changes also commonly accompany viral and bacterial infections (Dunn et al., 2005). However, it is erroneous to conclude that sickness behavioral and unipolar or bipolar depressive illnesses are one in the same (Dunn et al., 2005), and the psychological symptoms of sickness behavior appear to be more responsive to paroxetine than the vegetative symptoms (Capuron et al., 2002b).

Nevertheless, cytokines represent an important and feasible pathway for brain–body interactions in the etiopathogenesis of major depression. Recent data indicate that personality traits may be important premorbid determinates of vulnerability to cytokine-induced depression (Raison et al., 2006). Moreover, the subset of patients who become depressed while on cytokine therapy (i.e., IFN-α) tend to have prolonged decreases in tryptophan (a crucial precursor of serotonin) due to the therapy (Capuron et al., 2002a). These data suggest that even if cytokines are not the primary cause of idiopathic major depression, they could still act as disease modifiers in vulnerable individuals.

Animal models have taken a prominent position in elucidating the neurobehavioral basis of cytokine-induced sickness behavior (Dunn et al., 2005). Besides revealing the interactions of cytokines with neurotransmitters, animal experiments have helped clarify the interplay between proinflammatory cytokines and the hypothalamic-pituitary-adrenal (HPA) axis (Silverman et al., 2002). The regulation of proinflammatory cytokines by endogenous glucocorticoids also appears to play a role in anxiety-like behavior in virus-infected mice (Silverman et al., 2007).

Hyperactivity of the HPA axis in patients with major depression has been consistently demonstrated (Holsboer and Barden, 1996; Raison et al., 2006). The precise etiology of these HPA axis alterations remains obscure; though current evidence suggests that hypersecretion of corticotropin releasing hormone (CRH) – a neuropeptide with profound effects on depression and anxiety behavior that is also critical in HPA axis regulation – may be one mediator of such hypercortisolemia as well as depressive symptoms. By integrating data from animal and clinical studies, a pathophysiological picture can be brought into focus whereby interactions between cytokines and the HPA axis may contribute significantly to the etiology of at least some cases of major depression. Glucocorticoids represent the final product of HPA axis stimulation, and may represent a critical link between cytokines and major depression (Raison et al., 2006). Ligand-mediated activation of the glucocorticoid receptor (GR) by these hormones (e.g., cortisol) is pivotal to controlling the production of proinflammatory cytokines following infection, as well as regulating the homeostatic tone of the HPA axis. More specifically, GR activation in immune and connective tissues downregulates proinflammatory cytokines, and GR activation in the CNS mediates negative feedback on the HPA axis, in turn dampening production of CRH. Considering that impaired feedback inhibition by cortisol has been suggested to underlie the pathogenesis of depression (e.g., by allowing excess CRH production), it's not surprising that recent attention has turned to the role of defective GR activation as a possible instigator of major depression (Raison et al., 2006). Hence, a syndrome in which GR responds inadequately to cortisol may underlie some cases of major depression.

Along these lines, there is evidence that glucocorticoid resistance in immune tissues participates in the pathophysiology of inflammatory and autoimmune disorders (Raison et al., 2006). If proinflammatory cytokines play a role in eliciting the neurophysiological changes responsible for idiopathic major depression, then it is reasonable to question whether such patients have higher circulating levels of proinflammatory cytokines. The findings in this regard have been mixed, but by incorporating data on proinflammatory cytokines with other indices of immune activation (e.g., acute phase proteins), a convincing case can be made for an overactivation of the inflammatory response system in at least a subset of patients with major depression (Raison et al., 2006; Dantzer et al., 2008). It is also possible that the critical neuroimmunological abnormality in major depression is an aberration in the expression or local production of cytokine or their receptors within the CNS compartment. In such a case, the peripheral cytokine abnormalities observed in some patients may merely be a shadow-phenomenon of the causative neuroimmune abnormality in the CNS.

Viral-immune mechanisms have been proposed for bipolar disorder as well. Borna disease virus has been putatively implicated based on serology and the presence of viral sequences in the brain or blood of patients with psychotic and affective illnesses (Amsterdam et al., 1985; Bode et al., 1996; Salvatore et al., 1997). It remains unclear how such individuals could have contracted this virus, which is found mainly in veterinary settings (i.e., horses in Germany) or in research laboratories (Durrwald et al., 2007). The conflicting data and interpretations on Borna disease virus in humans is in danger of reaching an impasse, but new approaches integrating genetic and environmental risk factors for bipolar disorder are yielding promising results (Carter, 2007).

Recently, the endogenous retrovirus, HERV-W, was reported to have reduced expression in brain tissue from patients with bipolar disorder, though patients with schizophrenia and major depression also had lower expression than controls (Weis et al., 2007). The causal relationship of this finding is unknown. The human genome of every person has retrovirus sequences. On the whole, these do not represent de novo infections, but are stretches of retroviral DNA that have been retained over the course of vertebrate evolution. While many of these are incomplete and do not contain open reading frames, some proviral sequences capable of making viral proteins have been identified. Interestingly, the placenta appears to be a preferred site of expression (Frendo et al., 2003), and it is conceivable that the critical alteration in the pathogenesis of these disorders dates back to the womb, and the findings of abnormal HERV-W expression in the adult brain is just a marker for its abnormal expression in multiple tissues at the time of fetal development.

5.6 Obsessive-Compulsive Disorder (OCD)

Most of the interest concerning a possible infectious etiology of OCD centers around the Pediatric Autoimmune Neuropsychiatric Disorder associated with Streptococcus (PANDAS; Swedo and Grant, 2005). Since streptococcus is not a virus, it is not a topic of this chapter, but this literature holds important lessons for unraveling infectious mechanisms in neuropsychiatric illnesses, and the reader is referred to a recent review (Swedo and Grant, 2005). OCD may also arise in connection with viral infections. A neuropsychological study of complications following viral encephalitis found a high prevalence of cognitive deficits, depression, anxiety, and OCD (Pewter et al., 2007). Besides abnormalities in several cognitive domains, OCD scores were the most elevated.

5.7 Schizophrenia

In the last 30 years there have been hundreds of articles in peer-reviewed journals presenting evidence or positing theories to suggest that at least some cases of schizophrenia are due to infections with a virus or other microbe (Torrey and Peterson, 1973; Kirch, 1993; Mednick et al., 1994; Yolken and Torrey, 1995; Gilmore and Jarskog, 1997; Pearce, 2001; Brown and Susser, 2002; Patterson, 2007). The viral hypothesis for schizophrenia does not consist of a single line of reasoning, but rather is an accretion of data from diverse fields that are connected in various subtheories. These subtheories range from causality due to inheritance of endogenous retroviruses to autoimmunity triggered at the time of the psychotic episode (Pearce, 2003). Broadly speaking, if a virus is involved in schizophrenia, it could either act during neurodevelopment to initiate abnormalities that eventually give rise to psychotic symptoms, or the virus could act more proximally to the psychotic episode, as is seen in encephalitis. Most recent emphasis has been on infections during development.

The idea that schizophrenia originates in prenatal or neonatal brain development is supported by several converging lines of evidence derived from studies in epidemiology, neuropsychology, developmental biology, and pathology (Weinberger, 1987; Murray et al., 1992; Bloom, 1993; Walker et al., 1999). Besides viral infections, a variety of prenatal and delivery complications have been putatively implicated in schizophrenia, including preeclampsia, maternal malnutrition, low birth weight, preterm labor, placental abruption, premature rupture of membranes, chorioamnionitis, maternal hemorrhage, fetal asphyxia, and prenatal stressful events such as natural catastrophes and war (Parnas et al., 1982; Mednick et al., 1988; Eagles et al., 1990; Susser and Lin, 1992; Kendell et al., 1996; Sacker et al., 1996; Susser et al., 1996; Verdoux et al., 1997; Geddes et al., 1999; Hultman et al., 1999; Jablensky, 1999; Cannon et al., 2002). The effect size varies widely between studies (RR or OR of 1.3–15.1), and there are inconsistencies in exactly which obstetrical complications are the culprits (Geddes and Lawrie, 1995; Brown et al., 2001; Cannon et al., 2002). Still, there is a growing body of literature that leaves diminishing doubt that obstetric complications are in some way linked to schizophrenia pathogenesis, and gestational infections remain prime candidates as triggers.

Undoubtedly, genetic factors play a role in schizophrenia, but it has proven difficult to ascertain which genes are involved and how their contribution meshes with environmental antecedents such as infections (Murray et al., 1992; Petronis et al., 1999; Torrey and Yolken, 1999; Tsuang, 2000). It is becoming increasingly clear that schizophrenia is not usually caused by any single gene, but rather is underlaid by an ambiguous network of interacting genes that presumably confer susceptibility to environmental insults or psychosocial stressors. An intriguing idea is that relevant genes could exacerbate the impact of an obstetric complication in the fetal brain (or even select brain regions) while diminishing the deleterious consequences of the complication on other components of the maternal-fetal unit (Preti et al., 1998; Brune, 2004). A highly vigorous maternal immune response would fit well into this theory; i.e., successful maternal immune protection against an immediate infectious threat could be at the expense of subtle fetal brain injury that could ultimately be expressed as latent psychotic illness. This might explain the persistence of schizophrenia in the population despite reduced fecundity.

While a number of gestational infections have been proposed in the etiopathogenesis of schizophrenia, influenza has received the most attention. Influenza infection during pregnancy is relatively common and can be asymptomatic. A recent study found that 22% of women had rising titers against influenza A during the second trimester or third trimester, and 11% of these cases were confirmed to have intercurrent influenza infection via confirmatory testing (Irving et al., 2000). In a seminal ecological study, Mednick and colleagues found individuals exposed to the 1957 Helsinki influenza epidemic in their second trimester of gestation had an increased likelihood of being diagnosed with schizophrenia when assessed 25 years or more later (Mednick et al., 1988). Subsequent studies in Japan, Scotland, Australia, United Kingdom, and Denmark generally confirmed Mednick's original findings (Machon et al., 1995; Brown and Susser, 2002), but several studies were not confirmatory. Moreover, analysis of archived maternal plasma for influenza antibodies suggested that the first trimester rather than the second conferred risk for schizophrenia spectrum disorders in the offspring (Brown et al., 2004). Hence, the data on influenza and schizophrenia have not been entirely reconciled.

Wright et al. suggest that women genetically loaded for schizophrenia produce an over-active antibody response against second trimester influenza (Wright and Murray, 1993). Accordingly, women with this proposed immunogenetic anomaly might be expected to have less intense symptoms when exposed to the virus, and this could explain the disengagement between the various ecological studies of the influenza-schizophrenia connection and data derived from clinic or hospital records and/or maternal recall. In this theory, the maternal anti-influenza antibodies cross the placenta and traverse the undeveloped fetal blood–brain barrier, cross-reacting with fetal CNS tissue and impairing neuronal migration or neural development.

Convincing evidence for this mechanism is currently lacking in humans, though the idea of cross-reacting antibodies is not restricted to influenza. For example, sialic acid linked molecules such as gangliosides and the polysialylated form of the neural cell adhesion molecule (PSA-NCAM) are important determinants of brain development (Rosenberg and Noble, 1994; Barbeau et al., 1995). Antibodies that cross react with these molecules are proposed to be made against various bacterial and viral pathogens, and if these are able to reach the developing brain, they could cause subtle alterations in neurocircuitry (Nedelec et al., 1990; Nahmias et al., 2006). One appealing aspect of this hypothesis is the presence of several inherent mechanisms that would normally protect the immature brain from such antibodies, and these mechanisms may be defective in subset of pregnancies due to genetic or acquired abnormalities in the maternal B-cell repertoire, placenta, or fetal blood–brain barrier (Nahmias et al., 2006). Moreover, abnormalities in PSA-NCAM in schizophrenia have been suggested by genetic linkages and neurochemical findings (Barbeau et al., 1995; Tao et al., 2007). Thus, there could be multiple routes to schizophrenia that funnel though NCAM and interconnected molecules.

Over 10 years ago, Gilmore and Jarskog put forth the hypothesis that overproduction of proinflammatory cytokines may be a common link between infectious and non-infectious pregnancy complications and the development of adult schizophrenia (Gilmore and Jarskog, 1997). In the ensuing years there has been a burgeoning literature to support this idea, though the mechanisms remain undefined. Cytokines made by the mother and placenta can cross into the fetal circulation, and impact the fetal brain (Gilmore and Jarskog, 1997; Cai et al., 2000; Urakubo et al., 2001; du Plessis and Volpe, 2002). There is conflicting data as to what degree these cytokines can induce local inflammatory response in the fetal brain (reviewed in Ashdown et al., 2006), but it is clear that maternal viral stimuli can cause changes in neurochemistry and behavior in the offspring that are measurable postnatally (Urakubo et al., 2001; Gilmore et al., 2003; Meyer et al., 2005, 2007; Smith et al., 2007). Several studies have employed poly I:C as a virus-like stimulant, but Fatemi and colleagues have also shown that infection of pregnant mice with live influenza at mid-gestation induces abnormalities in the offspring, including aberrant corticogenesis and alterations in the levels of reelin (Fatemi et al., 1998, 1999, 2002; Shi et al., 2003). Reelin has been connected with schizophrenia at multiple levels (i.e., epigenetics, neurocircuitry, behavior; Fatemi, 2005). Also, poly I:C causes decrements in parvalbumin GABAergic interneurons in the hippocampus. Such a virus-induced disinhibition affecting hippocampal parvalbumin interneurons was previously reported in a rat model using neonatal infection with the live virus, lymphocytic choriomeningitis virus (LCMV; Pearce et al., 2000). This points to a biphasic disease mechanism in which a viral- or immune-mediated disruption of inhibitory (GABAergic) circuits occurs during brain development resulting in unbalanced excitatory neurotransmission, which in turn causes the gradual loss of principle cells due to excitotoxicity (Pearce et al., 1996, 2000). Additionally, a link between reelin alteration and deficits in inhibitory neurons in schizophrenia has been proposed (Costa et al., 2001).

Poly I:C, influenza virus, and bacterial immunostimulants given during rodent gestation have been reported to engender abnormalities in sensorimotor gating similar to those found in schizophrenia (Borrell et al., 2002; Shi et al., 2003; Zuckerman et al., 2003; Meyer et al., 2005; Shi et al., 2005). Recently, a mouse model was used to identify IL-6 as a key mediator of such behavioral abnormalities that follow in utero exposure into viral stimuli (Smith et al., 2007). There is also evidence that inflammatory stimuli during neurodevelopment can lead to a hyper-dopaminergic state manifested in the adults (Zuckerman et al., 2003; Ozawa et al., 2006). Another long-term consequence of viral infection during development is a decrease in hippocampal neurogenesis in the adult (Sharma et al., 2002). This effect appears to be mediated through the destruction or impairment of pluripotent neuronal progenitor cells that give rise to hippocampal dentate granule cells (Fig. 2). A reduction in adult neurogenesis has been tied to stress and depression, and to a lesser extent to schizophrenia (Schmidt and Duman, 2007; Toro and Deakin, 2007).

One strength of the cytokine hypothesis for the neurodevelopmental origins of schizophrenia is that proinflammatory cytokines amplify hypoxic brain damage and may constitute a common pathway for perinatal brain injury due to diverse insults in humans (Gilmore and Jarskog, 1997; Yoon et al., 2000; Dammann and Leviton, 2001; Kadhim et al., 2001; Dammann et al., 2002; du Plessis and Volpe, 2002). Moreover, there is considerable evidence that obstetric complications and prenatal stressors are partial determinants of immune responses of the adult offspring (Shanks and Lightman, 2001). Conceivably, this could represent a connection between a history of exposure to a pregnancy complications and immune abnormalities in adult patients with schizophrenia.

Studies indicating abnormal immune function in adult schizophrenia predate the use of anti-psychotic medications (Bruce and Peebles, 1904). Since then, a massive literature on the subject has evolved, leaving little doubt that the immune system is tied to schizophrenia (Ganguli et al., 1987, 1994a; Muller et al., 1999; Pearce, 2003). Nevertheless, this literature is also rife with confounds and inconsistencies.

Most studies have found an increase in IL-2 soluble receptor (IL-2sR) in schizophrenia, which is an indicator that T-cells are activated (Ganguli and Rabin, 1989; Rapaport et al., 1993, 1994; Muller et al., 2000) and is consistent with a background of autoimmunity (Steiner et al., 1995). Schizophrenia is associated with an increase in the proinflammatory cytokines TNF-α and IL-6; the latter often normalizing with remission (Ganguli et al., 1987, 1994b; Rapaport and Lohr, 1994; Naudin et al., 1996;

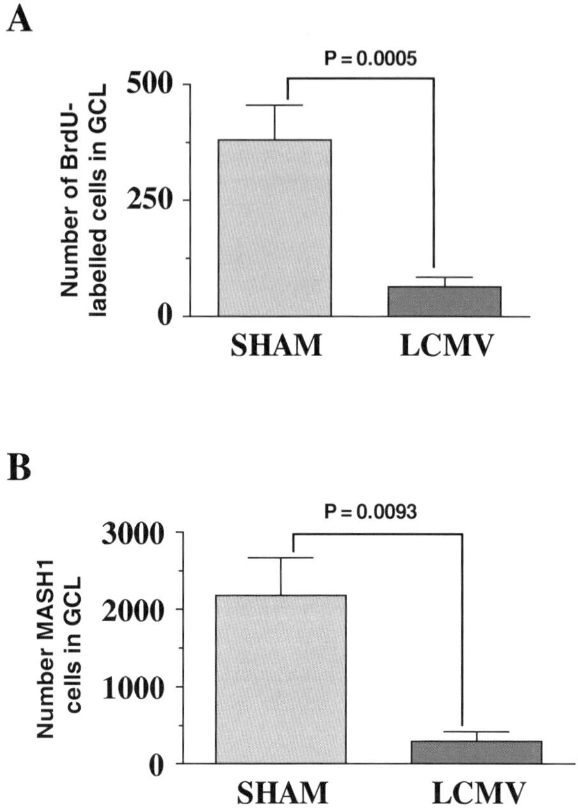

Fig. 2 Impact of neonatal lymphocyte choriomeningitis virus (LCMV) infection on adult neurogenesis in the hippocampus. *Panel A.* Proliferating cells labeled with BrdU in the dentate granule cell layer (GCL) of 8.5-month-old rats infected as neonates with LCMV, or sham injected. *Panel B.* Progenitor cells in the adult hippocampus as a function of neonatal LCMV infection. Reprinted from Sharma A, Valadi N, Miller AH, Pearce BD (2002) Neonatal viral infection decreases neuronal progenitors and impairs adult neurogenesis in the hippocampus. Neurobiol Dis 11:246–256

Monteleone et al., 1997; Muller et al., 1999; Maes et al., 2002). Moreover, some studies have shown that the addition of the anti-inflammatory agent, celecoxib, to an atypical antipsychotic medication produces at least a short-term benefit on psychopathology (Muller et al., 2002; Akhondzadeh et al., 2007). Also consistent with an autoimmune disease, several studies have shown in vitro production of IFN-γ is decreased in lymphocytes from schizophrenia patients (reviewed in Arolt et al., 2000).

Since CD4 lymphocytes are a major source of cytokines such as IL-2 and IFN-γ, the question has been raised whether schizophrenia is associated with an imbalance between the two types of CD4 cells – the TH1 cells which produce IFN-γ, IL-2, and IL-12, and the TH2 cells which produce IL-4, IL-5, and IL-10 (Schwarz et al., 2001). Although there is some evidence for elevated IL-4 and IL-10 in schizophrenia, the data have been inconsistent, perhaps because of medication effects or disease chronicity (Song et al., 2000; Schwarz et al., 2001).

Epidemiological studies have shown positive associations of schizophrenia with some autoimmune diseases (even prior to the psychiatric diagnosis) but negative associations with others (Torrey and Yolken, 2001; Eaton et al., 2006). This may reflect a predisposition toward certain HLA types or unusual acquired immune responses in patients with schizophrenia, but thus far the data have been inconclusive. Still, one cannot ignore the voluminous literature presenting evidence for autoantibodies against various CNS and peripheral antigens in schizophrenic patients (Ganguli et al., 1993; Borda et al., 2002; Wang et al., 2003). Viruses could plausibly instigate autoimmunity in psychotic disorders, but a specific mechanism has not been identified, and most studies have been met with the usual covariates involving antipsychotic medications, lifestyle factors (smoking, homelessness), and disease subtypes. Another possibility is that immune and environmental factors converge to disrupt the functional integrity of the blood–brain barrier in schizophrenia (Yolken and Torrey, 1995; Hanson and Gottesman, 2005). This could in theory be traced back to neurodevelopment, and even a slight delay in a crucial aspect of blood–brain barrier development could leave the immature brain vulnerable pathogens and other insults having the capacity to provoke a common proinflammatory cascade.

Although outside the scope of this chapter, it should be mentioned that several non-viral microbes have been posited in schizophrenia etiology. In particular, a role for toxoplasmosis has been put forth by several authors (Torrey and Yolken, 2003; Leweke et al., 2004; Brown et al., 2005; Skallova et al., 2005).

6 Conclusions

While it is clear that infections and immune responses can engender neuropsychiatric symptoms, we do not know what fraction of the major psychiatric disorders discussed above that can be attributed to viruses. A lesson from oncology can help put this into perspective. The belief that cancer could be caused by a contagious agent goes back centuries, but work showing the role of viruses in human cancer was constantly met with stringent criticism throughout most of the twentieth century (Epstein, 2001). The realization that cancer has a number of different causes, including viruses, has finally nudged its way into scientific dogma, and this has introduced new strategies for prevention and treatment (Epstein, 2001; Markowitz et al., 2007). The endeavor to find the causes of mental illness deserves no less thorough an investigation.

Acknowledgments The author gratefully acknowledges research support from the National Institutes of Health (NIMH, NIAAA, NINDS), the Theodore and Vada Stanley Foundation, and NARSAD.

References

Abeles AM, Pillinger MH, Solitar BM, Abeles M (2007) Narrative review: the pathophysiology of fibromyalgia. Ann Intern Med 146:726–734.

Ahmed R, Morrison LA, Knipe DM (1997) Viral persistance. In: Nathanson N, Ahmed R, (eds) Viral Pathogenesis. Lippincott-Raven, Philadelphia, pp 181–205.

Akhondzadeh S, Tabatabaee M, Amini H, Ahmadi Abhari SA, Abbasi SH, Behnam B (2007) Celecoxib as adjunctive therapy in schizophrenia: a double-blind, randomized and placebo-controlled trial. Schizophr Res 90:179–185.

Amsterdam JD, Winokur A, Dyson W, Herzog S, Gonzalez F, Rott R, Koprowski H (1985) Borna disease virus. A possible etiologic factor in human affective disorders? Arch Gen Psychiatry 42:1093–1096.

Appel S, Chapman J, Shoenfeld Y (2007) Infection and vaccination in chronic fatigue syndrome: myth or reality? Autoimmunity 40:48–53.

Arnason BGW (1998) Autoimmune diseases of the central and peripheral nervous systems. In: Rose NR, Mackay IR, (eds) The Autoimmune Diseases. Academic Press, San Diego, pp 571–602.

Arolt V, Rothermundt M, Wandinger KP, Kirchner H (2000) Decreased in vitro production of interferon-gamma and interleukin-2 in whole blood of patients with schizophrenia during treatment. Mol Psychiatry 5:150–158.

Ashdown H, Dumont Y, Ng M, Poole S, Boksa P, Luheshi GN (2006) The role of cytokines in mediating effects of prenatal infection on the fetus: implications for schizophrenia. Mol Psychiatry 11:47–55.

Ball MJ (1982) "Limbic predilection in Alzheimer dementia: is reactivated herpesvirus involved?" Can J Neurol Sci 9:303–306.

Ball LK, Ball R, Pratt RD (2001) An assessment of thimerosal use in childhood vaccines. Pediatrics 107:1147–1154.

Banks WA, Watkins LR (2006) Mediation of chronic pain: not by neurons alone. Pain 124:1–2.

Barbeau D, Liang JJ, Robitalille Y, Quirion R, Srivastava LK (1995) Decreased expression of the embryonic form of the neural cell adhesion molecule in schizophrenic brains. Proc Natl Acad Sci U S A 92:2785–2789.

Bautista JR, Rubin SA, Moran TH, Schwartz GJ, Carbone KM (1995) Developmental injury to the cerebellum following perinatal Borna disease virus infection. Developmental Brain Research 90:45–53.

Beffert U, Bertrand P, Champagne D, Gauthier S, Poirier J (1998) HSV-1 in brain and risk of Alzheimer's disease. Lancet 351:1330–1331.

Biederman J, Faraone SV (2002) Current concepts on the neurobiology of attention-deficit/hyperactivity disorder. J Atten Disord 6 Suppl 1:S7–S16.

Bilzer T, Stitz L (1996) Immunopathogenesis of virus diseases affecting the central nervous system. Crit Rev Immunol 16:145–222.

Biron CA (1994) Cytokines in the generation of immune responses to, and resolution of, virus infection. Curr Opin Immunol 6:530–538.

Blanchard MS, Eisen SA, Alpern R, Karlinsky J, Toomey R, Reda DJ, Murphy FM, Jackson LW, Kang HK (2006) Chronic multisymptom illness complex in Gulf War I veterans 10 years later. Am J Epidemiol 163:66–75.

Blennow K, de Leon MJ, Zetterberg H (2006) Alzheimer's disease. Lancet 368:387–403.

Bloom FE (1993) Advancing a neurodevelopmental origin for schizophrenia. Arch Gen Psychiatry 50:224–227.

Bode L, Durrwald R, Rantam FA, Ferszt R, Ludwig H (1996) First isolates of infectious human Borna disease virus from patients with mood disorders. Mol Psychiatry 1:200–212.

Bonthius DJ, Wright R, Tseng B, Barton L, Marco E, Karacay B, Larsen PD (2007) Congenital lymphocytic choriomeningitis virus infection: spectrum of disease. Ann Neurol 62:347–355.

Borda T, Perez Rivera R, Joensen L, Gomez RM, Sterin-Borda L (2002) Antibodies against cerebral M1 cholinergic muscarinic receptor from schizophrenic patients: molecular interaction. J Immunol 168:3667–3674.

Borrell J, Vela JM, Arevalo-Martin A, Molina-Holgado E, Guaza C (2002) Prenatal immune challenge disrupts sensorimotor gating in adult rats. Implications for the etiopathogenesis of schizophrenia. Neuropsychopharmacology 26:204–215.

Brown AS, Cohen P, Harkavy-Friedman J, Babulas V, Malaspina D, Gorman JM, Susser ES (2001) Prenatal rubella, premorbid abnormalities, and adult schizophrenia. Biol Psychiatry 49: 473–486.

Brown AS, Susser ES (2002) In utero infection and adult schizophrenia. Ment Retard Dev Disabil Res Rev 8:51–57.
Brown AS, Begg MD, Gravenstein S, Schaefer CA, Wyatt RJ, Bresnahan M, Babulas VP, Susser ES (2004) Serologic evidence of prenatal influenza in the etiology of schizophrenia. Arch Gen Psychiatry 61:774–780.
Brown AS, Schaefer CA, Quesenberry CP, Jr., Liu L, Babulas VP, Susser ES (2005) Maternal exposure to toxoplasmosis and risk of schizophrenia in adult offspring. Am J Psychiatry 162:767–773.
Bruce LC, Peebles AMS (1904) Quantitative and qualitative leukocyte counts in various forms of mental disease. J Ment Sci 50:409–417.
Brune M (2004) Schizophrenia-an evolutionary enigma? Neurosci Biobehav Rev 28:41–53.
Cai Z, Pan ZL, Pang Y, Evans OB, Rhodes PG (2000) Cytokine induction in fetal rat brains and brain injury in neonatal rats after maternal lipopolysaccharide administration. Pediatr Res 47:64–72.
Cannon M, Jones PB, Murray RM (2002) Obstetric complications and schizophrenia: historical and meta-analytic review. Am J Psychiatry 159:1080–1092.
Capuron L, Ravaud A, Neveu PJ, Miller AH, Maes M, Dantzer R (2002a) Association between decreased serum tryptophan concentrations and depressive symptoms in cancer patients undergoing cytokine therapy. Mol Psychiatry 7:468–473.
Capuron L, Gumnick JF, Musselman DL, Lawson DH, Reemsnyder A, Nemeroff CB, Miller AH (2002b) Neurobehavioral effects of interferon-alpha in cancer patients: phenomenology and paroxetine responsiveness of symptom dimensions. Neuropsychopharmacology 26: 643–652.
Carter CJ (2007) Multiple genes and factors associated with bipolar disorder converge on growth factor and stress activated kinase pathways controlling translation initiation: implications for oligodendrocyte viability. Neurochem Int 50:461–490.
Chess S (1977) Follow-up report on autism in congenital rubella. J Autism Child Schizophr 7:69–81.
Comi AM, Zimmerman AW, Frye VH, Law PA, Peeden JN (1999) Familial clustering of autoimmune disorders and evaluation of medical risk factors in autism. J Child Neurol 14:388–394.
Costa E, Davis J, Grayson DR, Guidotti A, Pappas GD, Pesold C (2001) Dendritic spine hypoplasticity and downregulation of reelin and GABAergic tone in schizophrenia vulnerability. Neurobiol Dis 8:723–742.
Courchesne E, Yeung-Courchesne R, Press GA, Hesselink JR, Jernigan TL (1988) Hypoplasia of cerebellar vermal lobules VI and VII in autism. New England J of Medicine 318:1349–1354.
Croen LA, Grether JK, Yoshida CK, Odouli R, Van de Water J (2005) Maternal autoimmune diseases, asthma and allergies, and childhood autism spectrum disorders: a case-control study. Arch Pediatr Adolesc Med 159:151–157.
Dalman C, Allebeck P, Gunnell D, Harrison G, Kristensson K, Lewis G, Lofving S, Rasmussen F, Wicks S, Karlsson H (2008) Infections in the CNS during childhood and the risk of subsequent psychotic illness: a cohort study of more than one million Swedish subjects. Am J Psychiatry 165:59–65.
Dammann O, Leviton A (2001) Possible strategies to protect the preterm brain against the fetal inflammatory response. Dev Med Child Neurol Suppl 86:18–20.
Dammann O, Kuban KC, Leviton A (2002) Perinatal infection, fetal inflammatory response, white matter damage, and cognitive limitations in children born preterm. Ment Retard Dev Disabil Res Rev 8:46–50.
Dantzer R, O'Connor JC, Freund GG, Johnson RW, Kelley KW (2008) From inflammation to sickness and depression: when the immune system subjugates the brain. Nature Reviews Neuroscience 9:46–56.
de la Torre JC, Oldstone MBA (1996) Anatomy of viral persistence: mechanisms of persistence and associated disease. Adv Virus Res 46:311–343.
du Plessis AJ, Volpe JJ (2002) Perinatal brain injury in the preterm and term newborn. Curr Opin Neurol 15:151–157.
Dunn AJ, Swiergiel AH, de Beaurepaire R (2005) Cytokines as mediators of depression: what can we learn from animal studies? Neurosci Biobehav Rev 29:891–909.

Durrwald R, Kolodziejek J, Herzog S, Nowotny N (2007) Meta-analysis of putative human bornavirus sequences fails to provide evidence implicating Borna disease virus in mental illness. Reviews in Medical Virology 17:181–203.
Eagles JM, Gibson I, Bremner MH, Clunie F, Ebmeier KP, Smith NC (1990) Obstetric complications in DSM-III schizophrenics and their siblings. Lancet 335:1139–1141.
Eaton WW, Byrne M, Ewald H, Mors O, Chen CY, Agerbo E, Mortensen PB (2006) Association of schizophrenia and autoimmune diseases: linkage of Danish national registers. Am J Psychiatry 163:521–528.
Eggenberger E (2007) Prion disease. Neurol Clin 25:833–842.
Elphick GF, Querbes W, Jordan JA, Gee GV, Eash S, Manley K, Dugan A, Stanifer M, Bhatnagar A, Kroeze WK, Roth BL, Atwood WJ (2004) The human polyomavirus, JCV, uses serotonin receptors to infect cells. Science 306:1380–1383.
Epstein MA (2001) Historical background. Philosophical Transactions of the Royal Society of London – Series B: Biological Sciences 356:413–420.
Fatemi SH, Sidwell R, Akhter P, Sedgewick J, Thuras P, Bailey K, Kist D (1998) Human influenza viral infection in utero increases nNOS expression in hippocampi of neonatal mice. Synapse 29:84–88.
Fatemi SH, Emamian ES, Kist D, Sidwell RW, Nakajima K, Akhter P, Shier A, Sheikh S, Bailey K (1999) Defective corticogenesis and reduction in Reelin immunoreactivity in cortex and hippocampus of prenatally infected neonatal mice. Mol Psychiatry 4:145–154.
Fatemi SH, Earle J, Kanodia R, Kist D, Emamian ES, Patterson PH, Shi L, Sidwell R (2002) Prenatal viral infection leads to pyramidal cell atrophy and macrocephaly in adulthood: Implications for genesis of autism and schizophrenia. Cell Mol Neurobiol 22:25–33.
Fatemi SH (2005) Reelin glycoprotein in autism and schizophrenia. Int Rev Neurobiol 71:179–187.
Fatemi SH, Pearce DA, Brooks AI, Sidwell RW (2005) Prenatal viral infection in mouse causes differential expression of genes in brains of mouse progeny: a potential animal model for schizophrenia and autism. Synapse 57:91–99.
Fombonne E (2008) Thimerosal disappears but autism remains. Arch Gen Psychiatry 65:15–16.
Frendo JL, Olivier D, Cheynet V, Blond JL, Bouton O, Vidaud M, Rabreau M, Evain-Brion D, Mallet F (2003) Direct involvement of HERV-W Env glycoprotein in human trophoblast cell fusion and differentiation. Mol Cell Biol 23:3566–3574.
Ganguli R, Rabin BS, Kelly RH, Lyte M, Ragu U (1987) Clinical and laboratory evidence of autoimmunity in acute schizophrenia. Ann N Y Acad Sci 496:676–685.
Ganguli R, Rabin BS (1989) Increased serum interleukin 2 receptor concentration in schizophrenic and brain-damaged subjects. Arch Gen Psychiatry 46:292.
Ganguli R, Brar JS, Chengappa KNR, Yang ZW, Nimgaonkar VL, Rabin BS (1993) Autoimmunity in schizophrenia: a review of recent findings. Ann Med 25:489–496.
Ganguli R, Brar JS, Rabin BS (1994a) Immune abnormalities in schizophrenia: evidence for the autoimmune hypothesis. Harv Rev Psychiatry 2:70–83.
Ganguli R, Yang Z, Shurin G, Chengappa KN, Brar JS, Gubbi AV, Rabin BS (1994b) Serum interleukin-6 concentration in schizophrenia: elevation associated with duration of illness. Psychiatry Res 51:1–10.
Garden GA, Moller T (2006) Microglia biology in health and disease. J Neuroimmun Pharmacol 1:127–137.
Geddes JR, Lawrie SM (1995) Obstetric complications and schizophrenia: a meta-analysis. Br J Psychiatry 167:786–793.
Geddes JR, Verdoux H, Takei N, Lawrie SM, Bovet P, Eagles JM, Heun R, McCreadie RG, T EM, O'Callaghan E, Stober G, Willinger U, Murray RM (1999) Schizophrenia and complications of pregnancy and labor: An individual patient data meta-analysis. Schizophr Bull 25:413–423.
Gerard HC, Dreses-Werringloer U, Wildt KS, Deka S, Oszust C, Balin BJ, Frey WH, 2nd, Bordayo EZ, Whittum-Hudson JA, Hudson AP (2006) Chlamydophila (Chlamydia) pneumoniae in the Alzheimer's brain. FEMS Immunol Med Microbiol 48:355–366.
Gilmore JH, Jarskog LF (1997) Exposure to infection and brain development: cytokines in the pathogenesis of schizophrenia [letter]. Schizophr Res 24:365–367.

Gilmore JH, Jarskog LF, Vadlamudi S (2003) Maternal infection regulates BDNF and NGF expression in fetal and neonatal brain and maternal-fetal unit of the rat. J Neuroimmunol 138:49–55.
Gosztonyi G, Koprowski H (2001) The concept of neurotropism and selective vulnerability ("pathoclisis") in virus infections of the nervous system – a historical overview. Curr Top Microbiol Immunol 253:1–13.
Gosztonyi G, Ludwig H (2001) Interactions of viral proteins with neurotransmitter receptors may protect or destroy neurons. Curr Top Microbiol Immunol 253:121–144.
Hanson DR, Gottesman, II (2005) Theories of schizophrenia: a genetic-inflammatory-vascular synthesis. BMC Medical Genetics 6:7.
Hemachudha T, Laothamatas J, Rupprecht CE (2002) Human rabies: a disease of complex neuropathogenetic mechanisms and diagnostic challenges. Lancet Neurol 1:101–109.
Hickie I, Davenport T, Wakefield D, Vollmer-Conna U, Cameron B, Vernon SD, Reeves WC, Lloyd A (2006) Post-infective and chronic fatigue syndromes precipitated by viral and non-viral pathogens: prospective cohort study. BMJ 333:575.
Holsboer F, Barden N (1996) Antidepressants and hypothalamic-pituitary-adrenocortical regulation. Endocr Rev 17:187–205.
Hornig M, Weissenbock H, Horscroft N, Lipkin WI (1999) An infection-based model of neurodevelopmental damage. Proc Natl Acad Sci U S A 96:12102–12107.
Hornig M, Lipkin WI (2001) Infectious and immune factors in the pathogenesis of neurodevelopmental disorders: epidemiology, hypotheses, and animal models. Ment Retard Dev Disabil Res Rev 7:200–210.
Hultman CM, Sparen P, Takei N, Murray RM, Cnattingius S (1999) Prenatal and perinatal risk factors for schizophrenia, affective psychosis, and reactive psychosis of early onset: case-control study. BMJ 318:421–426.
Irving WL, James DK, Stephenson T, Laing P, Jameson C, Oxford JS, Chakraverty P, Brown DW, Boon AC, Zambon MC (2000) Influenza virus infection in the second and third trimesters of pregnancy: a clinical and seroepidemiological study. BJOG 107:1282–1289.
Itzhaki RF, Wozniak MA, Appelt DM, Balin BJ (2004) Infiltration of the brain by pathogens causes Alzheimer's disease. Neurobiol Aging 25:619–627.
Jablensky A (1999) Schizophrenia: Epidemiology. Curr Opin Psychiatry 12:19–28.
Johnson RT (1998) Viral Infections of the Nervous System, 2nd Edition. Lippincott-Raven, Philadelphia.
Jones-Brando L, Torrey EF, Yolken R (2003) Drugs used in the treatment of schizophrenia and bipolar disorder inhibit the replication of Toxoplasma gondii. Schizophr Res 62:237–244.
Jyonouchi H, Geng L, Ruby A, Reddy C, Zimmerman-Bier B (2005) Evaluation of an association between gastrointestinal symptoms and cytokine production against common dietary proteins in children with autism spectrum disorders. J Pediatr 146:605–610.
Kadhim H, Tabarki B, Verellen G, De Prez C, Rona AM, Sebire G (2001) Inflammatory cytokines in the pathogenesis of periventricular leukomalacia. Neurology 56:1278–1284.
Kendell RE, Juszczak E, Cole SK (1996) Obstetric complications and schizophrenia: a case control study based on standardised obstetric records. Br J Psychiatry 168:556–561.
Kirch DG (1993) Infection and autoimmunity as etiologic factors in schizophrenia: a review and reappraisal. Schizophr Bull 19:355–370.
Korvatska E, Van de Water J, Anders TF, Gershwin ME (2002) Genetic and immunologic considerations in autism. Neurobiol Dis 9:107–125.
Koskiniemi M, Vaheri A (1989) Effect of measles, mumps, rubella vaccination on pattern of encephalitis in children. Lancet 1:31–34.
Lancaster K, Dietz DM, Moran TH, Pletnikov MV (2007) Abnormal social behaviors in young and adult rats neonatally infected with Borna disease virus. Behav Brain Res 176:141–148.
Lederman MM, Penn-Nicholson A, Cho M, Mosier D (2006) Biology of CCR5 and its role in HIV infection and treatment. JAMA 296:815–826.
Leweke FM, Gerth CW, Koethe D, Klosterkotter J, Ruslanova I, Krivogorsky B, Torrey EF, Yolken RH (2004) Antibodies to infectious agents in individuals with recent onset schizophrenia. Eur Arch Psychiatry Clin Neurosci 254:4–8.

Libbey JE, Sweeten TL, McMahon WM, Fujinami RS (2005) Autistic disorder and viral infections. J Neurovirol 11:1–10.

Lishman WA (1987) Organic Psychiatry. Blackwell, Oxford.

Machon RA, Mednick SA, Huttunen MO (1995) Fetal viral infection and adult schizophrenia: empirical findings and interpretation. In: Mednick SA, Hollister JM, (eds) Neural Development and Schizophrenia. Plenum, New York, pp 190–202.

Madsen KM, Hviid A, Vestergaard M, Schendel D, Wohlfahrt J, Thorsen P, Olsen J, Melbye M (2002) A population-based study of measles, mumps, and rubella vaccination and autism. N Engl J Med 347:1477–1482.

Maes M, Bocchio Chiavetto L, Bignotti S, Battisa Tura GJ, Pioli R, Boin F, Kenis G, Bosmans E, de Jongh R, Altamura CA (2002) Increased serum interleukin-8 and interleukin-10 in schizophrenic patients resistant to treatment with neuroleptics and the stimulatory effects of clozapine on serum leukemia inhibitory factor receptor. Schizophr Res 54:281–291.

Markowitz LE, Dunne EF, Saraiya M, Lawson HW, Chesson H, Unger ER, Centers for Disease C, Prevention, Advisory Committee on Immunization P (2007) Quadrivalent Human Papillomavirus Vaccine: Recommendations of the Advisory Committee on Immunization Practices (ACIP). MMWR Recomm Rep 56:1–24.

McGeer PL, Rogers J, McGeer EG (2006) Inflammation, anti-inflammatory agents and Alzheimer disease: the last 12 years. J Alzheimers Dis 9:271–276.

Mednick SA, Machon RA, Huttunen MO, Bonett D (1988) Adult schizophrenia following prenatal exposure to an influenza epidemic. Arch Gen Psychiatry 45:189–192.

Mednick SA, Huttunen MO, Machon RA (1994) Prenatal influenza infections and adult schizophrenia. Schizophr Bull 20:263–267.

Menninger KA (1926) Influenza and schizophrenia. An analysis of post-influenzal "dementia praecox" as of 1918, and five years later. Am J Psychiatry 5:469–529.

Meyer U, Feldon J, Schedlowski M, Yee BK (2005) Towards an immuno-precipitated neurodevelopmental animal model of schizophrenia. Neurosci Biobehav Rev 29:913–947.

Meyer U, Nyffeler M, Yee BK, Knuesel I, Feldon J (2007) Adult brain and behavioral pathological markers of prenatal immune challenge during early/middle and late fetal development in mice. Brain Behav Immun 22:469–486.

Millichap JG (1997) Encephalitis virus and attention deficit hyperactivity disorder. J R Soc Med 90:709–710.

Misery L (1997) Skin, immunity and the nervous system. Br J Dermatol 137:843–850.

Monteleone P, Fabrazzo M, Tortorella A, Maj M (1997) Plasma levels of interleukin-6 and tumor necrosis factor alpha in chronic schizophrenia: effects of clozapine treatment. Psychiatry Res 71:11–17.

Mrak RE, Griffin WS (2005) Glia and their cytokines in progression of neurodegeneration. Neurobiol Aging 26:349–354.

Muller N, Riedel M, Ackenheil M, Schwarz MJ (1999) The role of immune function in schizophrenia: an overview. Eur Arch Psychiatry Clin Neurosci 249 Suppl 4:62–68.

Muller N, Riedel M, Gruber R, Ackenheil M, Schwarz MJ (2000) The immune system and schizophrenia. An integrative view. Ann N Y Acad Sci 917:456–467.

Muller N, Riedel M, Scheppach C, Brandstatter B, Sokullu S, Krampe K, Ulmschneider M, Engel RR, Moller HJ, Schwarz MJ (2002) Beneficial antipsychotic effects of celecoxib add-on therapy compared to risperidone alone in schizophrenia. Am J Psychiatry 159:1029–1034.

Murray RM, Jones P, O'Callaghan E, Takei N, Sham P (1992) Genes, viruses and neurodevelopmental schizophrenia. J Psychiatr Res 26:225–235.

Nahmias AJ, Nahmias SB, Danielsson D (2006) The possible role of transplacentally-acquired antibodies to infectious agents, with molecular mimicry to nervous system sialic acid epitopes, as causes of neuromental disorders: prevention and vaccine implications. Clin Dev Immunol 13:167–183.

Nathanson N (1997) Introduction and history. In: Nathanson N, (ed) Viral Pathogenesis. Lippincott-Raven, Philadelphia, pp 3–11.

Naudin J, Mege JL, Azorin JM, Dassa D (1996) Elevated circulating levels of IL-6 in schizophrenia. Schizophr Res 20:269–273.

Nedelec J, Boucraut J, Garnier JM, Bernard D, Rougon G (1990) Evidence for autoimmune antibodies directed against embryonic neural cell adhesion molecules (N-CAM) in patients with group B meningitis. J Neuroimmunol 29:49–56.

Nelson PG, Kuddo T, Song EY, Dambrosia JM, Kohler S, Satyanarayana G, Vandunk C, Grether JK, Nelson KB (2006) Selected neurotrophins, neuropeptides, and cytokines: developmental trajectory and concentrations in neonatal blood of children with autism or Down syndrome. Int J Dev Neurosci 24:73–80.

Newschaffer CJ, Croen LA, Daniels J, Giarelli E, Grether JK, Levy SE, Mandell DS, Miller LA, Pinto-Martin J, Reaven J, Reynolds AM, Rice CE, Schendel D, Windham GC (2007) The epidemiology of autism spectrum disorders. Annu Rev Public Health 28:235–258.

Onami TM, Harrington LE, Williams MA, Galvan M, Larsen CP, Pearson TC, Manjunath N, Baum LG, Pearce BD, Ahmed R (2002) Dynamic regulation of T cell immunity by CD43. J Immunol 168:6022–6031.

Ozawa K, Hashimoto K, Kishimoto T, Shimizu E, Ishikura H, Iyo M (2006) Immune activation during pregnancy in mice leads to dopaminergic hyperfunction and cognitive impairment in the offspring: a neurodevelopmental animal model of schizophrenia. Biol Psychiatry 59:546–554.

Paquette M (2007) Morgellons: disease or delusions? Perspect Psychiatr Care 43:67–68.

Pardo CA, Vargas DL, Zimmerman AW (2005) Immunity, neuroglia and neuroinflammation in autism. International Review of Psychiatry 17:485–495.

Parnas J, Schulsinger F, Teasdale TW, Schulsinger H, Feldman PM, Mednick SA (1982) Perinatal complications and clinical outcome within the schizophrenia spectrum. Br J Psychiatry 140:416–420.

Patterson PH (2002) Maternal infection: window on neuroimmune interactions in fetal brain development and mental illness. Curr Opin Neurobiol 12:115–118.

Patterson PH (2007) Neuroscience. Maternal effects on schizophrenia risk. Science 318:576–577.

Pearce BD, Steffensen SC, Paoletti AD, Henriksen SJ, Buchmeier MJ (1996) Persistent dentate granule cell hyperexcitability after neonatal infection with lymphocytic choriomeningitis virus. J Neuroscience 16:220–228.

Pearce BD, Valadi NM, Po CL, Miller AH (2000) Viral infection of developing GABAergic neurons in a model of hippocampal disinhibition. Neuroreport 11:2433–2439.

Pearce BD (2001) Schizophrenia and viral infection during neurodevelopment: A focus on mechanisms. Mol Psychiatry 6:634–646.

Pearce BD (2003) Can a Virus Cause Schizophrenia: Facts and Hypotheses. Kluwer Academic Publishers, Boston.

Peterson BS, Leckman JF, Tucker D, Scahill L, Staib L, Zhang H, King R, Cohen DJ, Gore JC, Lombroso P (2000) Preliminary findings of antistreptococcal antibody titers and basal ganglia volumes in tic, obsessive-compulsive, and attention deficit/hyperactivity disorders. Arch Gen Psychiatry 57:364–372.

Petronis A, Paterson AD, Kennedy JL (1999) Schizophrenia: an epigenetic puzzle? Schizophr Bull 25:639–655.

Pewter SM, Williams WH, Haslam C, Kay JM (2007) Neuropsychological and psychiatric profiles in acute encephalitis in adults. Neuropsychological Rehabilitation 17:478–505.

Pickett J, London E (2005) The neuropathology of autism: a review. J Neuropathol Exp Neurol 64:925–935.

Pletnikov MV, Moran TH, Carbone KM (2002) Borna disease virus infection of the neonatal rat: developmental brain injury model of autism spectrum disorders. Front Biosci 7:d593–d607.

Plioplys AV, Greaves A, Kazemi K, Silverman E (1994) Lymphocyte function in autism and Rett syndrome. Neuropsychobiology 29:12–16.

Preti A, Miotto P, Zen T (1998) Perinatal complications, genetic risk and schizophrenia. Acta Psychiatr Scand 97:381–383.

Raison CL, Capuron L, Miller AH (2006) Cytokines sing the blues: inflammation and the pathogenesis of depression. Trends Immunol 27:24–31.

Ramirez JC, Fairen A, Almendral JM (1996) Parvovirus minute virus of mice strain i multiplication and pathogenesis in the newborn mouse brain are restricted to proliferative areas and to migratory cerebellar young neurons. J Virol 70:8109–8116.

Rapaport MH, Torrey EF, McAllister CG, Nelson DL, Pickar D, Paul SM (1993) Increased serum soluble interleukin-2 receptors in schizophrenic monozygotic twins. Eur Arch Psychiatry Clin Neurosci 243:7–10.

Rapaport MH, Lohr JB (1994) Serum-soluble interleukin-2 receptors in neuroleptic-naive schizophrenic subjects and in medicated schizophrenic subjects with and without tardive dyskinesia. Acta Psychiatr Scand 90:311–315.

Rapaport MH, McAllister CG, Kim YS, Han JH, Pickar D, Nelson DL, Kirch DG, Paul SM (1994) Increased serum soluble interleukin-2 receptors in Caucasian and Korean schizophrenic patients. Biol Psychiatry 35:767–771.

Rosen NJ, Yoshida CK, Croen LA (2007) Infection in the first 2 years of life and autism spectrum disorders. Pediatrics 119:e61–e69.

Rosenberg A, Noble EP (1994) Ethanol attenuation of ganglioside sialylation and neuritogenesis. Alcohol 11:565–569.

Rzhetsky A, Wajngurt D, Park N, Zheng T (2007) Probing genetic overlap among complex human phenotypes. Proc Natl Acad Sci U S A 104:11694–11699.

Sacker A, Done DJ, Crow TJ (1996) Obstetric complications in children born to parents with schizophrenia: a meta-analysis of case-control studies. Psychol Med 26:279–287.

Salvatore M, Morzunov S, Schwemmle M, Lipkin WI, Bunney WE, Cotman CW, Even C, Hatalski CG, Potkin SG, Portlace ML, Nichol ST, Tourtelloutte WW, Riederer P, Ter Meulen V (1997) Borna disease virus in brains of North American and European people with schizophrenia and bipolar disorder. Lancet 349:1813–1814.

Schechter R, Grether JK (2008) Continuing increases in autism reported to California's developmental services system: mercury in retrograde. Arch Gen Psychiatry 65:19–24.

Schiefermeier M, Kollegger H, Madl C, Schwarz C, Holzer M, Kofler J, Sterz F (2000) Apolipoprotein E polymorphism: survival and neurological outcome after cardiopulmonary resuscitation. Stroke 31:2068–2073.

Schmidt HD, Duman RS (2007) The role of neurotrophic factors in adult hippocampal neurogenesis, antidepressant treatments and animal models of depressive-like behavior. Behav Pharmacol 18:391–418.

Schwarz MJ, Chiang S, Muller N, Ackenheil M (2001) T-helper-1 and T-helper-2 responses in psychiatric disorders. Brain Behav Immun 15:340–370.

Shanks N, Lightman SL (2001) The maternal-neonatal neuro-immune interface: are there long-term implications for inflammatory or stress-related disease? J Clin Invest 108:1567–1573.

Sharma AK, Valadi N, Miller AH, Pearce BD (2002) Neonatal viral infection decreases neuronal progenitors and impairs adult neurogenesis in the hippocampus. Neurobiol Dis 11:246–256.

Shi L, Fatemi SH, Sidwell RW, Patterson PH (2003) Maternal influenza infection causes marked behavioral and pharmacological changes in the offspring. J Neurosci 23:297–302.

Shi L, Tu N, Patterson PH (2005) Maternal influenza infection is likely to alter fetal brain development indirectly: the virus is not detected in the fetus. Int J Dev Neurosci 23:299–305.

Silverman MN, Pearce BD, Miller AH (2002) Cytokines and HPA axis regulation. In: Kronfol Z, (ed) Cytokines and Mental Health. Kluwer, Norwell.

Silverman MN, Macdougall MG, Hu F, Pace TW, Raison CL, Miller AH (2007) Endogenous glucocorticoids protect against TNF-alpha-induced increases in anxiety-like behavior in virally infected mice. Mol Psychiatry 12:408–417.

Skallova A, Novotna M, Kolbekova P, Gasova Z, Vesely V, Sechovska M, Flegr J (2005) Decreased level of novelty seeking in blood donors infected with Toxoplasma. Neuro Endocrinol Lett 26:480–486.

Smith SE, Li J, Garbett K, Mirnics K, Patterson PH (2007) Maternal immune activation alters fetal brain development through interleukin-6. J Neurosci 27:10695–10702.

Song C, Lin A, Kenis G, Bosmans E, Maes M (2000) Immunosuppressive effects of clozapine and haloperidol: enhanced production of the interleukin-1 receptor antagonist. Schizophr Res 42:157–164.

Steiner G, Studnicka-Benke A, Witzmann G, Hofler E, Smolen J (1995) Soluble receptors for tumor necrosis factor and interleukin-2 in serum and synovial fluid of patients with rheumatoid arthritis, reactive arthritis and osteoarthritis. J Rheumatol 22:406–412.

Susser ES, Lin SP (1992) Schizophrenia after prenatal exposure to the Dutch hunger winter of 1944–1945. Arch Gen Psychiatry 49:983–988.
Susser E, Neugebauer R, Hoek HW, Brown AS, Lin S, Labovitz D, Gorman JM (1996) Schizophrenia after prenatal famine: Further evidence. Arch Gen Psychiatry 53:25–31.
Swedo SE, Grant PJ (2005) Annotation: PANDAS: a model for human autoimmune disease. J Child Psychol Psychiatry 46:227–234.
Szekely CA, Breitner JC, Fitzpatrick AL, Rea TD, Psaty BM, Kuller LH, Zandi PP (2008) NSAID use and dementia risk in the Cardiovascular Health Study: role of APOE and NSAID type. Neurology 70:17–24.
Tao R, Li C, Zheng Y, Qin W, Zhang J, Li X, Xu Y, Shi YY, Feng G, He L (2007) Positive association between SIAT8B and schizophrenia in the Chinese Han population. Schizophr Res 90:108–114.
Toro CT, Deakin JF (2007) Adult neurogenesis and schizophrenia: a window on abnormal early brain development? Schizophr Res 90:1–14.
Torrey EF, Peterson MR (1973) Slow and latent viruses in schizophrenia. Lancet 2:22–24.
Torrey EF, Peterson MR (1976) The viral hypothesis of schizophrenia. Schizophr Bull 2:136–146.
Torrey EF, Yolken RH (1999) Familial and genetic mechanisms in schizophrenia. Brain Research Reviews 31:113–117.
Torrey EF, Yolken RH (2001) The schizophrenia-rheumatoid arthritis connection: infectious, immune, or both? Brain Behav Immun 15:401–410.
Torrey EF, Yolken RH (2003) Toxoplasma gondii and schizophrenia. Emerg Infect Dis 9:1375–1380.
Tsuang M (2000) Schizophrenia: genes and environment. Biol Psychiatry 47:210–220.
Urakubo A, Jarskog LF, Lieberman JA, Gilmore JH (2001) Prenatal exposure to maternal infection alters cytokine expression in the placenta, amniotic fluid, and fetal brain. Schizophr Res 47:27–36.
van den Pol AN, Reuter JD, Santarelli JG (2002) Enhanced cytomegalovirus infection of developing brain independent of the adaptive immune system. J Virol 76:8842–8854.
van den Pol AN (2006) Viral infections in the developing and mature brain. Trends Neurosci 29:398–406.
van Gent T, Heijnen CJ, Treffers PD (1997) Autism and the immune system. J Child Psychol Psychiatry 38:337–349.
Vastag B (2001) Congressional autism hearings continue: no evidence MMR vaccine causes disorder. JAMA 285:2567–2569.
Verdoux H, Geddes JR, Takei N, Lawrie SM, Bovet P, Eagles JM, Heun R, McCreadie RG, McNeil TF, O'Callaghan E, Stober G, Willinger MU, Wright P, Murray RM (1997) Obstetric complications and age at onset in schizophrenia: an international collaborative meta-analysis of individual patient data. Am J Psychiatry 154:1220–1227.
Verma S, Cikurel K, Koralnik IJ, Morgello S, Cunningham-Rundles C, Weinstein ZR, Bergmann C, Simpson DM (2007) Mirtazapine in progressive multifocal leukoencephalopathy associated with polycythemia vera. J Infect Dis 196:709–711.
Vilensky JA, Foley P, Gilman S (2007) Children and encephalitis lethargica: a historical review. Pediatr Neurol 37:79–84.
Wakefield AJ (1999) MMR vaccination and autism. Lancet 354:949–950.
Walker EF, Diforio D, Baum K (1999) Developmental neuropathology and the precursors of schizophrenia. Acta Psychiatr Scand, Suppl 395:12–19.
Wang XF, Wang D, Zhu W, Delrahim KK, Dolnak D, Rapaport MH (2003) Studies characterizing 60 kda autoantibodies in subjects with schizophrenia. Biol Psychiatry 53:361–375.
Ward KN, Bryant NJ, Andrews NJ, Bowley JS, Ohrling A, Verity CM, Ross EM, Miller E (2007) Risk of serious neurologic disease after immunization of young children in Britain and Ireland. Pediatrics 120:314–321.
Weinberger DR (1987) Implications of normal brain development for the pathogenesis of schizophrenia. Arch Gen Psychiatry 44:660–669.
Weis S, Llenos IC, Sabunciyan S, Dulay JR, Isler L, Yolken R, Perron H (2007) Reduced expression of human endogenous retrovirus (HERV)-W GAG protein in the cingulate gyrus and hippocampus in schizophrenia, bipolar disorder, and depression. J Neural Transm 114:645–655.

Weiss LA, Shen Y, Korn JM, Arking DE, Miller DT, Fossdal R, Saemundsen E, Stefansson H, Ferreira MA, Green T, Platt OS, Ruderfer DM, Walsh CA, Altshuler D, Chakravarti A, Tanzi RE, Stefansson K, Santangelo SL, Gusella JF, Sklar P, Wu BL, Daly MJ (2008) Association between Microdeletion and Microduplication at 16p11.2 and Autism. N Engl J Med 358:667–675.

Whitley RJ (1997) Herpes simplex virus. In: Scheld WM, Whitley RJ, Durack DT, (eds) Infections of the Central Nervous System. Lippencott-Raven, Philadelphia, pp 73–89.

Williamson WD (1994) The longitudinal assessment of congenitally infected infants. Semin Pediatr Neurol 1:58–62.

Wright P, Murray RM (1993) Schizophrenia: prenatal influenza and autoimmunity. Ann Med 25:497–502.

Yolken RH, Torrey EF (1995) Viruses, schizophrenia, and bipolar disorder. Clin Microbiol Rev 8:131–145.

Yolken RH, Karlsson H, Yee F, Johnston-Wilson NL, Torrey EF (2000) Endogenous retroviruses and schizophrenia. Brain Research Reviews 31:193–199.

Yoon BH, Romero R, Park JS, Kim CJ, Kim SH, Choi JH, Han TR (2000) Fetal exposure to an intra-amniotic inflammation and the development of cerebral palsy at the age of three years. Am J Obstet Gynecol 182:675–681.

Zalcman SS, Siegel A (2006) The neurobiology of aggression and rage: role of cytokines. Brain Behav Immun 20:507–514.

Zerrate MC, Pletnikov M, Connors SL, Vargas DL, Seidler FJ, Zimmerman AW, Slotkin TA, Pardo CA (2007) Neuroinflammation and behavioral abnormalities after neonatal terbutaline treatment in rats: implications for autism. J Pharmacol Exp Ther 322:16–22.

Zimmerman AW, Connors SL, Matteson KJ, Lee LC, Singer HS, Castaneda JA, Pearce DA (2007) Maternal antibrain antibodies in autism. Brain Behav Immun 21:351–357.

Zuckerman L, Rehavi M, Nachman R, Weiner I (2003) Immune activation during pregnancy in rats leads to a postpubertal emergence of disrupted latent inhibition, dopaminergic hyperfunction, and altered limbic morphology in the offspring: a novel neurodevelopmental model of schizophrenia. Neuropsychopharmacology 28:1778–1789.

Microglial Cells and Inflammatory Cytokines in the Aged Brain

Amy F. Richwine and Rodney W. Johnson

Abstract Brain microglial cells are ordinarily quiescent, but when stimulated can transition to a "primed" or activated state. Both primed and activated microglia express markers that suggest activation, but only activated microglia produce appreciable levels of inflammatory cytokines. Primed microglia, however, are hyper-responsive to signals from the peripheral immune system and thus can produce an exaggerated cytokine response when provoked. The potential for primed microglia to mount an exaggerated response is important because inflammatory cytokines mediate the sickness behavior syndrome, induce deficits in cognition, and are involved in chronic neurodegenerative diseases. One event that may prime microglial cells for an exaggerated response is aging. Signs of neuroinflammation emerge in healthy aged subjects and new findings suggest that aged individuals suffer an exaggerated neuroinflammatory response during peripheral infection. The exaggerated neuroinflammatory response is accompanied by severe, longer-lasting behavioral deficits and may cause acute cognitive disorders that are often reported in elderly patients with peripheral infections.

Key words Aging · Behavior · Brain · Cytokines · Inflammation · Microglia

1 Introduction

During peripheral infection, the immune system conveys a message to the brain and microglial cells respond and produce inflammatory cytokines that help coordinate a behavioral response that is ordinarily adaptive (Dantzer et al., 2008). Excessive production of inflammatory cytokines in the brain, however, can produce severe

R.W. Johnson (✉)
Department of Animal Sciences, Integrative Immunology and Behavior, University of Illinois, 4 Animal Science Laboratory, 1207 West Gregory Drive, Urbana, IL 61801, USA
e-mail: rwjohn@uiuc.edu

Supported by NIH grants AG16710, AG023580, MH069148 and DA024443

A. Siegel, S.S. Zalcman (eds.), *The Neuroimmunological Basis of Behavior and Mental Disorders*, DOI 10.1007/978-0-387-84851-8_19,

behavioral deficits and promote neurotoxicity. In the murine ME7 model of prion disease, for example, activation of the peripheral innate immune system with lipopolysaccharide (LPS) induced an aberrant inflammatory cytokine response by brain microglial cells, excessive sickness behavior, and accelerated progression of the neurodegenerative disease (Combrinck et al., 2002; Cunningham et al., 2005). A similar interaction between peripheral infection and other chronic neurodegenerative diseases in mice has been reported (Nguyen et al., 2004) and consequently, it was proposed that elements unique to the neurodegenerative disease provide a priming stimulus for microglia and the signal from the peripheral immune system during an infection provides a secondary triggering stimulus (Perry et al., 2007). The end result of combining the two stimuli is an overall response that is greater than the sum of the responses to each stimulus alone. This interaction may explain why infection is a risk factor for relapse in multiple sclerosis patients or dementia in patients with Alzheimer's disease (Sibley et al., 1985; Holmes et al., 2003).

Similar to chronic neurodegenerative disease, aging was recently proposed to prime microglial cells (Perry, 2004; Godbout et al., 2005; Johnson and Godbout, 2006). Major histocompatibility complex (MHC) class II, a marker for activated microglia, was increased in brain of healthy aged mice (Godbout et al., 2005) and peripheral injection of LPS resulted in an exaggerated inflammatory cytokine response in the aged brain (Godbout et al., 2005; Chen et al., 2008). Similar findings have been reported in older rats inoculated with *Escherichia coli* (Barrientos et al., 2006). The exaggerated inflammatory cytokine response in the brain is not due to the message from the peripheral immune system, as intracerebroventricular (ICV) injection of LPS also elicited a heightened response by the central inflammatory cytokine compartment in old mice (Huang et al., 2008). Consistent with an exaggerated inflammatory cytokine response in the brain, anorexia, depression-like behavior, and deficits in hippocampal-dependent learning and memory are more evident in old mice than in young adults after peripheral LPS administration (Godbout et al., 2005; Chen et al., 2007; Godbout et al., 2008). Importantly, we recently found IL-1ra given ICV to old mice reduced the depression in social behavior caused by peripheral LPS and facilitated recovery (unpublished data).

These observations from rodent studies are noteworthy because acute cognitive impairments are common in elderly human patients and often occur in association with an infection that is unrelated to the central nervous system (CNS; Wofford et al., 1996; Chiovenda et al., 2002). Demented elderly patients are routinely screened for bladder infections because of the close association between infection and cognitive disorders. And a striking characteristic of pneumonia in the elderly is that it commonly presents clinically as delirium (Janssens and Krause, 2004). Thus, the inflammatory response which is usually adaptive, may elicit severe behavior deficits in elderly individuals which prolong recovery. How aging affects the central cytokine compartment and behavior is critical given the rapid growth of the 65 and older segment of the population. The effects of aging on neuroinflammation and behavior have been recently reviewed (Godbout and Johnson, 2006; Johnson and Godbout, 2006), although little attention was given to microglial cell regulation.

Therefore, the purpose of this chapter is to provide a brief background on microglial cell regulation, discuss evidence that suggests aging "primes" microglial cells, and present important new information that links microglial cell activity to behavioral pathology in the aged.

2 Microglial Cell Activity

The communication between the immune system and the CNS and the propagation of the inflammatory response through the brain parenchyma is dependent on resident mononuclear phagocytic cells in the brain (i.e., microglial cells) that continuously search for pathogens or injuries throughout the microenvironment of the CNS and respond to signals from the peripheral immune system (Davalos et al., 2005; Nimmerjahn et al., 2005; Dantzer et al., 2008). Activation of microglia, whether by brain trauma or infection, initiates the brain's innate immune response and inflammatory cytokine production. Before activation, microglia in the normal healthy brain have a highly branched morphology and have low levels of cell surface antigens such as CD45, CD68, and MHC class I and II. Despite being considered in the "resting-state" when these cell surface antigens are downregulated, microglia are still rather active and move about the brain parenchyma in a random manner. Using in vivo two-photon imaging, processes from resting microglia were reported to continuously extend and retract with filopodia-like protrusions (Nimmerjahn et al., 2005). Thus microglia are actively moving in search of pathogens or injuries within the CNS even when in a resting state.

Microglia are sensitive to changes in the CNS and can rapidly become activated. When provoked, microglia upregulate expression of cell surface antigens, including complement receptors, MHC class I and II, and Toll Like Receptors (TLRs; Perry and Gordon, 1988). Additionally, morphology of microglia change upon activation with processes that become shorter and stouter while the soma increases in size (Long et al., 1998). Along with these physical changes in shape, it has been proposed that phenotypic changes occur when microglia become activated (such as switching between inflammatory state to an anti-inflammatory state) and that different insults trigger the release of different cytokines and other inflammatory proteins (Perry et al., 2007). These phenotypic changes, however, are not accompanied with noticeable changes in morphology.

One proposed mechanism for maintaining microglia in a resting state is through cell–cell communication with neurons, and recent studies have begun to unravel the mechanisms behind this interaction. It is thought that proteins on or secreted by neurons interact with cell surface molecules expressed on resting microglia to suppress activation (Hoek et al., 2000; Wright et al., 2000; Mott et al., 2004). For example, the CD200 receptor is located on neurons while its ligand is present on microglia, and in mice deficient in CD200, microglia have less ramified morphology and increased expression of CD11b and CD45 suggesting an activated state (Hoek et al., 2000). During experimental autoimmune encephalomyelitis, mice deficient in CD200 receptor have an increase in inducible nitric oxide synthase compared to wild-type mice (Wright et al., 2000). Also, CD45 has been shown to suppress inflammatory cytokine production by reducing microglia activation demonstrated by a decrease in tumor necrosis factor-α (TNFα) production and mitogen-activated protein kinase activation, an important cell signaling molecule that is involved in the synthesis of cytokines (Tan et al., 2000b, a). Recently it was shown that neurons secrete CD22, a ligand for CD45 (Mott et al., 2004), and when primary microglia cultures were treated with neuron conditioned media, LPS-induced production of TNFα by microglia was inhibited which was attributed to the presence of CD22 secreted by neurons. Importantly, antibodies against CD22 without neuron conditioned media had no effect on TNFα production. Furthermore, microglial cells cultured from CD45-deficient mice had greater production of TNFα, nitric oxide synthesis, and neuronal injury after treatment with amyloid β (Tan et al., 2000b). These studies suggest an important role for communication between neurons and microglia to maintain a quiescent environment and a decrease in these cell surface molecules may play an important role in the unique inflammatory gene transcription profile seen during aging.

3 Microglia in the Aged Brain

In the adult brain there is balance between inflammatory and anti-inflammatory mediators, but aging appears to tip the scale towards inflammation. A heightened inflammatory state may partially explain why age is a risk factor for certain neurodegenerative diseases and cognitive impairment. Studies investigating cellular changes in the brain have reported age-related morphological changes in microglia (Streit, 2004; Streit et al., 2004). By using flow cytometry to sort microglia expressing green florescent protein, microglia from aged mice were characterized by having decreased processes, presence of lipofuscins, and altered granularity (Sierra et al., 2007). Similar dystrophic characteristics have also been reported in brains from human subjects (Streit et al., 2004). By comparing a 68-year-old human brain to 38-year-old brain, a 10-fold increase in the number of microglia with abnormalities was seen in the aged brain compared to the young adult. As described above, these changes are different from activated microglia that have short, stouter processes and a larger soma. The dystrophic microglia in the aged brain suggests a reactive state, and are therefore, thought to be more responsive to a secondary stimulus compared to microglia from a young brain.

Along with changes in morphology, expression of cell surface molecules, similar to those increased on activated microglia, are increased on microglia from healthy aged brains of humans, nonhuman primates, canines, and rodents (Perry et al., 1993; Tafti et al., 1996; Godbout et al., 2005; Frank et al., 2006). For example, MHC class II, a molecule responsible for presenting foreign antigens to CD4+ T cells, and complement receptors, which allow phagocytic cells to identify complement proteins attached to a pathogen to be phagocytized, have been reported to increase (Perry et al., 1993; Streit and Sparks, 1997; Godbout et al., 2005; Frank et al., 2006). Furthermore, TLR4 and CD14 mRNA, important cell surface molecules for detecting LPS, have also been reported to be elevated in the aged brain (Huang et al., 2008; Letiembre et al., 2007). Despite the increases in multiple cell surface molecules involved in the inflammatory response, primed or aged microglia do not appear to produce substantial levels of inflammatory cytokines unless further provoked.

4 Are Microglia "Primed" in Aged Brain?

Microglia from an aged brain are similar to primed microglia in the sense that both produce low levels of inflammatory cytokines and have an increase in some cell surface molecules compared to those in a resting state. However, a classical definition of a primed cell, whether it be macrophage or microglia, is when a host-derived factor causes a cell to respond more rapidly and to a greater degree to a secondary stimulus, and in the absence of this factor the cell would either be unresponsive or have only a modest response to the secondary stimulus (Schroder et al., 2006). For example, primary murine bone marrow-derived macrophages had minimal nitric oxide production when cultured with LPS alone, but when cultured 2–16 h with interferon-γ (IFNγ) followed by LPS, there was a substantial increase in nitric oxide (Sester et al., 2005). The function of IFNγ in this case is to sensitize macrophages to inflammatory insults, and in fact, INFγ was originally named macrophage-activating factor (Schroder et al., 2006). IFNγ is produced by several inflammatory cells including natural killer cells, Th1 cells, macrophages, and microglia, which allow it to send a priming signal to several cells throughout the host.

Microglia from aged mice appear to be similar to a primed microglia because of the heightened inflammatory response in the brain after a peripheral injection of LPS compared to young cohorts (Godbout et al., 2005; Chen et al., 2008). Furthermore, glial cells and brain slices from aged mice produce more IL-6 in response to LPS in vitro than glia and brain slices from younger adults (Ye and Johnson, 1999, 2001a). Aged mice have been reported to have detectable mRNA and protein levels of IFNγ in endothelial cells of the choroid plexus and microvessels of the cortex and cerebellum while levels were undetectable in adult mice (Wei et al., 2000). However, further studies are needed to truly define by immunological standards if microglia in the aged brain are primed. For example, the heightened inflammatory cytokine response could be due to an increase in the number of microglial cells, as opposed to the responsiveness of the individual cell. We recently reported increased microglial staining in brains of old mice (Chen et al., 2008). Furthermore, peritoneal macrophages collected from aged rats produced significantly less TNFα in response to IFNγ and LPS compared to macrophages collected from adult rats (Davila et al., 1990). This experiment suggests that in the periphery the priming effect of IFNγ is actually disrupted during aging.

5 Does Peripheral Infection Lead to Behavioral Pathology in the Aged?

Even if microglia in the aged brain are determined not to be primed, they still have an amplified response after activation of the peripheral innate immune system, which would suggest that aging sensitizes microglia (or other cytokine-producing cells) to inflammatory insults. This is of great importance because systemic infections and inflammation can increase the risk for neurodegenerative diseases suggesting that the brain becomes more vulnerable to neurotoxicity (Sheffield and Berman, 1998). For example, in the ME7 murine model of prion disease which causes neurodegeneration, microglia have an increase in CD68 and ramified morphology (Cunningham et al., 2003). Following a peripheral injection of LPS, there was a greater decrease in locomotor activity in ME7 mice compared to control mice injected with LPS (Combrinck et al., 2002; Cunningham et al., 2005). This was associated with heightened brain levels of IL-1β in ME7-LPS mice compared to ME7-saline and control-LPS mice. Furthermore, in ME7-LPS animals, the systemic challenge increased neuronal apoptosis and synaptic loss (Cunningham et al., 2005). These data indicate that chronic neurodegenerative diseases can sensitize microglia and that if the peripheral immune system is activated, brain inflammation is amplified and neurotoxicity can occur.

Because microglia in the aged brain have increased expression of cell surface molecules similar to microglia in mice injected with ME7 (Betmouni and Perry, 1999), it is reasonable to postulate that microglia become sensitized with age and have a similar heightened response to peripheral LPS. In a microarray study, old mice were found to have a unique gene expression profile in the brain that indicated increased inflammation (Godbout et al., 2005). Quantitative real-time PCR was used to measure MHC class II expression, which was increased due to age and was accompanied with an increase in IL-6 and IL-1β mRNA. After a peripheral injection of LPS in adult and aged mice, MHC class II expression was unchanged, but IL-6 and IL-1β were increased 43-fold in the aged brain, while the adults had a 22-fold increase 4 h after an injection. To more specifically investigate the microenvironment of hippocampal neurons after LPS, laser capture microdissection was used to separate the neuronal layers of the hippocampus from surrounding tissue. It was found that 4 h after activation of the peripheral innate immune system with LPS, IL-1β, IL-6, and TNFα were markedly higher in the neuronal cell layers of the Cornu Ammonis (CA) regions and dentate gyrus in old mice compared to young adults (Chen et al., 2007). Other studies have demonstrated that the LPS-induced increase in inflammatory cytokines is prolonged in hippocampus of aged mice (Richwine et al., submitted).

Accompanying the heightened neuroinflammation, LPS induced a prolonged depression in social investigation and locomotor activity in aged mice (Godbout et al., 2005). At 24 h, adult mice had fully recovered whereas aged mice still showed reductions in these behaviors. Also, aged mice were anorectic longer and lost more body weight after LPS. A recent study reported aged mice to express depression-like behavior 3 days after peripheral LPS administration when other signs of illness had dissipated (Godbout et al., 2008). In another study, LPS was injected ICV in young and old mice to determine if the prolonged sickness behavior was caused by amplified cytokine production from the brain and not the peripheral response (Huang et al., 2008). Similar to a peripheral injection of LPS, aged mice receiving LPS ICV had a prolonged decrease in food intake, locomotor activity, and social exploration which was accompanied with a greater induction of IL-1β, IL-6, and TNFα in the cerebellum and hippocampus. This study shows that direct activation of the brain's innate immune system in aged mice resulted in a heightened inflammatory response and the peripheral innate immune system is not necessary for the increased inflammatory response seen in the aged brain. Thus, peripheral infection causes a more intense, longer-lasting inflammatory cytokine response in the aged brain, perhaps due to an age-related increase in microglial cell sensitivity.

It has long been known that a peripheral infection and the consequent activation of immune cells can lead to a disruption in cognitive processing in healthy humans and animals (Gibertini et al., 1995; Gibertini, 1998; Pugh et al., 1998; Arai et al., 2001; Reichenberg et al., 2001; Barrientos et al., 2006). For example, in rats, peripheral LPS administration after conditioning caused deficits in contextual fear conditioning which was blocked when the anti-inflammatory cytokine IL-1ra was administered immediately following LPS (Pugh et al., 1998). Similarly, animals infected with gram negative bacterium *Legionella pneumophila* (Lp) showed significantly longer latencies and distance to find the platform in the Morris water maze compared to noninfected animals, and when neutralizing antibodies against IL-1β were administered, these deficits caused by Lp were eliminated (Gibertini, 1996). In a more recent study, mice lacking a functional IL-6 gene (IL-6$^{-/-}$ mice) were tested in the matching-to-place version of the Morris water maze to investigate the role of IL-6 in cognitive impairments associated with peripheral infections (Sparkman et al., 2006). After LPS treatment, IL-6$^{+/+}$ mice exhibited deficits in this task while IL-6$^{-/-}$ mice did not. Interestingly, IL-6$^{+/+}$ mice had a greater expression of the inflammatory cytokines IL-1β and TNFα in the hippocampus compared to IL-6$^{-/-}$ mice. These studies and others demonstrate a connection between inflammatory cytokines and disruption in hippocampal-dependent learning and memory.

Because neuroinflammation can influence cognitive processing, it is not surprising that the age-associated changes in cytokines during inflammatory insults result in cognitive impairments (Barrientos et al., 2006; Chen et al., 2008). When adult and aged mice were tested in a version of the water maze that required mice to integrate new information with a pre-existing schema, aged mice were more vulnerable to disruption in cognitive processing compared to young adults after a peripheral injection of LPS (Chen et al., 2008). This study is supported by another that showed older rats inoculated with *Escherichia coli* (Barrientos et al., 2006) showed impairments in a spatial reference version of the water maze and impaired contextual fear conditioning compared to younger rats. In both studies, activation of the peripheral innate immune system led to higher levels of inflammatory cytokines in the hippocampus in aged rodents compared to adults (Barrientos et al., 2006; Chen et al., 2008). These are important observations because cognitive impairments can hinder

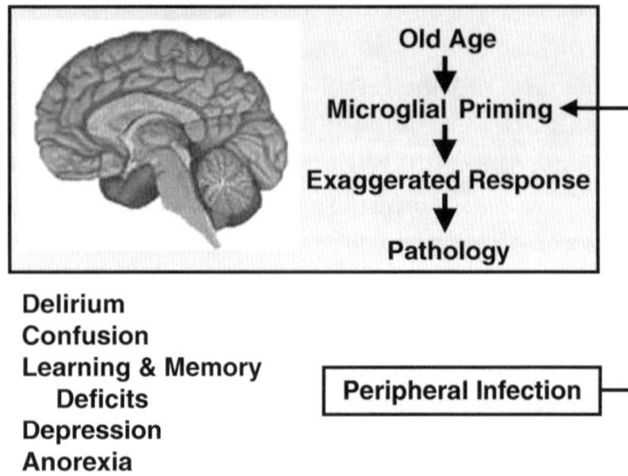

Fig. 1 Aging may "prime" microglial cells so they overreact to signals from the peripheral immune system. Excessive production of inflammatory cytokines can cause behavioral pathology, further complicating healthcare

self-care and, in turn, increase hospitalization and delay recovery (Johnston et al., 1987). Thus, age-associated changes in brain cytokines during a systemic infection can cause more severe cognitive disorders that impact on the overall health of elderly patients (Fig. 1).

6 Does a Deficit in Interleukin-10 Sensitize the Brain Cytokine Compartment in the Aged?

Originally described as a cytokine synthesis inhibitory factor, IL-10 is capable of inhibiting inflammation by reducing synthesis of inflammatory cytokines, suppressing cytokine receptor expression, and decreasing receptor activation (Moore et al., 2001; Strle et al., 2001). IL-10 can suppress synthesis of IL-6, IL-1β, and TNF-α by degrading cytokine mRNA or by inhibiting the activation of NFκB (Bogdan et al., 1992; Heyen et al., 2000). In primary murine microglia cultures, addition of recombinant IL-10 inhibited LPS-induced activation of NFκB and suppressed IL-6 production (Heyen et al., 2000). IL-10 has been reported to increase other anti-inflammatory cytokines like IL-1ra and SOCS (suppressor of cytokine synthesis; Howard et al., 1992; Kasai et al., 1997; Strle et al., 2001). Furthermore, IL-10 can suppress expression of inflammatory cell surface molecules present on activated microglia, such as MHC class II and co-stimulatory B7–2 signals (Menendez Iglesias et al., 1997; O'Keefe et al., 1999). Altogether, these studies indicate that IL-10 is a critical anti-inflammatory protein that can reduce inflammation via multiple paths.

By suppressing inflammatory cytokines, IL-10 can mediate sickness behaviors, fever, and increase survival during inflammatory insults (Gerard et al., 1993; Howard et al., 1993; Di Santo et al., 1995; Bluthe et al., 1999; Leon et al., 1999). For example, when IL-10 was administered ICV prior to a peripheral LPS injection it dose-dependently attenuated deficits in social exploration (Bluthe et al., 1999); and after ICV LPS injection it completely inhibited brain IL-1β and TNFα production

(Di Santo et al., 1995). Conversely, in IL-$10^{-/-}$ mice which lack a functional IL-10 gene, IL-6, IL-1β, and TNFα in the periphery and in discrete brain regions were higher after peripheral LPS injection (Leon et al., 1999; Krzyszton et al., submitted). The fatigue, deficits in motor coordination, and sickness behavior were also exacerbated in IL-$10^{-/-}$ mice after LPS injection (Krzyszton et al., submitted), not unlike what is seen in old mice treated with LPS (Godbout et al., 2005; Johnson and Godbout, 2006; Huang et al., 2008). Taken together, these studies demonstrate the importance of IL-10 in mediating inflammatory cytokines in the brain and behavior during peripheral infection.

During healthy aging, high IL-10 production may increase the capacity to live to maximum lifespan. In a study of centenarians, the polymorphism of the IL-10 gene promoter that causes IL-10 to be produced at high levels was more prominent compared to the polymorphism that leads to low levels of IL-10 (Caruso et al., 2004). Since centenarians are the best example of successful aging, the high levels of IL-10 might be vital to healthy aging. However, our lab and others have demonstrated that IL-10 decreases in the brain of aged rodents (Smith et al., 1999; Ye and Johnson, 2001a; Frank et al., 2006). Coronal brain slices taken from old mice secreted approximately 25% less IL-10 than coronal brain slices taken from adult mice, and after treating with LPS, the tissue from old mice secreted approximately 35% less IL-10 compared to the tissue taken from adults (Ye and Johnson, 2001a). A similar pattern of IL-10 production was observed in primary glial cell cultures established from young and old mice (Ye and Johnson, 2001a); and IL-10 mRNA was found to be reduced in the hippocampus of aged rats (Frank et al., 2006). This decrease in IL-10 may play an important role in the age-related increase in brain inflammatory cytokines since recombinant IL-10 reduced NFκB DNA binding activity, IL-6 mRNA expression, and protein secretion in glial cells from aged mice (Ye and Johnson, 2001b). Since IL-10 can inhibit the synthesis of inflammatory cytokines, the decrease in IL-10 during aging may permit increased production of inflammatory cytokines in the healthy brain.

7 Conclusion

The elderly are often immunosuppressed and susceptible to infection. Thus, the idea that aging may sensitize or prime microglia so that an exaggerated inflammatory cytokine response occurs in the brain when the peripheral immune system is stimulated is vitally important. The magnitude and duration of the behavioral response appears to be proportional to the level of cytokines produced in the brain so the immune-to-brain signaling that induces "normal" sickness behavior in the young, may induce behavioral pathology in the old (Fig. 1). A discordant response of brain microglial cells to signals from the peripheral immune system may underlie some of the acute cognitive disorders that complicate health care for geriatric patients. Therefore, treatments to mitigate the brain's inflammatory cytokine response during peripheral infection should be investigated in geriatric patients.

References

Arai K, Matsuki N, Ikegaya Y, Nishiyama N (2001) Deterioration of spatial learning performances in lipopolysaccharide-treated mice. Jpn J Pharmacol 87:195–201.

Barrientos RM, Higgins EA, Biedenkapp JC, Sprunger DB, Wright-Hardesty KJ, Watkins LR, Rudy JW, Maier SF (2006) Peripheral infection and aging interact to impair hippocampal memory consolidation. Neurobiol Aging 27:723–732.

Betmouni S, Perry VH (1999) The acute inflammatory response in CNS following injection of prion brain homogenate or normal brain homogenate. Neuropathol Appl Neurobiol 25:20–28.

Bluthe RM, Castanon N, Pousset F, Bristow A, Ball C, Lestage J, Michaud B, Kelley KW, Dantzer R (1999) Central injection of IL-10 antagonizes the behavioural effects of lipopolysaccharide in rats. Psychoneuroendocrinology 24:301–311.

Bogdan C, Paik J, Vodovotz Y, Nathan C (1992) Contrasting mechanisms for suppression of macrophage cytokine release by transforming growth factor-beta and interleukin-10. J Biol Chem 267:23301–23308.

Caruso C, Lio D, Cavallone L, Franceschi C (2004) Aging, longevity, inflammation, and cancer. Ann N Y Acad Sci 1028:1–13.

Chen, J, Buchanan JB, Buchanan, Sparkman NL, Godbout JP, Freund GG, and Johnson RW (2008) Neuroinflammation and disruption in working memory in aged mice after acute stimulation of the peripheral innate immune system. Brain Behav. Immun 22:301–311.

Chiovenda P, Vincentelli GM, Alegiani F (2002) Cognitive impairment in elderly ED patients: need for multidimensional assessment for better management after discharge. Am J Emerg Med 20:332–335.

Combrinck MI, Perry VH, Cunningham C (2002) Peripheral infection evokes exaggerated sickness behaviour in pre-clinical murine prion disease. Neuroscience 112:7–11.

Cunningham C, Deacon R, Wells H, Boche D, Waters S, Diniz CP, Scott H, Rawlins JN, Perry VH (2003) Synaptic changes characterize early behavioural signs in the ME7 model of murine prion disease. Eur J Neurosci 17:2147–2155.

Cunningham C, Wilcockson DC, Campion S, Lunnon K, Perry VH (2005) Central and systemic endotoxin challenges exacerbate the local inflammatory response and increase neuronal death during chronic neurodegeneration. J Neurosci 25:9275–9284.

Dantzer R, O'Connor JC, Freund GG, Johnson RW, Kelley KW (2008) From inflammation to sickness and depression: when the immune system subjugates the brain. Nat Rev Neurosci 9:46–56.

Davalos D, Grutzendler J, Yang G, Kim JV, Zuo Y, Jung S, Littman DR, Dustin ML, Gan WB (2005) ATP mediates rapid microglial response to local brain injury in vivo. Nat Neurosci 8:752–758.

Davila DR, Edwards CK, 3rd, Arkins S, Simon J, Kelley KW (1990) Interferon-gamma-induced priming for secretion of superoxide anion and tumor necrosis factor-alpha declines in macrophages from aged rats. FASEB J 4:2906–2911.

Di Santo E, Sironi M, Pozzi P, Gnocchi P, Isetta AM, Delvaux A, Goldman M, Marchant A, Ghezzi P (1995) Interleukin-10 inhibits lipopolysaccharide-induced tumor necrosis factor and interleukin-1 beta production in the brain without affecting the activation of the hypothalamus-pituitary-adrenal axis. Neuroimmunomodulation 2:149–154.

Frank MG, Barrientos RM, Biedenkapp JC, Rudy JW, Watkins LR, Maier SF (2006) mRNA upregulation of MHC II and pivotal pro-inflammatory genes in normal brain aging. Neurobiol Aging 27:717–732.

Gerard C, Bruyns C, Marchant A, Abramowicz D, Vandenabeele P, Delvaux A, Fiers W, Goldman M, Velu T (1993) Interleukin 10 reduces the release of tumor necrosis factor and prevents lethality in experimental endotoxemia. J Exp Med 177:547–550.

Gibertini M (1996) IL1 beta impairs relational but not procedural rodent learning in a water maze task. Adv Exp Med Biol 402:207–217.

Gibertini M (1998) Cytokines and cognitive behavior. Neuroimmunomodulation 5:160–165.

Gibertini M, Newton C, Klein TW, Friedman H (1995) Legionella pneumophila-induced visual learning impairment reversed by anti-interleukin-1 beta. Proc Soc Exp Biol Med 210:7–11.
Godbout JP, Chen J, Abraham J, Richwine AF, Berg BM, Kelley KW, Johnson RW (2005) Exaggerated neuroinflammation and sickness behavior in aged mice following activation of the peripheral innate immune system. FASEB J 19:1329–1331.
Godbout JP, Johnson RW (2006) Age and neuroinflammation: a lifetime of psychoneuroimmune consequences. Neurol Clin 24:521–538.
Godbout, J. P., M. Moreau, J. Lestage, J. Chen, N. L. Sparkman, N. Castanon, K. W. Kelley, R. Dantzer, and R. W. Johnson. 2008. Aging exacerbates depressive-like behavior in mice in response to activation of the peripheral innate immune system. Neuropsychopharmacology 33:2341-2351.
Heyen JR, Ye S, Finck BN, Johnson RW (2000) Interleukin (IL)-10 inhibits IL-6 production in microglia by preventing activation of NF-kappaB. Brain Res Mol Brain Res 77:138–147.
Hoek RM, Ruuls SR, Murphy CA, Wright GJ, Goddard R, Zurawski SM, Blom B, Homola ME, Streit WJ, Brown MH, Barclay AN, Sedgwick JD (2000) Down-regulation of the macrophage lineage through interaction with OX2 (CD200). Science 290:1768–1771.
Holmes C, El-Okl M, Williams AL, Cunningham C, Wilcockson D, Perry VH (2003) Systemic infection, interleukin 1beta, and cognitive decline in Alzheimer's disease. J Neurol Neurosurg Psychiatry 74:788–789.
Howard M, Muchamuel T, Andrade S, Menon S (1993) Interleukin 10 protects mice from lethal endotoxemia. J Exp Med 177:1205–1208.
Howard M, O'Garra A, Ishida H, de Waal Malefyt R, de Vries J (1992) Biological properties of interleukin 10. J Clin Immunol 12:239–247.
Huang, Y., C. J. Henry, R. Dantzer, R. W. Johnson, and J. P. Godbout. 2008. Exaggerated sickness behavior and brain proinflammatory cytokine expression in aged mice in response to intracerebroventricular lipopolysaccharide. Neurobiol. Aging 29:1744–1753.
Janssens JP, Krause KH (2004) Pneumonia in the very old. Lancet Infect Dis 4:112–124.
Johnson RW, Godbout JP (2006) Aging, neuroinflammation and behavior. In: Psychoneuroimmunology, 4th Edition (Heijnen CJ, Dantzer R, Irwin M, Glaser R, Sheridan J, Padgett D, eds). Burlington, MA: Elsevier.
Johnston M, Wakeling A, Graham N, Stokes F (1987) Cognitive impairment, emotional disorder and length of stay of elderly patients in a district general hospital. Br J Med Psychol 60 (Pt 2):133–139.
Kasai T, Inada K, Takakuwa T, Yamada Y, Inoue Y, Shimamura T, Taniguchi S, Sato S, Wakabayashi G, Endo S (1997) Anti-inflammatory cytokine levels in patients with septic shock. Res Commun Mol Pathol Pharmacol 98:34–42.
Krzyszton, C. P. N. L. Sparkman, R. W. Grant, J. B. Buchanan, S. R. Broussard, J. Woods, and R. W. Johnson (2008). Exacerbated fatigue and motor deficits in interleukin-10 deficient mice after peripheral immune stimulation. Am. J. Physiol. doi:10.1152/ajpregu.90302.2008.
Leon LR, Kozak W, Rudolph K, Kluger MJ (1999) An antipyretic role for interleukin 10 in LPS fever in mice. Am J Physiol 276:R81–R89.
Letiembre M, Hao W, Liu Y, Walter S, Mihaljevic I, Rivest S, Hartmann T, Fassbender K (2007) Innate immune receptor expression in normal brain aging. Neuroscience 146:248–254.
Long JM, Kalehua AN, Muth NJ, Calhoun ME, Jucker M, Hengemihle JM, Ingram DK, Mouton PR (1998) Stereological analysis of astrocyte and microglia in aging mouse hippocampus. Neurobiol Aging 19:497–503.
Menendez Iglesias B, Cerase J, Ceracchini C, Levi G, Aloisi F (1997) Analysis of B7–1 and B7–2 costimulatory ligands in cultured mouse microglia: upregulation by interferon-gamma and lipopolysaccharide and downregulation by interleukin-10, prostaglandin E2 and cyclic AMP-elevating agents. J Neuroimmunol 72:83–93.
Moore KW, de Waal Malefyt R, Coffman RL, O'Garra A (2001) Interleukin-10 and the interleukin-10 receptor. Annu Rev Immunol 19:683–765.
Mott RT, Ait-Ghezala G, Town T, Mori T, Vendrame M, Zeng J, Ehrhart J, Mullan M, Tan J (2004) Neuronal expression of CD22: novel mechanism for inhibiting microglial proinflammatory cytokine production. Glia 46:369–379.

Nguyen MD, D'Aigle T, Gowing G, Julien JP, Rivest S (2004) Exacerbation of motor neuron disease by chronic stimulation of innate immunity in a mouse model of amyotrophic lateral sclerosis. J Neurosci 24:1340–1349.

Nimmerjahn A, Kirchhoff F, Helmchen F (2005) Resting microglial cells are highly dynamic surveillants of brain parenchyma in vivo. Science 308:1314–1318.

O'Keefe GM, Nguyen VT, Benveniste EN (1999) Class II transactivator and class II MHC gene expression in microglia: modulation by the cytokines TGF-beta, IL-4, IL-13 and IL-10. Eur J Immunol 29:1275–1285.

Perry VH (2004) The influence of systemic inflammation on inflammation in the brain: implications for chronic neurodegenerative disease. Brain Behav Immun 18:407–413.

Perry VH, Cunningham C, Holmes C (2007) Systemic infections and inflammation affect chronic neurodegeneration. Nat Rev Immunol 7:161–167.

Perry VH, Gordon S (1988) Macrophages and microglia in the nervous system. Trends Neurosci 11:273–277.

Perry VH, Matyszak MK, Fearn S (1993) Altered antigen expression of microglia in the aged rodent CNS. Glia 7:60–67.

Pugh CR, Kumagawa K, Fleshner M, Watkins LR, Maier SF, Rudy JW (1998) Selective effects of peripheral lipopolysaccharide administration on contextual and auditory-cue fear conditioning. Brain Behav Immun 12:212–229.

Reichenberg A, Yirmiya R, Schuld A, Kraus T, Haack M, Morag A, Pollmacher T (2001) Cytokine-associated emotional and cognitive disturbances in humans. Arch Gen Psychiatry 58:445–452.

Richwine, A. F., A. O. Parkin, J. B. Buchanan, J. Chen, J. A. Markham, J. M. Juraska, and R.W. Johnson. (2008). Architectural changes to CA1 pyramidal neurons in adult and aged mice after peripheral immune stimulation. Psychoneurendocrinology doi:1016/j.psyneuen.2008.08.003.

Schroder K, Sweet MJ, Hume DA (2006) Signal integration between IFNgamma and TLR signalling pathways in macrophages. Immunobiology 211:511–524.

Sester DP, Trieu A, Brion K, Schroder K, Ravasi T, Robinson JA, McDonald RC, Ripoll V, Wells CA, Suzuki H, Hayashizaki Y, Stacey KJ, Hume DA, Sweet MJ (2005) LPS regulates a set of genes in primary murine macrophages by antagonising CSF-1 action. Immunobiology 210:97–107.

Sheffield LG, Berman NE (1998) Microglial expression of MHC class II increases in normal aging of nonhuman primates. Neurobiol Aging 19:47–55.

Sibley WA, Bamford CR, Clark K (1985) Clinical viral infections and multiple sclerosis. Lancet 1:1313–1315.

Sierra A, Gottfried-Blackmore AC, McEwen BS, Bulloch K (2007) Microglia derived from aging mice exhibit an altered inflammatory profile. Glia 55:412–424.

Smith EM, Cadet P, Stefano GB, Opp MR, Hughes TK, Jr. (1999) IL-10 as a mediator in the HPA axis and brain. J Neuroimmunol 100:140–148.

Sparkman NL, Buchanan JB, Heyen JR, Chen J, Beverly JL, Johnson RW (2006) Interleukin-6 facilitates lipopolysaccharide-induced disruption in working memory and expression of other proinflammatory cytokines in hippocampal neuronal cell layers. J Neurosci 26:10709–10716.

Streit WJ (2004) Microglia and Alzheimer's disease pathogenesis. J Neurosci Res 77:1–8.

Streit WJ, Sammons NW, Kuhns AJ, Sparks DL (2004) Dystrophic microglia in the aging human brain. Glia 45:208–212.

Streit WJ, Sparks DL (1997) Activation of microglia in the brains of humans with heart disease and hypercholesterolemic rabbits. J Mol Med 75:130–138.

Strle K, Zhou JH, Shen WH, Broussard SR, Johnson RW, Freund GG, Dantzer R, Kelley KW (2001) Interleukin-10 in the brain. Crit Rev Immunol 21:427–449.

Tafti M, Nishino S, Aldrich MS, Liao W, Dement WC, Mignot E (1996) Major histocompatibility class II molecules in the CNS: increased microglial expression at the onset of narcolepsy in canine model. J Neurosci 16:4588–4595.

Tan J, Town T, Mori T, Wu Y, Saxe M, Crawford F, Mullan M (2000b) CD45 opposes beta-amyloid peptide-induced microglial activation via inhibition of p44/42 mitogen-activated protein kinase. J Neurosci 20:7587–7594.

Tan J, Town T, Mullan M (2000a) CD45 inhibits CD40L-induced microglial activation via negative regulation of the Src/p44/42 MAPK pathway. J Biol Chem 275:37224–37231.
Wei YP, Kita M, Shinmura K, Yan XQ, Fukuyama R, Fushiki S, Imanishi J (2000) Expression of IFN-gamma in cerebrovascular endothelial cells from aged mice. J Interferon Cytokine Res 20:403–409.
Wofford JL, Loehr LR, Schwartz E (1996) Acute cognitive impairment in elderly ED patients: etiologies and outcomes. Am J Emerg Med 14:649–653.
Wright GJ, Puklavec MJ, Willis AC, Hoek RM, Sedgwick JD, Brown MH, Barclay AN (2000) Lymphoid/neuronal cell surface OX2 glycoprotein recognizes a novel receptor on macrophages implicated in the control of their function. Immunity 13:233–242.
Ye SM, Johnson RW (1999) Increased interleukin-6 expression by microglia from brain of aged mice. J Neuroimmunol 93:139–148.
Ye SM, Johnson RW (2001a) An age-related decline in interleukin-10 may contribute to the increased expression of interleukin-6 in brain of aged mice. Neuroimmunomodulation 9:183–192.
Ye SM, Johnson RW (2001b) Regulation of interleukin-6 gene expression in brain of aged mice by nuclear factor kappaB. J Neuroimmunol 117:87–96.

Index